Allylboration of Carbonyl Compounds

Allylboration of Carbonyl Compounds

DENNIS G. HALL and HUGO LACHANCE

A JOHN WILEY & SONS, INC., PUBLICATION

Published by John Wiley & Sons, Inc., Hoboken, New Jersey
Published simultaneously in Canada.

For general information on our other products and services or for technical support, please contact our Customer Care Department within the United States at (800) 762-2974, outside the United States at (317) 572-3993 or fax (317) 572-4002.

Wiley also publishes its books in a variety of electronic formats. Some content that appears in print may not be available in electronic formats. For more information about Wiley products, visit our web site at www.wiley.com.

Library of Congress Catalog Card Number: 42-20265
ISBN 978-1-118-34445-3

10 9 8 7 6 5 4 3 2 1

CONTENTS

FOREWORD

Chemical synthesis is an intellectually and technically challenging enterprise. Over the many decades of progress in this discipline, spectacular advances in methods have made once intimidating transformations now routine. However, as the frontier advances and the demands for ready access to greater molecular complexity increases, so does the sophistication of the chemical reactions needed to achieve these goals. With this greater sophistication (and the attendant expectation of enhanced generality, efficiency, and selectivity) comes the challenge of adapting these technologies to the specific applications needed by the practitioner. In its 69-year history, *Organic Reactions* has endeavored to meet this challenge by providing focused, scholarly, and comprehensive overviews of a given transformation.

The impact of organoboron chemistry in organic synthesis can hardly be overstated. Two chemistry Nobel Prizes have been awarded for the use of organoboron reagents (H. C. Brown, 1979 and A. Suzuki, 2010), not to mention the prize for boron hydride structures (W. N. Lipscomb, 1976). The remarkable diversity of transformations mediated by boron combined with the high selectivities observed make this element the envy of the main group. Among the myriad transformations, the allylation of carbonyl compounds is one of the most well-recognized reactions mediated by boron. This process delivers homoallylic alcohols in high yield and with a high degree of stereoselectivity. Moreover, the reactions are mechanistically intriguing and their utility stimulates a synergy between fundamental studies of stereochemistry and applications in target-oriented synthesis.

The *Organic Reactions* series is fortunate to have published a comprehensive chapter on this important process that constituted Volume 73. This timely chapter was authored by one of the internationally recognized leaders in this field, Professor Dennis G. Hall together with his student and coauthor Hugo Lachance. Although many reviews and book chapters have been written on the allylboration transform, this massive chapter constitutes the definitive work in the field. Thus, in keeping with our educational mission, the Board of Editors of *Organic Reactions* has decided to publish this chapter as a separate, soft cover book to make the work available to a wider audience of chemists. In addition, to keep pace with the rapid development of this field, Prof. Hall has provided updated references that bring the literature coverage up to December 2010. These references are appended at the end of the original reference section and organized by the Tabular presentation of the different classes of allylation reagents.

The publication of this book represents the fourth soft cover reproduction of single-volume *Organic Reactions* chapters. The success of the first three soft cover books ("The Stille Reaction" (Volume 50), "Handbook of Nucleoside Synthesis" (Volume 55), and "Ionic and Organometallic-Catalyzed Organosilane Reductions" (Volume 71) has convinced us that the availability of low-cost, high-quality publications that cover broadly useful transformations is addressing an unmet need in the organic synthesis community. Thus we will continue to identify candidates for the compilation of such individual volumes as opportunities present themselves.

Scott E. Denmark
Urbana, Illinois

PREFACE

Allylic boron compounds have gained a prominent position as a truly practical class of synthetic reagents over the past 25 years. Their main application as a method for the stereoselective synthesis of homoallylic alcohols by allyl transfer to carbonyl compounds in essentially unmatched. In this process, a new carbon-carbon bond is formed, and up to two new stereogenic centers are created in a predictably controlled manner. Highly effective and general enantioselective variants are available, and countless syntheses of polyacetate and polypropionate natural products have been achieved using stereoselective allylborations as key steps. Furthermore, the residual allylic unit can be manipulated through a number of different transformations such as oxidative cleavage, olefin metathesis, hydrometalation, and many other reactions. Cyclic reagents, as well as propargylic and allenic reagents, have also been developed to further extend the realm of applications of this chemistry. The popularity of carbonyl allylboration chemistry stems largely from its scope and reliability as a C–C bond-forming process affording high and predictable control of chemo-, site-, diastereo-, and enantioselectivity. The recent discovery of Lewis and Brønsted acid catalyzed variants has opened new doors for further development of this important reaction. Further expanding the scope of this transformation is the continuous development of mild and efficient methods for the preparation of functionalized allylic boron reagents. This topic is also captured in this monograph.

Experimental reaction conditions are discussed and numerous specific experimental procedures are provided from the literature for a wide spectrum of representative functional group transformations and synthetic applications using this method. A tabular survey of all examples of carbonyl allylboration used in organic synthesis is presented in nine tables organized by type of allylic boron reagent with listings within each table according to increasing carbon number of the carbonyl substrates.

The literature is covered through December 2010, supplementing by more than 180 literature citations the 834 references found in the earlier hard-cover chapter. These latest literature references have been collected in separate sections according to the sequence of the tables in the tabular survey section. In each of the sections, the individual citations have been arranged in alphabetic order of the author names.

Carbonyl Allylboration is intended to be a useful, easily read tool for all practitioners of organic synthesis.

CHAPTER 1

ALLYLBORATION OF CARBONYL COMPOUNDS

Hugo Lachance and Dennis G. Hall

*Department of Chemistry, University of Alberta, Edmonton,
Alberta, T6G 2G2 Canada*

CONTENTS

dennis.hall@ualberta.ca
Allylboration of Carbonyl Compounds, by Dennis G. Hall and Hugo Lachance.
© 2012 Organic Reactions, Inc. Published 2012 by John Wiley & Sons, Inc.

ACKNOWLEDGMENTS

The authors are grateful to Prof. Scott Rychnovsky, Prof. Robert Bittman, and Dr. Danielle Soenen for their help in copy-editing and proofreading of this chapter.

INTRODUCTION

Allylic boron compounds have gained a prominent position as a useful class of synthetic reagents in the past 25 years. Their general structures, **1** and **2**, and their utility in carbonyl additions are shown in Eq. 1. The main use of these reagents is in the stereoselective synthesis of homoallylic alcohols **3** by an allyl-transfer reaction to carbonyl compounds. In this process, a new carbon–carbon bond is formed, and up to two new stereogenic centers are created. Moreover, the residual allylic unit can be manipulated through a number of different transformations such as oxidative cleavage, olefin metathesis, and many others. Although less prevalent, the propargyl and allenyl reagents typified by **4** and **5** have also been described (Eqs. 2 and 3). Most examples of allylic boron reagents used in carbonyl additions belong to one of two main classes, boranes (structure **1**, Y = alkyl) and boronate derivatives (structure **2**, Y = OR or NR$_2$ for bis(sulfonamide) derivatives). This chapter focuses on describing and comparing both classes, and when needed, they will be discussed separately. A chart of ligand structures with the acronyms used in this text can be found preceding the Tables.

1 Y = alkyl
2 Y = O-alkyl(aryl)
      or N-alkyl(aryl)

3

(Eq. 1)

4

(Eq. 2)

$$\text{(Eq. 3)}$$

Allylic boron reagents react with several classes of carbonyl compounds and their derivatives, including imines. Their most common use, however, is in nucleophilic additions to aldehydes to produce homoallylic secondary alcohols (i.e., the process of Eq. 1 where $R^5 = H$). This reaction process was discovered by Mikhailov and Bubnov in 1964, who observed the formation of a homoallylic alcohol product from the reaction of triallylborane with aldehydes.[1] The first example of allylboration using an allylic boronate was disclosed by Gaudemar and coworkers in 1966.[2] The true beginnings of this chemistry in terms of practical synthetic applications can be traced back to the late 1970's, with Hoffmann's crucial realization of the regio- and diastereospecific nature of the additions of both crotylboronate isomers to aldehydes (see "Mechanism and Stereochemistry").[3] In the following two decades, the work of Hoffmann, Brown, Roush, and Corey was determinant in maturing carbonyl allylboration chemistry into one of the primary methodological tools in stereoselective synthesis. In particular, numerous syntheses of polyacetate and polypropionate natural products feature stereoselective allylborations as key steps (see "Applications to the Synthesis of Natural Products"). The discovery of the Lewis acid catalyzed manifold by Hall and Miyaura recently opened new doors for further development of this important reaction.

In addition to the predominant allyl and crotyl reagents, a large number of allylic borane **1** and boronate derivatives **2** (Eq. 1) with various substituents (R^1-R^4) have been reported. Interested readers can refer to the comprehensive Tabular Survey at the end of this monograph, which covers the literature up to the end of 2005. Several reviews on allylic boron compounds and other allylmetal reagents and their additions to carbonyl compounds and imines have been written prior to this one,[4-14] and these sources may be consulted if a more in-depth historical perspective is desired.

MECHANISM AND STEREOCHEMISTRY

Thermal Uncatalyzed Reactions

Mechanistically, allylic boron reagents belong to the Type I class of carbonyl allylation reagents.[15] Type I reactions proceed through a rigid, chairlike transition structure which requires a synclinal orientation of reacting π systems. Although catalytic variants for additions of allylic boronates have been reported recently, the reaction between most allylic boron reagents and aldehydes is spontaneous, irreversible, and requires no external activator. The uncatalyzed reactions of these Type I reagents proceed by way of a six-membered, chairlike transition structure that features a dative bond between the boron and the carbonyl oxygen of the aldehyde (Scheme 1).[16-20] Competitive kinetics in the addition of

R^1, R^2 = H, alkyl, OR, NR_2, SiR_3, Cl
R^3 = H, alkyl, CO_2R, halogen, $B(OR)_2$
R^4 = H, alkyl, aryl, OR, halogen

1 Y = alkyl
2 Y = O-alkyl(aryl)
   or N-alkyl(aryl)

Transition Structure

Scheme 1. Overall aldehyde allylboration process.

an allylboronate to benzaldehyde and deuterobenzaldehyde in different solvents revealed a negative secondary deuterium kinetic isotope effect, which rules out a single-electron transfer mechanism.[20] Theoretical studies using ab initio MO and DFT methods strongly suggest that the strength of the dative bond between the boron and the aldehyde carbonyl oxygen in the transition structure is the dominant factor in determining the rate of the reaction.[19,20] Indeed, whereas the B–O bond is short (~1.6 Å) and very advanced in the transition state, the incipient C–C bond between the carbonyl carbon and the allylic unit has merely initiated (~2.4 Å) (Fig. 1). A weakly bound B–O coordinated complex was detected as an early intermediate in the calculated reaction pathway.[19] Not surprisingly, allylboration reactions tend to proceed faster in non-coordinating solvents, and the most electrophilic boron reagents react faster.[21]

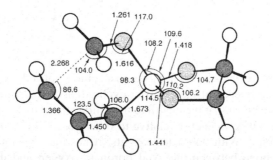

Figure 1. Calculated transition structure for the allylboration reaction between *B*-allyl-1, 3,2-dioxaborolane ($CH_2{=}CHCH_2B[OCH_2]_2$) and formaldehyde at the MP2/6-31G** level of theory. Indicated bond distances are in angstrom. Reprinted with permission from *J. Org. Chem.* **1998**, *63*, 8331.[19] Copyright 1998 American Chemical Society.

The immediate products of additions between carbonyl substrates and allylic boranes **1** or boronate derivatives **2** are borinate or borate esters, respectively. To cleave the covalent B–O bond in these intermediates (structure **6**, Scheme 1) and to obtain the desired free alcohol, a hydrolytic or oxidative work-up is required. This issue is discussed in detail in the section "Work-Up Conditions". In the interest of simplifying chemical equations, specific work-up conditions are not included in most of the examples highlighted in this chapter.

One of the main reasons for the popularity of allylic boron reagents in stereocontrolled synthesis is that their additions to aldehydes are reliably highly stereoselective and the outcome is predictable. The diastereospecificity of the reaction was first recognized by Hoffmann and Zeiss using both E- and Z-crotylboronates **7** (Eqs. 4 and 5).[3,22] For both crotylboranes and crotylboronates, the additions generally proceed with near-perfect reflection of the olefin geometry of the reagent into the configuration of the product. Specifically, the E-crotylboron reagents (Scheme 1, $R_E = Me$, $R_Z = H$) lead to anti-propanoate products and the Z-crotyl reagents (Scheme 1, $R_E = H$, $R_Z = Me$) lead to syn-products. In both instances, of the two possible chairlike transition structures the favored one places the aldehyde substituent (R^5) in a pseudo-equatorial orientation. This model, which has been reproduced computationally for both crotylborane[17] and crotylboronate[18] derivatives, accounts for the high diastereospecificity of most allylborations. Even highly functionalized reagents and most intramolecular additions follow the same trend.

$$(E)\text{-}\mathbf{7}\ \ 93{:}7\ E{:}Z \qquad\qquad anti \qquad\qquad (Eq.\ 4)$$
$$(>85\%)\ 93{:}7\ d.r.$$

$$(Z)\text{-}\mathbf{7}\ \ >95{:}5\ Z{:}E \qquad\qquad syn \qquad\qquad (Eq.\ 5)$$
$$(>85\%)\ >96{:}4\ d.r.$$

It is well accepted that the high diastereospecificity of aldehyde allylboration reactions is a consequence of the compact cyclic transition structure. Theoretical calculations have shown that the chairlike transition structure shown in Scheme 1 and Fig. 1 is the lowest in energy relative to other possibilities such as the twist-boat conformation.[16] With boronate reagents, it has also been suggested that a weak hydrogen bond between the axial boronate oxygen and the hydrogen of the polarized formyl unit contributes to the preference for the transition structure with the aldehyde substituent in the pseudo-equatorial position.[23]

Lewis Acid Catalyzed Reactions

As described above, allylic boron reagents are self-activating, Type I reagents where the allylation is effected by coordination of the aldehyde carbonyl oxygen to the boron atom within a cyclic six-membered transition structure. Because of this self-activation mechanism, there would appear to be no advantage to using an external promoter such as a Lewis or Brønsted acid. Furthermore, one could expect an added Lewis acid to compete with the boron atom for the basic aldehyde oxygen, potentially leading to a switch from the highly diastereospecific Type I mechanism to a less selective, open-chain Type II mechanism.[15] Recent publications, however, show that the additions of allylic boronates can be efficiently and beneficially catalyzed by several Lewis acids.[24–28] Allylic boranes are not subject to this catalytic effect.[29] For additions of allylic boronates, the rate enhancements observed in the presence of these catalysts are quite dramatic.[24,29] For example, the addition of 2-ethoxycarbonyl allylboronate **8** to benzaldehyde to give an exo-methylene butyrolactone requires almost two weeks at room temperature, but only 12 hours in the presence of a catalytic amount of Sc(OTf)$_3$ (Eq. 6).[24] Note that in this example the resulting homoallylic alcohol intermediate cyclizes in situ with the carboxy ester group to form a lactone product.

(Eq. 6)

It is noteworthy that the stereospecificity observed in the thermal reaction is fully preserved under this new catalytic manifold. Furthermore, the presence of a 2-alkoxycarbonyl substituent on the allylic boronate is not necessary for the metal-promoted activation to occur (Eq. 7).[24,25] According to mechanistic studies, a chairlike bimolecular transition structure similar to the thermal additions can be proposed for these catalyzed allylborations.[29] Control experiments have confirmed the inefficiency of Lewis acids with dialkylallylboranes (Eq. 8). Hence, the catalytic effect is thought to derive from an increase in the electrophilicity of the boron atom through binding of the metal ion to one of the boronate oxygens [transition structure (T.S.) **9**], as opposed to coordination of the carbonyl oxygen (T.S. **10**).[29]

(Eq. 7)

(Eq. 8)

Coordination of the Lewis acid to the boronate oxygens is thought to decrease the overlap between the oxygen lone electron pairs and the vacant p-orbital of the boron atom. Consequently, the boron center is rendered more electron-deficient, and compensates by strengthening the key boron–carbonyl oxygen interaction and concomitantly lowering the activation energy of the reaction. This idea is consistent with theoretical calculations of the non-catalyzed allylboration[19] and from an experimental study consisting of a quantitative survey of steric and electronic effects in the reactions of different allylboronates.[21] The latter results led to the conclusion that the rate of a given allylboration can "be rationalized in terms of the relative availability of lone pairs of electrons on the oxygen atoms attached to the boron".

Brønsted Acid Catalyzed Reactions

In line with the effect of Lewis acids, the huge rate acceleration in the additions of allylic boronates to aldehydes by strong protic acids such as triflic acid was recently reported.[30] In the example of Eq. 9, the reaction yield is almost quantitative after 16 hours at 0° under conditions where the use of $Sc(OTf)_3$ gives less than 5% of the product. Interestingly, additions with geometrically defined allylic boronates are not always stereospecific. Although definitive mechanistic studies have not yet appeared, it can be presumed that the origin of the acceleration could be similar to that of Lewis acid catalysis, with activation by protonation of a boronate oxygen in the cyclic chairlike transition structure.

(Eq. 9)

(99%)

(<5%) with $Sc(OTf)_3$

SCOPE AND LIMITATIONS

Preparation of Allylic Boron Reagents

Unlike aldehydes and ketones, allylic boron compounds are not ubiquitous, commercial organic substrates. There are several methods for the preparation of allylic boronates, however, and many of these have been developed in the past decade. This topic has been reviewed recently[14] so only the most common methods are emphasized in this section. As a result of the lesser stability of allylic boranes, methods to access these reagents are more limited and it is generally easier to prepare allylic boronates with a wide range of functional groups.

From Hard Allylic Organometallics. The most common preparation of allylic boranes and boronates is the addition of a reactive allylic metal species to a borinic or boric ester, respectively (Eqs. 10 and 11). Preparations from allyllithium,[31-34] allylmagnesium,[32,35] and allylpotassium[22,31,36-40] reagents are all well known. These methods are popular because the required allylic anions are quite easy to prepare, and because they generally lead to high yields of products.

$$\text{(Eq. 10)}$$

$$\text{(Eq. 11)}$$

One potential drawback to this approach is that the allylic metal precursor may not be configurationally stable and may be subject to facile metallotropic rearrangement leading to either constitutional or geometrical isomers (Eq. 12).[41-43] This is in fact a major problem even with 3-substituted allylic boranes, whose geometrical integrity can be compromised according to Eq. 12 even at temperatures well below 0° (see section "Stability and Handling"). Because of their potential configurational instability, and also because of their high sensitivity to air, allylic boranes are not isolated from the process depicted in Eq. 10, but rather are used in situ for their addition to carbonyl substrates. In contrast, allylic boronates are much more stable and can be isolated and purified prior to subsequent carbonyl allylation reactions.

$$\text{(Eq. 12)}$$

Another potential problem with the preparations from allylic organometallics is that they often involve highly reactive reagents that can lead to incompatibilities with many functional groups. While these impediments constitute no real problem for simple allyl- or crotylboron reagents, they can lead to poor regio- and stereoselectivities in more substituted examples. Significantly, as shown with the diisopropyl tartrate (DIPT) derivatives (*E*)-**11** and (*Z*)-**11** of Eqs. 13 and 14,[37] the use of Schlosser's "superbase" conditions[43] to prepare the configurationally stable crotylpotassium anions allows the preparation of both E- and Z-crotylboronates with very high selectivity from the respective (*E*)- and (*Z*)-2-butene. Likewise, the bis(isopinocampheyl)crotylboranes [bis(Ipc)crotylboranes] **12** are prepared by anion trapping with the requisite borinic ester [MeOB(Ipc)$_2$] at −78° (Eqs. 15 and 16).[44] Because of the strong Lewis acidity of the resulting dialkylborane, however, the addition of BF$_3$ is required in order to remove the methoxide anion and form the desired "free" borane. This additional operation is not required with the less acidic crotylboronates. Moreover, crotyldialkylboranes are too sensitive to be isolated and must be treated immediately with carbonyl substrates (see section on "Stability and Handling"). These approaches using crotylpotassium anions remain the methods of choice for preparing boron-based crotylation reagents, which are very useful in the synthesis of polypropionate natural products (see section "Applications to the Synthesis of Natural Products").

$$(Eq. 13)$$

$$(Eq. 14)$$

$$(Eq. 15)$$

$$(Eq. 16)$$

MeOB[(+)-Ipc]$_2$　　　　MeOB[(–)-Ipc]$_2$

A number of other functionalized allylic boron reagents can be made using allylmetal intermediates, including E-3-silyl-substituted reagents,[32,38–40,45] exemplified by the bis(Ipc) derivative **13**[45] (Eq. 17), and the useful Z-3-alkoxy-substituted reagents,[46–48] exemplified by **14**[48] (Eq. 18). The double bond geometry of **14** arises from the chelated alkyllithium intermediate employed in its preparation (Eq. 18).[33,46] Like most allylic dialkylboranes, reagents **13** and **14** are not isolated and rather are allowed to react in situ with carbonyl substrates (see section "Stability and Handling").

$$Me\text{-}Si(Me)\text{-}N(i\text{-}Pr)_2 \xrightarrow[\text{Et}_2\text{O, } 0°]{n\text{-BuLi, TMEDA}} \left[(i\text{-Pr})_2N\text{-}Si(Me)(Me)\diagup\diagup\text{Li} \right] \xrightarrow[\text{2. BF}_3\bullet\text{OEt}_2]{\text{1. MeOB(Ipc)}_2}$$

$$(i\text{-Pr})_2N\text{-}Si(Me)(Me)\diagup\diagup\text{B(Ipc)}_2$$

13 (—)   (Eq. 17)

$$MeO\diagup\diagup \xrightarrow[\text{THF, } -78°]{n\text{-BuLi}} \left[MeO\cdots\text{Li} \right] \xrightarrow[\text{2. BF}_3\bullet\text{OEt}_2]{\text{1. MeOB(Ipc)}_2} MeO\diagup\diagup\text{B(Ipc)}_2$$

14 (—)   (Eq. 18)

If the presence of sensitive functional groups poses problems of chemoselectivity in the use of hard allylic metal reagents, allylboronate derivatives also can be accessed by a milder transmetalation of allylic tin species with boron halides.[49] This approach has been used by Corey in the synthesis of chiral bis(sulfonamido)boron reagents such as the methallyl reagent **15** (Eq. 19) (see section "Chiral Boronate Derivatives").[50]

$$\diagup\diagdown\text{Sn(Bu-}n)_3 \xrightarrow[\text{CH}_2\text{Cl}_2, 0° \text{ to rt}]{} \text{(Eq. 19)}$$

15 (—)

Recently, the enantiomerically enriched α-carbamoyloxycrotylboronate **16** has been synthesized using a similar anti-S_E' reaction with in situ generated 2-chloro-1,3-bis(toluenesulfonyl)-1,3,2-diazaborolidine (Eq. 20). Following a ligand-exchange with pinacol, the resulting chiral α-substituted boronate **16** is isolated in a good overall yield with minimal erosion of enantiomeric purity.[51]

$$(n\text{-Bu})_3\text{Sn}\diagup\diagdown\text{OC(O)N(Pr-}i)_2 \xrightarrow[\text{CH}_2\text{Cl}_2, 0°]{}$$

88% ee

(Eq. 20)

$$\xrightarrow[\text{2. pinacol, pTSA, MgSO}_4, \text{CH}_2\text{Cl}_2]{\text{1. HCl, H}_2\text{O}}$$

16 (83%, 2 steps) 85% ee

From Alkenylmetal Precursors. The reaction of an alkenylmetal with a halomethyl boronic ester also leads to allylic boronates (Eq. 21).[52] This strategy takes advantage of the configurational stability of vinylic carbanions and circumvents the rearrangements that can plague allylic metal intermediates. For example, an alkenylmagnesium reagent reacts readily with pinacol chloromethylboronate to give the corresponding isoprenylboronate in good yield (Eq. 22).[53]

$$\begin{array}{c}\underset{R^3}{\overset{R^4}{R^2}} M \xrightarrow{XCH(R^5)B(OR^1)_2} \underset{R^3 \quad R^5}{\overset{R^4 \quad OR^1}{R^2}} B\text{-}OR^1\end{array}$$ (Eq. 21)

$$M = Li, Mg, Al, K \qquad \qquad \textbf{2}$$

$$\text{(isopropenyl-MgBr)} + \text{Cl}\text{-B(pinacol)} \xrightarrow[-78° \text{ to rt}]{THF} \text{(isoprenyl-B(pinacol))}$$ (Eq. 22)

(73%)

Alkenylmetal reagents are generally more configurationally stable than allylic metal species and so alkylations with these anions are usually more stereoselective than with allylic anions. The low reactivity of many alkenylmetal species, however, can sometimes bring about poor yields in the alkylation step. While the reactions of alkenyllithium[54–56] and alkenylmagnesium[53] reagents with halomethylboronates are well established, the high reactivity of these organometallics limits the type of functional groups that may be present. In this regard, less reactive alkenylaluminum[57,58] (Eq. 23) and alkenylcopper[26,59,60] reagents (Eq. 24) have been used to produce more sensitive, functionalized allylic boronates such as 2-ethoxycarbonyl derivatives of types **17** and **19** from the corresponding alkyne precursors. In the generic example in Eq. 24, the cis-carbocupration of alkynoic esters provides a weakly nucleophilic and configurationally unstable 1-alkoxycarbonylalkenylcopper intermediate **18**, and the presence of hexamethylphosphoric triamide (HMPA) as an additive was found to be crucial to avoid erosion of the E/Z stereoselectivity in the alkylation step with the iodomethylboronate electrophile.[26,59,60] Under these conditions, a variety of unsymmetrically substituted 3,3-dialkylated allylboronates **19** have been prepared in over 20:1 selectivity, and these reagents have been subsequently employed in the stereoselective preparation of quaternary carbon centers by reaction with aldehydes.

$$=\!\!-CO_2Me \xrightarrow[THF, 0°]{DIBAL\text{-}H, HMPA} \left[\underset{Al(Bu\text{-}i)_2}{\overset{CO_2Me}{=}} \right] \xrightarrow[rt]{Cl\text{-}B(pinacol)} MeO_2C\text{-B(pinacol)}$$

17 (80%)

(Eq. 23)

(Eq. 24)

One limitation of this preparative method is the concurrent formation of an alkenylboronate, which is formed presumably via α-elimination pathways involving the intermediate ate adduct of addition between the organometallic reagent and the α-haloalkylboronate (Eq. 25).[52,61]

(Eq. 25)

The use of enantiomerically pure α-chloroalkyl boronic esters as electrophiles, for example the dicyclohexylboronate **20** obtained from Matteson's asymmetric homologation method,[62] provides access to α-alkyl-substituted allylic boronates. For example, reagent **21** is isolated in very high diastereomeric purity by way of a net inversion of configuration (Eq. 26).[63,64] Likewise, the Zn(II)-promoted reaction of alkenylmetal fragments with chiral dichloromethylboronate **22** leads to optically pure α-chloroallylboronates such as **23** (Eq. 27).[65] This reagent has a small tendency to isomerize so it is combined directly with aldehydes. Addition of all these α-substituted reagents to aldehydes is highly diastereoselective (see Schemes 8 and 9 under "Enantioselective Allylations").[66,67] Furthermore, the chloride substituent can be displaced with a variety of nucleophiles including metal alkoxides and alkylmetal reagents to provide other useful allylation reagents.[68]

(Eq. 26)

(Eq. 27)

From Other Organic Fragments.   Other direct methods to access allylic boron reagents, including chiral ones such as 2-cyclohexenylborane **24**, are based on the hydroboration of 1,3-dienes[69] (Eq. 28) and allenes[70] (Eq. 29). Allylic boronates can also be accessed by the transition-metal-catalyzed diboration[71] (Eq. 30) and silaboration[72] (Eq. 31) of dienes and allenes. With allenes as substrates, the use of a chiral phosphoramidite ligand leads to high enantioselectivities in the preparation of chiral 2-borylated α-substituted reagents of type **25** (Eq. 32).[73]

$$(Ipc)_2BH \quad + \quad \bigcirc \quad \xrightarrow[-25°,\ 12\ h]{THF} \quad \text{(Eq. 28)}$$

from (+)-2-pinene

24 (—) >94% ee

(Eq. 29)

1.5 eq   (86%)

(Eq. 30)

1.5 eq   (93%) >99:1 Z:E

(Eq. 31)

2 eq   (90%) >99:1 Z:E

(Eq. 32)

(pin₂B₂)   **25** (61-75%) 87-92% ee

Allylic boronates with a wide variety of functional groups can also be obtained by the treatment of allylic acetates with pinacolatodiboron [(pin)₂B₂], although side products of oxidative homodimerization can be observed in some instances (Eq. 33).[74]

(Eq. 33)

(73%)   (13%)

Very recently, it was shown that allylic trifluoroborates such as **26** can be prepared stereospecifically from allylic alcohols using a palladium complex made from a pincer ligand (Eq. 34).[75] A copper-catalyzed, site selective and stereospecific substitution of allylic carbonates with $(pin)_2B_2$ has been optimized to afford several examples of α-substituted allylic boronates.[76] When using the E-isomer of an enantiomerically pure secondary allylic carbonate, the resulting chiral α-substituted allylic boronate **27** is obtained with minimal erosion of enantiomeric purity (Eq. 35). The corresponding Z-configured allylic carbonate affords the other enantiomer, and this stereochemical outcome has been explained with a mechanism involving anti attack of the borylcopper reagent in a conformation minimizing allylic 1,3-strain. Allylic halides are also suitable substrates with pinacolborane as the borylating agent under platinum catalysis.[77] As shown in a palladium-catalyzed Stille cross-coupling reaction with soft alkenylmetal electrophiles (Eq. 36),[78] transition-metal-catalyzed processes can even tolerate unprotected acidic functionalities. Related Negishi-type coupling strategies have also been employed.[79,80]

$n\text{-Pr}$⁀⁀OH + $[B(OH)_2]_2$ → $n\text{-Pr}$⁀⁀$BF_3^-K^+$

1. PhSe—Pd—SePh, Cl (5 mol%), DMSO/MeOH, 50°, 16 h
2. aq KHF_2

26 (94%) (Eq. 34)

$n\text{-Bu}$⁀⁀OCO_2Me, 98% ee

Cu(OBu-t) (10 mol%), Xantphos (10 mol%), $(pin)_2B_2$ (2.2 eq), THF, 0°, 23 h

(Eq. 35)

27 (95%) 96% ee

$Pd_2(dba)_3 \cdot CHCl_3$ (3 mol%), PPh_3 (12 mol%), HMPA, 50°

(63%) (Eq. 36)

From Boron-Containing Fragments. A number of indirect methods to synthesize more highly functionalized allylic boronates are based on the isomerization or derivatization of organoboron substrates. Selected examples include the homologation of alkenylboronates with chloromethyllithium[56,81] (Eq. 37),[82] and the allylic rearrangement of 3-siloxyalkenylboronates to give chiral α-chloro derivatives such as **28** (Eq. 38).[83] A variant of Eq. 37 using lithiated allyl chloride has also been described.[84] The Johnson orthoester Claisen rearrangement of a chiral 3-hydroxypropenylboronic ester leads to an epimeric, separable mixture of α-substituted allylic boronates **29** and **30** (Eq. 39).[85,86] A related intermolecular method based on the S_N2' addition of Grignard reagents has been reported (Eq. 40).[87] These reagents are not isolated, rather, they are immediately combined with aldehydes. Other methods of preparation of allylic boronates involving boronic ester fragments include the isomerization of alkenylboronates (Eq. 41),[88] cycloadditions of dienylboronates (Eq. 42),[89] and olefin cross-metathesis (Eq. 43).[90]

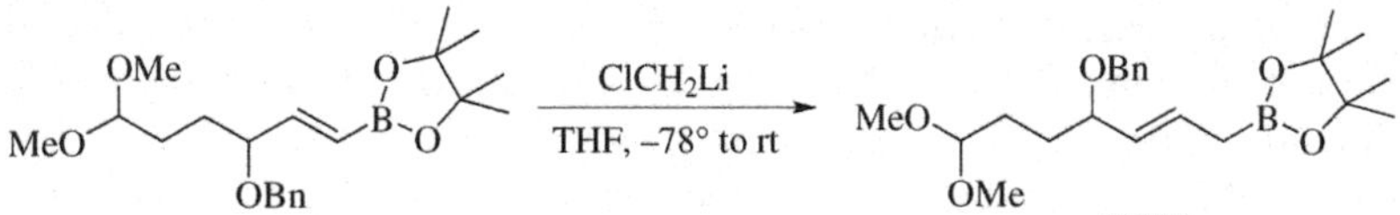

(77%)   (Eq. 37)

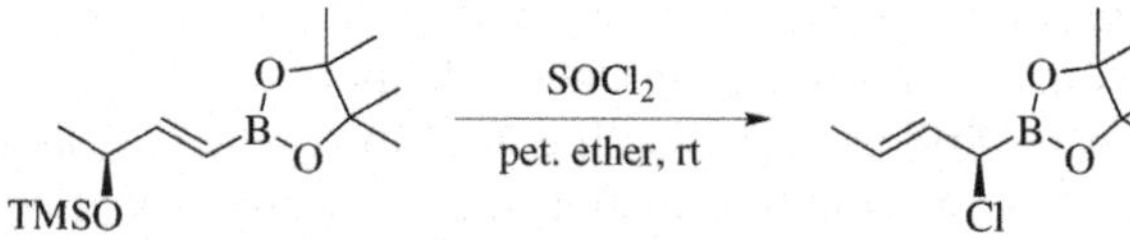

28 (75%)   (Eq. 38)

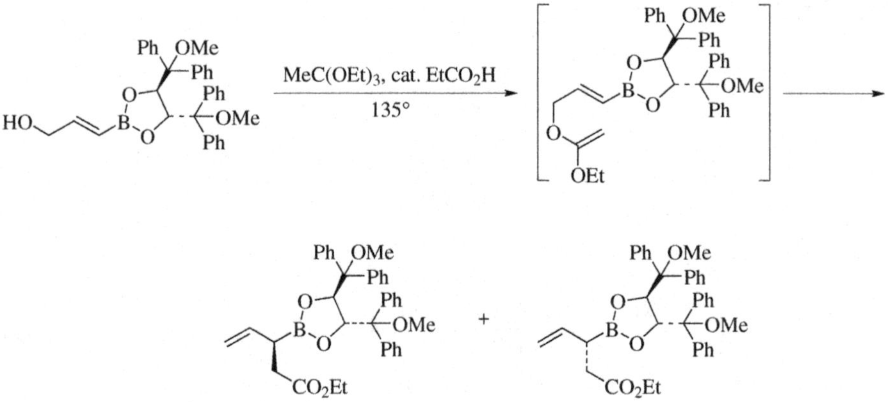

29   (71%) 1:1 (separable)   **30**

(Eq. 39)

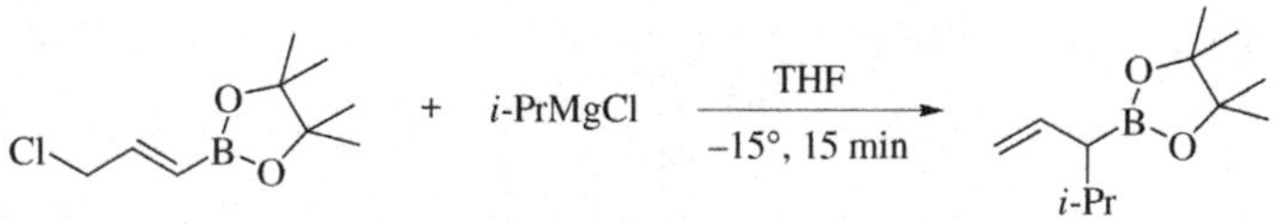

i-Pr

(—)   (Eq. 40)

[IrH$_2$(AcOEt)$_2$(PPh$_2$Me)$_2$]PF$_6$
(3 mol%)

AcOEt, rt, 10 min

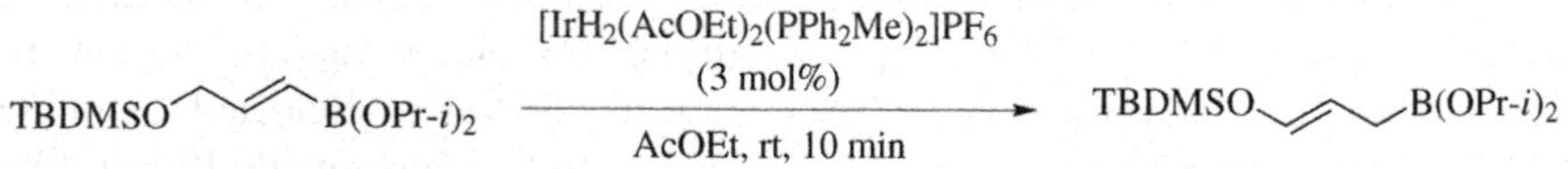

(94% conv.) 98:2 E:Z

(Eq. 41)

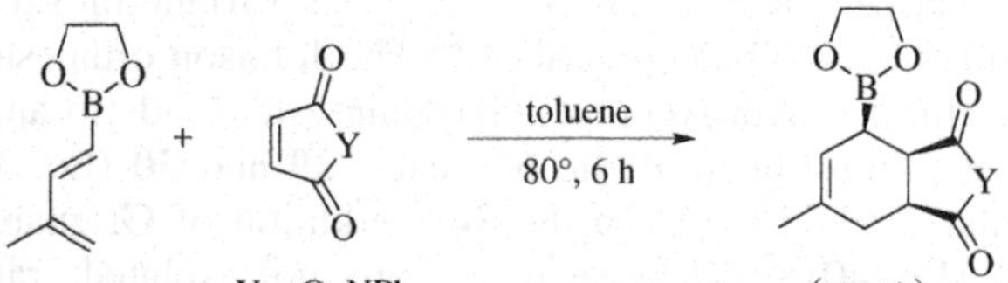

Y = O, NPh   (quant.)   (Eq. 42)

$$\text{(Eq. 43)}$$

Stability and Handling of Allylic Boron Reagents

Allylic boronates are more stable to atmospheric oxidation and are thus much easier to handle than the corresponding allylic boranes. The stability of the boronate reagents arises from the partial donation of the lone pairs of electrons on the oxygen atoms into the empty p-orbital of boron. This mesomeric effect is responsible for the upfield shift of the boron atom in ^{11}B NMR compared to that of allylic boranes (compare allylboronate **31** and allylborane **32**).[21]

31
^{11}B NMR (200 MHz, CH$_2$Cl$_2$) δ 33 ppm

32
^{11}B NMR (200 MHz, THF) δ 82 ppm

As a consequence of their superior stability, many types of allylic boronates can be isolated and purified. It should be noted that most pinacol allylic boronic esters and other bulky esters are stable to hydrolysis and can be conveniently purified by chromatography on silica gel. A potential pitfall of all allylic boron compounds is their stereochemical integrity, and substituted allylic boranes are known to undergo reversible borotropic rearrangements at temperatures above −45° (see Eq. 12, M = BR$_2$).[22,91,92] For this reason, allylic boranes are normally not handled under atmospheric conditions, and are rather prepared and employed in situ at low temperatures. These borotropic rearrangements are the bane of stereoselective syntheses since they can scramble the double bond geometry of the reagent. Thus, a major advantage in using allylic boronates compared to boranes is that they are much less prone to such rearrangement. For example, whereas *B*-crotyl-9-borabicyclo[3.3.1]nonane (*B*-crotyl-9-BBN) (**33**) exists as an equilibrating mixture of E- and Z-isomers at room temperature,[92] the two isomers of pinacol crotylboronate (**7**) are sufficiently stable to be independently prepared, isolated, and employed at that temperature (Fig. 2).[22,54] The allylic borotropic rearrangement has been studied in detail with a few distinct reagents, including the α-methyl allylic reagents of generic structure **34** (Eq. 44). The dialkylboranes are fluxional at room temperature and exist almost exclusively as the thermodynamically favored crotylborane with a more highly substituted double bond.[91-93]

33
inseparable E:Z mixture
at room temperature

(E)-**7**

(Z)-**7**

E:Z isomers readily prepared
and handled at room temperature

Figure 2. Comparison of the stereochemical stability of crotylboranes vs crotylboronates.

34 (Eq. 44)

R = alkyl   fast at room temperature
Y = OR   stable at room temperature
Y = NMe$_2$ stable up to 200°

The speed of the rearrangement depends on the Lewis acidity of the boron atom.[94] Electron-donating substituents that reduce the acidity of boron slow down the process. Thus, whereas a monoamino derivative slowly isomerizes at 150°,[95] the corresponding diamino derivative is stable up to a temperature of 200°.[96] The corresponding boronic esters are less stabilized towards the borotropic rearrangement, but they are stable enough to be conveniently handled at room temperature. The presence of Lewis acids, however, can promote or accelerate the rearrangement.[49] These allylic borotropic rearrangements obey first-order kinetics and the absence of crossover products in control experiments demonstrates their intramolecular nature.[92] In the case of the B-allyl 9-BBN reagent, the coalescence temperature is 10°, which corresponds to a rate of exchange of 330 sec^{-1} ($\Delta G^{\neq}$ 13.3 kcal/mol).[92] Unsurprisingly, bases such as pyridine completely stop the rearrangement.[92,94] Activation parameters of several allylic dialkylboranes have been determined by dynamic NMR spectroscopy.[91]

Reactivity of Allylic Boron Reagents

The boron–oxygen mesomeric effect described in the previous section explains the lower reactivity of allylic boronates towards carbonyl compounds compared to that of allylic boranes. The use of Lewis acids, however, allows boronate derivatives, including hindered ones, to react at temperatures comparable to the analogous boranes. As described above (see section "Mechanism and Stereochemistry"), the most reactive allylic boronates are those with the most electrophilic boron centers.[19,21] The nucleophilicity of the γ-position of an allylic boron reagent (the position that forms the new C–C bond with the aldehyde) is also important to the reactivity of the reagent. For example, allylic boronates with faster   slower

reactivity towards carbonyl compounds

Figure 3. Scale of reactivity for selected allylboron derivatives toward carbonyl compounds.

substituents that reduce electron density at this position, such as 2-ethoxycarbonyl groups,[58,59] are correspondingly less reactive than similar allylboronates that lack these groups. Between the crotylboronate isomers, the E-isomer reacts noticeably faster than the Z counterpart.[22,34] The electronic and steric effects of different allylic borane and boronate derivatives has been studied by qualitative kinetic analysis.[21] The resulting scale of reactivity depicted in Fig. 3 emphasizes the importance of both effects in the reactivity of these boron-based reagents. In general, acyclic boronates react faster than cyclic ones. Moreover, the reaction is quite sensitive to electronic effects. For example, N,N,N',N'-tartramide derivatives react at a much slower rate than the corresponding esters.

Allylic dihaloboranes have been isolated and characterized.[97] The highly electrophilic difluoroallylborane is predicted to be extremely reactive in carbonyl allylation.[19] Allylic difluoroboranes are postulated intermediates in the Lewis acid promoted additions of allylic trifluoroborate salts.[98,99] These reagents combine stability and reactivity as desirable attributes in a single class of allylic boron reagents. They are prepared by reaction of the corresponding allylic boronic acids with aqueous KHF$_2$, and they can be purified by recrystallization and stored as air- and water-stable solids for extended periods of time at room temperature. For example, the crotyl trifluoroborates (E)- and (Z)-**35** do not interconvert. Allylic trifluoroborate salts are inert to carbonyl compounds until activated by an external Lewis acid, at which point they become efficient allylating agents that produce homoallylic alcohols in high yields. Instead of aldehyde coordination, this new method of activation is thought to involve a mechanism whereby the electrophilic additive (usually BF$_3$•OEt$_2$) strips off a fluoride anion from the trifluoroborate salt. The expected product of this equilibration is a highly electrophilic tricoordinate allylic difluoroborane, which is thought to react with aldehydes through the usual Type I mechanism. Consistent with this hypothesis, the crotyl reagents (E)- and (Z)-**35** provide, in over 90% yield, the respective anti and syn addition products expected from a cyclic, chairlike transition structure (Eq. 45). Although the reaction is usually carried out using two equivalents of BF$_3$•OEt$_2$ at $-78°$, it can also be performed with as low as 5 mol% of the same additive, albeit at room temperature. A phase-transfer procedure has also been developed.[100]

$$R^3 = \text{Ph} \quad (>90\%) \quad >98{:}2 \text{ d.r.}$$

(Eq. 45)

(E)-35 $R^1 = \text{Me}, R^2 = \text{H}$ $R^3 = \text{Ph}$ (>90%) >98:2 d.r.
(Z)-35 $R^1 = \text{H}, R^2 = \text{Me}$ $R^3 = \text{Ph}$ (>90%) >98:2 d.r.

BF$_3$ (0.05 eq), rt, 3-6 h; or
BF$_3$ (2.0 eq), −78°, 15 min

Work-Up Conditions

The addition of allylic boron reagents to carbonyl compounds first leads to homoallylic alcohol derivatives **36** or **37** that contain a covalent B–O bond (Eqs. 46 and 47). These adducts must be cleaved at the end of the reaction to isolate the free alcohol product from the reaction mixture. To cleave the covalent B–O bond in these intermediates, a hydrolytic or oxidative work-up is required. For additions of allylic boranes, an oxidative work-up of the borinic ester intermediate **36** ($R^1 = $ alkyl) with basic hydrogen peroxide is preferred. For additions of allylic boronate derivatives, a simpler hydrolysis (acidic or basic) or triethanolamine exchange[22] is generally performed as a means to cleave the borate intermediate **37** ($Y = $ O-alkyl). The facility with which the borate ester is hydrolyzed depends primarily on the size of the substituents, but this operation is usually straightforward. For sensitive carbonyl substrates, the choice of allylic derivative, borane or boronate, may thus be dictated by the particular work-up conditions required.

(Eq. 46)

(Eq. 47)

1 $R^1 = $ alkyl   **36**

2 $Y = $ O-alkyl(aryl)   **37**
       or N-alkyl(aryl)

Carbonyl Substrate Generality

Additions of allylic boron reagents have been reported on a very wide range of classes of functionalized aldehydes. Some types of aldehydes, however, are very reactive and may lead to side-reactions. For example, β,γ-unsaturated aldehydes are notoriously difficult substrates but an indirect procedure for their in situ generation leads to clean products of allylboration.[101] Although most examples of additions of allylic boron reagents involve aldehydes as substrates, additions to ketones are also possible. Additions to simple ketones are much slower than similar reactions with aldehydes, and forcing conditions may be required. The stereoselectivity observed in these additions is variable depending on the difference in size between the two substituents on the ketone. As exemplified in Eq. 48, ketones react slowly with allylic boronates, under high pressure, to yield tertiary homoallylic alcohols as products.[102] Additions of allylic boranes are also much slower with ketones. For example, additions of methallyl-BBN are performed at room temperature, compared to −78° for aldehydes (Eq. 49), and they were found to be highly sensitive to the steric bulk of the ketone.[103] Because of the high temperature required, these additions are accompanied by reversible allylic rearrangement in reactions of 3-monosubstituted allylic boranes, which renders them impractical.

$$\text{(Eq. 48)}$$

anti   syn

R = Et (79%) 1:1 anti:syn, ~1:2 E:Z
R = i-Pr (—) >99:1 anti:syn, ~1:3 E:Z

$$\text{(Eq. 49)}$$

(95%)

The additions of allylic boronates to ketones are greatly enhanced if an appropriate chelating group is present on the ketone (such as an α- or β-hydroxy or carboxylic acid group).[104–109] For example, both (E)- and (Z)-diisopropyl crotylboronate exchange 2-propanol and add to α-hydroxyacetone within reasonable time frames to give the expected products in high diastereoselectivity (Eq. 50).[109] In the proposed transition structure **38**, the ketone residue bearing the α-hydroxy group occupies an axial position due to the formation of a cyclic boronate complex.

38 R_E = Me, R_Z = H (72%) 97:3 d.r.
          R_E = H, R_Z = Me (65%) 94:6 d.r.

$$\text{(Eq. 50)}$$

Moderate levels of diastereomeric differentiation are observed in the additions of achiral allylboronates to β-hydroxy ketones, as exemplified in Eq. 51a.[110] The major product is tentatively assigned as the anti diastereomer. Few reports describe the use of carbonyl substrates other than aldehydes and ketones, and these reactions have not led to applications. For instance, low reactivity leading to modest yields and stereoselectivity are observed in the additions of allylic boranes to acylsilanes (Eq. 51b).[111]

(Eq. 51a)

(Eq. 51b)

Acid chlorides (which react vigorously), esters, and N,N-dimethylamides react with two equivalents of the reagent to give diallyl tertiary alcohols after work-up (Eq. 52).[103]

(Eq. 52)

The Lewis acid catalyzed reactions expand the scope of aldehyde substrates with certain boronate reagents. For example, whereas cyclohexanecarboxaldehyde is unreactive under thermal (uncatalyzed) conditions, it does react with allylic boronate **8** under Sc(OTf)$_3$ catalysis at room temperature (Eq. 53, see also Eq. 6).[26]

(Eq. 53)

Diastereoselective Additions to Chiral α-Substituted Carbonyl Substrates

The presence of a stereogenic center on the aldehyde can strongly influence the diastereoselectivity in allylboration reactions, especially if this center is in the α-position. Predictive rules for nucleophilic addition on such α-substituted carbonyl substrates such as the Felkin model are not always suitable for closed transition structures.[112] For α-substituted aldehydes devoid of a polar substituent, Roush has established that the minimization of "gauche-gauche" ("syn-pentane") interactions can overrule the influence of stereoelectronic effects.[113] This model is valid for any 3-monosubstituted allylic boron reagent. For example, although crotylboronate (E)-**7** adds to aldehyde **39** to afford as the major product the diastereomer predicted by the Felkin model (Scheme 2),[114,115] it is proposed that the dominant factor is rather the minimization of syn-pentane interactions between the γ-substituents of the allyl unit and the α-carbon of the aldehyde. With this

Scheme 2. Stereoinduction model for additions of (*E*)-**7** to α-chiral aldehyde **39**.

model, **40** is the favored transition structure leading to the Felkin product because the two smallest substituents on the α-carbon (H, Me) are aligned respectively with a methyl and a hydrogen of the E-crotyl unit. This arrangement leaves the largest group (Et) of the aldehyde "outside" the chair, away from the reactive centers and almost orthogonal with the plane of the carbonyl unit. This conformer, which in this instance also happens to be the reactive one predicted by the Felkin model, best minimizes the so-called gauche-gauche (syn-pentane) interactions. In the alternative transition structure **41** that leads to the minor diastereomer, the ethyl group induces slightly more important gauche-gauche interactions compared to those found in **40** (Et–H for **41** vs. Me–H for **40**). Other transition structures with the smallest group (H) outside the chair are less favorable and are not considered in this analysis. These transition structures feature Me–Me or Et–Me "syn-pentane" interactions, and thus are significantly higher in energy.

The value of this model is more apparent in rationalizing the results of crotyl-boronate (*Z*)-**7** for which the possible transition structures, **42**–**45**, are shown in Scheme 3.[113] In these examples, the major diastereomer is the anti-Felkin product.[115] The Felkin transition structure **45** shows a severe Me–Me syn-pentane interaction. Structure **43**, analogous to **40** in Scheme 2, best minimizes these interactions by aligning the smallest groups together. With respect to the minor diastereomer, structure **44** is preferable over **45**, but not quite as effective as **43** in minimizing the overall effect of syn-pentane interactions because it places the methyl group outside the chair instead of the larger ethyl group.

The stereoinductive effect of the α-substituent is often more powerful when this substituent is a polar group such as an alkoxy or amino group. In this instance as well, minimization of syn-pentane interactions with 3-monosubstituted reagents is important, and the selectivity can be amplified if it acts in con-cert with other effects. Known models include the stereoelectronic preference

Scheme 3. Stereoinduction model for additions of (Z)-**7** to α-chiral aldehyde **39**.

for the orthogonal polar group dictated by the Felkin–Anh model,[116] and the minimization of dipoles (i.e., the Cornforth model[117]). With such aldehydes as glyceraldehyde acetals (e.g., **51** in Scheme 4), Z-crotylboron reagents tend to be more selective than the E-isomers.[118] In the absence of any γ-substituents, or with 3,3-disubstituted reagents, syn-pentane interactions are not expected to play a key role. As shown in Scheme 4, the generic addition of pinacol allylboronate (**31**) to aldehydes substituted with a polar substituent can proceed through four reasonable transition structures **46–49** (high-energy rotamers with the small group, the hydrogen atom, in the "outside" position are not considered).[115,118] Comparison of the selectivity among aldehydes **39, 50**, and **51** shows an increase in the anti product with a polar aldehyde substituent. By presenting the polar group almost

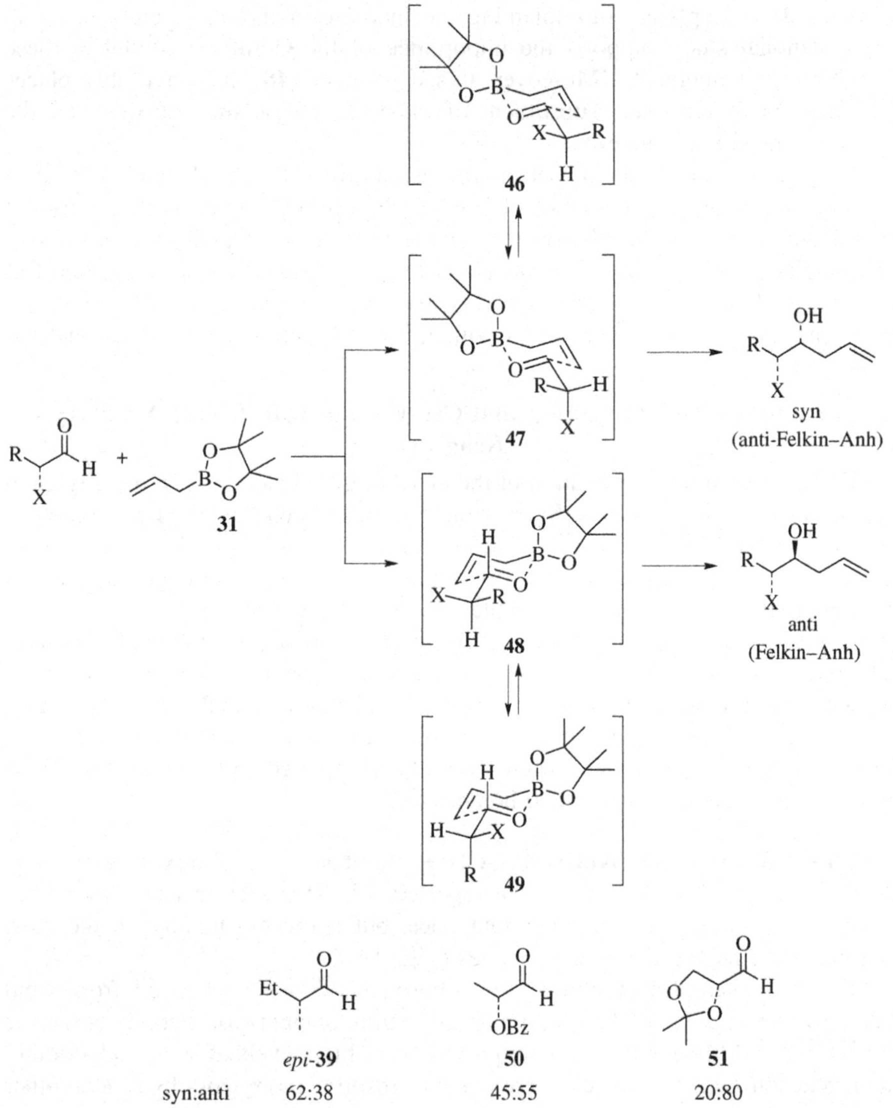

Scheme 4. Stereoinduction model for additions of allylboronate **31** to aldehydes with a polar α-substituent.

perpendicular to the plane of the aldehyde carbonyl group, structures **46** and **49** are stereoelectronically favorable and probably comparable in energy even if structure **49** is formally the reactive conformer according to the Felkin–Anh interpretation. The results of reactions of allylboronate **31** and the corresponding crotyl reagents suggest that with aldehyde **51**, the Cornforth-like conformer **48** (minimizing dipole repulsion between the α-alkoxy substituent and the carbonyl) is particularly important in explaining the increase of the anti stereoisomer. A computational study supports the importance of the Cornforth model in these allylboration reactions.[119] Moreover, this conformer (48) also favorably places the large methylene(oxy) substituent of aldehyde 51 on the "outside" on the chairlike transition structure.

To improve the levels of selectivity in additions to chiral aldehydes, it is possible to resort to the tactic of double diastereoselection with the use of chiral allylic boranes and boronates (see section "Double Diastereoselection"). Bis(isopinocampheyl) allylic boranes and the tartrate allylic boronates (see following section), in particular, are very useful in the synthesis of polypropionate natural products by reaction with α-methyl and α-alkoxy functionalized aldehydes.

Enantioselective Allylations and Crotylations with Chiral Auxiliary Reagents

The most common application of the allylboration reaction is in the allylation and crotylation of aldehydes. Many strategies have been devised for controlling the absolute stereoselectivity.[10,11] These strategies involve two general classes of chiral reagents: (1) allylic boronates or boranes containing a stereogenic α-carbon on the allylic unit; (2) allylic boronates fitted with a chiral auxiliary that creates the cyclic boronate, or that is attached to the non-allylic substituents in boranes. The use of chiral auxiliary-based reagents is more popular because it is generally easier to manipulate the non-allylic substituents than it is to make an enantiomerically pure allylic boronate with a stereogenic α-carbon on the allylic unit. Several examples of both classes of reagents have been reported and reviewed recently, both for boronate derivatives and boranes.[10,11]

Chiral Boronate Derivatives.   A large number of chiral auxiliary reagents based on allylic boronates has been reported.[10,11] This section provides a brief overview of the historically important ones, but it focuses mainly on the most popular systems and the emerging ones (Fig. 4).

The first examples of chiral allylic boronates[120] were prepared from rigid camphor-derived 1,2-diols, providing allylation reagents of generic structure 52.[114,120–122] Although they did not provide very high levels of stereoinduction in their reactions with aldehydes, these reagents inspired more work by several other groups, and these efforts led to significantly improved systems. For example, the class of tartrate-derived allenyl-, allyl- 53, and crotylboronates 11 has evolved into one of the most recognizable class of reagents in organic synthesis.[37,123,124] These reagents are very reactive, allowing additions to aldehydes to proceed readily at −78°. Their reactivity comes at a price, however, as they are hydrolytically unstable and must be employed in conjunction with molecular sieves to rigorously eliminate adventitious traces of water.[125] Nonetheless, when appropriately stored at −20°, the crotylboronates 11 can be kept for months without appreciable deterioration. As expected, reagents (E)-11 and (Z)-11 give the respective anti- and syn-propionate units in a diastereospecific manner and with very high diastereoselectivity (Eqs. 54 and 55).[37] The enantioselectivity tends to be higher

Figure 4. Common allylic boronate derivatives used as chiral auxiliary reagents in enantioselective carbonyl additions. (Only one stereoisomer is shown for simplicity.)

with (E)-**11**, but values are typically less than 85% ee for most aldehydes. In this regard, toluene is the best solvent for aliphatic aldehydes whereas THF is optimal for aromatic ones.[125] Although simple stereoselection with these reagents does not provide practical levels of enantioselectivity, their use in double diastereoselection with α-substituted aldehydes[126] has been amply demonstrated in the context of numerous total syntheses of complex natural products.[11]

$$\text{(Eq. 54)}$$

(R,R)-(E)-**11**   anti
(70–95%) >98% d.r., up to 88% ee

$$\text{(Eq. 55)}$$

(R,R)-(Z)-**11**   syn
(70–95%) >98% d.r., up to 86% ee

Originally, it was proposed that lone pair repulsions between one of the tartrate ester carbonyl oxygens and the aldehyde oxygen in transition structure **60** were responsible for the preference for transition structure **59** and the consequent enantiofacial selectivity (Scheme 5). Recent theoretical calculations,

R = Cy (97%) 87% ee
R = t-Bu (56%) 86% ee

Scheme 5. Model for absolute stereoinduction in additions of tartrate allylic boronate **53** to aldehydes.

however, point to an attractive n_O-$\pi^*_{C=O}$ interaction between the basic oxygen atom of one of the two ester groups and the boron-activated aldehyde carbonyl as the main factor favoring transition structure **59** in these stereoselective allylations.[127]

Significantly higher levels of enantioselectivity can be obtained with aromatic and unsaturated aldehydes by making use of the corresponding metal carbonyl complexes (Eq. 56).[128,129] The selectivity is higher in toluene as solvent, and the resulting product can be easily decomplexed oxidatively to afford enantioenriched material. Likewise, dicobalt hexacarbonyl complexes of propargylic aldehydes provide much improved selectivities.[128,130]

1. 4 Å MS, toluene, −78°
2. CH_3CN, O_2, hv

(>90%) 83% ee

56% ee with PhCHO

(Eq. 56)

Cyclic derivatives **54** and **55** lead to higher levels of enantioselectivity with several types of aldehydes but their preparation requires more effort than the simpler parent reagents **53** and **11** (Fig. 4).[131,132] The mixed O/N boronate derivative **56** provides high enantioselectivities with aliphatic aldehydes (Eq. 57).[133] Its preparation, however, is also tedious and requires five steps from enantiomerically pure camphorquinone.

$$(Eq. 57)$$

(47-92%) up to 96% ee

The bis(sulfonamide)-derived reagents (e.g., **57, 58**, and **15** in Fig. 4) are based on a 1,2-diamino-1,2-diphenylethane auxiliary.[50] These hydrolytically unstable boronate derivatives are typically prepared by Sn-to-B exchange of allylic tin precursors (e.g., Eq. 19). When employed at $-78°$ in toluene or methylene chloride, unsubstituted allyl and methallyl reagents **57** and **15** provide very high enantioselectivities (>95% ee).[50] The 2-bromoallyl reagent is also quite efficient but the crotylation reagents **58** provide slightly lower enantioselectivities (90–95% ee). It is thought that in the favored transition state **61** involving a tetracoordinate boron atom, the tolylsulfonyl groups orient away from the phenyl substituents of the chiral auxiliary (Scheme 6). It was proposed that the high enantioselectivity can be rationalized by the presence of a steric interaction between the methylene group of the allyl unit and the sulfonyl substituent on the pseudo-equatorial nitrogen atom in the disfavored transition state **62**.[50] Several 2-substituted derivatives of these bis(sulfonamide) allylation reagents have been employed in the total synthesis of complex natural products (see section "Applications to the Synthesis of Natural Products").

There are only a few examples where chiral allylic boronates react successfully with ketones, and the enantioselectivities are modest.[134] A recent report, however,

R = Ph (>90%) 95% ee
R = Cy (>90%) 97% ee

61
favored diastereomeric T.S.

62

Scheme 6. Model for absolute stereoinduction in additions of bis(sulfonamide) reagent **57** to aldehydes.

describes the use of very reactive binaphthol-derived allylboronates. In particular, high levels of enantioselection (>96% ee) are obtained in the reaction of the 3,3'-(CF$_3$)$_2$-BINOL reagent **63** with several aromatic ketones, as exemplified with 4-chloroacetophenone (Eq. 58).[135]

(Eq. 58)

Chiral Dialkylboranes. Several allylic boranes have been developed as chiral auxiliary reagents (Fig. 5). The introduction of terpene-based reagents such as **12** and **64–68** has been pioneered by H.C. Brown, and the most popular class remains the bis(isopinocampheyl) derivatives (structures **12, 64–66**).[136] A wide variety of substituted analogs have been reported,[9,12] including the popular crotylboranes **12**,[44,137] but also a number of other reagents bearing heteroatom-substituents. All are made from the inexpensive precursor α-pinene, readily available in both enantiomeric forms.

64 R^1, R^2, R^3 = H
(*E*)-**12** R^1 = Me, R^2, R^3 = H
(*Z*)-**12** R^1 = H, R^2 = Me, R^3 = H
65 R^1, R^2 = H, R^3 = Me
66 R^1, R^2 = Me, R^3 = H

67

68

69

70 R^1, R^2, R^3 = H
(*E*)-**71** R^1 = Me, R^2, R^3 = H
(*Z*)-**71** R^1 = H, R^2 = Me, R^3 = H

Figure 5. Chiral allylic boranes used as chiral auxiliary reagents in enantioselective additions to carbonyl compounds. (Only one isomer is shown for simplicity. For reagents **12** and **64–66**, (−)-Ipc is shown.)

Most of the bis(isopinocampheyl) reagents, which are prepared from hard allylic organometallic precursors (e.g., Eqs. 15 and 16), are very reactive and add rapidly to aldehydes at $-78°$. Like other boranes, they are oxidatively unstable in ambient atmosphere and must be reacted with aldehydes in situ. Moreover, because of the borotropic rearrangement discussed above, the crotylboranes must be preserved and used at a low temperature to avoid geometrical isomerization with consequent formation of mixtures of diastereomeric products (see section on "Stability and Handling"). Not surprisingly, the 3,3-disubstituted reagent **66**, which provides quaternary carbon centers in the products, reacts more slowly and necessitates a higher reaction temperature.[138] The alcohol products from these additions are most often isolated through a basic hydrolytic oxidative work-up of the borinate intermediate (see Eq. 46). The unsubstituted reagent **64**, allyl(Ipc)$_2$borane, is very enantioselective.[136] It provides enantioselectivies over 95% for a wide range of aldehyde substrates. The removal of residual Mg^{2+} salts from the preparation of **64** leads to a dramatic increase of the reagent's reactivity, allowing reactions to be carried out at $-100°$ with improved enantioselectivities (up to 98% ee).[139] Simple allylation reactions have been performed successfully in good to high yields with a wide variety of aliphatic, aromatic and heteroaromatic aldehydes (see section "Applications to the Synthesis of Natural Products"). Crotylations of aldehydes with reagents **12** are less selective, providing enantioselectivities in the range of 88–92% ee at $-78°$.[137] The bis(isopinocampheyl)boranes behave poorly with ketones as substrates. For example, the simple allylation of acetophenone with reagent **64** gives the addition product with only 5% ee.[140]

The enantioselectivity of these reagents is explained by comparison of transition structures **72** and **73** shown in Scheme 7. The disfavored transition structure **73** leading to the minor enantiomer displays a steric interaction between the methylene of the allylic unit and the methyl group of one of the pinane units. Unlike the tartrate boronates described above, the directing effect of the bis(isopinocampheyl) allylic boranes is extremely powerful, giving rise to high reagent control in double diastereoselective additions (see section on "Double Diastereoselection").

Other chiral dialkylallylboranes have been reported, including the borolanes **69** (Fig. 5).[141,142] Although the high reactivity and outstanding enantioselectivity of these borolanes make them very attractive reagents, they are notorious for their tedious preparation and have thus remained rather under-utilized. Recently, the 10-TMS-9-borabicyclo[3.3.2]decane allylborane and crotylboranes (**70**) and (**71**) were shown to provide remarkable stereoselectivities (>98% d.r., 94–99% ee) and good to high yields for a wide range of aldehydes.[143] These reagents require reaction times of just a few hours at $-78°$, and are said to be robust and recyclable. They can be prepared using allylic carbanions in three simple steps that includes, however, a resolution of the borinic ester precursor as a stable and crystalline pseudoephedrine complex. The corresponding allenylborane has been reported.[144] Reagent **74**, the 10-phenyl analog of reagent **70**, is particularly effective in allylations of ketones.[145] As exemplified in the reaction between (*R*)-**74** and acetophenone, enantioselectivities are excellent for aromatic ketones

Scheme 7. Model for absolute stereoinduction in additions of $(-)$-bis(isopinocampheyl) allylic boranes to aldehydes.

(Eq. 59), and even surprisingly high for aliphatic ketones such as 2-butanone, a substrate that offers very little steric discrimination (Eq. 60). Reagent **74** is less effective than **70** in allylations of aldehydes (e.g., 90% ee vs. >98% ee for **70** in the allylation of benzaldehyde). The superior reactivity and selectivity of **74** with ketones is ascribed in part to the lesser steric bulk of the phenyl substituent compared to the trimethylsilyl unit of reagent **70**. The smaller phenyl substituent of **74** would provide a better fit for ketones in the chiral "pocket" of the reagent.

Enantioselective Allylations and Crotylations with Chiral α-Substituted Reagents

One major advantage of chiral auxiliary reagents over chiral α-substituted reagents is the fact that the chiral diol or diamine unit is not modified in the bond-making process and is thus potentially recyclable. The preparation of enantiomerically pure α-substituted reagents requires a stereoinductive transformation such as the Matteson asymmetric homologation (see Eqs. 26 and 27), and their addition to aldehydes leads to destruction of the stereogenic α-center. Despite these disadvantages, many such reagents have found extensive use in the total synthesis of complex natural products.[11] Examples of chiral α-substituted allylic boronates are shown in Fig. 6.[4,5] The possibility for allylic rearrangement precludes the use of chiral α-substituted allylic boranes except in special instances such as the cyclohexenyl derivative **24**, which can be prepared by the asymmetric hydroboration of cyclohexadiene (Eq. 28). Enantiomerically enriched reagent **24** (94% ee) adds to aldehydes at −78° to give the corresponding homoallylic alcohols with high enantioselectivities (94% ee).[70]

From the useful α-chloro allylic boronates **23** and **28**,[65–67,83,146] a number of other α-substituted reagents can be obtained by nucleophilic substitution of the chloride substituent (Fig. 6). For example, the α-methoxy reagent **75**[68,147] and α-methyl substituted reagent **21**[63–65] provide excellent enantiofacial discrimination with levels of enantiocontrol over 95% ee. These reagents provide interesting insight into the important steric and electronic factors involved in the allyboration transition structure. For all of these α-substituted reagents, two competing chairlike transition structure models can be proposed where the α substituent is positioned either in a pseudo-equatorial or in a pseudo-axial orientation.[146] These two competing structures lead to opposite configurations for the resulting homoallylic alcohols, and their relative energy difference depends on the nature of the substituent on the α-carbon, on the nature of the diol boronate auxiliary, and also on the presence of γ-substituents on the allylic boronate. For example, in reactions with α-chloro reagent **23**, the chloroalkenyl homoallylic alcohol product (Z)-**78** is largely predominant (>99:1) over the E-configured product **80**, and it is obtained in a very high level of enantioselectivity (Scheme 8).[65,66] This very high enantioselectivity of the product reflects the high enantiopurity of reagent **23**. The predominance of stereoisomer **78** and its Z-olefin geometry can be explained in

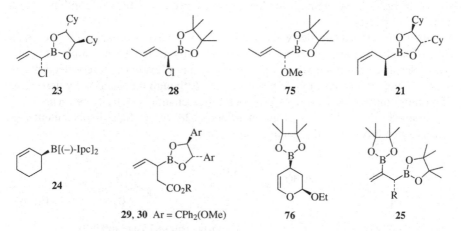

Figure 6. Chiral α-substituted allylic boron reagents.

Scheme 8. Stereoinduction model for the additions of chiral α-chloro allylboronate **23**.

terms of one preferred transition structure **77** where the chloro substituent adopts a pseudo-axial orientation to minimize dipole repulsion, and to avoid Coulombic repulsions with the oxygen atoms of the boronate. In transition structure **79**, unfavorable steric interactions and electronic repulsions are unavoidable between the pseudo-equatorial chloro substituent and the dioxaborolane substituents. It had been shown previously that the 1,2-dicyclohexyl ethanediol (DICHED) auxiliary had little effect on the stereochemical outcome of the reaction.[64] Thus, the stereogenic α-chloro center and not the chiral boronate unit is the main contributor to the highly efficient enantiofacial differentiation in these allylations. The size of the boronate auxiliary, however, influences the relative proportions of isomers by further disfavoring transition structures like **79** where the α-substituent occupies the pseudo-equatorial position.[148] Reagent **23** also performs very well in double diastereoselection with chiral α-alkoxy aldehydes.[65,66] This reagent, prepared as shown in Eq. 27, has a small tendency to isomerize and thus is combined with aldehydes directly.

The E-crotylboronate derivatives **28** and **75** behave similarly. As shown in Eqs. 61 and 62, these reagents were successfully tested with simple aliphatic aldehydes and benzaldehyde and provide very high levels of stereoselectivity in the formation of anti-propionate products.[67,68] Although the α-methoxy derivative **75** is more diastereoselective, it provides lower enantioselectivity because it can be obtained only in 90% ee from enantiopure **28** via an S$_N$2 displacement that causes some erosion of the enantiomeric purity.

(Eq. 61)

28 95% ee

R = Et (47-65%) 95:5 d.r., 96% ee
R = Ph (53-68%) 95:5 d.r., 98% ee

epi-**75** 90% ee

(Eq. 62)

R = Et (81%) >99:1 d.r., 90% ee
R = Ph (79%) >99:1 d.r., 90% ee

The α-methyl Z-crotylboronate derivative **21** can be obtained in high enantiomeric purity, and its additions to aldehydes are highly diastereoselective and occur with almost perfect enantiofacial selectivity (Scheme 9).[63,64] In this instance, the unfavored transition structure **83** (leading to Z-product **84**) with the pseudo-axial α-methyl group develops a strong pseudo-1,3-diaxial interaction between the two methyl substituents of the reagent. The favored transition structure **81** is devoid of such an interaction, and provides the E-configured syn-propionate adducts **82** with very high stereoselectivity. Reagents **21**, **28**, and **75** (Fig. 6) have also been used extensively in double diastereoselective synthesis with chiral aldehydes, and they generally lead to high levels of reagent control (see section "Double Diastereoselection").[63,64,67,68,83,149]

The enantiomerically pure α-substituted reagents of type **29/30** (Fig. 6, Eq. 39) react with benzaldehyde to give the expected homoallylic alcohol products such as **86** with high enantioselectivity (Eq. 63).[85,86] By analogy with the above-described reagents, the addition is presumed to occur through transition structure **85** with the pseudo-axial α-alkyl substituent. The competing transition structure with the pseudo-equatorial substituent would experience severe non-bonding interactions between this group and the bulky boronic ester moiety.

R = Et (79%) 98% ee
R = Ph (71%) 99% ee

Scheme 9. Stereoinduction model for the additions of chiral α-methyl crotylboronate **21**.

29
R = CPh$_2$(OMe)

85

86 (91%) 96% ee

(Eq. 63)

Recently, the first examples of catalytic enantioselective preparations of chiral α-substituted allylic boronates have appeared. Cyclic dihydropyranylboronate **76** (Fig. 6) is prepared in very high enantiomeric purity by an inverse electron-demand hetero-Diels–Alder reaction between 3-boronoacrolein pinacolate (**87**) and ethyl vinyl ether catalyzed by chiral Cr(III) complex **88** (Eq. 64). The resulting boronate **76** adds stereoselectively to aldehydes to give 2-hydroxyalkyl dihydropyran products **90** in a "one-pot" process.[150–152] The diastereoselectivity of the addition is explained by invoking transition structure **89**. Key to this process is the fact that the possible "self-allylboration" between **76** and **87** does not take place at room temperature. Several applications of this three-component reaction to the synthesis of complex natural products have been described (see section on "Applications to the Synthesis of Natural Products").

The catalytic asymmetric diboration of allenes provides α-substituted 2-boronyl allylic boronates of type **25** (see Eq. 32). One of them, **91**, adds to benzaldehyde, albeit with a slight erosion of stereoselectivity (Eq. 65).[73] The major β-hydroxy ketone stereoisomer, isolated after an oxidative work-up, originates from the putative chairlike transition structure **92**.

87

88 (1 mol%)

4 Å MS, rt, 1 h
(solvent)

76 96% ee

RCHO, 45°

89

90
(80-90%) >99:1 d.r., >95% ee

(Eq. 64)

(Eq. 65)

Enantioselective Additions with Chiral Propargyl Reagents

As demonstrated by the example in Eq. 66, propargylboron reagent **93** affords allenic alcohols in very high enantioselectivities.[153] Although this reagent has not been widely employed, the selectivity it provides with all classes of aldehydes is truly remarkable (>99% ee for model aliphatic, unsaturated, and aromatic aldehydes).

(Eq. 66)

Enantioselective Additions with Chiral Allenyl Reagents

Although they were among the first examples to illustrate the use of tartrate boronate auxiliaries, allenylboronate reagents are seldom employed. The tartrate-derived reagent **94** provides homopropargylic alcohols in high enantioselectivity from aliphatic aldehydes (Eq. 67).[123] This reagent provides optimal selectivity when substituted with the bulkier diisopropyl tartrate ester, and even in this situation aromatic aldehydes react sluggishly and with low stereoselectivity. The efficacy and scope of bis(sulfonamide) allenylboron reagent **95** has been investigated in detail.[153] As exemplified in Eq. 68, this reagent adds to most types of aldehydes with high enantioselectivities. A recently described chiral dialkylallenylborane provides comparable selectivities.[144]

R = n-C$_5$H$_{11}$ (63%) >95% ee
R = Ph (43%) 79% ee

(Eq. 67)

(76%) 94% ee

(Eq. 68)

Enantioselective Additions with Chiral 3-Heterosubstituted Reagents

A large number of chiral 3-substituted allylic borane and boronate reagents have been described (Fig. 7). To be effective, these reagents must be prepared in geometrically pure form. Like the corresponding crotyl reagents, additions to aldehydes are diastereospecific as the anti/syn configuration of the resulting homoallylic alcohol products is dependent upon the geometry of the original allyl unit.

Preparation of 1,2-Diols. The most direct method of preparation of 1,2 diols from allylic boron reagents involves the use of 3-alkoxy-substituted reagents. Although these reagents are prepared easily from the corresponding lithiated 3-alkoxypropene, only the Z-isomers are prepared directly in this fashion as a result of the preference for a cis-configuration in the lithium-chelated allylic carbanion (see Eq. 18, section "Preparation of Allylic Boron Reagents"). The resulting Z-configured reagents provide the syn-1,2-diol products in excellent diastereoselectivity in agreement with the usual cyclic chairlike transition structure. These reagents are very effective in double diastereoselection with α-substituted aldehydes.[11] Several chiral analogues such as **14**, **96**, **97**,[48,154] and **98**,[155] are sufficiently effective for use in absolute stereocontrol (Fig. 7). In particular, the bis(isopinocampheyl)-derived 3-alkoxyallylboranes such as **96** react highly enantioselectively with a wide range of aldehyde substrates, including the notoriously troublesome acrolein (Eq. 69).[156]

14 R = Me
96 R = CH₂OMe
97 R = CH₂O(CH₂)₂TMS

98

99

100

13

101 Ar = Ph
102 Ar =

103

104

Figure 7. Chiral 3-substituted allylic boron reagents.

$$\text{(Eq. 69)}$$

(66%) 95% ee

To access anti-1,2-diols, indirect methods are required for the preparation of geometrically pure, chiral E-3-alkoxy reagents. To this end, the isomerization of alkenylboronic esters described above (Eq. 41), provides a reliable route to tartrate-derived E-3-siloxy allylboronate **99** (Fig. 7). The latter shows variable enantioselectivities in additions to aldehydes, with cyclohexanecarboxaldehyde affording the highest selectivity (Eq. 70).[88]

$$\text{(Eq. 70)}$$

(R,R)-**99**   (85%) >99:1 d.r., 91% ee

Alternatively, masked hydroxy groups may be used as the γ-substituent. For example, following allylation with the E-3-boronyl reagent **100** (Fig. 7), the second hydroxy group is generated upon an oxidative work-up that transforms the alkylboronate unit of intermediate **105** with full retention of configuration (Eq. 71).[157] Likewise, provided it can be protodesilylated in the presence of an allylsilane moiety, a suitable silyl substituent can be subjected to stereospecific Tamao–Fleming oxidative conditions to afford the desired anti 1,2-diol. The requisite E-3-silyl-substituted reagents are conveniently prepared by trapping of the corresponding allylic carbanions[154] with a borate ester (Eq. 17).[38–40,45,46] Both borane and boronate reagents, exemplified by the respective pinene- and tartrate-derived reagents **13**[45] and **102**[158] (Fig. 7, Eqs. 72 and 73), afford diol products in moderate yields but with high enantioselectivies after a stereospecific Tamao–Fleming protodesilylation/oxidation of the β-hydroxy silyl intermediates (e.g., **106**). As with other substituted reagents, however, only the bis(isopinocampheyl) allylic boranes **100** and **13** lead to practical levels of enantioselectivity suitable for absolute stereocontrol. Reagent **100**, in particular, has found useful application in the synthesis of complex natural products (see section on "Applications to the Synthesis of Natural Products").

100   **105**

(63-80%)
>97.5:2.5 d.r., >90% ee

$$\text{(Eq. 71)}$$

(Eq. 72)

(Eq. 73)

Preparation of 1,4-Diols, Epoxides, and Amino Alcohols. Reagent **101** (Fig. 7) illustrates the complications typically encountered in the design of 3-silyl-substituted allylic boron reagents, which require a selective protolytic desilylation of the resulting allylic silane to afford a 1,2-diol product. As opposed to the closely related 2-menthofuryl-derived reagent **102**, this selective desilylation is not possible with reagent **101** (Fig. 7). Epoxidation of the aldehyde addition products of allylic silane (R,R)-**101** with dimethyldioxirane, however, followed by an acid-catalyzed Peterson rearrangement, provides E-4-hydroxy-3-alkenol products **107** in good enantioselectivities (Eq. 74).[40] Reagent **101** is also very efficient in double diastereoselection with several types of chiral α-substituted aldehydes. The Z-3-chloro-substituted bis(Ipc) reagent **103** (Fig. 7), prepared by lithiation of 3-chloropropene, exists in equilibrium with the α-chloro isomer **108** (Eq. 75). Isomer **103**, however, adds faster to a wide range of aldehydes to give syn-chlorohydrins **109** with high enantioselectivities.[159] The diastereoselectivity is explained by invoking the usual cyclic chairlike transition structure. Treatment of the chlorohydrins under basic oxidative conditions affords synthetically useful vinyl epoxides. Several aldehyde types (aliphatic, aromatic, unsaturated) have been used successfully with reagent **103**. A 3-imino reagent of E-geometry, **104**, adds to aldehydes to afford anti amino alcohols with good enantioselectivities (Eq. 76).[160] However, even the simplest aldehydes tested (benzaldehyde, cyclohexanecarboxaldehyde) give modest yields of products (ca. 50–60%).

(Eq. 74)

103

108

109

(60-65%)
>94:6 d.r., 90-97% ee

(Eq. 75)

104

1. RCHO, −78°
2. H$_2$O$_2$, aq NaOH
3. MeONH$_3$+Cl$^-$

(50-60%)
>90% d.r., 90-93% ee

(Eq. 76)

Lewis and Brønsted Acid Catalyzed Enantioselective Additions

The recent advent of the Lewis acid catalyzed allylboration reactions opened new doors for enantioselective allylborations and motivated the reexamination of a number of chiral auxiliary systems. In this perspective, reagents **110–112**[121] are extremely enantioselective at −78° in the presence of Sc(OTf)$_3$ as catalyst (Eq. 77).[27,28] These hydrolytically stable reagents can be conveniently purified by silica gel chromatography. Consistently high levels of absolute stereoinduction (>95% ee) are observed with a broad range of aldehydes for the unsubstituted allyl reagent **110** as well as the methallyl reagent **111**, and both crotyl reagents (*E*)- and (*Z*)-**112**. Although this method requires a stoichiometric amount of the chiral diol auxiliary, this diol is readily recovered after the hydrolytic work-up. Aromatic aldehydes and, unless they are very hindered, aliphatic aldehydes (including functionalized ones) react efficiently to give good to high yields of products. Although α,β-unsaturated aldehydes react poorly, acetylenic aldehydes are very competent substrates even with the crotyl reagents **112** (Eq. 78).[28]

Sc(OTf)$_3$ (2-10 mol%)
CH$_2$Cl$_2$, −78°

110 R^2, R^3, R^4 = H
111 R^2, R^3 = H, R^4 = Me
(*E*)-**112** R^2 = Me, R^3, R^4 = H
(*Z*)-**112** R^2, R^4 = H, R^3 = Me

113

(50-95%)
>98% d.r., up to 98% ee

(Eq. 77)

(Eq. 78)

On the basis of mechanistic studies,[29] a closed bimolecular transition structure **113** involving activation of the boronate unit via coordination of the scandium to one of the dioxaborolane oxygen atoms has been proposed (Eq. 77). A possible interpretation to explain the enantioselectivity of this allylation system originates from the accepted stereoinduction model for the non-catalyzed reaction arising from a π_{phenyl}-$\pi^*_{C=O}$ attraction.[121,161] It has been proposed that the Sc(III) ion coordinates to the least hindered lone pair (syn to H) of the pseudo-equatorial oxygen, thereby suppressing n_O-p_B conjugation and maximizing boron-carbonyl bonding.[29]

The real promise of this catalytic reaction is the eventual development of an efficient enantioselective allylboration catalyzed by chiral Lewis acids. A stereoselective reaction using a substoichiometric amount of a chiral director has been reported, but only modest levels of stereo-induction were achieved with an aluminum–BINOL catalyst system (Eq. 79).[25] Recently, a chiral Brønsted acid catalyzed system has been devised based on a diol–tin(IV) complex (Eq. 80).[162] In this approach, aliphatic aldehydes provide enantioselectivities (up to 80% ee) higher than those of aromatic aldehydes when using the optimal complex **114**.[163] Although the levels of absolute stereoselectivity of this method remain too low for practical uses, promising applications are possible in double diastereoselection (see section on "Double Diastereoselection").

(Eq. 79)

(Eq. 80)

A copper-catalyzed reaction using a chiral diphosphine ligand, DuPHOS, with an added lanthanide salt, provides good levels of enantioselectivity (67–91% ee) in additions of the simple allylboronate **31** to both aromatic and aliphatic ketones that present a large difference of steric bulk on the two sides of the carbonyl group.[164] One such example is shown in Eq. 81. On the basis of ^{11}B NMR experiments and on the lack of diastereoselectivity in crotylation reactions, the mechanism of this allylation is believed to involve transmetalation of the boron to an allylcopper species.

$$\text{31 (1.2 eq)} + \xrightarrow[\substack{\text{La(OPr-}i)_3 \text{ (4.5 mol\%)} \\ \text{DMF, } -40°, \text{ 1 h}}]{\substack{\text{CuF}_2\bullet\text{2H}_2\text{O (3 mol\%),} \\ (R,R)\text{-}i\text{-Pr-DUPHOS (6 mol\%)}}} \quad\quad (87\%) \text{ 90\% ee} \quad\quad \text{(Eq. 81)}$$

Double Diastereoselection with Chiral Carbonyl Substrates

Double diastereoselective additions between chiral allylic boron reagents and enantiopure chiral carbonyl compounds can be highly advantageous because the resulting diastereomers are separable (in theory, though not always in practice).[165] The most desirable situation is where there is very high reagent control to override the intrinsic effect of the stereogenic center of the aldehyde or ketone on the diastereofacial selectivity. The success of this approach depends heavily on the structure of the particular chiral aldehyde used and the nature of its substituents, including protecting groups. Fortunately, the directing effect of the bis(isopinocampheyl) allylic boranes is extremely powerful, giving rise to high reagent control in double diastereoselective additions. As demonstrated with the simple allylations of chiral, α-substituted aldehydes 39 and 115, reagent-controlled additions are very effective (Eqs. 82–85).[166] Both examples involving aldehyde 39 (Eqs. 82 and 83) are very favorable in comparison to the modest selectivity of the achiral pinacol allylboronates (compare to Scheme 2). Reagent control is observed even in mismatched situations such as the difficult case of α-phenyl aldehyde 115 (Eq. 85).[166,167]

By using a judicious combination of the appropriate double bond geometry and pinene enantiomer, the corresponding crotylboranes present a very nice solution to the problem of full stereocontrol of acyclic dipropionate units when combined with chiral α-methyl β-alkoxy aldehydes such as **116** (Eqs. 86–89).[167] In this situation as well, reagent-controlled additions are very selective even in mismatched cases, affording stereotriads in high yields and >90% d.r. Simple α-alkoxy aldehydes provide comparable levels of selectivites but exceptions have been reported with stereogenic β-alkoxy centers and more complex aldehydes.[168]

(Eq. 86)

(84%) 95:5 d.r.

(Eq. 87)

(87%) 98:2 d.r.

(Eq. 88)

(78%) 95:5 d.r.

(Eq. 89)

(83%) 92:8 d.r.

In comparison, additions of chiral allylic boronates such as tartrate-based reagents tend to be less selective in double diastereoselective additions to α-methyl aldehydes, especially in mismatched cases (Eqs. 90–92).[169] Nonetheless, these reagents are advantageous as they can increase, even in the mismatched manifold, the intrinsic levels of selectivities dictated by the chiral aldehyde substrate, and the resulting diastereomeric products are usually separable by chromatography. When used with α-alkoxy aldehydes such as glyceraldehyde acetonide (**51**, Scheme 4), tartrate-based reagents are highly effective.[126,170] Several applications of double diastereoselective additions of tartrate-based allylic boronate reagents have been described in the course of the synthesis of complex natural products (see section on "Applications to the Synthesis of Natural Products").

(Eq. 90)

anti-syn   syn-syn

(—) 55:29 anti-syn:syn-syn
+ other diastereomers (16%)

(71%) 95:1 anti-syn:syn-syn
+ other diastereomer (4%)

(Eq. 91)

(—) 45:41 anti-syn:syn-syn
+ other diastereomers (14%)

(Eq. 92)

The use of chiral Brønsted acids is illustrated in Eq. 93 as a method for catalyst-controlled double diastereoselective additions of pinacol allylic boronates.[163] Aside from circumventing the need for a chiral boronate, these additions can lead to very good amplification of facial stereoselectivity. For example, compared to both non-catalyzed (room temperature, Eq. 90) and $SnCl_4$-catalyzed variants, the use of the "matched" diol-$SnCl_4$ enantiomer at a low temperature leads to a significant improvement in the proportion of the desired anti-syn diastereomer in the crotylation of aldehyde 117 with pinacolate reagent (Z)-7 (Eq. 93). Moreover, unlike reagent (Z)-11 (Eq. 91) none of the other diastereomers arising from Z- to E-isomerization is observed.

Conditions		anti-syn:syn-syn
$SnCl_4$ alone	(81%)	66:34
$SnCl_4$, (R,R)-diol (11 mol%)	(77%)	95:5
$SnCl_4$, (S,S)-diol (11 mol%)	(50%)	68:32

(Eq. 93)

The chiral α-substituted reagents **21**, **23**, **28**, and **75** (Fig. 6) are very powerful controllers of double diastereoselective additions and perform admirably well even in mismatched situations.[63–66,68,83,147,149] Reagent **75** tends to be more selective than the α-chloro analog **28**, and the high stereocontrol displayed by reagent **75** in the examples of Eqs. 94 and 95 can be explained using a transition structure model similar to that of Scheme 8.[68,147] Remarkably, even the mismatched pair (Eq. 95) affords outstanding selectivity. Several other examples have been described, including applications to the synthesis of complex polypropionate natural products (see section on "Applications to the Synthesis of Natural Products": Scheme 19).

syn-anti   anti-anti

(59%) >95:5 syn-anti:anti-anti   (Eq. 94)

(—) 5:>95 syn-anti:anti-anti   (Eq. 95)

Intramolecular Reactions

Intramolecular additions generally follow the same trends of stereoselectivity as observed in the bimolecular reactions.[80,82,171,172] For example, allylic boronates (*E*)- and (*Z*)-**118** provide the respective trans- and cis-fused products of intramolecular allylation.[173] As shown with allylboronate (*E*)-**118**, a Yb(OTf)$_3$-catalyzed hydrolysis of the acetal triggers the intramolecular allylboration and leads to isolation of the trans-fused product **119** in agreement with the usual cyclic transition structure (Eq. 96).

119
(77%) >92% trans

(Eq. 96)

Intramolecular allylations can also provide five-membered rings with ease, and the entropic benefits facilitate additions to ketone substrates. For example, the allylic boronate **121**, formed by a Negishi-type coupling between alkenyl bromide **120** and a methylzinc dialkoxyboron reagent, cyclizes in situ into a cyclopentanone to provide a [3.3.0] bicyclic alcohol product (Eq. 97).[174] This example constitutes a formal 5-exo-trig cyclization, and seven-membered ring systems can also be elaborated using a similar approach.[79,80,174]

(Eq. 97)

Reagents for Multiple Additions

A tandem double-allylation strategy, based on the E-3-boronyl allylborane reagent **100**,[157] has been optimized for the synthesis of 1,5-diol products from two different aldehyde substrates.[175] Specifically, reagent **100** undergoes allylation with a limiting amount of an aldehyde, R^1CHO, and the resulting α-substituted allylboronate **122** can then add to a second added aldehyde (R^2CHO) (Eq. 98). The first allylation with the bis(isopinocampheyl)allylic borane unit is highly enantioselective and the resulting configuration of **122** controls the fate of the second allylation. Thus, from intermediate **122**, transition structure **123** featuring a pseudo-equatorial α-substituent explains the stereocontrolled formation of 1,5-diol **124**. The lower reactivity of **122** compared to **100**, as well as a tight control of reagent stoichiometry, helps to minimize the formation of the double allylation product of the first aldehyde (R^1CHO).

(Eq. 98)

With the analogous reagent **125**, however, the corresponding allylboronate intermediate **126** is thought to favor a transition structure **127** where the α-substituent is positioned in a pseudo-axial orientation in order to escape non-bonding interactions with the bulky tetraphenyl dioxaborolane (Eq. 99). This way, a Z-configured allylic alcohol unit of opposite configuration is obtained in diol product **128**. This type of steric control with chiral α-substituted allylboronates had been demonstrated before (see section on α-substituted reagents). The usefulness of this powerful tandem allylation/allylation strategy is demonstrated by the preparation of several examples of both types of 1,5-diols **124** and **128**, and by a number of successful applications to the synthesis of complex natural products (see section on "Applications to the Synthesis of Natural Products").

(Eq. 99)

A double Brown-type allylation reagent, **129**, gives C_2-symmetric 3-methylenepentane-1,5-diols **130** in modest yields but high enantioselectivities with both aliphatic and aromatic aldehydes (Eq. 100).[176] This reagent can be prepared by double deprotonation of 2-methylpropene followed by condensation with either antipodes of $(Ipc)_2BCl$. For example, (S,S)-**129** originates from $[(+)-Ipc]_2BCl$. Double allylboration of an excess of aldehyde with (S,S)-**129** affords C_2-symmetric diol product **130** accompanied with approximately 5–15% proportion of "monoallylboration" alcohol after the oxidative work-up.

(Eq. 100)

Tandem Reactions

Many of the recent advances in synthetic applications of allylic boron reagents have focused on the use of these reagents as key components of tandem reactions and "one-pot" sequential processes, including multicomponent reactions. The following examples briefly illustrate the range of possibilities. Most cases involve masked allylboronates as substrates, and the tandem process is usually terminated by the allylboration step.

Intramolecular Allylboration. In one rare but impressive example involving a masked aldehyde, a domino hydroformylation/allylboration/hydroformylation reaction cascade has been designed to generate bicyclic annulated tetrahydropyrans[177] and nitrogen heterocycles.[178,179] For example, treatment of allylboronate **131** under hydroformylation conditions first leads to aldehyde intermediate **132** (Eq. 101). With a poised allylic boronate, formation of **132** triggers the key intramolecular allylation to give intermediate **134**.[179] This intermediate then undergoes a second hydroformylation, followed by a final cyclization to give the bicyclic lactol product **135** in 83% yield as a 1 : 1 mixture of anomers. The final products are obtained in this one-pot process with a very high diastereoselectivity (97 : 3 ratio) as a result of simple diastereocontrol in the allylboration step involving the putative transition structure **133**. Similar approaches are employed to access pyran derivatives.[177]

(Eq. 101)

A masked allylic boron unit can be revealed through a transition-metal-catalyzed borylation reaction. For example, a one-pot borylation/allylation tandem process based on the borylation of various ketone-containing allylic acetates has been developed.[180] The intramolecular allylboration step is very slow in DMSO, which is the usual solvent for these borylations of allylic acetates (see Eq. 33). The use of a non-coordinating solvent like toluene is more suitable for the overall process provided that an arsine or phosphine ligand is added to stabilize the active Pd(0) species during the borylation reaction. With cyclic ketones such as **136**, the intramolecular allylation provides cis-fused bicyclic products in agreement with the involvement of the usual chairlike transition structure, **137** (Eq. 102).

(Eq. 102)

Intermolecular Allylboration. A tandem aza[4+2] cycloaddition/allylboration three-component reaction[181,182] has been designed based on the precedented carbocyclic [4+2] cycloaddition/allylboration[89] and a subsequent one-pot variant.[183] Thus, the thermal reaction between hydrazonobutadienes **138**, N-substituted maleimides, and aldehydes provides polysubstituted α-hydroxyalkylpiperidines **141** via the cyclic allylboronate intermediate **139** and the proposed chairlike transition structure **140** (Eq. 103).[181] Monoactivated dienophiles like acrylates fail to react with heterodienes **138** but the scope of aldehydes is very broad; both aliphatic and aromatic aldehydes are suitable, including electron-rich ones.[182] An inverse electron-demand variant to access the corresponding dihydropyran derivatives via the intermediacy of enantiomerically enriched pyranyl allylic boronate **76** has been subsequently developed (see Eq. 64).[150–152]

(Eq. 103)

As described above in Eq. 43, simple allylboronates can be transformed into more elaborated ones using olefin cross-metathesis.[184] Treatment of pinacol allylboronate **31** with a variety of olefin partners in the presence of Grubbs' second-generation catalyst **142** smoothly leads to formation of 3-substituted allylboronates **143** as cross-metathesis products (Eq. 104). Unfortunately, these new allylic boronates are formed as mixtures of geometrical isomers with modest E/Z selectivity. They are not isolated but rather are treated directly with benzaldehyde to give the corresponding homoallylic alcohol products in good yields (Table A).

(Eq. 104)

Table A. Functionalized homoallylic alcohols from olefin cross metathesis and subsequent allylboration reactions with benzaldehyde

Entry	Cross Partner	Product	Yield	d.r.
1			(73%)	3.8:1
2			(78%)	4.9:1
3			(58%)	>20:1
4			(66%)	—
5			(60%)	>20:1

The main advantage of this one-pot sequence is that it is exceptionally tolerant of sensitive functional groups. Entry 3 (Table A) shows an example with an unprotected alcohol, and entries 1, 2, and 5 all show examples with halogenated groups that are delivered directly from an allylboronate intermediate. These groups, which would not survive the strongly basic conditions or the active metals used in many other preparative methods, are carried through this procedure without incident. Entry 4 is also noteworthy because it shows that quaternary carbon centers can be made with this tandem process. A serious limitation of this approach, however, is that the diastereoselectivity seen in the formation of allylation products is quite variable as a consequence of the low geometric (E/Z) selectivity in the cross-metathesis. Olefin partners with large allylic substituents (entries 3–5) react to give exclusively the anti product shown in Table A. Unfortunately, olefins with smaller substituents (entries 1 and 2) show a much lower preference. Furthermore, both E- and Z-olefins afford the same stereoisomer (compare entries 1 and 2).

A mild one-pot procedure based on a platinum-catalyzed diborylation of 1,3-butadienes (see Eq. 30) gives doubly allylic boronate **144**, which adds to an aldehyde to form a quaternary carbon center in the intermediate **145** (Eq. 105).[185] The use of a tartrate auxiliary in this process leads to good levels of enantioselectivity in the final diol product, which is obtained after oxidation of the primary alkylboronate intermediate. Although examples of aliphatic, aromatic, and unsaturated aldehydes have been described, enantioselectivities vary widely (33 to 74% ee), and are good only for aliphatic aldehydes. An intramolecular variant of this interesting tandem reaction is also known.[186]

(72%) >19:1 syn:anti, 74% ee

(Eq. 105)

The diboration of allenes can also be employed in a similar tandem process. By taking advantage of an improved, more enantioselective variant for the preparation of allylboronates of type **25** (Eq. 32), a one-pot sequence involving allene diboration, aldehyde allylation, and Suzuki cross-coupling has been developed (Eq. 106). It is noteworthy that no change of solvent, nor any addition of palladium catalyst is needed in the last stage involving the cross-coupling of alkenylboronate intermediate **146** to give final product **147**.[187]

(Eq. 106)

APPLICATIONS TO THE SYNTHESIS OF NATURAL PRODUCTS

Enantioselective Additions

The bis(isopinocampheyl)borane reagents described in the sections on enantioselective additions have found extensive use in the total synthesis of complex, bioactive natural products. A synthesis of the potent anticancer agent epothilone

Scheme 10

exemplifies particularly well the suitability of the Brown allylation with highly functionalized and hindered aldehydes (Scheme 10). Thus, using the salt-free, low-temperature conditions, reagent **64** is added to unsaturated heterocyclic aldehyde **148** to afford the desired secondary homoallylic alcohol in very high yield and excellent enantioselectivity. The synthesis also featured another Brown allylation on the hindered α-dimethylated β-keto aldehyde **149**, thus providing two of the key fragments necessary to assemble the structure of the anticancer agents epothilones.[188]

The applicability of the bis(isopinocampheyl) reagents is not limited to simple allylations and crotylations. For example, reagent **97** affords syn vicinal diols in very high enantioselectivity (Scheme 11).[189] Compared with reagents **14** and **96** (Fig. 7), reagent **97** possesses the added advantage of yielding the alcohol product with an easily removable (2-trimethylsilylethoxy)methyl (SEM) protecting group. In this example, the resulting monoprotected diol has been subsequently converted into the eight-membered ring ether laurencin.

With only a small number of substrate types are the tartrate-based allylic boronate reagents enantioselective enough for applications in the control of

Scheme 11

Scheme 12

absolute stereochemistry. One such example involves iron dienyl complexes such as the dialdehyde **150**, which reacts with crotylboronate reagent (*E*)-**11** to give the anti-propionate product in the form of complex **151** in over 98% ee (Scheme 12). After a few transformations and decomplexation, the resulting diene is further elaborated into a triene precursor that undergoes an intramolecular Diels–Alder reaction leading to the tricyclic core of ikarugamycin.[190]

The Corey allylation system based on a chiral bis(sulfonamide) auxiliary was put to use with success in a number of synthetic efforts, including the total synthesis of the anticancer agent leucascandrolide (Scheme 13).[191] Chiral reagent **152** is added to an achiral aldehyde, 3-(*p*-methoxybenzyloxy)propanal, affording intermediate **153** in high stereoselectivity. The latter is transformed into a pyranyl aldehyde, which is subjected to a second allylation (this time, a doubly diastereoselective addition) en route to the completion of leucascandrolide.

Double Diastereoselection

The powerful directing effect of bis(isopinocampheyl) allylic boranes has been put to great use in the context of several applications of double diastereoselective allylations in the total synthesis of natural products. As discussed in a previous section, the Brown allylation can be exploited to overcome the stereodirecting effect of chiral α-stereogenic aldehydes, including α-alkoxy substituted ones. Thus, the simple allylation of aldehyde **154** provides as major product the desired diastereomer needed towards a total synthesis of brasilenyne (Scheme 14).[192] The yield and stereoselectivity is even increased to over 97 : 3 under the low-temperature, magnesium-free conditions described before.[139]

A synthesis of the naturally occurring phosphatase inhibitor calyculin C showcased several other interesting examples of doubly diastereoselective additions involving the Brown reagents.[168] The addition of tetrasubstituted reagent **155** to 2,3-*O*-isopropylidene-D-glyceraldehyde (**51**) provides intermediate **156** in very high selectivity despite a substrate-reagent mismatch, which is overruled by the borane reagent (Scheme 15). The latter is transformed in several steps into aldehyde **157**, and a crotylboration of this β-alkoxy aldehyde provided a very high diastereoselectivity for the desired anti-propanoate unit. With the subsequent aldehyde intermediate **158**, a third example involves a difficult mismatched case that

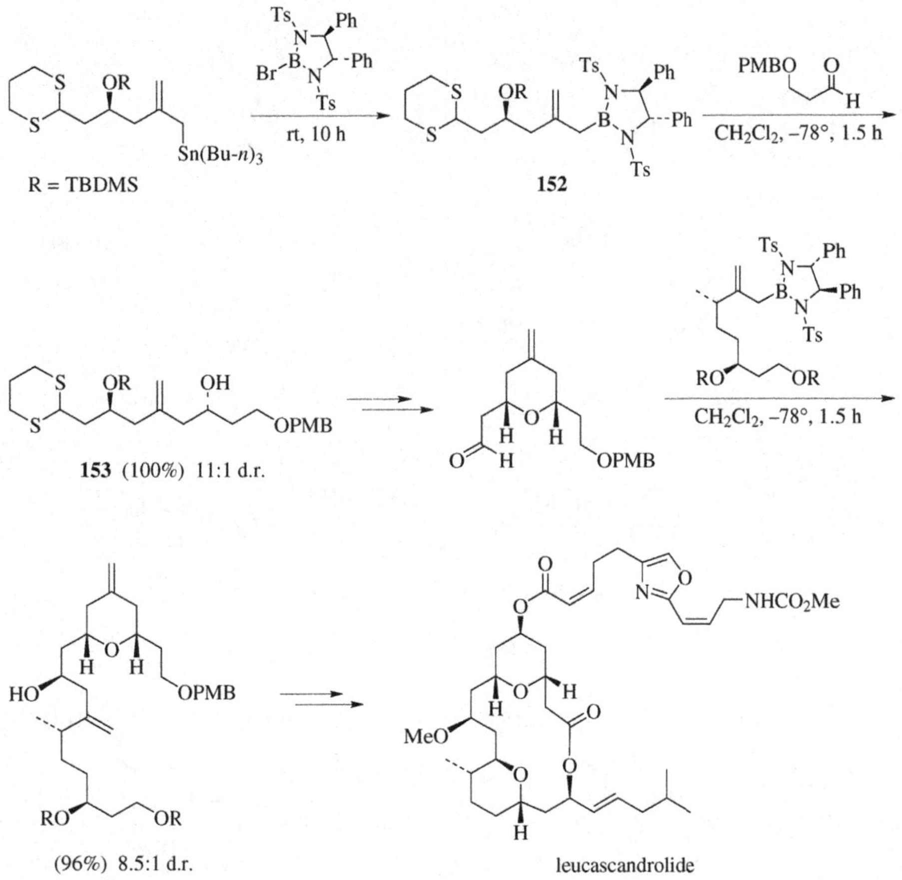

Scheme 13

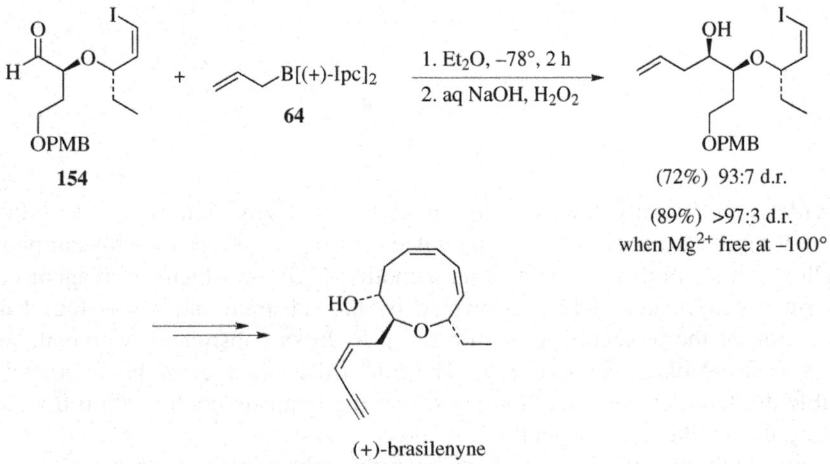

(+)-brasilenyne

Scheme 14

Scheme 15

provides an unusually low level of diastereoselectivity (almost $1:1$), indicating that high selectivity is never guaranteed even with the bis(isopinocampheyl) allylic boranes. In this example of an α-methyl-β-alkoxy aldehyde, reagent control by crotylborane (E)-**12** is overruled by the substrate and it was found that the nature of the protecting group on the β-hydroxy substituent is crucial, with the most favorable group a benzoyl. The minor diastereomer of this mismatched double diastereoselective addition possesses the requisite configuration for completing the synthesis of calyculin C.

Several allylic boron reagents have been recently employed in a series of doubly diastereoselective additions toward a synthesis of mycalamide.[193,194] Thus,

Scheme 16

the addition of tartrate-derived reagent **159** to 2,3-O-isopropylidene-D-glyceralde-hyde provides diastereomer **161** with very high selectivity (Scheme 16). This result contrasts with the formation of **156** using a bis(isocampheyl)-borane reagent (Scheme 15). As described in a previous section, double diastereoselective additions of allylic boron reagents to the glyceraldehyde acetals usually proceed with high selectivity in both matched and mismatched cases due to the lack of a strong bias among the four main rotamers in the transition structures minimizing syn-pentane interactions (see Scheme 4). Unlike the additions of crotyl reagents (see Schemes 2 and 3), transition structures from additions of 3,3-disubstituted reagents cannot escape from syn-pentane interactions and all four of the structures of Scheme 4 display one such interaction between a methyl group and the diox-olane ring. Thus, as substrate control is somewhat mitigated in these instances, reagent control is more effective, which explains the formation of epimeric alcohols with a judicious choice of the right enantiomer of reagents **155** or **159**. In the formation of product **161** from reagent (R,R)-**159**, the stereoselectivity of this matched manifold can be explained by a Felkin–Anh transition structure **160**. Such double diastereoselective additions between tartrate-derived reagents and glyceraldehyde ketals usually proceed with very high stereoselectivity.[170] The intermediate **161** is transformed into the hindered α-dimethylated aldehyde **162**, which undergoes another doubly diastereoselective addition with Brown's reagent **64** that results in stereomerically pure product **163**. The corresponding tartrate-derived reagent [(S,S)-**53**, Fig. 4] gives a much lower selectivity with the same aldehyde. Further transformations of the intermediate **163** leads to α-methoxy aldehyde **164**, whose reaction with E-3-trimethylsilyl tartrate reagent **165** (similar to reagent **101**, Fig. 7) occurred with high selectivity in the matched manifold to give product **166**. The stereoselectivity of this addition can be explained by minimization of syn-pentane interactions just as observed with an E-crotyl reagent (see structure **40** in Scheme 2). As with reagent **101** (see Eq. 74), the allylsilane unit of **166** is oxidized with rearrangement to give an advanced intermediate towards mycalamide A.

As demonstrated in the course of a total synthesis of the macrolide bafilomycin,[195] double diastereoselective additions can be useful even in the mismatched manifold. For example, the crotylation of chiral α-substituted aldehyde **167** with (E)-**11** affords an 85 : 15 ratio of diastereomers favoring the desired anti-anti product (Scheme 17). Without a chiral tartrate reagent, the undesired anti-syn diastereomer would be intrinsically favored from aldehyde **167**. The use of the appropriate tartrate reagent, the (R,R) unit in this instance, overturns this preference to afford an acceptable ratio of the two separable diasteomers.

Although the class of bis(isopinocampheyl)allylboranes often leads to better levels of double diastereoselectivity, the basic oxidative work-up required in these allylations is not compatible with all substrates. This is true for the crotylation of the aldehyde derived from the terminal alkene of **168** (Scheme 18),

167 (R,R)-(E)-**11**

3,4-anti-4,5-anti (70%)   3,4-anti-4,5-syn (—)

85:15

Scheme 17

Scheme 18

which requires the use of boronate (E)-**11** because of the milder hydrolytic workup associated with this class of reagents.[196] This matched case of double diastereoselection is very efficient, giving intermediate **170** as the exclusive diastereomer, which was further elaborated to reach the targeted natural product, 13-deoxytedanolide. The high stereoselectivity in favor of the Felkin product **170** can be explained through chairlike transition structure **169** using the same stereoinduction model described in Scheme 2, with the largest chain (R) occupying the "outside" position.

As demonstrated in a total synthesis of the denticulatins A and B, the class of chiral α-substituted allylic boronates is very effective in double diastereoselection (Scheme 19).[197] From aldehyde **171**, a simple E-crotylation provides the Felkin product **173**. The high stereoselectivity of this crotylation can be explained by invoking transition structure **172** using the same model described in Scheme 2, which minimizes syn-pentane interactions with the largest of the three α-substituents positioned at the outside. Intermediate **173** is then transformed into aldehyde **174**, which is subjected to another E-crotylation, however this time with α-methoxy reagent **75**. The use of reagent **75** is necessary to optimize the diastereomeric ratio in favor of the desired anti-Felkin stereoisomer **175**, which results from a mismatched situation. Thus, the 4:1 diastereoselectivity of this reagent-controlled addition is remarkable as additions to α-methyl β-branched aldehydes like **171** and **174** intrinsically favor the Felkin product (e.g., **173** from **171**). Here, the stereocontrolling effect of reagent **75** is powerful enough to reverse this trend and to favor diastereomer **175**. The latter cyclizes spontaneously into a hemiketal intermediate en route to the synthesis of the denticulatins A and B.

Scheme 19

Tandem Reactions

Allylic boronates are very suitable for the design of tandem reactions that can be applied in natural product synthesis. Using the strategy illustrated in Eq. 64, an inverse electron-demand hetero-Diels–Alder reaction between **87** and a 2-substituted enol ether, catalyzed by chiral Cr(III) complex **88**, leads in high enantioselectivity to cyclic dihydropyranyl boronate **176** (Scheme 20).[198] The latter adds stereoselectively, in a one-pot sequential process, to unsaturated aldehyde **177** to give 2-hydroxyalkyl dihydropyran product **178**. Interestingly, the large, chiral catalyst **88** promotes the cycloaddition of the requisite Z-enol ether faster than that of the corresponding E-isomer. The remarkable selectivity of this tandem

$R^1 = CH_2CH=CMe_2$
$R^2 = (E)-MeC=CHCO_2Et$

178 (76%) 98% d.r., 95% ee

thiomarinol H

Scheme 20

reaction allowed an expedient enantioselective synthesis of a potent antibiotic of the thiomarinol family. A related aza[4+2] cycloaddition/allylboration has been applied to the enantioselective synthesis of palustrine alkaloids.[199,200]

The double allylation reagent **125** described in a previous section (see Eq. 99) has been put to use with remarkable efficiency in the context of natural product synthesis. In the construction of the pyran-containing C43-C67 fragment of the complex marine natural product amphidinol 3, a one-pot double allylation with two different acetonide-protected chiral α-substituted aldehydes is performed (Scheme 21).[201] The first allylation involved a stereochemically mismatched case of double diastereoselection, which is resolved by the strong directing power of the bis(isopinocampheyl)boryl unit and the optimal choice of protecting group. The desired intermediate **179** then reacts with the second aldehyde to give product **180** in high yield and with good overall stereoselectivity (9:1 d.r.). Functional group manipulations and a subsequent cyclization afford the pyran core of amphidinol 3. Reagent **125** is also used in the construction of the C1-C25 fragment of the same natural product.[202] The related reagent **100** (see Eq. 98) has been put to use toward the synthesis of peloruside.[203]

180 (70-80%) 9:1 d.r.

pyran core of amphidinol 3

Scheme 21

Transformations of Residual Groups

The residual alkenyl unit obtained in the products of addition of allylic boron reagents offers myriad possibilities for further derivatization in the synthesis of natural products. For instance, earlier sections (see sections on Preparation of 1,2-Diols, and 1,4-Diols) describe oxidative processes involving allylic silane units originating from additions of 3-silyl-substituted reagents. With the recent advent of alkene metathesis reactions, the possibilities of manipulations of the residual alkenyl unit have grown further. One attractive application in the synthesis of γ-lactones is exemplified in a synthesis of goniothalamin.[204] Thus, a simple allylation of cinnamaldehyde followed by acryloylation provide intermediate **181** (Scheme 22). The latter then undergoes a ring-closing metathesis using Grubbs' first-generation ruthenium catalyst to afford the desired natural product.

(72%) 92% ee

181

cat. (10 mol%)

CH₂Cl₂, 40°

(76%)

goniothalamin

Scheme 22

Scheme 23

A similar strategy has been employed in the synthesis of a dihydropyran required to achieve the synthesis of (−)-laulimalide (Scheme 23).[205] Thus, allylation of the secondary alcohol resulting from a methallylation of 3-tri-*n*-butylstannylacrolein provided unsymmetrical ether **182**, which is closed to a pyran with a ring-closing metathesis using Schrock's molybdenum catalyst.

The class of 3-silyl-substituted reagents provides, upon addition with aldehydes, allylic silanes that offer many options for further derivatization. Oxidative processes are described in previous sections (see the sections on Preparation of 1,2-Diols and 1,4-Diols). If the appropriate silicon substituents are chosen, formal [3+2] cycloadditions with aldehydes can be promoted under Lewis acid catalysis. For example, the mismatched addition of the Z-3-propyl-3-benzhydryldimethyl allylsilane **183** to an α-benzyloxy aldehyde proceeds with low diastereofacial selectivity in favor of product **184**; however, after protection of the secondary alcohol, an efficient [3+2] annulation provides the polysubsubstituted furan **185** in good yield and acceptable stereoselectivity (Scheme 24).[206] The latter is brought forward to a tricyclic unit found in the antitumor natural product angelmicin B.

A special α-substituted allylic boronate, reagent **187**, has been developed as part of a tandem method for synthesizing trienes through an extended homologation of aldehydes.[207] This process, a one-pot aldehyde allylation/dioxene thermolysis, has been applied to a total synthesis of phenalamide A_2 (Scheme 25).[208] Thus, reagent **187** is added to unsaturated aldehyde **186** to give dioxene intermediate **188**. Upon warming to 120°, **188** undergoes a formal retro[4+2] cycloaddition to give the desired dienal as a mixture of geometrical isomers. The addition of iodine promotes a complete Z- to E- isomerization to give the E,E-dienal intermediate **189**. A dehydration of the latter then provides the tetraenal **190**, an advanced intermediate subsequently transformed into phenalamide A_2.

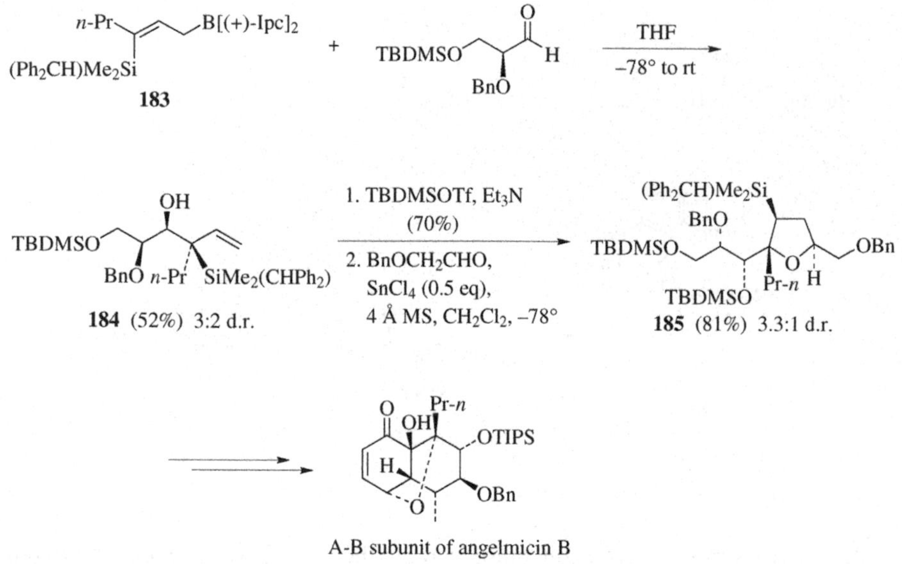

A-B subunit of angelmicin B

Scheme 24

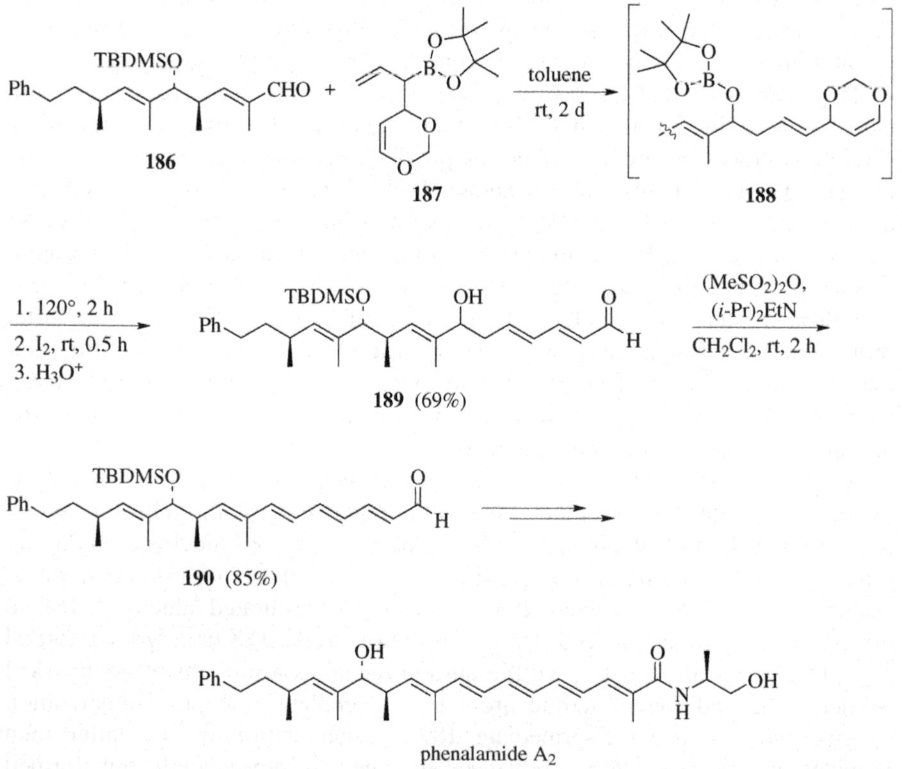

phenalamide A$_2$

Scheme 25

COMPARISON WITH OTHER METHODS

Despite the extensive efforts by numerous research groups over the course of more than two decades, there is still no carbonyl allylation methodology that possesses all of the following attributes: mildness and chemoselectivity, substrate generality (for both allylmetal reagent and aldehyde substrate), high levels of diastereo- and enantioselectivity, and high practicality (ease of use, low cost, non-toxicity, and low environmental impact). Very importantly, the ideal enantioselective allylation methodology would circumvent the use of a chiral auxiliary through a simple and efficient chiral catalyst.[209] While several systems based on allylic boron reagents do possess many of the above attributes, several other allylation systems based on other metals (mainly tin, titanium, and silicon) demonstrate very interesting properties. It is noteworthy that a wide body of aldol-based reaction methods have been developed to afford chiral secondary alcohols, propionate units, and other useful intermediates closely related to those afforded by allylation processes. These methods are not described in this section but the interested reader will find several excellent reviews and monographs on the topic of aldol reactions.[210–212] This section focuses on presenting a brief comparison of allylic boron reagents with other enantioselective allylation reagents.

Methods Employing Allylic Tin Reagents

Allylic trialkyltin reagents can react at high temperatures with aldehydes[213] but because of the low acidity of the trialkyltin center, they are always employed as Type II allylation reagents that normally require activation of the carbonyl substrate with an external Lewis acid.[15] This behavior contrasts the cyclic, organized transition structure of allylboration reactions. Consequently, additions of allylic tin reagents proceed through open transition structures that tend to demonstrate lower levels of diastereoselectivity (with 3-substituted reagents) compared with the corresponding boron reagents. Allylic trialkyltin reagents are highly nucleophilic in comparison to the corresponding trialkylsilanes. Their reactivity has contributed largely to their popularity. One major drawback of organotin-based reagents, however, is their high toxicity. Moreover, crotylation reagents are rather difficult to synthesize in geometrically pure form, albeit, this point is rather inconsequential because of the stereoconvergent (Type II) behavior of 3-substituted tin-based reagents. Catalytic enantioselective methods that make use of achiral reagents constitute the most attractive carbonyl allylation processes. In this respect, the Keck allylation remains the most efficient and most popular system based on achiral allylic tin reagents (Eq. 107).[214,215] This system has been employed mainly for simple allylations and methallylations of aldehydes. It makes use of a chiral BINOL-based titanium Lewis acid, which provides enantiomeric excesses as high as 94–96% for several types of aldehydes, both aliphatic and aromatic. The catalyst can be employed in a loading as low as 1 mol% when i-PrSSiMe$_3$ is added.[216] Efficient variants of these systems have been described for the allylation of ketones.[217] A BINAP–AgOTf complex catalyzes the simple allylation of aldehydes with enantiomeric excesses as high as

96% for benzaldehyde.[218] Methallylations and crotylations of aldehydes under this catalytic system provide good enantioselectivities.[219]

$$\text{RCHO} \quad + \quad \diagup\!\!\diagdown\!\!\diagup\text{SnBu}_3 \xrightarrow[\text{4 Å MS, CH}_2\text{Cl}_2, -20°]{} \qquad \text{(Eq. 107)}$$

(78-98%) 89-96% ee

Additions of enantiomerically enriched allenyltin reagents provide homopropargylic alcohol products with high levels of stereoselectivity (Eq. 108).[220] These reagents also function through Lewis acid catalysis. Although their preparation requires several steps through the intermediacy of enantiomerically pure propargylic alcohols, allenyltin reagents are effective with a wide range of aldehydes, including double diastereoselective additions with chiral α-substituted aldehydes.[220]

$$\text{RCHO} \quad + \quad \xrightarrow[\text{CH}_2\text{Cl}_2]{\text{BF}_3\cdot\text{OEt}_2} \qquad \text{(Eq. 108)}$$

(>80%) up to 99:1 d.r.

Methods Employing Allylic Silicon Reagents

Allylic trialkylsilicon reagents are less nucleophilic than the corresponding tin reagents. However, they are cheap and safer in terms of toxicity. One of the early successes in the area of catalytic enantioselective carbonyl allylation chemistry makes use of chiral acyloxy borane (CAB) catalysts and methallyltrimethylsilane (Eq. 109).[221,222] This system works for aromatic, unsaturated, and aliphatic aldehydes but it gives the highest enantioselectivities in the reactions of 3-alkyl substituted reagents and aromatic aldehydes. Aliphatic aldehydes give noticeably lower selectivities, which probably accounts for the low level of adoption of this allylation method.

$$\text{RCHO} \quad + \quad \diagup\!\!\diagdown\!\!\diagdown\text{TMS} \xrightarrow[\text{EtCN}, -78°]{\text{(10–20 mol\%)}} \qquad \text{(Eq. 109)}$$

Ar = 3,5-(CF$_3$)$_2$C$_6$H$_3$ (55-99%) 54-88% ee

A method using dual activation has been developed in which a Lewis acid activates the aldehyde with concomitant nucleophilic activation of the allylic silicon reagent with fluoride anion. Thus, by using a BINOL-based titanium tetrafluoride complex, both aromatic and aliphatic aldehydes can be reacted with allyltrimethylsilane with enantioselectivities up to 94% (Eq. 110).[223] Aromatic aldehydes provide lower selectivity but this method is particularly efficient with hindered α,α-dialkylated aldehydes such as pivalaldehyde.[224]

$$
\text{RCHO} \quad + \quad \diagup\!\!\diagdown\!\!\diagup\text{TMS} \quad \xrightarrow[\text{CH}_2\text{Cl}_2/\text{CH}_3\text{CN, 0}^\circ]{\substack{\text{TiF}_4 \text{ (10 mol\%),} \\ (S)\text{-BINOL (10 mol\%)}}} \quad \underset{R}{\overset{\text{OH}}{\diagup\!\!\diagdown\!\!\diagup\!\!\diagdown}} \qquad \text{(Eq. 110)}
$$

$$(69\text{-}93\%) \ 60\text{-}94\% \ ee$$

Allylic trialkoxysilanes are effective nucleophiles under the bifunctional Ag(I)–fluoride catalytic system when using a combination of KF and 18-crown-6 as catalytic source of soluble fluoride anion (Eq. 111).[225] Although this method requires the use of three molar equivalents of allyltrimethoxysilane, only 5 mol% of AgOTf with 2 mol% of BINAP are sufficient to afford enantiomeric excesses in the range of 93–97% for aromatic aldehydes. Aliphatic aldehydes display lower selectivities, and require a higher loading of the catalyst. The corresponding crotylation reactions are stereoconvergent, as both (E)- and (Z)-crotyltrimethoxysilane provide the anti-propanoate product in 95% ee.[225]

$$
\text{RCHO} \quad + \quad \underset{\text{3 eq}}{\diagup\!\!\diagdown\!\!\diagup\text{Si(OMe)}_3} \quad \xrightarrow[\text{THF, }-20^\circ, \text{ 4 h}]{\substack{(R)\text{-BINAP (2 mol\%), AgOTf (5 mol\%),} \\ \text{KF (5 mol\%), 18-crown-6 (5 mol\%)}}} \quad \underset{R}{\overset{\text{OH}}{\diagup\!\!\diagdown\!\!\diagup\!\!\diagdown}}
$$

$$(61\text{-}95\%) \ 87\text{-}97\% \ ee$$

$$\text{(Eq. 111)}$$

A similar Cu(I)-catalyzed reaction has been reported, and one example represents a modestly enantioselective addition to a ketone (Eq. 112).[226]

$$
\underset{\text{Ph}}{\overset{\text{O}}{\|}}\!\!\text{Me} \quad + \quad \underset{\text{1.5 eq}}{\diagup\!\!\diagdown\!\!\diagup\text{Si(OMe)}_3} \quad \xrightarrow[\text{THF, rt}]{\substack{\text{CuCl/}(R)\text{-tol-BINAP (15 mol\%)} \\ \text{TBAT (15 mol\%)}}} \quad \underset{\text{Ph}}{\overset{\text{HO}}{\diagup\!\!\diagdown\!\!\diagup\!\!\diagdown}}
$$

$$(100\%) \ 56\% \ ee$$

$$\text{(Eq. 112)}$$

Another approach toward activating allylic silanes exploits the ability of some basic solvents and additives to accelerate the additions of allylic halosilanes. In the latest advance, treatment of an allylic trichlorosilane with a dimeric chiral phosphoramide catalyst generates a highly reactive silane species which then allylates aldehydes to give homoallylic alcohols in high stereoselectivity (Eq. 113).[227] Aromatic and unsaturated aldehydes react smoothly with the simple allyl reagent and both crotylsilanes to give homoallylic alcohols in good yields and high diastereo- and enantioselectivities.[228] The mechanism of these allylations has been studied in detail. The high diastereoselectivity and the dependence of product stereochemistry on the geometry of γ-substituted silanes indicate that these allyltrichlorosilanes react like allylic boron reagents through the Type I pathway. A significant limitation to this approach is that it is not applicable to aliphatic aldehydes.

$$R^1CHO + \quad R^2 \diagdown \diagup SiCl_3 \qquad \xrightarrow[\text{CH}_2\text{Cl}_2, -78°]{\text{(5 mol%)}} \qquad R^1 \diagup \diagup \text{OH} \qquad \text{(Eq. 113)}$$

$$R^2, R^3 = \text{H or Me}$$

(57-92%) 80-97% ee

Non-racemic α-substituted allylic silanes, in particular crotylsilanes, are very attractive reagents despite their rather tedious preparation. They were found to provide very high transfer of chirality in their additions to achiral aldehydes under Lewis acid catalysis (Eq. 114).[229] These reagents have been tested several times in the context of natural product synthesis. Their diastereoselectivity (syn/anti) depends on several factors, including the nature of the aldehyde substrate, the reagent, and the nature of the Lewis acid employed. For example, the syn product can be obtained predominantly in the reaction of Eq. 114 by switching to the use of a monodentate Lewis acid such as BF_3.

anti
(59%) 12:1 d.r.

(Eq. 114)

Another approach for the activation of allylic silanes takes advantage of the fact that constraining silicon in a small ring increases its Lewis acidity. By reacting allyltrichlorosilane with 1,2-diols, diamines, or amino alcohols, allylchlorosilanes are generated in which the silicon is part of a strained heterocyclic 5-membered ring. The most interesting aspect of this recent work comes from allylsilanes bearing chiral 1,2-amino alcohols and 1,2-diamines as substituents. The optimal reagents are derived from C_2-symmetrical cyclohexanediamines, and are exceptionally selective for a broad range of aldehydes, including unsaturated and aromatic ones (Eq. 115).[230] These novel reagents are stable and easy to handle, and possess a long shelf-life. Interestingly, the corresponding crotylsilanes are also efficient reagents, and behave as Type I reagents by providing stereospecific crotylations of aldehydes.[231]

$$RCHO + \qquad \xrightarrow[-10°, 20\,h]{\text{CH}_2\text{Cl}_2} \qquad R \diagup \diagup \text{OH} \qquad \text{(Eq. 115)}$$

(61-93%) 95-98% ee

Methods Employing Other Metals

Several other metals have been examined in allyl transfer reactions to carbonyl compounds. Only modest success has been obtained except for a class of cyclopentadienyl titanium reagents based on a tartrate-derived auxiliary (Eq. 116).[232] These reagents display a very high level of enantioselectivity in the simple allylation of several types of aldehydes, including hindered ones like pivalaldehyde (63% yield, 97% ee). A number of 3-substituted derivatives have been examined, including the E-crotyl reagent, which provides the anti propanoate products in excellent yields and over 98% ee for model aromatic and aliphatic aldehydes. These reagents are remarkably effective in double diastereoselective additions. Their inherent disadvantages are their preparation and their stoichiometric mode of stereochemical induction.

$$\text{(86-93\%) 94-97\% ee} \qquad \text{(Eq. 116)}$$

Allylic chromium species can also add to aldehydes. In this regard, an efficient catalytic enantioselective variant using allylic halides as substrates and manganese as co-oxidant has been described recently (Eq. 117).[233] This method provides high enantiomeric excesses in the simple allylation of a wide range of aliphatic, aromatic, and heteroaromatic aldehydes. Crotylation examples are also very enantioselective, albeit with modest anti/syn diastereoselectivity.

$$\text{RCHO} + \text{\qquad Br} \xrightarrow[\substack{\text{Mn, TESCl,} \\ \text{DME/CH}_3\text{CN (3:1), rt, 24 h}}]{\text{(3 mol\%)}} \text{(81-93\%) 93-98\% ee}$$

$$\text{(Eq. 117)}$$

EXPERIMENTAL CONDITIONS

Additions of allylic boron reagents are typically performed under experimentally simple conditions, and under a wide range of temperatures that is dictated by the reactivity of the particular reagent employed. Work-up conditions are discussed in a previous section. For the reaction itself, uncatalyzed additions can be performed in a wide variety of aprotic solvents. Non-coordinating solvents usually lead to shorter reactions times,[21] but the identification of an optimal solvent in stereoselective additions is rather unpredictable and may require coordinating solvents for higher selectivity. The use of molecular sieves is strongly recommended when using water-sensitive allylic boronates. Lewis or Brønsted acid catalyzed additions are also better performed with non-coordinating solvents, typically toluene or methylene chloride. The literature describes only a few examples of allylations with a polymer-supported reagent,[234] and with supported aldehydes.[235,236]

EXPERIMENTAL PROCEDURES

$$\overset{\displaystyle \diagup\!\!\!\diagdown\!\!\!\diagup}{} MgBr \ + \ MeOB[(+)\text{-Ipc}]_2 \xrightarrow[\text{2. pentane}]{\text{1. THF, 0° to rt}} \overset{\displaystyle \diagup\!\!\!\diagdown\!\!\!\diagup}{} B[(+)\text{-Ipc}]_2 \quad (\sim100\%)$$

(+)-*B*-Allyl bis(Isopinocampheyl)borane [General Procedure for Preparation of the Reagent Free of MgBr(OMe)].[139] Allylmagnesium bromide in ether (48 mL, 1.0 M, 48 mmol) was added dropwise to a well-stirred solution of (+)-*B*-methoxy bis(isopinocampheyl)borane (15.8 g, 50 mmol) in THF (50 mL) at 0°. After completion of addition, the reaction mixture was vigorously stirred for 1 hour at room temperature, and the solvents were removed under vacuum (14 mm Hg, 2 hours). The residue was extracted with pentane (2 × 100 mL) under nitrogen, and stirring was discontinued to permit the MgBr(OMe) salt to settle. The clear supernatant pentane extract (free from the magnesium salts) was transferred to another flask using a double-ended needle through a Kramer filter. Evaporation of pentane (14 mm Hg, 1 hour; 2 mm Hg, 1 hour) afforded the pure title product in nearly quantitative yield.

$$\overset{\displaystyle \diagup\!\!\!\diagdown\!\!\!\diagup}{} B[(+)\text{-Ipc}]_2 \ + \ \overset{O}{\underset{H}{\diagup\!\!\!\diagdown\!\!\!\diagup}} \xrightarrow[\text{2. NaOH, H}_2\text{O}_2]{\text{1. ether, }-100°} \overset{OH}{\diagdown\!\!\!\diagup\!\!\!\diagdown\!\!\!\diagup} \quad (82\%)$$

(*R*)-1,5-Hexadien-3-ol [Typical Procedure for the Simple Allylation of a Representative Aldehyde with the AllylB(Ipc)$_2$ Reagent].[139] Anhydrous ether (100 mL) was added to allylB[(+)-Ipc]$_2$, and the resulting solution was cooled to −100°. A solution of acrolein (2.8 g, 50 mmol) in Et$_2$O (50 mL), maintained at −78°, was slowly added along the side of the flask, kept at −100°. The reaction mixture was stirred for 0.5 hour at −100°, and MeOH (1 mL) was added. The reaction mixture was then brought to room temperature (1 hour) and treated with 3 N NaOH (20 mL) and 30% H$_2$O$_2$ (40 mL). Heating the reaction mixture at reflux for 3 hours ensured the completion of the oxidation. The organic layer was separated, washed with water (30 mL) and brine (30 mL), dried over anhydrous MgSO$_4$, and filtered and concentrated under reduced pressure. Distillation afforded 4.17 g of the title product (82% yield), bp 54° at 20 mm Hg; ^{1}H NMR (300 MHz, CDCl$_3$) δ 5.85 (m, 2H), 5.20 (m, 4H), 4.19 (m, 1H), 2.33 (m, 2H), 1.72 (br, 1H); ^{13}C NMR (300 MHz, CDCl$_3$) δ 139.3, 133.4, 117.9, 115.2, 72.1, 41.0. GC analysis of its Mosher ester on a capillary Supelcowax column (15 m × 0.25 cm) showed the enantiomeric excess to be 96%.

(2R,3R)-3-Methyl-4-penten-2-ol [Preparation of (−)-(Z)-Crotyl Bis(isopinocampheyl)borane and a Representative Reaction with Acetaldehyde].[137]

To a stirred mixture of t-BuOK (2.8 g, 25 mmol, dried at 80°/0.5 mm Hg for 8 hours), cis-2-butene (4.5 mL, 50 mmol), and THF (7 mL) at −78° was added n-BuLi (2.3 M in THF, 25 mmol). After complete addition of n-BuLi, the mixture was stirred at −45° for 10 minutes. The resulting solution was cooled to −78°, and methoxydiisopinocampheylborane in ether [1 M, 30 mmol, derived from (+)-α-pinene] was added dropwise. After the reaction mixture was stirred at −78° for 30 minutes, BF$_3$•OEt$_2$ (4 mL, 33.5 mmol) was added dropwise. Acetaldehyde (2 mL, 35 mmol) was then added dropwise at −78°. The reaction solution was stirred at −78° for 3 hours and treated with 3 N NaOH (18.3 mL, 55 mmol) and 30% H$_2$O$_2$ (7.5 mL). The resulting mixture was heated at reflux for 1 hour. The organic layer was separated, washed with water (30 mL) and brine (30 mL), dried over anhydrous MgSO$_4$, and filtered and concentrated under reduced pressure to afford the title product (75% yield, 90% ee, ≥99% d.r.), bp 78°/85 mm Hg; [α]$^{23}_D$ +19.40° (neat); ^{1}H NMR (CDCl$_3$) δ 6.00−5.68 (m, 1H), 5.21−4.91 (m, 2H), 3.70 (p, J = 6 Hz, 1H), 2.25 (sextet, J = 7 Hz, 1H), 2.07 (s, br, 1H), 1.16 (d, J = 6 Hz, 3H), 1.04 (d, J = 7 Hz, 3H). ^{13}C NMR (CDCl$_3$, Me$_4$Si) δ 140.62, 115.06, 70.78, 44.87, 19.96, 14.85.

(2R,3S)-3-Methyl-4-penten-2-ol [Preparation of (−)-(E)-Crotyl Bis(isopinocampheyl)borane and a Representative Reaction with Acetaldehyde].[137]

To a stirred mixture of t-BuOK (2.8 g, 25 mmol, dried at 80°/0.5 mm for 8 h), trans-2-butene (4.5 mL, 50 mmol), and THF (7 mL) at −78°, was added n-BuLi (2.3 M in THF, 25 mmol). The mixture was stirred at −45° for 10 minutes. The resulting solution was recooled to −78°, and a solution of methoxydiisopinocampheylborane (1.0 M in ether, 30 mmol, derived from (+)-α-pinene) was added. After the reaction mixture was stirred at −78° for 30 minutes, BF$_3$•OEt$_2$ (4 mL,

33.5 mmol) was added dropwise. Acetaldehyde (35 mmol) was added dropwise at $-78°$. The reaction mixture was stirred at $-78°$ for 4 hours and treated with 18.3 mL (55 mmol) of 3 N NaOH and 30% H_2O_2 (7.5 mL). After the mixture was heated at reflux for 1 hour, the organic layer was separated, washed with water (30 mL) and brine (30 mL), dried over anhydrous $MgSO_4$, and filtered and concentrated under reduced pressure to afford the title product (78% yield, 90% ee, $\geq$99% d.r.); bp $78°/85$ mm Hg; $[\alpha]^{23}_D$ +9.141° (neat); ^1H NMR (CDCl$_3$) δ 6.00–5.55 (m, 1 H), 5.30–4.95 (m, 2H), 3.4 (p, $J = 7$ Hz, 1H), 2.15 (sextet, $J = 7$ Hz, 1H), 1.70 (s, 1 H), 1.20 (d, $J = 7$ Hz, 3H), 1.05 (d, $J = 7$ Hz, 3H); ^{13}C NMR (CDCl$_3$, Me$_4$Si) δ 140.73, 115.76, 70.72, 45.64, 19.94, 15.72.

(R,R)-[(E)-2-Butenyl]diisopropyl Tartrate Boronate [Preparation of the (E)-Crotyl Diisopropyl Tartrate Boronate Reagent].[37]   An oven-dried 1-L three-neck round-bottom flask equipped with a magnetic stir bar and a $-100°$ thermometer was charged with anhydrous THF (472 mL) and t-BuOK (48.0 g, 425 mmol). This mixture was flushed with Ar and cooled to $-78°$. trans-2-Butene (42.0 mL, 450 mmol), condensed from a gas lecture bottle into a rubber-stoppered 250-mL graduated cylinder immersed in a $-78°$ dry ice–acetone bath, was added via cannula. n-BuLi (2.5 M in hexane, 170 mL, 425 mmol) was then added dropwise via cannula at a rate such that the internal temperature did not rise above $-65°$. After completion of the addition (roughly 2 hours on this scale), the cooling bath was removed and the reaction mixture was allowed to warm until the internal temperature reached $-50°$. The solution was maintained at $-50°$ for exactly 15 minutes and then was immediately cooled to $-78°$. Triisopropylborate (80.0 g, 98.2 mL, 425 mmol) was added dropwise via cannula to the (E)-crotylpotassium solution at a rate such that the internal temperature did not rise above $-65°$. On this scale the addition time was approximately 2 hours. After the addition was complete, the reaction mixture was maintained at $-78°$ for 10 minutes and then rapidly poured into a 2-L separatory funnel containing 800 mL of 1 N HCl saturated with NaCl. The aqueous layer was adjusted to pH 1 by using aqueous 1 N HCl, and a solution of (R,R)-diisopropyl tartrate (100 g, 425 mmol) in 150 mL of Et$_2$O was added. The phases were separated, and the aqueous layer was extracted with additional Et$_2$O (4 × 200 mL). The combined extracts were dried with anhydrous MgSO$_4$ for at least 2 hours at room temperature and then vacuum filtered through a fritted glass funnel under a N$_2$ blanket into an oven-dried round-bottom flask. The filtrate was concentrated under reduced pressure to afford a colorless thick liquid, which was then pumped to constant weight (125 g) at 0.5–1.0 mm Hg. It was necessary for the neat reagent to be stirred to remove residual volatile materials, especially THF. Analysis of the crude product by capillary GC [50 m × 0.25 mm fused quartz SE-54 column (0.2 μm polydiphenylvinyl-dimethylsiloxane), 170° isotherm; t_r of the E-isomer, 10.9 min vs. t_r of the Z-isomer, 11.3 min; t_r of butylboronate, 10.5 min] indicated that the isomeric purity of this batch of reagent was 99%; 12% of tartrate butylboronate was also present, reflecting incomplete metalation of (E)-2-butene in this experiment. The reagent is moisture sensitive, and it is better handled as a stable solution in toluene or THF. The (S,S)-reagent derived from (S,S)-DIPT had $[\alpha]^{23}_D$ +36.5° (neat); ^{1}H NMR (300 MHz, CDCl$_3$) δ 5.63–5.53 (m, 1H), 5.48–5.37 (m, 1H), 5.00–4.83 (m, 2H), 4.89 (s, 2H), 1.86 (br d, $J = 6.4$ Hz, 2H), 1.53 (br d, $J = 6.3$ Hz, 3H), 0.90 (d, $J = 6.3$ Hz, 12H); ^{13}C NMR (75.4 MHz, C$_6$D$_6$) δ 169.0, 126.1, 125.3, 78.4, 69.6, 21.6, 18.1; ^{11}B NMR (115.8 MHz, C$_6$D$_6$) δ 34.4; CIMS (NH$_3$) (m/z): [M + 1]$^+$ 299; HRMS (m/z): calcd for C$_{14}$H$_{23}$^{10}BO$_6$, 297.1617; found, 297.1618; HRMS (m/z): calcd for C$_{14}$H$_{23}$^{11}BO$_6$, 298.1581; found, 298.1683.

(70-75%)

(R,R)-[(Z)-2-Butenyl]diisopropyl Tartrate Boronate [Preparation of the (Z)-Crotyl Diisopropyl Tartrate Boronate Reagent].[37]

The preparation of Z-crotylboronate from (Z)-2-butene is analogous to that described for the corresponding E-reagent with the following modification: on completion of the n-BuLi addition, the reaction mixture was warmed to −20° to −25° for 30–45 minutes before being cooled to −78°. This ensured near quantitative formation of the (Z)-crotylpotassium. Temperature control is less critical here since the (Z)-crotylpotassium species is highly favored at equilibrium (99 : 1). The remainder of this preparation was the same as that described above for the synthesis of the E-crotylboronate reagent. On a 100-mmol scale, the yield of Z-crotylboronate was 70–75% (1–2% of butylboronate) and the isomeric purity was >98% (generally >99%). The (R,R)-(Z)-reagent prepared from (R,R)-DIPT had $[\alpha]^{23}_D$ −80.2°(c 2.42, CHCl$_3$); IR (thin film) 3500 (br), 3220 (m), 2980 (s), 1745 (s), 1375 (s), 1220 (s), 1100 (s) cm^{-1}; ^{1}H NMR (300 MHz, CDCl$_3$) δ 5.76–5.67 (m, 1H), 5.55–5.48 (m, 1H), 5.05–4.99 (m, 2H), 4.92 (s, 2H), 1.91 (d, $J = 7.8$ Hz, 2H), 1.56 (d, $J = 6.6$ Hz, 3H), 0.94 (d, $J = 6.3$ Hz, 12H); ^{13}C NMR (75.4 MHz, C$_6$D$_6$) δ 169.0, 124.6, 124.4, 78.4, 69.6, 21.3, 12.7; ^{11}B NMR (115.8 MHz, C$_6$D$_6$) δ 34.8; CIMS (NH$_3$) (m/z): [M$^+$] 298; HRMS (m/z): calcd for C$_{14}$H$_{23}$^{11}BO$_6$, 298.1581; found, 298.1624.

(1S,2R)-1-Cyclohexyl-2-methyl-3-buten-1-ol [Representative Procedure for Additions of (E)-Crotyl Diisopropyl Tartrate Boronate to Aldehydes].[37] A solution of (R,R)-isopropyl tartrate (E)-crotylboronate (3.19 g, 10.7 mmol, crude reagent, 99.3% isomeric purity) in 32 mL of dry toluene under N_2 was treated with powdered 4 Å molecular sieves (600 mg) and cooled to −78°. A solution of freshly distilled cyclohexanecarboxaldehyde (1.00 g, 8.92 mmol) in 10 mL of toluene was added dropwise over 30 minutes. The reaction mixture was stirred at −78° for 3 hours and then was treated with 10 mL of 2 N NaOH to hydrolyze the DIPT borate. The two-phase mixture was warmed to 0° and stirred for 20 minutes before being filtered through a pad of Celite. The aqueous layer was extracted with Et_2O (4 × 10 mL), and the combined organic extracts were dried over anhydrous K_2CO_3 and filtered. Flash chromatographic purification on silica gel (60 × 150 mm column, 6 : 1 hexane/ether) provided 1.41 g of the title product (94% yield) in >99% d.r. (capillary GC analysis: Carbowax 20 column; 70° for 4 minutes then 10°/min to a final temperature of 150°) and 86% ee as determined by chiral capillary GC analysis of the methyl ether. The product had $[\alpha]^{25}_D$ −15.8° (c 1.02, CHCl$_3$, data obtained on a sample with >99% d.r. and 86% ee); ^{1}H NMR (300 MHz, CDCl$_3$) δ 5.80 (ddd, J = 18.0, 10.9, 7.2 Hz, 1H), 5.15–5.10 (m, 2H), 3.11 (dd, J = 5.1, 5.1 Hz, 1H), 2.39 (m, 1H), 1.85–1.60 (m, 5H), 1.48–1.35 (m, 2H), 1.28–1.06 (m, 5H), 1.04 (d, J = 7.2 Hz, 3H); ^{13}C NMR (75.4 MHz, CDCl$_3$) δ 140.4, 116.0, 78.8, 40.5, 40.3, 30.0, 27.0, 26.5, 26.4, 26.1, 17.0; Anal. Calcd for $C_{11}H_{20}O$: C, 78.52; H, 11.98. Found: C, 78.22; H, 12.20.

(1R,2R)-1-Cyclohexyl-2-methyl-3-buten-1-ol [Representative Procedure for Additions of (Z)-Crotyl Diisopropyl Tartrate Boronate to Aldehydes].[37] The title product was prepared by the procedure described above using (R,R)-isopropyl tartrate (Z)-crotylboronate (91% yield); $[\alpha]^{25}_D$ +28.0° (c 0.61, CHCl$_3$, data obtained on a sample with >99% d.r.); ^{1}H NMR (400 MHz, CDCl$_3$) δ 5.80 (ddd, J = 17.3, 10.0, 7.0 Hz, 1H), 5.08 (d, J = 17.3 Hz, 1H), 5.06 (d, J = 10.0 Hz, 1H), 3.18 (ddd, J = 5.0, 5.0, 4.3 Hz, 1H), 2.38 (m, 1H), 1.91 (d, J = 13.0 Hz, 1H), 1.73 (m, 2H), 1.64 (m, 1H), 1.58 (m, 1H), 1.41 (m, 1H), 1.35 (d, J = 4.3 Hz, 1H), 1.30–1.00 (m, 5H), 0.97 (d, J = 6.8 Hz, 3H); ^{13}C NMR (75.4 MHz, CDCl$_3$) δ 142.0, 114.4, 78.6, 40.2, 39.8, 29.6, 27.8, 26.4, 26.2, 25.9, 13.1; MS (m/z): [M$^+$ − crotyl] 113. Anal. Calcd for $C_{11}H_{20}O$: C, 78.51; H, 11.98. Found: C, 78.33; H, 12.03.

(R)-(+)-1-Phenyl-3-buten-1-ol [Preparation of (1R,2R)-N,N'-Bis-(p-Toluenesulfonamide)-1,2-diamino-1,2-diphenylethane Allyldiazaborolidine and a Representative Reaction with Benzaldehyde].[50] (+)-(1R,2R)-N,N'-Bis(p-toluenesulfonamide)-1,2-diamino-1,2-diphenylethane (0.75 g, 1.44 mmol) was placed in a dry 25-mL round-bottom flask equipped with a magnetic stir bar and sealed with a septum. The flask was evacuated and flushed with argon three times. Freshly distilled CH_2Cl_2 (12 mL) was added and the resulting solution was cooled to 0°, at which point BBr_3 (1 M in CH_2Cl_2, 1.4 mL, 1.4 mmol) was added. The resulting solution was stirred at 0° for 10 minutes and then warmed to room temperature and stirred for 40 minutes. The solution was concentrated under vacuum (2 mm Hg) and rigorously protected from moisture. Freshly distilled CH_2Cl_2 (5 mL) was added and removed under vacuum to give the bromoborane intermediate; ^{11}B NMR (96 MHz, $CDCl_3$) δ 25.7; ^{1}H NMR (300 MHz, $CDCl_3$) δ 7.32–7.28 (m, 6H), 7.26–7.22 (m, 4H), 7.20–7.09 (m, 4H), 7.02–6.98 (m, 4H), 5.11 (s, 2H), 2.33 (s, 6H). Freshly distilled CH_2Cl_2 was added and the homogeneous solution was cooled to 0° and allyltri-n-butyltin (435 μL, 464 mg, 1.4 mmol) was added dropwise. The reaction mixture was stirred at 0° for 10 minutes, followed by 2 hours at room temperature. A purified aliquot displayed the following spectroscopic data: ^{1}H NMR (300 MHz, $CDCl_3$) δ 7.52–6.92 (m, 18H), 6.21–6.09 (m, 1H), 5.34–5.11 (m, 2H), 4.78 (s, 2H), 2.77 (dd, $J = 8.9$, 1.9 Hz, 2H), 2.32 (s, 3H).

The resulting solution was cooled to −78° and a solution of benzaldehyde (1.28 mL, 133.6 mg, 1.26 mmol) in CH_2Cl_2 (1 mL) was added dropwise. The reaction mixture was stirred at −78° for 2.5 hours, and aqueous buffer (pH 7.5) was added. The aqueous layer was extracted with CH_2Cl_2 (5 mL), the combined organic phases were washed with brine, and the solvent was evaporated. The residue was taken up in Et_2O/hexane (2 : 1, 10 mL), and the resulting solution was cooled to 0° for 20 minutes to complete precipitation of the solid. After removal of the (R,R)-bis(sulfonamide) (0.694 g, 92%) by filtration through a sintered glass funnel, the filtrate was vigorously stirred with aqueous KF (50%, 15 mL) at room temperature for 40 minutes. The organic layer was separated and dried with anhydrous $MgSO_4$, filtered, and the solvent was removed under reduced pressure. The product was purified by silica gel chromatography (hexane/Et_2O, 5 : 1) to afford 170 mg (1.12 mmol) of the title product (91% yield) as a colorless liquid; [α]$_D$ +44.78° (c 1.26, benzene); ^{1}H NMR (300 MHz, $CDCl_3$) δ 7.41–7.28

(m, 5H), 5.92–5.73 (m, 1H), 5.22–5.13 (m, 2H), 4.80–4.61 (m, 1H), 2.51–2.30 (m, 2H), 2.05 (d, $J = 3.0$ Hz, 1H).

(1R)-2-[1-Chloro-2-(E)-butenyl]pinacol Boronate [Preparation of a Chiral α-Chloro Allylic Boronate].[83,237] (R)-3-Hydroxy-1-butyne (16.58 g, 236.5 mmol) and hexamethyldisilazane (19.09 g, 118 mmol) were heated to 110° for 12 hours. The mixture was then filtered through a short pad of silica gel to give 33.30 g (99%) of 3-trimethylsiloxy-1-butyne as an almost colorless liquid. Under an atmosphere of N_2, borane dimethylsulfide complex (8.4 mL, 84 mmol) was dissolved in DME (120 mL). The solution was cooled to 0° and cyclohexene (13.80 g, 168 mmol) was added. The mixture was stirred at 0° for 15 minutes and then warmed to room temperature and stirred for another 1 hour. The solution was cooled to 0° and 3-trimethylsiloxy-1-butyne (15.0 g, 84 mmol) was added. The reaction mixture was warmed to room temperature to dissolve the dicyclohexylborane, and stirred for an additional 1 hour. Anhydrous trimethylamine oxide (12.65 g, 84 mmol) was added in small portions to maintain the solution at reflux. The mixture was cooled to room temperature and stirred for an additional 1 hour. Pinacol (10.0 g, 84 mmol) was added and the mixture was stirred for 12 hours. The solution was filtered and the solvent was removed under reduced pressure. The resulting liquid was distilled (0.3 Torr) to give 18.60 g of pinacol (R)-(E)-2-(3-trimethylsiloxy-1-butene)boronate (83% yield), bp 68°/0.1 Torr; ^{1}H NMR (400 MHz, CDCl$_3$) δ 6.58 (dd, $J = 17.9$, 4.3 Hz, 1H), 5.58 (dd, $J = 17.9$, 1.7 Hz, 2H), 4.30 (ddq, $J = 6.5$, 4.3, 1.7 Hz, 1H), 1.25 (s, 12H), 1.20 (d, $J = 6.5$ Hz, 3H), 0.09 (s, 9H); ^{13}C NMR (125 MHz, CDCl$_3$) δ 156.5, 83.0, 69.6, 24.7, 23.5, 0.0. Anal. Calcd for C$_{13}$H$_{27}$BO$_3$Si: C, 57.78; H, 10.07. Found: C, 57.71; H, 10.16.

The excess of trimethylamine was removed by washing the pinacol (R)-(E)-2-(3-trimethylsiloxy-1-butene)boronate (5.0 g) in petroleum ether (50 mL) using 5% aqueous AcOH (10 mL), 4% aqueous NaHCO$_3$ (10 mL), and saturated aqueous Na$_2$SO$_4$ solution. The aqueous phases were back extracted each time with petroleum ether (10 mL). The combined organic layers were dried over anhydrous MgSO$_4$, filtered, and the solvent was removed under reduced pressure to afford pure pinacol (E)-2-(3-trimethylsiloxy-1-butene)boronate (4.84 g). Co(NO$_3$)$_2$·6H$_2$O (20 mg, 0.069 mmol) was added to a stirred solution of pinacol (R)-(E)-2-(3-trimethylsiloxy-1-butene)boronate (5.70 g, 21.1 mmol) in petroleum ether (130 mL). Freshly distilled SOCl$_2$ (2.75 g, 23.1 mmol) was added, and slow evolution of SO_2 was observed. After 4 hours, when no more bubbling was observed, the mixture was filtered, and the solvent was removed under reduced pressure to afford 4.39 g of the crude title product (96% yield), which was distilled, bp 30°/0.1 Torr; ^{1}H NMR (400 MHz, CDCl$_3$) δ 5.78 (ddq, $J = 15.1$, 6.4, 0.8 Hz, 1H), 5.65 (ddq, $J = 15.1$, 9.2, 1.5 Hz, 1H), 3.94 (d, $J = 9.2$ Hz, 1H), 1.71 (ddd, $J = 6.4$, 1.5, 0.8 Hz, 3H), 1.28 (s, 12H); ^{13}C NMR (125 MHz, CDCl$_3$) δ 129.0, 128.1, 84.4, 24.4, 17.7. Anal. Calcd for $C_{10}H_{18}BClO_2$: C, 55.47; H, 8.38. Found: C, 55.50; H, 8.64.

(1S,2S)-(Z)-4-Chloro-2-methyl-1-phenylbut-3-en-1-ol   [Representative Addition of Pinacol (1R)-2-[1-Chloro-2-(E)-butenyl]boronate to Aldehydes].[67]

To a solution of crude pinacol (1R)-2-[1-chloro-2-(E)-butenyl]boronate (1.80 g, 8.3 mmol) in petroleum ether (15 mL) was added benzaldehyde (0.59 g, 5.5 mmol) at 0°. After 18 hours at room temperature, the mixture was evaporated, the residue was dissolved in Et$_2$O, and triethanolamine (1.24 g, 1.10 mL, 8.3 mmol) was added under vigorous strirring. After 4 hours, the mixture was filtered and the solvent was removed by evaporation. The crude product was purified by flash column chromatography (7:3 petroleum ether/ether) to afford 0.66 g of the title product (61% yield); $[\alpha]^{20}_D$ +44.5° (c 10, CDCl$_3$); ^{1}H NMR (400 MHz, CDCl$_3$) δ 7.34 (m, 5H), 6.13 (dd, $J = 0.9$, 7.2 Hz, 1H), 5.75 (dd, $J = 7.2$, 9.4 Hz, 1H), 4.50 (d, $J = 7.3$ Hz, 1H), 3.14 (dddq, $J = 0.9$, 6.9, 7.3, 9.4 Hz, 1H), 2.02 (br s, 1H), 0.90 (d, $J = 6.9$ Hz, 3H); ^{13}C NMR (100 MHz, CDCl$_3$) δ 142.3, 133.6, 128.2, 127.7, 126.6, 119.2, 77.8, 39.8, 16.0. Anal. Calcd for $C_{11}H_{13}ClO$: C, 67.18; H, 6.66. Found: C, 67.27; H, 6.74.

(1R,2S,3R,4S)-2,3-O-[(E)-2-Butenylboryl]-2-phenyl-1,7,7-trimethylbornanediol [Preparation of the (E)-Crotyl Camphordiol Boronate Reagent].[28]

A 200-mL three-neck round-bottom flask equipped with a magnetic stir bar and a thermometer was charged with 110 mL of THF and t-BuOK (4.65 g, 41.5 mmol). The mixture was flushed with Ar and cooled to −78°. trans-2-Butene (2.46 g, 43.9 mmol), condensed into a rubber-stoppered 10-mL round-bottom flask kept at −78°, was added via cannula. n-BuLi (1.43 M in hexane, 29 mL, 41.5 mmol) was added dropwise over 1 hour with a syringe pump, so that the internal temperature did not rise at all. After completion of the addition, the reaction mixture was allowed to warm until the internal temperature reached −52°. The solution was maintained at −52° for 15 minutes, and then cooled back to −78°. Triisopropyl borate (10.53 mL, 45.6 mmol) was added dropwise over 30 minutes through a syringe pump. The reaction mixture was maintained at −78° for 2 hours, and was sealed under Ar and stored in a freezer for a few weeks without any noticeable change of its quality. The stored solution (10 mL) was reacted with 15 mL of aqueous 1 N HCl, and extracted with ether three times. The combined organic layers were mixed with (1R,2S,3R,4S)-2-phenyl-1,7,7-trimethylbornanediol (220 mg, 0.89 mmol), stirred at ambient temperature for 45 minutes, and then concentrated under reduced pressure. Flash chromatography (5% EtOAc/hexanes, SiO$_2$ pre-treated with 5% Et$_3$N/hexanes) yielded 277 mg of the title product as a colorless oil (99% yield); [α]^{25}$_D$ +10.5° (c 1.84, CHCl$_3$); ^{1}H NMR (500 MHz, CDCl$_3$) δ 7.42–7.41 (m, 2H), 7.34–7.24 (m, 3H), 5.43–5.30 (m, 2H), 4.71 (s, 1H), 2.13 (d, $J = 5.5$ Hz, 1H), 1.84–1.78 (m, 1H), 1.61–1.52 (m, 5H), 1.20–1.13 (m, 5H), 1.04–0.94 (m, 1H), 0.93 (s, 3H), 0.92 (s, 3H); ^{13}C NMR (125 MHz, CDCl$_3$) δ 141.8, 127.4, 127.2, 126.7, 125.7, 125.2, 95.6, 88.5, 52.0, 50.2, 48.8, 29.6, 24.8, 23.6, 20.7, 18.0, 9.3; ^{11}B NMR (64 MHz, CDCl$_3$) δ 34.2; HRMS-EI (m/z): calcd for C$_{20}$H$_{27}$O$_2$B, 310.21042; found, 310.20994. Anal. Calcd for C$_{20}$H$_{27}$O$_2$B: C, 77.43; H, 8.77. Found: C, 77.24; H, 8.86.

(75%)

(3R,4R)-3-Methyl-1-undecen-5-yn-4-ol [Representative Procedure for Scandium-Catalyzed E-Crotylation of Aldehydes].[28] Scandium trifluoromethanesulfonate (238 mg, 0.48 mmol) and CH$_2$Cl$_2$ (5 mL) were introduced in a 50-mL round-bottom flask, and the mixture was cooled to −78°. 2-Octynal (1.03 mL, 7.25 mmol) was added, followed by a solution of (1R,2S,3R,4S)-2,3-O-[(E)-2-butenylboryl]-2-phenyl-1,7,7-trimethylbornanediol (1.50 g, 4.83 mmol) in CH$_2$Cl$_2$ (7 mL) dropwise over 30 minutes. The resulting mixture was stirred at −78° for 24 hours, then DIBAL-H (1.0 M in toluene, 14.5 mL, 14.5 mmol) was added. The mixture was stirred at −78° for 1 hour, then carefully poured into a 250-mL separatory funnel containing aqueous 1 N NaOH (50 mL). After the resulting layers were separated, the aqueous layer was extracted with EtOAc (3 × 25 mL). The combined organic layers were dried over anhydrous MgSO$_4$, filtered, and the solvent was removed under reduced pressure. Flash chromatography (5% EtOAc/hexanes) afforded 594 mg of the title product as a colorless oil (75% yield); [α]^{25}$_D$ +16.55° (c 2.00, CHCl$_3$); ^{1}H NMR (500 MHz, CDCl$_3$) δ 5.82 (ddd, $J = 17.7$, 10.4, 7.6 Hz, 1H), 5.17–5.12 (m, 2H), 4.19 (ddd, $J = 6.2$, 3.8, 1.9 Hz, 1H), 2.44–2.40 (m, 1H), 2.21 (ddd, $J = 9.0$, 7.1, 1.9 Hz, 2H), 1.82 (br s, 1H), 1.54–1.48 (m, 2H), 1.40–1.29 (m, 4H), 1.12 (d, $J = 7.1$ Hz, 3H), 0.90 (t, $J = 7.1$ Hz, 3H); ^{13}C NMR (125 MHz, CDCl$_3$) δ 139.6, 116.4,

86.6, 79.5, 66.4, 44.7, 31.0, 28.3, 22.1, 18.6, 15.2, 13.9; HRMS-ES (m/z): calcd for $C_{12}H_{20}ONa$, 203.14064; found, 203.14084. ^{19}F NMR analysis of the Mosher ester derivative indicated 97% ee: (376 MHz, $CDCl_3$) δ − 71.88 (major), −72.14 (minor).

The fractions containing the diol auxiliary and those containing borate derivatives were concentrated, and treated with a solution of THF (2 mL), 1 N NaOH (1 mL), and H_2O_2 (1 mL of a 30% aqueous solution) for 16 hours. The resulting mixture was diluted with water (5 mL) and extracted with EtOAc (3 × 10 mL). The combined organic layers were dried over anhydrous $MgSO_4$, filtered, and concentrated. Flash chromatography (5% EtOAc/hexanes) afforded 891 mg of recovered $(1R,2S,3R,4S)$-2-phenyl-1,7,7-trimethylbornanediol (75% yield).

$(1R,2S,3R,4S)$-2,3-O-[(Z)-2-Butenylboryl]-2-phenyl-1,7,7-trimethylbornanediol [Preparation of the (Z)-Crotyl Camphordiol Boronate Reagent].[28]
A 300-mL three-neck round-bottom flask equipped with a magnetic stir bar and a thermometer was charged with 110 mL of THF and t-BuOK (2.86 g, 25.5 mmol). After the mixture was flushed with Ar and cooled to −78°, cis-2-butene (1.50 g, 26.8 mmol), condensed into a rubber-stoppered 10-mL round-bottom flask kept at −78°, was added via cannula. n-BuLi (1.43 M in hexane, 17.9 mL, 25.5 mmol) was added dropwise over 1 hour such that the internal temperature did not rise above −70°. After completion of the addition, the cooling bath was removed and the reaction mixture was allowed to warm up until the internal temperature reached −25°. The solution was maintained at −25° for 30 minutes, and then cooled back to −78°. Triisopropyl borate (6.48 mL, 28.1 mmol) was added dropwise over 30 minutes. The reaction mixture was maintained at −78° for 2 hours, and was then rapidly poured into a 500-mL separatory funnel containing 200 mL of aqueous 1 N HCl. The layers were separated, and the aqueous layer was extracted with Et_2O (2 × 100 mL). The combined organic layers were dried over anhydrous $MgSO_4$, filtered, and concentrated on the rotary evaporator to a volume of about 60 mL. The resulting solution was approximately 0.4 M in (Z)-2-butenylboronic acid, and could be kept at 4° for a few weeks without any noticeable change in its concentration or its reactivity with diols. To this solution (3.00 mL, 1.20 mmol) was added $(1R,2S,3R,4S)$-2-phenyl-1,7,7-trimethylbornanediol (246 mg, 1.00 mmol). The resulting mixture was stirred at ambient temperature for 30 minutes, and concentrated under reduced pressure. Flash chromatography (10% EtOAc/hexanes, SiO_2 pre-treated with 5% Et_3N/hexanes) yielded 307 mg of the title product as a colorless oil (99% yield); $[\alpha]^{25}_D$ +12.8° (c 1.73, $CHCl_3$); ^1H NMR (300 MHz, $CDCl_3$) δ 7.44–7.38 (m,

2H), 7.35–7.22 (m, 3H), 5.52–5.34 (m, 2H), 4.70 (s, 1H), 2.13 (d, $J = 5.2$ Hz, 1H), 1.87–1.74 (m, 1H), 1.66–1.61 (m, 2H), 1.54–1.50 (m, 3H), 1.21 (s, 3H), 1.21–1.10 (m, 2H), 1.07–0.96 (m, 1H), 0.94 (s, 3H), 0.91 (s, 3H); ^{13}C NMR (125 MHz, CDCl$_3$) δ 141.8, 127.4, 127.2, 126.7, 124.9, 123.5, 95.7, 88.6, 52.0, 50.2, 48.9, 29.6, 24.7, 23.6, 20.7, 12.5, 9.3; ^{11}B NMR (128 MHz, CDCl$_3$) δ 34.1; HRMS-EI (m/z): calcd for C$_{20}$H$_{27}$O$_2$B, 310.21042; found, 310.21036. Anal. Calcd for C$_{20}$H$_{27}$O$_2$B: C, 77.43; H, 8.77. Found: C, 77.41; H, 8.77.

(72%)

(3S,4R)-3-Methyl-1-undecen-5-yn-4-ol [Representative Procedure for Scandium-Catalyzed Z-Crotylation of Aldehydes].[28]   The title compound was prepared in 72% yield by using a procedure analogous to that used to prepare (3R,4R)-3-methyl-1-undecen-5-yn-4-ol; $[\alpha]^{25}_D$ +22.7° (c 1.27, CHCl$_3$); ^{1}H NMR (300 MHz, CDCl$_3$) δ 5.94–5.80 (m, 1H), 5.20–5.17 (m, 1H), 5.15–5.12 (m, 1H), 4.26 (ddd, $J = 4.8, 2.0, 2.0$ Hz, 1H), 2.50–2.38 (m, 1H), 2.22 (ddd, $J = 6.9, 6.9, 2.0$ Hz, 2H), 1.72 (br s, 1H), 1.57–1.46 (m, 2H), 1.44–1.26 (m, 4H), 1.11 (d, $J = 6.9$ Hz, 3H), 0.91 (t, $J = 6.9$ Hz, 3H); ^{13}C NMR (125 MHz, CDCl$_3$) δ 139.2, 116.8, 86.7, 79.2, 66.3, 44.5, 31.0, 28.4, 22.1, 18.6, 15.6, 13.9; HRMS-EI (m/z): calcd for C$_{12}$H$_{20}$O, 180.15141; found, 180.15092. ^{19}F NMR analysis of the Mosher ester derivative indicated an optical purity of 95% ee: (376 MHz, CDCl$_3$) δ –71.90 (major), –72.19 (minor).

The fractions containing the diol auxiliary and the ones containing diol-borate derivatives were concentrated, and treated with a solution of THF (2 mL), 1 N NaOH (1 mL), and H$_2$O$_2$ (1 mL of a 30% aqueous solution) for 16 hours. The resulting mixture was diluted with water (5 mL) and extracted with EtOAc (3 × 10 mL). The combined organic layers were dried over anhydrous MgSO$_4$, filtered, and concentrated. Flash chromatography (5% EtOAc/hexanes) afforded 891 mg of recovered (1R,2S,3R,4S)-2-phenyl-1,7,7-trimethylbornanediol (75% yield).

(85%)

(3S)-1-Phenylhex-5-en-3-ol [Representative Allylation Catalyzed by a Chiral Diol-SnCl$_4$ Complex].[163]   Into a flame-dried 25-mL round-bottom flask equipped with a rice needle stir bar were successively added (R,R)-1,2-dinaphthylethanediol (8.65 mg, 0.025 mmol, 0.11 eq), anhydrous Na$_2$CO$_3$ (5.6 mg,

0.05 mmol, 0.05 eq), activated 4 Å molecular sieves (50 mg), and freshly distilled toluene (0.5 mL). To the stirred mixture under argon at room temperature was added 25 μL (0.025 mmol, 0.10 eq) of a 1.0 M solution of $SnCl_4$
in CH_2Cl_2; in order to obtain reproducible results, it is necessary to use a
gas tight 25 μL syringe, and that all of the $SnCl_4$ drops into the diol solution without touching the side walls. The resulting mixture was stirred for 5
minutes, cooled to −78°, and maintained at this temperature for 15 minutes,
after which allyl boronic pinacol ester (46.2 mg, 1.1 eq) dissolved in 0.5 mL
of toluene was added via syringe and this mixture was maintained at −78°
for 15 minutes. Freshly distilled hydrocinnamaldehyde (32.6 μL, 0.25 mmol,
1.0 eq) was added and the reaction mixture was stirred at −78° for 12 hours.
DIBAL-H in toluene (0.5 mL, 0.50 mmol, 2.0 eq) was added to reduce any
remaining aldehyde. After 30 minutes, 2 mL of aqueous 1 N HCl was added,
after which the reaction mixture was brought to room temperature and stirred
for 1 hour. The dark brown biphasic mixture was extracted with Et_2O (2 ×
25 mL) to give a clear etheral solution, which was washed with brine and
dried over anydrous Na_2SO_4, filtered, and concentrated under reduced pressure. Purification of the residue by flash chromatography (5% EtOAc/hexanes)
afforded 37.4 mg of the title product (85% yield, 78% ee); HPLC (chiracel-
OD, 5% i-PrOH/hexane, 0.5 mL/min, 254 nm) t_r (major, 89%) = 26.19 min,
t_r (minor, 11%) = 37.62 min; $[\alpha]^{25}_D$ −4.54° (c 0.41, $CHCl_3$); IR (thin film)
3370, 1640 cm^{-1}; ^{1}H NMR (300 MHz, $CDCl_3$) δ 7.32–7.28 (m, 2H), 7.23–7.18
(m, 3H), 5.90–5.78 (m, 1H), 5.18 (m, 1H), 5.14 (m, 1H), 3.72–3.65 (m, 1H),
2.87–2.79 (m, 1H), 2.74–2.67 (m, 1H), 2.37–2.32 (m, 1H), 2.24–2.16 (m, 1H),
1.86–1.75 (m, 2H), 1.68–1.66 (s, 1H); ^{13}C NMR (125 MHz, $CDCl_3$) δ 142.05,
134.6, 128.43, 128.38, 125.8, 118.3, 69.9, 42.1, 38.5, 32.05; HRMS-EI (m/z):
calcd for $C_{12}H_{16}O$, 176.12011; found, 176.12011. Anal. Calcd for $C_{12}H_{16}O$: C,
81.77; H, 9.15. Found: C, 81.80; H, 9.14.

**(R)-[($2R,6S$)-6-Ethoxy-5,6-dihydro-2H-pyran-2-yl]phenylmethanol [Rep-
resentative Example of a Tandem Catalytic Enantioselective [4+2] Cyclo-
addition/Allylboration].[150]** A mixture of 3-boronoacrolein pinacolate (364 mg,
2.00 mmol) and ethyl vinyl ether (1.90 mL, 20.0 mmol) was placed in an oven-
dried 10-mL round-bottom flask with a stir bar. To this solution was added
the Jacobsen tridentate chromium(III) catalyst **88** (9.60 mg, 0.020 mmol) and
powdered BaO (400 mg). After the mixture had been stirred for 1 hour at room

temperature, benzaldehyde (424 mg, 4.00 mmol) was added. The reaction mixture was stirred at 40–45° for 24 hours, and diluted with EtOAc and filtered through Celite. The EtOAc solution was stirred for 30 minutes with an aqueous saturated solution of NaHCO$_3$. The organic layer was separated and the aqueous layer was extracted with EtOAc (2 × 20 mL). The combined organic layers were washed with an aqueous saturated solution of NaCl (20 mL), dried over anhydrous MgSO$_4$, filtered, and concentrated to afford the crude product. Purification by flash column chromatography (deactivated silica gel, 9 : 1 hexanes/Et$_2$O) afforded 384 mg of the pure title product as a clear oil (82% yield, 96% ee); HPLC (chiralpak AD-RH, 50% i-PrOH/H$_2$O, 0.300 mL/min, 210.8 nm) t$_r$ (major) = 8.60 min, t$_r$ (minor) = 10.64 min; [α]$^{23}_D$ +24.9° (c 1.0, CHCl$_3$); IR (CH$_2$Cl$_2$, cast) 3451, 3035, 2877, 1640, 1257 cm^{-1}; ^{1}H NMR (500 MHz, CDCl$_3$) δ 7.40–7.26 (m, 5H), 5.79–5.75 (m, 1H), 5.40–5.36 (m, 1H), 4.78 (dd, J = 5.9, 5.9 Hz, 1H), 4.56 (dd, J = 7.4, 1.5 Hz, 1H), 4.32–4.28 (m, 1H), 3.96 (dq, J = 9.7, 7.1 Hz, 1H), 3.58 (dq, J = 9.7, 7.1 Hz, 1H), 3.11 (br s, 1H), 2.12–2.10 (m, 2H), 1.25 (t, J = 7.1 Hz, 3H); ^{13}C NMR (125 MHz, CDCl$_3$) δ 139.8, 128.3, 128.0, 127.2, 125.4, 124.8, 98.5, 78.7, 76.8, 64.5, 31.1, 15.3; HRMS-ESI (m/z): [M + Na]$^+$ calcd for C$_{14}$H$_{18}$O$_3$Na, 257.1154; found, 257.1152.

TABULAR SURVEY

Within each table, entries are organized in the order of increasing number of carbons in the carbonyl substrate, not including protecting groups. This system was adopted so as to have similar substrates grouped together. The tables cover entries from the literature up until December 31, 2005. See pages 573–580 for supplemental references for Tables 1–9 and for recent information about experimental methods, reviews, and mechanisms. In each table, for the same aldehyde, allyl transfer reagents are grouped by the type of boron reagent involved; boranes are followed by boronates, and then the others such as borinates and borate salts. In all tables, entries with the same experimental conditions but differing only in the results (yield and selectivities) are listed under a single entry. In these instances, the results noted correspond to the best reported example. Table 1A contains entries of simple allyl (C$_3$H$_5$) transfer to non-aromatic carbonyl substrates. Table 1B contains entries for aromatic and heteroaromatic substrates. Tables 2 and 3 cover entries of Z-crotylation and E-crotylation respectively (cis and trans C$_4$H$_7$). In the instances where a mixture of E/Z crotyl reagents is used, these entries are found in Table 2. It should be noted that readers interested specifically in the propionate products of crotylation reactions should also consult Table 4, which contains several entries of α-substituted crotylation reagents. Table 4 includes all entries with reagents having a substituent in the α position of the allylic group. Table 5 lists entries involving allyl transfer where the β position is substituted with a variety of functional groups. Table 6 reports entries where the γ position of the reagent is functionalized with a carbon or heteroatom-containing group. Table 7 presents examples of allenyl and propargyl group transfer. Table 8 consists of entries of cyclic allylic boron reagents, where the allylic unit is part of a ring. Table 9 contains entries of intramolecular allylboration reactions. For multiple substitution patterns, the entries are ordered by least to most substituted; entries with the same number of substitutions are prioritized by $\alpha > \gamma > \beta$. The authors would like to mention that some entries employing the Ipc reagents are inconsistent with respect to the nomenclature used for the absolute configuration of the reagent. These inconsistencies originate from the source references.

The following abbreviations were used in the text and the Tables:

Ac	acetyl
acac	acetylacetone
Adm	adamantyl
AIBN	2,2'-azo(bis)isobutyronitrile
Ald	aldehyde
All	allyl
$B_2(cat)_2$	bis(catecholato)diboron
BBN	9-borabicyclo[3.3.1]nonanyl
BEC	2-bromoethyloxycarbonyl
bmim	butylmethylimidazolium
Bn	benzyl
Boc	*tert*-butoxycarbonyl
BPS	*tert*-butyldiphenylsilyl
Bu	butyl
Bz	benzoyl
Cbz	carbobenzyloxy
Conc.	concentration
Cy	cyclohexyl
dba	dibenzylideneacetone
DBU	1,8-diazabicyclo[5.4.0]undec-7-ene
DDQ	2,3-dichloro-5,6-dicyano-1,4-benzoquinone
DEIPS	diethylisopropylsilyl
DIPT	diisopropyl tartrate
DMB	3,4-dimethoxybenzyl
DMF	dimethylformamide
DMPM	3,4-dimethoxybenzyl
DMSO	dimethyl sulfoxide
dppe	diphenylphosphinoethane
dppf	diphenylphosphinoferrocene
d.r.	diastereomeric ratio
ee	enantiomeric excess
emim	ethylmethylimidazolium
equiv or eq.	equivalent
e.r.	enantiomeric ratio
Et	ethyl
HQ	hydroquinone
8-HQ	8-hydroquinoline

Icr	isocaranyl
IL	ionic liquid
Ipc	isopinocampheyl
Ket	ketone
L.A.	Lewis acid
LDA	lithium diisopropylamide
MEM	2-methoxyethoxymethyl
Mes	mesityl
min	minute
MOM	methoxymethyl
MS	molecular sieves
Nph	naphthyl
Ph	phenyl
Phth	phthalyl
Piv	pivaloyl
PMB	*p*-methoxybenzyl
PMP	*p*-methoxyphenyl
Pr	propyl
PTC	phase-transfer catalyst
pTSA	*p*-toluenesulfonic acid
pyr	pyridine
rt	room temperature
SEM	2-(trimethylsilyl)ethoxymethyl
siamyl	*sec*-isoamyl
TBAT	tetrabutylammonium triphenylsilyldifluorosilicate
TBDMS	*tert*-butyldimethylsilyl
TBDPS	*tert*-butyldiphenylsilyl
TEAN	tetraethylammonium nitrate
TES	triethylsilyl
Thex	thexyl (2,3-dimethyl-2-butyl)
Tf	trifluoromethanesulfonyl
TFA	trifluoroacetic acid
TfOH	trifluoromethanesulfonic acid
THF	tetrahydrofuran
THP	tetrahydropyranyl
TIPS	triisopropylsilyl
TMP	2,2,6,6-tetramethylpiperidine
TMS	trimethylsilyl
TMSE	2-(trimethylsilyl)ethyl
Tol	*p*-tolyl
Ts	*p*-toluenesulfonyl
TPS	triphenylsilyl
Tr	trityl

CHART 1. ACRONYMS FOR LIGANDS USED IN TABLES

(R)-Tol-BINAP

BIPHEPHOS

(R)-BINOL

(R,R)-i-Pr-DUPHOS

85

TABLE 1. ADDITION OF UNSUBSTITUTED ALLYL REAGENTS
A. NON-AROMATIC CARBONYL SUBSTRATES

Carbonyl Substrate	Allylborane Reagent	Conditions	Product(s) and Yield(s) (%), % ee	Refs.
C_2 CH$_3$CHO	BEt$_2$	Neat, 0–10°	(88)	238
	B[(−)-limonene]	Ether, −78° to rt	(72), 7 **I**	239, 136a
	B[(−)-Ipc]$_2$	THF, −78°, 3 h	**I** (—), 92	136, 239, 240
	"	Ether, −78° to rt	**I** (74), 93	240, 136a
	"	Ether, −100° [a]	**I** (—), >99	139, 241
	B(4-d-Icr)$_2$	Ether, −78° to rt	**I** (72), 99	239, 136a
	"	THF, −78°, 3 h	**I** (—), 98	240
	"	Ether, −100° [a]	**I** (—), >99	139
	(TMS B structure)	Ether, 3 h	**I** Temp: rt (—), 94; 0° (—), 93; −25° (—), 98; −78° (71), 96	143
	B[(+)-Ipc]$_2$	Pentane, ether, −100°	(70), 94 **I**	204
	B(2-d-Icr)$_2$	THF, −78°, 3 h	**I** (—), 98	240
	"	Ether, −100° [a]	**I** (—), >99	139, 242

Reagent	Conditions	Product	Refs.
B[(+)-β-pinene]$_2$	Ether, −78° to rt	I (65), 11	239, 136a
B[(+)-10-methyl-Ipc]$_2$	Ether, −78° to rt	I (72), 93	239, 136a
B[(+)-longifolene]$_2$	Ether, −78° to rt	I (67), 34	239, 136a
	Hexane, −40°	(86), 86	122, 243
"	Hexane, −40°, 1 h; rt, 12 h	I (92), 65	244
	Hexane, −40°	I (82), 33	122
	Hexane, −40°	(92), 38	122
	Ether, −78° to rt, 2 h	I (47), 96	133

TABLE 1. ADDITION OF UNSUBSTITUTED ALLYL REAGENTS (Continued)
A. NON-AROMATIC CARBONYL SUBSTRATES (Continued)

Carbonyl Substrate	Allylborane Reagent	Conditions	Product(s) and Yield(s) (%), % ee	Refs.
C₂				
CH₃CHO	(polymer-supported sulfonamide boronate reagent)	Ether, −78°, 7 h	OH (51), 92	245, 246
	B[(−)-Ipc]₂	Pentane, ether, −100°	OH (71), 94	247
O=C–R	BBN	Pentane, rt, 2 h	OH)₂ R: Cl (95), OAc (95), OEt (90), NMe₂ (50–57), OBBN (84)	103
CF₃CHO	B[(+)-Ipc]₂	Pentane, ether, −100°	OH / CF₃ I (50), 99	248, 249
	B(2-d-Icr)₂	Pentane, ether, −100°	I (—), >99	248

88

Substrate	Reagent	Conditions	Product(s), (%) de/ee, (% yield)	Refs.
$RO\text{-}CH_2\text{-}CHO$, R = Bn	allyl–B[(–)-Ipc]$_2$	—	(—)	250
	allyl–B[(–)-Ipc]$_2$	Pentane, ether, –100°, 1 h	**I** (74), 92	251
R = PMB	allyl–B[(+)-Ipc]$_2$	Ether, –78° to rt, 12 h	(95), 71 **I**	252
R = Bn, TBDPS	allyl–B(2-d-Icr)$_2$	Ether, –78°, 4 h	**I** (75), 94	253
R = Bn, TBDPS	allyl tartrate boronate	4 Å MS, toluene, –78°	**I** R: Bn (42–70), 59; TBDPS (42–70), 56	170
R = Bn, TBDMS, TBDPS	allyl boronate	Sc(OTf)$_3$ (10 mol%), CH$_2$Cl$_2$, –78°, 12 h	**I** R: Bn (62), 77; TBDMS (76), 90; TBDPS (61), 90	27, 28
R = Bn	allyl diamine boronate	4 Å MS, toluene, –78°, 46 h	(45), 86 **I**	131
R = TBDMS	allyl boronate	4 Å MS, toluene, –78°	**I** (42–70), 59	170

TABLE 1. ADDITION OF UNSUBSTITUTED ALLYL REAGENTS (*Continued*)

A. NON-AROMATIC CARBONYL SUBSTRATES (*Continued*)

Carbonyl Substrate	Allylborane Reagent	Conditions	Product(s) and Yield(s) (%), % ee	Refs.
C$_2$ (glyoxal-type, OHC–CHO)	B[(–)-Ipc]$_2$	THF, –100°, 4 h	(80), >98, 97:3 d.r.	254, 255
	B[(+)-Ipc]$_2$	THF, –100°, 4 h	(82), >98, 97:3 d.r.	254
(acetyl, OBu-*n*)	B(Bu-*n*)$_2$	Neat, 0°	(85)	238
(acetyl, OEt)	BC$_6$H$_{13}$ (reagent)$_2$	Neat, 0°	I (88)	238
(acetyl, OH)	B (reagent)$_3$	Neat, 0°	I (70)	238
	B (reagent)$_3$	Neat, 0–20°	I (80)	238
(BrCH$_2$CHO)	B[(+)-Ipc]$_2$	Ether, –110° to 20°, 15 min	(50), 89	256
	B(4-*d*-Icr)$_2$	Ether, –110° to 20°, 15 min	(50), 92	256
(MeO$_2$C–CHO)	B[(–)-Ipc]$_2$	Ether, –100°, 1 h	(82), 94	257

C₃

Aldehyde	Reagent	Conditions	Product, (yield), % ee	Refs.
(acrolein / propenal)	B[(−)-Ipc]₂ (allyl)	THF, −78°, 3 h	(−), 92	240
	"	Ether, −100° [a]	**I** (−), 96	139
	B(4-d-Icr)₂ (allyl)	Ether, −78° to rt	**I** (79), 86	239, 136a
	"	THF, −78°, 3 h	**I** (−), 93	240
	"	Ether, −100° [a]	**I** (−), 98	139
	B(2-d-Icr)₂ (allyl)	THF, −78°, 3 h	(−), 95	240
	"	Ether, −100° [a]	**I** (−), >99	139
	(TMS-substituted allylborabicyclononane)	Ether, −78°, 3 h	**I** (71), >99	143
	(Ph boronate)	Hexane, −40°	(86), 50	122
(E-iodo-enal)	B[(+)-Ipc]₂ (allyl)	Ether, −90°	(49), 95	258
(propynal)	B[(−)-Ipc]₂ (allyl)	Ether, −78°	**I** (49), 90	259

91

TABLE 1. ADDITION OF UNSUBSTITUTED ALLYL REAGENTS (*Continued*)

A. NON-AROMATIC CARBONYL SUBSTRATES (*Continued*)

Carbonyl Substrate	Allylborane Reagent	Conditions	Product(s) and Yield(s) (%), % ee	Refs.
C₃				
(TMS-propynal)	B[(+)-Ipc]₂	Ether, –100°, 1 h	(48), 98	261
(propanal)	(cis-propenyl)₃B	–70° to 130°	(71)	262, 1
	B(Bu-*n*)₂	Neat, 0–10°	**I** (92)	238
	allyl-B (Pr-*n* bicyclic)	Neat, 0–10°	**I** (90)	238
	TMS allyl-B (pyrrolidine)	Ether, –100°, 3 h	(80), 96	141
	B[(–)-Ipc]₂	THF, –78° to rt	**I** (71), 86	136, 136a
	"	Ether, –78°, 3 h	**I** (—), 86	240
	B(4-*d*-Icr)₂	Ether, –78° to rt	**I** (76), 91	239, 136a
	"	THF, –78°, 3 h	**I** (—), 91	240
	B(2-*d*-Icr)₂	THF, –78°, 3 h	(—), 94	240

	Hexane, −78°	(91), 77	122, 244
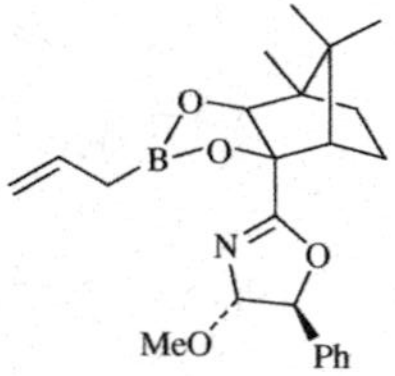	Hexane, −40°	I (85), 44	122
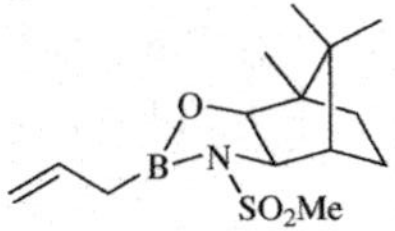	Ether, −78° to rt, 2 h	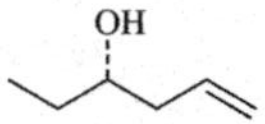 (92), 92	133
		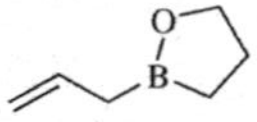 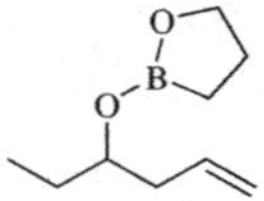 (85)	263
	Neat, 80-100°	(91)	238
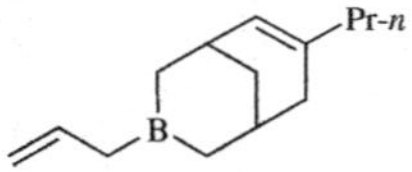	Neat, 0-10°	I (92)	238
	Neat, 0-10°	I (99)	103
	Pentane, rt, 2 h	(96)	263
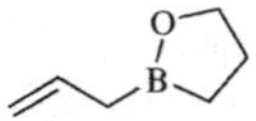	Neat, 80-100°	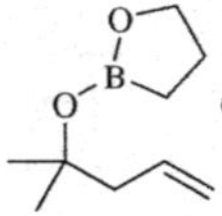	

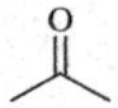

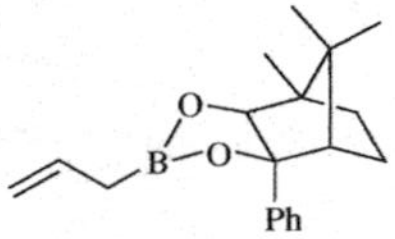

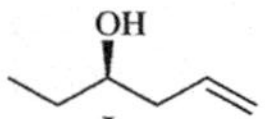

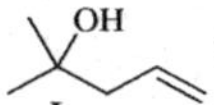

Ph

MeO Ph

SO$_2$Me

BEt$_2$

Pr-n

BBN

93

TABLE 1. ADDITION OF UNSUBSTITUTED ALLYL REAGENTS (*Continued*)

A. NON-AROMATIC CARBONYL SUBSTRATES (*Continued*)

Carbonyl Substrate	Allylborane Reagent	Conditions	Product(s) and Yield(s) (%), % ee	Refs.
C₃				
		Neat, 80-100°	(70)	263
	B[(+)-Ipc]₂	Pentane, ether, –100°	(65), >99	248
	B(2-*d*-Icr)₂	Pentane, ether, –100°	**I** (—), >99	248
	BEt₂	Neat, 0-10°	(78)	238
		Neat, 0-10°	**I** (83)	238
	B[(+)-Ipc]₂	Ether, –78° to rt, 3 h	HO, SiMe₂Ph (66), 6	111
		Ether, –100°, 3 h	(92), 84	141
R = Bn	B[(–)-Ipc]₂	Ether, –78°, 4 h	**I** (90), 85	264
R = TBDMS	"	Ether, –100° to –78°	**I** (82), 85	265

94

R	Reagent	Conditions	Product (yield), % ee	Refs.
R = TBDPS	allyl–B[(−)-Ipc]₂	THF, −78°	**I** (99), 88	266
R = PMB	"	Ether, −78°	**I** (97), 91	267
R = Bn	allyl–B[(−)-Ipc]₂	Ether, −78° to rt, 4 h	**I** (90), 85	253, 268
R = TBDPS	allyl–B[(+)-Ipc]₂	Ether, −78° to rt, 3 h	**I** (83), 94	269, 270
	"	THF, −78°, 3 h	RO—CH₂…CH(OH)—CH₂CH=CH₂ **I** (90), 94	264
R = Bn	allyl–B(2-d-Icr)₂	Toluene, −78°	**I** (98), 84	271
	"	Ether, THF	**I** (87), >95	272
R = PMB	allyl–B(2-d-Icr)₂	Ether, −78°	**I** (96)	273
R = Bn, TBDPS	tartrate allylboronate (OPr-i)	4 Å MS, toluene, −78°	**I** — R = Bn (−), 66 ; R = TBDPS (−), 63	170
R = TBDMS	tartrate allylboronate (OPr-i)	4 Å MS, toluene, −78°	RO—CH₂…CH(OH)—CH₂CH=CH₂ (−), 66	170
R = TBDPS	camphor-derived allylboronate (Ph)	Sc(OTf)₃ (10 mol%), CH₂Cl₂, −78°, 12 h	RO—CH₂…CH(OH)—CH₂CH=CH₂ (86), 93	27, 28

TABLE 1. ADDITION OF UNSUBSTITUTED ALLYL REAGENTS (*Continued*)

A. NON-AROMATIC CARBONYL SUBSTRATES (*Continued*)

Carbonyl Substrate	Allylborane Reagent	Conditions	Product(s) and Yield(s) (%), % ee	Refs.
C₃				
(O, H, HN–Boc)	B[(−)-Ipc]₂	—	(75), 97:3 d.r.	274
(O, H, OR; R = Bn)	B[(−)-Ipc]₂	Ether, −78°, 3 h	(80), 94:6 d.r.	167, 166
	B[(+)-Ipc]₂	Ether, −78°, 3 h	(75), 96:4 d.r. **I**	167, 166
R = TBDMS	B[(+)-Ipc]₂	Ether, −78°, 3 h	**I** (82), 92:8 d.r.	167, 166
R = Bn	(pinacol allylboronate)	—	**I** (—), 55:45 d.r.	118
	"	THF, −78° to rt, 12 h	**I** (97), 65:35 d.r.	275
	(tartrate allylboronate, OPr-i)	4 Å MS, toluene, −78°, 4 h	(—), 72:28 d.r.	170
	(tartrate allylboronate, OPr-i)	4 Å MS, toluene, −78°, 4 h	(16), 84:16 d.r.	170

Substrate	Reagent	Conditions	Product	Refs.
(aldehyde, OTBDMS)	BF$_3^-$K$^+$ (allyl)	BF$_3$·OEt$_2$ (0.05 mol%), CH$_2$Cl$_2$, rt, 3–6 h	(OH) (69), 70:30 d.r.	99
"	"	n-Bu$_4$NI (10 mol%), CH$_2$Cl$_2$/H$_2$O, rt, 30 min	OTBDMS **I** (95), 70:30 d.r.	276
(oxazolidine aldehyde, Boc)	B[(−)-Ipc]$_2$ (allyl)	Ether	(OH) (52), 97:3 d.r.	281
"	B[(+)-Ipc]$_2$ (allyl)	Ether	(OH) **I** (—), 88:12 d.r.	277, 278
"	"	—	(OH) **I** (80), 96:4 d.r.	279, 280
"	(allyl boronate, OPr-i diester)	CH$_2$Cl$_2$, rt	(OH) (—), 66:34 d.r.	277
"	"	4 Å MS, −78°	(OH) **I**	277
"	"	Toluene, −78°, 3 h	**I** (84), 83:17 d.r.	282
(oxazolidine aldehyde, Boc)	(allyl boronate, OEt diester)	Toluene	**I** (77), 89:11 d.r.	278

For the entry (allyl boronate, OPr-i diester), CH$_2$Cl$_2$, rt:

Solvent	d.r.
toluene	(77) 89:11
THF	(—) 85:15
ether	(—) 85:15

97

Carbonyl Substrate	Allylborane Reagent	Conditions	Product(s) and Yield(s) (%), % ee		Refs.

C3

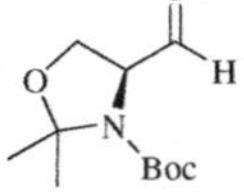

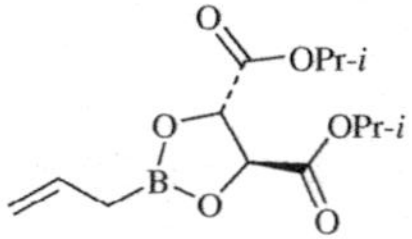

		4 Å MS, −78°	**I**	Solvent d.r.	277
				toluene (—) 85:15	
				THF (—) 87:13	
				Et$_2$O (84) 90:10	
		Toluene, −78°, 3 h	**I** (78), 82:18 d.r.		282
	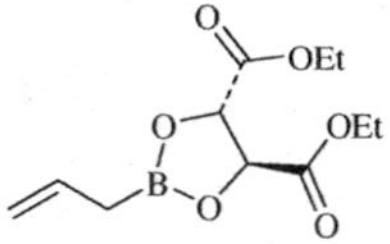	Ether	**I** (84), 90:12 d.r.		278
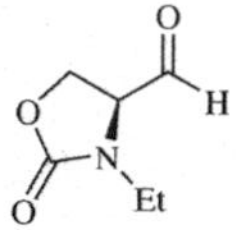	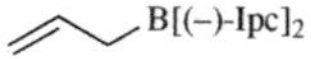 B[(−)-Ipc]$_2$	THF, −78° to rt	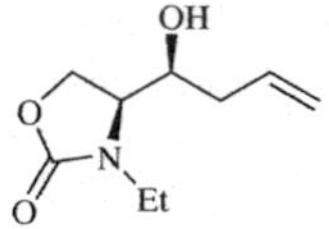 (75)		283, 284
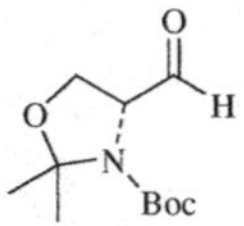	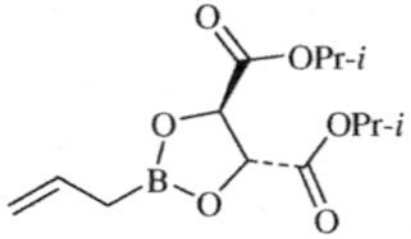	Toluene, −78°, 3 h	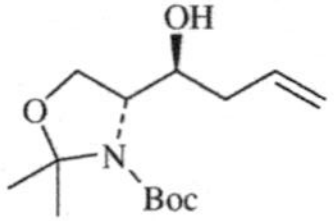 (89)		282
	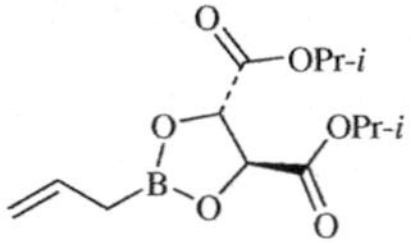	Toluene, −78°, 3 h	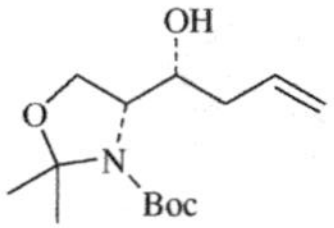 (84)		282
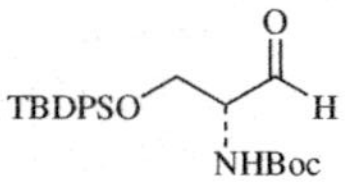	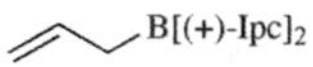 B[(+)-Ipc]$_2$	Ether, −78°, 3.5 h	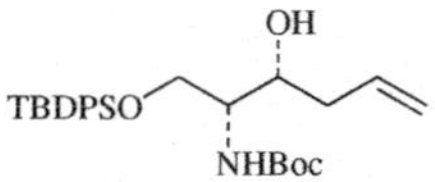 (68), 12:1 d.r.		285

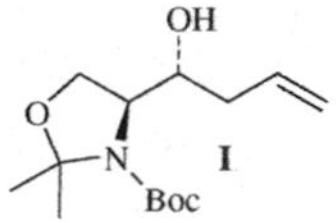

98

Substrate	Reagent	Conditions	Product	Ref.
	B[(−)-Ipc]₂	Ether, −78°, 3.5 h	(—), 12:1 d.r.	286
	B(OPr-*i*)₂	Et₃N, CH₂Cl₂, −10° to rt, 13 h	(86)	105
	B(OPr-*i*)₂	Et₃N, CH₂Cl₂, −10° to rt, 15 h	(84)	109
	B[(+)-Ipc]₂	Ether, −100°, 6 h	(88), 97:3 d.r. **I**	287
		Ether, −78°	**I** (—), 95:5 d.r.	143
		Ether, −78°	(—), >99:1 d.r.	143
		Petroleum ether, −78° to rt, 13 h	**I** (60), 78:22 d.r.	288
	"	CH₂Cl₂, −78°, 24–48 h	**I** (75), 80:20 d.r.	124
	"	Toluene, −78°, 24–48 h	**I** (—), 71:29 d.r.	124

TABLE 1. ADDITION OF UNSUBSTITUTED ALLYL REAGENTS (*Continued*)

A. NON-AROMATIC CARBONYL SUBSTRATES (*Continued*)

Carbonyl Substrate	Allylborane Reagent	Conditions	Product(s) and Yield(s) (%), % ee	Refs.
C₃		See table	(—) Solvent / Temp / d.r.: CH₂Cl₂ / –78° / 80:20 CH₂Cl₂ / rt / 77:23 toluene / –78° / 71:29 toluene / rt / 73:27 hexane / –78° / 75:25 hexane / rt / 79:21 ether / rt / 77:23 THF / rt / 71:29 DMF / 0° to rt / 58:42	118
		CH₂Cl₂, –78°, 24-48 h	**I** (—), 81:19 d.r.	124
	"	Toluene, –78°, 24-48 h	**I** (—), 76:24 d.r.	124
		CH₂Cl₂, –78°, 24-48 h	**I** (—), 68:32 d.r.	124
		4 Å MS, toluene	(—) Temp / d.r.: rt / 64:36 0° / 70:30 –23° / 77:23 –78° / 93:7	125

100

			Ref.
" x M	4 Å MS, toluene, −78°	**I** (—)	125

x	d.r.
0.1	94:6
0.2	93:7
1	89:11

			Ref.
"	4 Å MS	**I** (—)	125

Solvent	Temp	d.r.
toluene	−78°	97:3
CH₃CN	−30°	72:28
CCl₄	−23°	71:29
THF	−78°	70:30
CH₂Cl₂	−78°	61:39
ether	−78°	64:36
CHCl₃	−63°	25:75

			Ref.
"	4 Å MS, toluene, −78°	**I** (—), 93:7 d.r.	170
"	CH₂Cl₂, −78°, 24-48 h	**I** (—), 61:39 d.r.	124
"	Toluene, −78°, 24-48 h	**I** (79-83), 92:8 d.r.	124
	4 Å MS, toluene, −78°, 8 h	**I** (86), 95:5 d.r.	132
	CH₂Cl₂, −78°, 24-48 h	(91), 96:4 d.r.	124
"	Toluene, −78°, 24-48 h	**I** (86), 93:7 d.r.	124

TABLE 1. ADDITION OF UNSUBSTITUTED ALLYL REAGENTS (Continued)

A. NON-AROMATIC CARBONYL SUBSTRATES (Continued)

Carbonyl Substrate	Allylborane Reagent	Conditions	Product(s) and Yield(s) (%), % ee	Refs.
C₃		4 Å MS, −78°	I (—) Solvent / d.r. toluene 98:2 CH₂Cl₂ 96:4	125, 170
	"	4 Å MS, toluene	I (—) Temp / d.r. rt 93:7 0° 94:6 −23° 96:4 −78° 98:2	125
		4 Å MS, toluene, −78°, 8 h	I (73), >99:1 d.r.	132
		4 Å MS, toluene, −78°, 43 h	I Reagent / % Conv. / d.r. R,R 81 99.7:0.3 S,S 84 2:98	131
		CH₂Cl₂, H₂O, rt, 5 h	I (81), 67:33 d.r.	234

C$_4$

	Conditions	Product	Reference
	Petroleum ether, −78° to rt, 13 h	**I** (87), 96:4 d.r.	288
	t-BuCHO (0.1 equiv), 4 Å MS, toluene, −78°, 36 h	(53), >99:1 d.r.	132
	Pentane, rt, 2 h	(91)	103
	Ether, −100°, 3 h	(85), 97	141
	Ether, −100° ^{a}	(—), >99	241
	Hexane, −40°	(84), 30	122
	CH$_2$Cl$_2$, H$_2$O, rt, 3 h	(70)	234
	Neat, 80-100°	(87)	263

TABLE 1. ADDITION OF UNSUBSTITUTED ALLYL REAGENTS (Continued)

A. NON-AROMATIC CARBONYL SUBSTRATES (Continued)

Carbonyl Substrate	Allylborane Reagent	Conditions	Product(s) and Yield(s) (%), % ee	Refs.
C4				
	B[(−)-Ipc]$_2$	THF, −78° to rt	(76), 79	136a
	BEt$_2$	Neat, 0-10°	(77)	238
	BBN	Pentane, rt, 2 h	I (94)	103
	B[(−)-Ipc]$_2$	THF, −78° to rt	(79), 35	136a
	(Ph-substituted allyl-BBN)	Ether, −78°	I (77), 81	145
	BEt$_2$	Neat, 0-10°	(89)	238
	B[(−)-Ipc]$_2$	Ether, −100° [a]	(—), 96	139, 289
	"	Ether, −78° to rt	I (72), 87	136, 136a
				240
	"	THF, −78° to rt	I (88), 83	166
	(allyl-B-TMS)	Ether, −78°, 3 h	I (79), >99	143

104

B(4-d-Icr)$_2$	Ether, −100° [a]	**I** (—), 98	139
"	Ether, −78° to rt	**I** (73), 89	239, 166, 240
B(2-d-Icr)$_2$	THF, −78°, 3 h	**I** (—), 88	240
"	Ether, −100° [a]	[OH / n-Pr structure] **I** (—), >99	139, 241, 242
	THF, −78°, 3 h	**I** (—), 94	240
B(OBu-n)$_2$	—	OB(OBu-n)$_2$ (50-60) [structure]	240
[diamide NHBn structure]	Toluene, −78°, 18 h	[OH / n-Pr structure] (—), 95	2
[SO$_2$Me sultam structure]	Ether, −78° to rt, 2 h	**I** (78), 96	290
[Ph structure]	Hexane, −40°	[OH / n-Pr structure] (93), 72	133
"	Hexane, −40°, 1 h; rt, 12 h	**I** (93), 72	122
			244

TABLE 1. ADDITION OF UNSUBSTITUTED ALLYL REAGENTS (*Continued*)

A. NON-AROMATIC CARBONYL SUBSTRATES (*Continued*)

Carbonyl Substrate	Allylborane Reagent	Conditions	Product(s) and Yield(s) (%), % ee	Refs.
C₄				
		Ether, −78°	(80), 87	145
		Neat, 80-100°	(86)	263
		Ether, −100°, 3 h	(85), 96	141
		Ether, −100°, 3 h	**I** (82), 71	141
		Ether, −100°, 3 h	**I** (87), 24	141
		Ether, −100°, 3 h	**I** (95), 72	141
	—B[(−)-Ipc]₂	Ether, −78° to rt	**I** (86), 90	136, 136a
	"	Ether, −78°, 3 h	**I** (—), 88	240
	"	Ether, −100° ᵃ	**I** (—), 96	139

106

Reagent	Conditions	Product (yield), %	Refs.
$\diagup\!\!\diagdown$ B(4-d-Icr)$_2$	Ether, −78° to rt	**I** (73), 97	239, 136a
"	THF, −78°, 3 h	**I** (—), 95	240
"	Ether, −100° ^{a}	**I** (—), 98	139, 241
$\diagup\!\!\diagdown$ B(2-d-Icr)$_2$	Ether, −100° ^{a}	(—), >99	139
"	THF, −78°, 3 h	**I** (—), 94	240
[allyl boronate, Ph]	Ether, −78°, 3 h	(70), >99	143
[TMS allylborane]	Hexane, −40°	(88), 70	122, 244
[B–N SO$_2$Me]	Ether, −78° to rt, 2 h	(51), 94	133
$\diagup\!\!\diagdown$ B[(+)-Ipc]$_2$	Pentane, ether, −100°	(55), 99	248
$\diagup\!\!\diagdown$ B(2-d-Icr)$_2$	Pentane, ether, −100°	**I** (—), >99	248
$\diagup\!\!\diagdown$ B[(+)-Ipc]$_2$	Ether, −78° to rt, 3 h	(65)	111

Carbonyl Substrate	Allylborane Reagent	Conditions	Product(s) and Yield(s) (%), % ee	Refs.
C₄				
	B(OPr-*i*)₂	Et₃N, CH₂Cl₂, −20° to rt, 13 h	(90)	105
	BEt₂	Neat, 0–10°	(73)	238
	B[(−)-Ipc]₂	Ether, −100°, 0.5 h	(91)	291
	B[(−)-Ipc]₂	Ether, −100° to −78°, 3 h	(68), 91	292
		THF, reflux, 2 h	(90)	293, 294
R = TBDMS, TBDPS, Bn		4 Å MS, toluene, −78°	R: TBDMS (−), 77; TBDPS (−), 74; Bn (−), 78	170
R = TBDPS		4 Å MS, toluene, −78°	I (−), 82	131

Reagent	Conditions	Product (yield)	Ref.
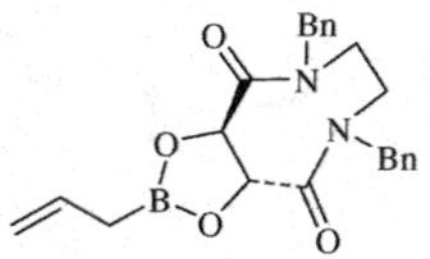 (Bn-N, Bn-N allyl boronate)	4 Å MS, toluene, −78°, 37 h	**I** (58), 94	131
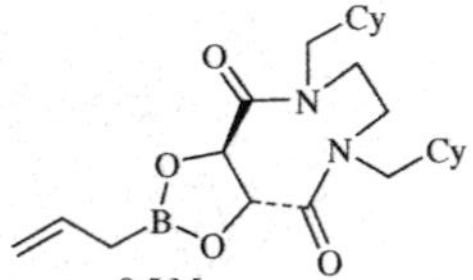 (Cy, Cy allyl boronate) 0.5 M	t-BuCHO (0.1 equiv), 4 Å MS, toluene, −78°, 36 h	**I** (72), 94	132
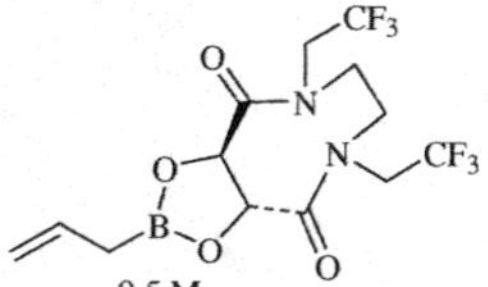 (CF$_3$, CF$_3$ allyl boronate) 0.5 M	t-BuCHO (0.1 equiv), 4 Å MS, THF, −78°, 5 h	**I** (80), 95	132
B[(−)-Ipc]$_2$	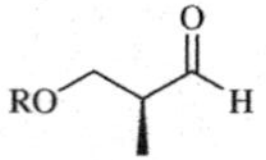 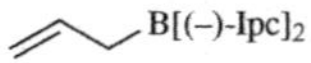 Ether, −110°, 3 h [a]	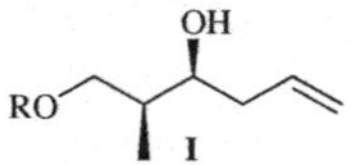 (92)	295
"	Ether, −100°	**I** (70)	296
 B[(−)-Ipc]$_2$	Ether, −78°, 3 h	**I** (80), 96:4 d.r.	167
 B[(+)-Ipc]$_2$	Ether, −78°, 3 h	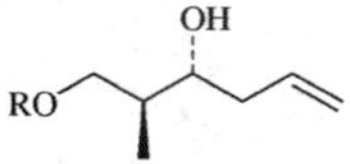 (78), 98:2 d.r.	167
B[(−)-Ipc]$_2$	—	 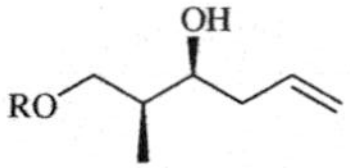 (76), 99:1 d.r.	280

Substrates:

R = PMB

R = Bn

R = TBDPS

Carbonyl Substrate	Allylborane Reagent	Conditions	Product(s) and Yield(s) (%), % ee	Refs.
C_4				
R = TBDMS, TBDPS, Bn	(pinacol allylboronate)	Toluene, rt	(—) **I** — R / d.r.: TBDMS 52:48; TBDPS 54:46; Bn 54:46	169
	(OPr-i tartrate allylboronate)	4 Å MS, toluene, −78°	**I** — R / d.r.: TBDMS (71) 89:11; TBDPS (—) 79:21; Bn (—) 83:17	169, 297, 170
R = TBDPS	(Cy diamide allylboronate, 0.5 M)	4 Å MS, THF, −78°, 36 h	**I** (71), 95:5 d.r.	132
R = TBDPS, TBDMS	(CF$_3$ diamide allylboronate)	4 Å MS, THF, −78°	**I** — R / Time / d.r.: TBDPS 8 h (82) 92:8; TBDMS 9 h (72) 95:5	132
R = TBDMS	(CF$_3$ diamide allylboronate)	4 Å MS, THF, −78°, 9 h	(76), 94:6 d.r.	132

R = TBDMS, TBDPS, Bn

4 Å MS, toluene, −78°

R		d.r.	
TBDMS	(—)	81:19	169, 297,
TBDPS	(72)	87:13	170
Bn	(—)	80:20	

R = TBDPS

4 Å MS, THF, −78°, 8 h

I (83), 97:3 d.r.   132

B[(−)-Ipc]₂

Ether, −78°, 3 h

(—), 98:2 d.r.   167

B[(+)-Ipc]₂

Ether, −78°, 3 h

(—), 95:5 d.r.   167

R = TBDMS, TBDPS, Bn

4 Å MS, toluene

131

Reagent	R	Temp	Time		d.r.
S,S	TBDMS	−78°	37 h	(46)	97:3
S,S	TBDMS	−50°	21 h	(76)	95:5
R,R	TBDMS	−78°	36 h	(43)	3:97
S,S	TBDPS	−78°	44 h	(48)	95:5
R,R	TBDPS	−78°	25 h	(35)	3:97
R,R	TBDPS	−50°	21 h	(68)	5:95
S,S	Bn	−50°	23 h	(56)	93:7
R,R	Bn	−50°	23 h	(53)	7:93

R = Bn

111

TABLE 1. ADDITION OF UNSUBSTITUTED ALLYL REAGENTS (*Continued*)
A. NON-AROMATIC CARBONYL SUBSTRATES (*Continued*)

Carbonyl Substrate	Allylborane Reagent	Conditions	Product(s) and Yield(s) (%), % ee	Refs.
C$_4$				
R = TBDMS	B[(−)-Ipc]$_2$	Ether, −78°, 1 h	(67)	298
R = Bn	B[(−)-Ipc]$_2$	Ether, −78° to 0°, 3 h	(67), 92:8 d.r. **I**	253
R = TBDMS	B[(+)-Ipc]$_2$	Ether, −78°, 1 h	**I** (67)	298
R = TBDMS	B[(+)-Ipc]$_2$	Ether, −78°, 1 h	(67)	298
R = TBDMS	B[(−)-Ipc]$_2$	Ether, −78°, 1 h	(67)	298
		1. (OMe B OMe with methyl) THF, −78° to rt, 1.5 h; 2. allyl–MgBr, −78° to rt, 3 h	(62), 70:30 d.r.	299
	(allyl–B bis-NHBn bis-oxazolidine)	Toluene, −78°, 18 h	(—), 95	290

112

		Hexane, −40°	(83), 70	122, 244
	B[(−)-Ipc]₂	Ether, THF, −78° to −25°, 5 h	**I** (79), 85:15 d.r.	300
	B[(+)-Ipc]₂	Ether, THF, −78° to −25°, 5 h	**I** (79), 85:15 d.r.	300
	B[(−)-Ipc]₂	Ether, THF, −78° to −25°, 5 h	(53), 75:25 d.r.	300
	B[(+)-Ipc]₂	Ether, THF, −78° to −25°, 5 h	(55), 97:3 d.r.	300
	B[(+)-Ipc]₂	0° to rt	R d.r. Boc (96) 9:1 Cbz (94) 19:1	301

TABLE 1. ADDITION OF UNSUBSTITUTED ALLYL REAGENTS (*Continued*)

A. NON-AROMATIC CARBONYL SUBSTRATES (*Continued*)

Carbonyl Substrate	Allylborane Reagent	Conditions	Product(s) and Yield(s) (%), % ee	Refs.
C₄				
		4 Å MS, toluene, −78°	(95), >96, 96:4 d.r.	302
		4 Å MS, toluene, −78°	(86), >97, 74:26 d.r.	302
		4 Å MS, toluene, −78°	(83), 96:4 d.r.	170
		4 Å MS, toluene, −78°	(79), 84:16 d.r.	170
		4 Å MS, toluene, −78°	(—), 75:25 d.r.	302
		4 Å MS, toluene, −78°	(82), >97, 70:30 d.r.	302

Conditions	Product	Ref.
4 Å MS, toluene, −78°	(96), >98, 99:1 d.r.	302
See table	(−)	118

Solvent	Temp	d.r.
CH₂Cl₂	−78°	90:10
CH₂Cl₂	rt	87:13
toluene	−78°	90:10
toluene	rt	87:13
n-BuLi, THF	−78° to rt	79:21

CH₂Cl₂, −78°, 24-48 h	**I** (85), 90:10 d.r.	124
Toluene, −78°, 24-43 h	**I** (−), 90:10 d.r.	124
CH₂Cl₂, −78°, 24-48 h	**I** (−), 86:14 d.r.	124
4 Å MS, toluene, −78°	**I** (−), >300:1 d.r.	170
CH₂Cl₂, −78°, 24-48h	**I** (94), 98:2 d.r.	124
Toluene, −78°, 24-48 h	(63), 68:32 d.r.	124

TABLE 1. ADDITION OF UNSUBSTITUTED ALLYL REAGENTS (*Continued*)

A. NON-AROMATIC CARBONYL SUBSTRATES (*Continued*)

Carbonyl Substrate	Allylborane Reagent	Conditions	Product(s) and Yield(s) (%), % ee	Refs.
C₄				
	(OPr-*i*)(OPr-*i*) allylborane	CH₂Cl₂, −78°, 24-48 h	(—), 71:29 d.r. **I**	124
	"	Toluene, −78°, 24-48 h	**I** (—), 64:36 d.r.	124
	"	4 Å MS, toluene, −78°	**I** (63), 73:27 d.r.	170
	(O(Adm-1))(O(Adm-1)) allylborane	Toluene, −78°, 24-48 h	**I** (—), 60:40 d.r.	124
	(OEt)(OEt) allylborane	4 Å MS, CH₂Cl₂, −78° to rt, 24 h	(93), >99:1 d.r.	303, 304
	B[(+)-Ipc]₂	Ether, −78° to rt	(72), 6:1 d.r.	305
	B[(−)-Ipc]₂	−70°	(60-65), 9:1 d.r.	306
	B[(−)-Ipc]₂	Ether, −75°, 2 h	(40), 9:1 d.r.	309
	B(2-*d*-Icr)₂	Ether, −75°, 2 h	**I** (52), 93:7 d.r.	309

Substrate	Reagent	Conditions	Product	Refs.
	B[(+)-Ipc]₂	Toluene, −78° to rt, 2 h	(75), 95:5 d.r.	307, 308
	(allyl diisopropyl tartrate boronate) OPr-*i*	4 Å MS, toluene	I (—), 88:12 d.r.	170
	(allyl diisopropyl tartrate boronate) OPr-*i*	4 Å MS, THF	(—), 87:13 d.r.	170
	B[(−)-Ipc]₂	Ether, −100°, 1 h	(82), >99	257
	B[(−)-Ipc]₂	Ether, −78°	(93), >95:5 d.r.	310
	(crotyl)₃B	THF, reflux, 15 h	(90)	293

TABLE I. ADDITION OF UNSUBSTITUTED ALLYL REAGENTS (*Continued*)
A. NON-AROMATIC CARBONYL SUBSTRATES (*Continued*)

Carbonyl Substrate	Allylborane Reagent	Conditions	Product(s) and Yield(s) (%), % ee	Refs.
C₄		THF, reflux, 15 h	(90)	294
C₅	B[(−)-Ipc]₂	Ether, −100° [a]	(66), 99	311
	"	Ether, −78°	**I** (66), 95	312
	BBN	Pentane, rt, 2 h	(97)	103
		Ether, −100°, 3 h	(90), 97	141
	B[(−)-Ipc]₂	THF, −78°, 3 h	**I** (—), 83	240
	"	THF, −78° to rt	**I** (88), 83	136, 136a
	"	Ether, −100° [a]	**I** (—), >99	139
		—	**I** (60), 99	313
	B(2-d-Icr)₂	THF, −78°, 3 h	**I** (—), 99	240
	"	Ether, −100° [a]	**I** (—), >99	139
		Ether, −78°, 3 h	**I** (79), >98	143

118

Reagent	Conditions	Product (yield %, ee %)	Refs.
B[(+)-Ipc]₂	Ether, −100° a	(—), 99	241
B(4-d-Icr)₂	THF, −78° to rt	I (80), 88	239, 136a
"	THF, −78°, 3 h	I (—), 88	240
"	Ether, −100° a	I (—), >99	139
(tartrate bis-benzylamide, NHBn / NHBn)	Toluene, −78°, 18 h	I (—), 98	290
(tartrate diisopropyl ester, OPr-i / OPr-i)	4 Å MS, toluene, −78°	(56), 86	131, 170
(N,N'-dibenzyl tartramide, Bn–N / N–Bn)	4 Å MS, toluene, −78°, 67 h	I (—), 96	131
(pinanediol-derived Ph borolane)	Hexane, −40°	I (85), 45	122, 244
(pinanediol-derived oxazoline, MeO, Ph)	Hexane, −40°	I (87), 48	122

119

TABLE 1. ADDITION OF UNSUBSTITUTED ALLYL REAGENTS (Continued)
A. NON-AROMATIC CARBONYL SUBSTRATES (Continued)

Carbonyl Substrate	Allylborane Reagent	Conditions	Product(s) and Yield(s) (%), % ee	Refs.
C$_5$				
t-Bu CHO		CH$_2$Cl$_2$, H$_2$O, rt, 4 h	OH t-Bu (49)	234
	(SO$_2$Me)	Ether, −78° to rt, 2 h	OH t-Bu **I** (80), 88	133
		Ether, −78°, 7 h	**I** (84), 84	245, 246
	BF$_3^-$K$^+$	n-Bu$_4$NI (10 mol%), CH$_2$Cl$_2$, H$_2$O, rt, 15 min	OH t-Bu (95)	100
n-Bu CHO		CH$_2$Cl$_2$, H$_2$O, rt, 3 h	OH n-Bu (82)	234
i-Pr CHO	(SO$_2$Me)	Ether, −78° to rt, 2 h	OH i-Pr (69), 90	133

120

Substrate	Reagent	Conditions	Product	Refs.
cyclopentanone	allyl–BE$_2$	Neat, 0–10°	(72)	238
	allyl–BBN	Pentane, rt, 2 h	**I** (100)	103
	n-Pr–BBN (allyl)	Neat, 0–10°	**I** (77)	238
(*E*)-pent-3-en-2-one	allyl–BBN	Pentane, rt, 2 h	(96)	103
i-Pr methyl ketone	allyl–B (Ph-9-BBN)	Ether, −78°	(74), 92	145
n-Pr methyl ketone	allyl oxaborolane	Neat, 80–100°	(87)	263
Et–CH(CH$_3$)CHO	allyl–B[(−)-Ipc]$_2$	THF, −78°, 3 h	(81), 96:4 d.r.	166, 167
	allyl–B[(+)-Ipc]$_2$	THF, −78°, 3 h	(83), 95:5 d.r.	166, 167
	allyl–B(pinacol)	Neat, rt, 3 d	(—), 62:38 d.r.	115

TABLE 1. ADDITION OF UNSUBSTITUTED ALLYL REAGENTS (*Continued*)

A. NON-AROMATIC CARBONYL SUBSTRATES (*Continued*)

Carbonyl Substrate	Allylborane Reagent	Conditions	Product(s) and Yield(s) (%), % ee	Refs.
C₅				
	B[(+)-Ipc]₂	Ether, −78° to rt, 3 h	R TMS (72), 89; TES (32), 59; TPS (61), 92; SiMe₂Ph (53), 81; SiMe₂Bn (44), 79	111
	(allyl)₃B	THF, reflux, 2 h	(79)	293, 294
	B[(−)-Ipc]₂	Ether, −78°	(61), 93:7 d.r.	314
	B[(+)-Ipc]₂	Ether, −78°	(61), 92:8 d.r.	314
	pinacol allylboronate	Neat, rt, 3 d	(—) R: TBDMS d.r. 79:21; MOM 79:21	115
	pinacol allylboronate	Neat, rt, 3 d	(—) R: TBDMS d.r. 49:51; MOM 61:39	115

122

Substrate	Reagent	Conditions	Product	Yield (%), d.r.	Refs.
R = Bn	⟍B[(−)-Ipc]₂	Ether, −100°	**I**	(75), 96:4 d.r	315
R = PMB	"	THF, −100°	**I** (75), 96:4 d.r.		316
	B[(−)-Ipc]₂	Ether, −100° [a]	**I** (88), 96		317
	"	Ether, −100° [a]	**I** (93), 95:5 d.r.		318
TBDMSO	B[(−)-Ipc]₂	—		(92), 88	319, 320
	B[(−)-Ipc]₂	THF, −100°, 4 h		(97), >98, 95:5 d.r.	254, 321
	B[(+)-Ipc]₂	THF, −78° to rt		(60), >95:5 d.r.	322, 323
	B[(+)-Ipc]₂	Ether, −78°		(65), 10:1 d.r.	274
	B[(−)-Ipc]₂	—		(70), 2:1 d.r.	324

123

TABLE 1. ADDITION OF UNSUBSTITUTED ALLYL REAGENTS (*Continued*)

A. NON-AROMATIC CARBONYL SUBSTRATES (*Continued*)

Carbonyl Substrate	Allylborane Reagent	Conditions	Product(s) and Yield(s) (%), % ee	Refs.
C₅				
	B[(–)-Ipc]₂	Ether, –78° to rt, 2 h	(92), 83	270
	B[(+)-Ipc]₂	Ether, –78°, 4 h	(75), 96:4 d.r	264
	B[(+)-Ipc]₂	Ether, –78° to rt, 16 h	(44)	325
	B[(–)-Ipc]₂	Ether, –78° to rt	(94)	326
	B[(–)-Ipc]₂	Ether, –78°	(66), 91.5:8.5 d.r.	327
	B[(–)-Ipc]₂	Ether, –78°	(74)	328, 329
		Toluene, –78°	(74), 80, 5.5:1 d.r.	330

124

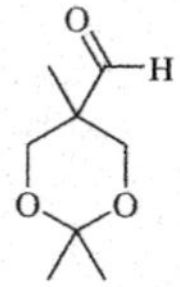

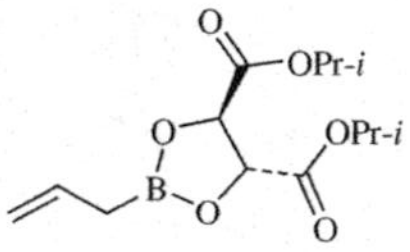

	4 Å MS, toluene, –78° to rt	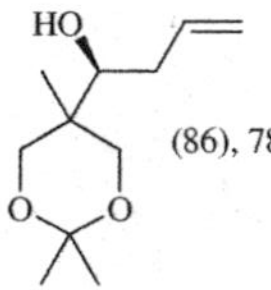 (86), 78	328, 329	
	4 Å MS, toluene, –78° to rt	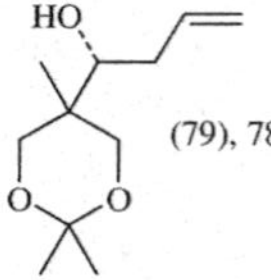 (79), 78	331	
	Ether, –93°	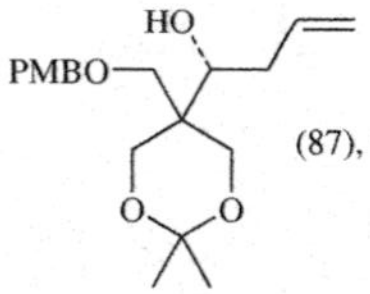 (87), 92	332, 333	
	Neat, rt, 3 d	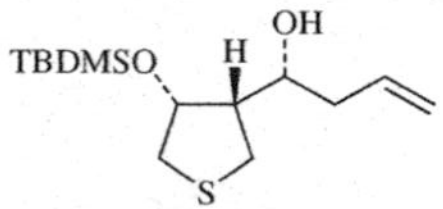 (—), 73:27 d.r.	115	
	—	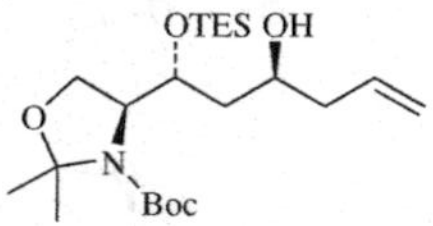 (76)	280, 280a	
	—	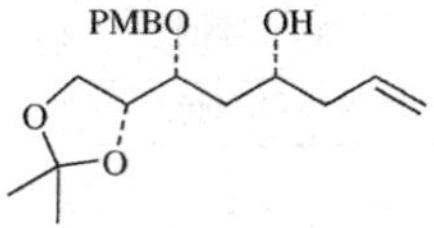 (82)	280, 280a	

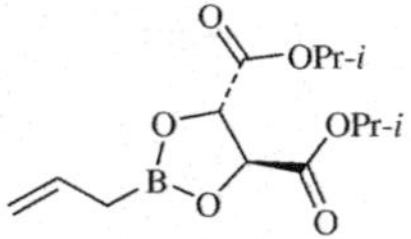

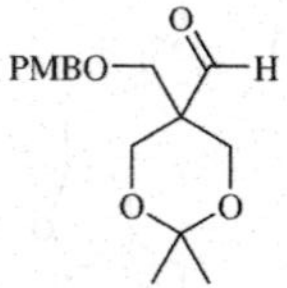

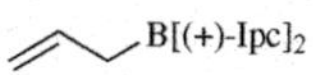

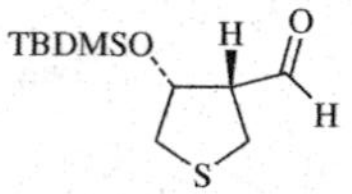

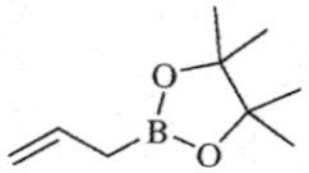

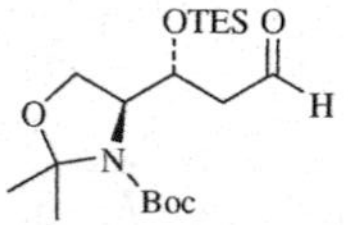

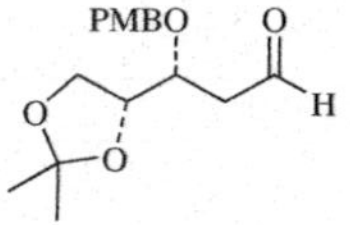

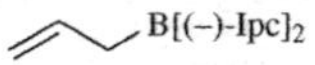

B[(+)-Ipc]₂
B[(+)-Ipc]₂
B[(–)-Ipc]₂

TABLE 1. ADDITION OF UNSUBSTITUTED ALLYL REAGENTS (*Continued*)

A. NON-AROMATIC CARBONYL SUBSTRATES (*Continued*)

Carbonyl Substrate	Allylborane Reagent	Conditions	Product(s) and Yield(s) (%), % ee	Refs.
C$_5$				
	B[(−)-Ipc]$_2$	Ether, −78°	(75)	335
	B[(−)-Ipc]$_2$	Ether, −78°	(71), 83:17 d.r.	336
	B[(−)-Ipc]$_2$	Ether, −78°	(72)	335
	B(4-*d*-Icr)$_2$	Ether, −78°	(95), 78:8 d.r.	337
	B[(+)-Ipc]$_2$	Ether, −78°	(80), 94	338
	B[(−)-Ipc]$_2$	Ether, −100°, 1 h	(88), 96	257

126

Substrate	Reagent	Conditions	Product(s) (%), % ee	Refs.
C₆ $n\text{-}C_5H_{11}CHO$	BBN (allyl)	Pentane, rt, 2 h	OH $n\text{-}C_5H_{11}$ (92)	103
	B[(−)-Ipc]₂ (allyl)	Pentane, ether, −100°	OH $n\text{-}C_5H_{11}$ (71), 97	204
	B[(+)-Ipc]₂ (allyl)	—	OH $n\text{-}C_5H_{11}$ **I** (63), 96	250
	"	—	**I** (91), 88	
			OH $n\text{-}C_5H_{11}$ **I** (93), 97	339
	Cy-tartrate allylboronate, 0.5 M	t-BuCHO (0.1 equiv), 4 Å MS, toluene, −78°, 7 h		132
	CF₃-tartrate allylboronate, 0.5 M	t-BuCHO (0.1 equiv), 4 Å MS, THF, −78°, 5 h	**I** (95), 95	132
	Ts-diamine allylboron (Ph, Ph)	CH₂Cl₂, −78°, 2 h	OH $n\text{-}C_5H_{11}$ **I** (>90), 90	50
	"	Toluene, −78°, 2 h	**I** (>90), 95	50
$n\text{-Bu}$ (2-hexanone)	BBN (allyl)	Pentane, rt, 2 h	OH $n\text{-Bu}$ (100)	103

TABLE 1. ADDITION OF UNSUBSTITUTED ALLYL REAGENTS (*Continued*)

A. NON-AROMATIC CARBONYL SUBSTRATES (*Continued*)

Carbonyl Substrate	Allylborane Reagent	Conditions	Product(s) and Yield(s) (%), % ee	Refs.
C$_6$				
		Ether, −78°	(70), 99	145
		THF, −78° to −40°, 48 h	**I** (75), 90	135
		(*R,R*)-*i*-Pr-DuPHOS, CuF, La(OPr-*i*)$_3$, DMF, −40°, 1 h	(99), 91	164
	BEt$_2$	Neat, 0-10°	(92)	238
	BBN	Pentane, rt, 2 h	**I** (94)	103
	BEt$_2$	Neat, 0-10°	(75)	238
	BBN	Pentane, rt, 2 h	**I** (100)	103
		Neat, 0-10°	**I** (88)	238

128

		Neat, 80-100°	(80)	263
	B[(−)-Ipc]₂	Ether, −78° to rt, 2 h	(24), 10	340
	B[(−)-Ipc]₂	Ether, −78° to rt, 18 h	(4), 41 (35), 41	340
	OPr-*i* OPr-*i*	4 Å MS, toluene, −78°	(85–90), 99:1 d.r.	129
	OPr-*i* OPr-*i*	4 Å MS, toluene, −78°	(85–90), 75:25 d.r.	129
	B[(+)-Ipc]₂	Ether, −78° to rt, 3 h	(70), 42	111
	B[(+)-Ipc]₂	Ether, −78° to rt, 3 h	(34), 29	111

129

TABLE 1. ADDITION OF UNSUBSTITUTED ALLYL REAGENTS (*Continued*)

A. NON-AROMATIC CARBONYL SUBSTRATES (*Continued*)

Carbonyl Substrate	Allylborane Reagent	Conditions	Product(s) and Yield(s) (%), % ee	Refs.
C_6	$(\text{allyl})_3\text{B}$	THF, reflux, 2 h	(50)	293, 294
	B[(−)-Ipc]$_2$	—	(93)	341
		4 Å MS, toluene, −78°	(82), 45:1 d.r.	129
		4 Å MS, toluene, −78°	(—), >98, >50:1 d.r.	342
	B[(−)-Ipc]$_2$	Ether, −100° to −78°, 3 h	(—) R / d.r.: TBDMS (79), 89 >99:1; PMB (83), 88 65:35	292
	B[(−)-Ipc]$_2$	—	(—), 90	343

Aldehyde	Allyl reagent	Conditions	Product	Refs.
(structure) OH, i-Pr	(structure) MeO–B–O, i-Pr, MgBr⁺, OMe	1. i-Pr, THF, −78° to rt, 15 h; OMe, OMe, B, O, i-Pr; 2. MgBr, −78° to rt, 3 h	(structure) OH OH, i-Pr (83), 72:28 d.r.	299
(structure) TBDMSO, i-Pr	$BF_3^-K^+$	$n\text{-}Bu_4NI$ (10 mol%), CH_2Cl_2/H_2O, rt, 30 min	(structure) TBDMSO OH, i-Pr **I** (97), 65:35 d.r.	276
"	"	$BF_3\cdot OEt_2$ (5 mol%), CH_2Cl_2, rt, 3–6 h	**I** (90), 65:35 d.r.	99
(structure) O, Pr-i, O, HO	$B(OPr\text{-}i)_2$	Ether, −10°, 18 h	(structure) HO Pr-i, O HO (81)	108
(structure) Fe(CO)₃, MeO	(structure) OPr-i, OPr-i, O, O, O–B–O	4 Å MS, toluene, −78°	(structure) OH, Fe(CO)₃, MeO (85–90), 98.5:1.5 d.r.	129
(same)	(structure) OPr-i, OPr-i, O, O, O–B–O	4 Å MS, toluene, −78°	(structure) OH, Fe(CO)₃, MeO (85–90), 82:18 d.r.	129
(structure) O, EtO	$B((-)\text{-}Ipc)_2$	Ether, −100°	(structure) OH, EtO (87), 95.5:4.5 d.r.	318

TABLE 1. ADDITION OF UNSUBSTITUTED ALLYL REAGENTS (*Continued*)

A. NON-AROMATIC CARBONYL SUBSTRATES (*Continued*)

Carbonyl Substrate	Allylborane Reagent	Conditions	Product(s) and Yield(s) (%), % ee	Refs.
C₆	$B[(+)\text{-Ipc}]_2$	Ether, −78° to 0°, 18 h	(91), 92:8 d.r.	253
	$B[(-)\text{-Ipc}]_2$	Ether, −78° to rt, 2 h	(33-42), 55 + (28), 52	344, 340, 345
	$B[(-)\text{-Ipc}]_2$	Ether, −78° to rt, 2 h	(82-86), >95:5 d.r.	344, 340, 345
	$B[(-)\text{-Ipc}]_2$	—	I (77)	346
	"	Ether, −78° to rt, 14 h	I (88)	347
	$B[(-)\text{-Ipc}]_2$	Ether, −100°, 1 h	(80), 92	257
	$B(OPr\text{-}i)_2$	Et₃N, CH₂Cl₂, −10° to rt, 13 h	(87)	105

C_7

Reagent	Conditions	Product	Ref.
B[(+)-Ipc]$_2$	Ether, −78°	(44), 86	314
(TMS-pyrrolidinyl-B-allyl)	Ether, −100°, 3 h	(92), 96	141
B[(+)-Ipc]$_2$	THF, −85°, 18 h	(66)	348
(pinacol allylboronate)	L.A. (10 mol%), toluene, −78°, 16 h	L.A. AlCl$_3$ (48), Sc(OTf)$_3$ (74)	25
(tartrate-derived allylboronate)	4 Å MS, toluene, −78°	(97), 87 **I**	125
(bis-Bn diamide allylboronate)	4 Å MS, toluene, −78°	**I**	131
"	4 Å MS, toluene, −73°, 47 h 0.03 M	**I** (40), 97	132

Table for ref. 131:

Temp	Time	% Conv.	
−78°	47 h	80	(40), 97
−50°	17 h	84	(—), 94
rt	2 h	97	(—), 87

TABLE 1. ADDITION OF UNSUBSTITUTED ALLYL REAGENTS (*Continued*)
A. NON-AROMATIC CARBONYL SUBSTRATES (*Continued*)

Carbonyl Substrate	Allylborane Reagent	Conditions	Product(s) and Yield(s) (%), % ee	Refs.
C₇ (cyclohexanecarbaldehyde)	(reagent with N,N-Cy diazaborolidine) 0.5 M	4 Å MS, toluene, −78°	**I** (OH, cyclohexyl, allyl) — Time: 5 h (—), 62; 3 h (—), 91	132
	" 0.5 M	MeCHO (0.1 equiv), 4 Å MS, toluene, −78°, 5 h	**I** (—), 90	132
	" x M	t-BuCHO (0.1 equiv), 4 Å MS, −78°	**I** — x / Solvent / Time: 0.5 toluene 7 h (82), 95; 0.1 toluene 1 h (71), 94; 0.5 THF 5 h (91), 94	132
	(tartrate-derived OPr-i boronate)	4 Å MS, −78°	**I** — Solvent: toluene (—), 87; ether (—), 82; THF (—), 78; CH₂Cl₂ (—), 59	131, 125
	"	Toluene	**I** — Temp: −78° (—), 87; −50° (—), 82; 0° (—), 57; rt (—), 50	131
	(tartrate-derived OR boronate)	4 Å MS, toluene, −78°	**I** — R: Pr-i (—), 86; Et (—), 86; Me (—), 84; C₁₀H₁₉-c (—), 86; CH(Pr-i)₂ (—), 86	125

Reagent	Conditions	Product(s) and Yield(s) (%)	Refs.
	Toluene, −78°, 18 h	**I** (—), 99	290
	OAc, Pd$_2$(dba)$_3$, DMSO, toluene, 40°, 66 h	**I** (58), 48	349
	THF, −78°, 1 h	**I** (90), 76	135
	Sc(OTf)$_3$ (10 mol%), CH$_2$Cl$_2$, −78°, 12 h	**I** (53), 92	28
	CH$_2$Cl$_2$, H$_2$O, rt, 3 h	(92)	234

TABLE 1. ADDITION OF UNSUBSTITUTED ALLYL REAGENTS (*Continued*)

A. NON-AROMATIC CARBONYL SUBSTRATES (*Continued*)

Carbonyl Substrate	Allylborane Reagent	Conditions	Product(s) and Yield(s) (%), % ee	Refs.
C_7				
0.05 M		–78°	I	350
			Solvent / Time / % Conv. / % ee	
			toluene / 0.25 h / 95 / 87	
			THF / 1.5 h / 70 / 78	
			CH$_2$Cl$_2$ / — / — / 59	
x M		–78°	I	350
			x / Solvent / Time / % Conv. / % ee	
			0.08 / toluene / 16 h / 27 / 21	
			0.08 / ether / 20 h / 14 / 32	
			0.1 / CH$_2$Cl$_2$ / 18 h / 80 / 53	
0.05 M		–78°	I	350
			Solvent / Time / % Conv. / % ee	
			toluene / 16 h / 24 / 27	
			THF / 16 h / 8 / 30	
			CH$_2$Cl$_2$ / 18 h / 40 / 46	
x M		–25°, 40 h	I	350
			x / Solvent / % Conv. / % ee	
			0.06 / toluene / 18 / 37	
			0.07 / THF / 10 / 29	
			0.07 / CHCl$_3$ / 30 / 56	

0.06 M —78°, 16 h

Solvent	% Conv.	% ee	350
toluene	50	15	350
CH$_2$Cl$_2$	67	6	

Product: OH cyclohexyl allyl (**I**)

0.08 M — See table

Solvent	Temp	Time	% Conv.	% ee	350
toluene	—25°	40 h	75	25	350
THF	—25°	40 h	91	23	
CH$_2$Cl$_2$	—25°	40 h	94	27	
CH$_2$Cl$_2$	—78°	15 h	21	30	

Product: OH cyclohexyl allyl (**I**)

x M — See table

Solvent	Temp	Time	% Conv.	% ee		350
toluene	—25°	20 h	46	1	S	350
CH$_2$Cl$_2$	—25°	20 h	46	4	S	
CH$_2$Cl$_2$	—78°	15 h	2	13	R	
CHCl$_3$	—25°	19 h	80	7	S	

x
0.1
0.1
0.2
0.15

x M — See table

Solvent	Temp	Time	% Conv.	% ee		350
toluene	—78°	5 h	21	11	R	350
CH$_2$Cl$_2$	—78°	16 h	64	6	R	
CHCl$_3$	—58°	21 h	85	13	S	

x
0.08
0.07
0.08

0.05 M — —78°

Solvent	Time	% Conv.	% ee		350
toluene	15 h	82	11	R	350
CH$_2$Cl$_2$	18 h	95	6	S	

Carbonyl Substrate	Allylborane Reagent	Conditions	Product(s) and Yield(s) (%), % ee	Refs.
C₇ 0.1 M		Toluene	I $\begin{array}{ccc} \text{Temp} & \text{Time} & \text{\% Conv.} \\ -78° & 72\text{ h} & 4 \\ -25° & 20\text{ h} & 46 \end{array}$	350
0.1 M		Toluene	I $\begin{array}{ccc} \text{Temp} & \text{Time} & \text{\% Conv.} \\ -78° & 15\text{ h} & 15 \\ -25° & 20\text{ h} & 73 \end{array}$	350
		−78°, 2 h	I $\begin{array}{ccc} \text{Solvent} & & \\ \text{CH}_2\text{Cl}_2 & (>90), 93 \\ \text{toluene} & (>90), 97 \end{array}$	50
		—	I (84), 92	351
	BF₃⁻K⁺	n-Bu₄NI (10 mol%), CH₂Cl₂, H₂O, rt, 15 min	I (97)	100
		n-Bu₃P, PhSeCl, AIBN, n-Bu₃SnH, THF, reflux	I (56)	352

138

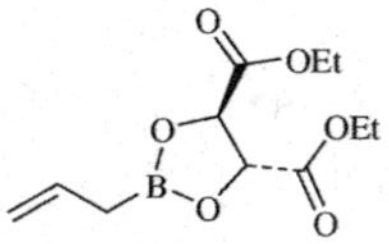

352

103

263

103

103

352

353

(30), 8

(98)

(65)

(94)

I + II (100), I:II = 95:5

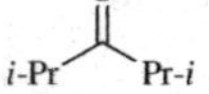

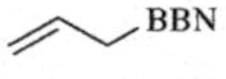

I + **II**

(79)

(82)

n-Bu$_3$P, PhSeCl, Et$_3$B, n-Bu$_3$SnH, THF, −78°

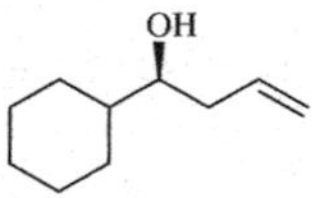

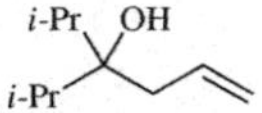

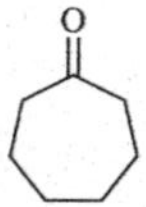

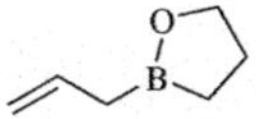

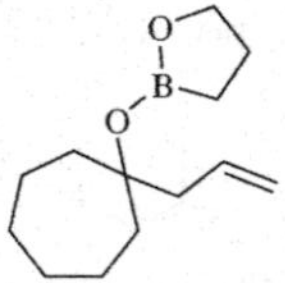

Pentane, rt, 2 h

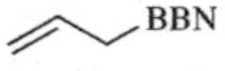

Neat, 80-100°

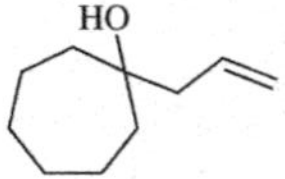

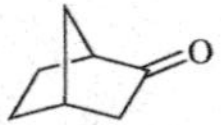

Pentane, rt, 2 h

Pentane, rt, 2 h

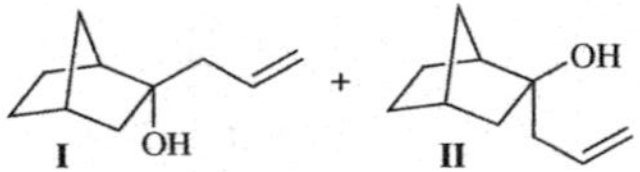

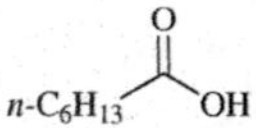

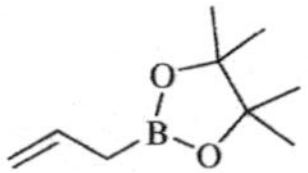

n-Bu$_3$P, PhSeCl, AIBN, n-Bu$_3$SnH, THF, reflux

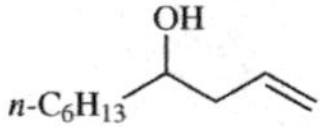

bmimBr, rt, 3 h

BBN

BBN

BBN

B(OPr-i)$_2$

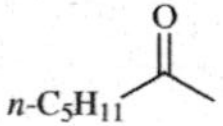

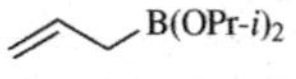

i-Pr

n-C$_6$H$_{13}$

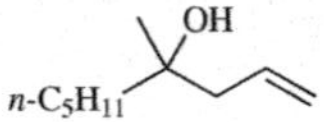

n-C$_5$H$_{11}$

TABLE 1. ADDITION OF UNSUBSTITUTED ALLYL REAGENTS (*Continued*)

A. NON-AROMATIC CARBONYL SUBSTRATES (*Continued*)

Carbonyl Substrate	Allylborane Reagent	Conditions	Product(s) and Yield(s) (%), % ee	Refs.
C₇	B[(+)-Ipc]₂	Ether, –100°, 2 h	(95), 10:1 d.r.	354, 355, 356
	B[(–)-Ipc]₂	4 Å MS, toluene, –78°	(98), 9:1 d.r.	357
	"	CH₂Cl₂, –78°	**I** (73), 2:1 d.r.	357
	B[(+)-Ipc]₂	CH₂Cl₂, –78°	(77), 2:1 d.r.	357
	(allyl dioxaborolane-OEt)	4 Å MS, toluene, –78°	(76), 1:1 d.r.	357
	B(OH)₂	CH₂Cl₂, rt	(84), 68:32 d.r.	110
	B(OH)₂	CH₂Cl₂, rt	(86), 70:30 d.r.	110
	B[(–)-Ipc]₂	Ether, –100°	(60), 80	358

Aldehyde	Reagent	Conditions	Product	Ref.
		4 Å MS, toluene, −78°	(95), 8:1 d.r.	359
	B[(−)-Ipc]₂	Ether, −78°	(91), 97:3 d.r.	360
	B[(−)-Ipc]₂	Ether, −100°	(75), 95:5 d.r.	361
	B[(−)-Ipc]₂	THF, −78°, 1 h	(80), 96:4	362
	B[(+)-Ipc]₂	Ether, −100°, 2 h [a]	(80), >90	363
	B[(−)-Ipc]₂	Ether, −100°, 0.5 h	(74), >98	188, 671
	B[(−)-Ipc]₂	Ether, −100°	(62), 86	358
	"	Ether, −78° [a]	I (75), 88	285

TABLE 1. ADDITION OF UNSUBSTITUTED ALLYL REAGENTS (*Continued*)

A. NON-AROMATIC CARBONYL SUBSTRATES (*Continued*)

Carbonyl Substrate	Allylborane Reagent	Conditions	Product(s) and Yield(s) (%), % ee	Refs.
C7				
	(OPr-*i*, OPr-*i*)	4 Å MS, toluene, –78°	(85), 4:1 d.r.	364
	B[(–)-Ipc]₂	—	**I** (48), 8.6:1 d.r.	364
	B[(–)-Ipc]₂	Ether, –78°	(76), 10:1 d.r.	365
	B[(+)-Ipc]₂	Ether, –78°	(65), 10:1 d.r.	365
racemic	B[(+)-Ipc]₂	1. Allylation 2. K₂CO₃, MeOH	(78), >95, 42:36 d.r.	366
	B[(–)-Ipc]₂	Ether, –78°, 1 h	(88), 95:5 d.r.	367
	B[(–)-Ipc]₂	Ether, –78° to rt, 2 h	(95)	270

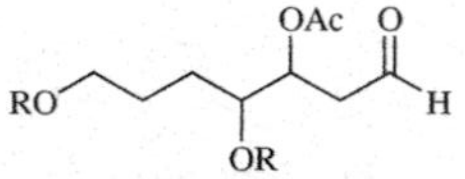

368   (71–79), 5:1–8:1 d.r.

369   (48)

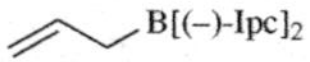

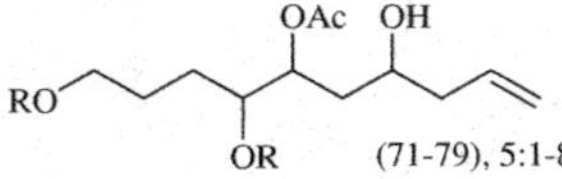

370   (60), 90:10 d.r.

193   (89), >98:2 d.r.

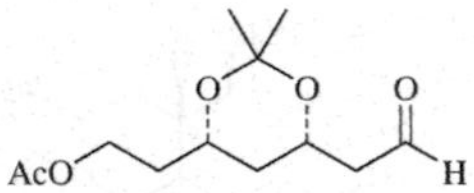

338   (70), 92

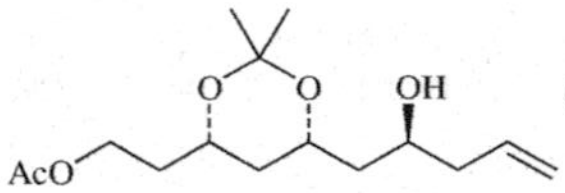

310   (58), 90

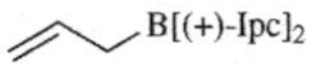

340   (35), 62   +   (17), 54

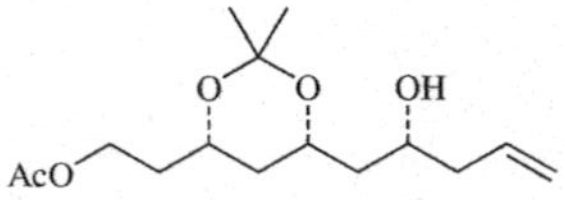

—

Ether, −78°

Ether, −78°, 1 h

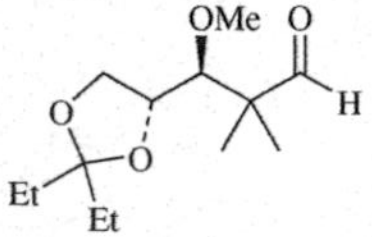

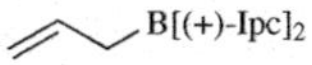

Ether, −95°

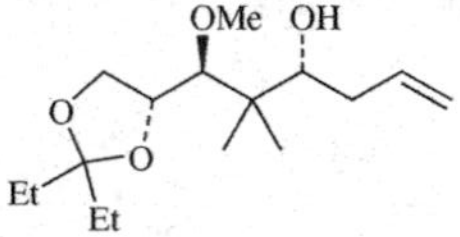

Ether, −78°

Ether, −78°

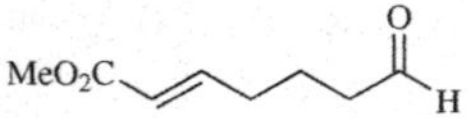

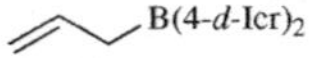

Ether, −78° to rt, 18 h

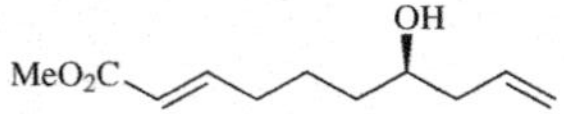

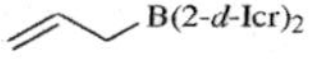

B[(−)-Ipc]$_2$

B[(−)-Ipc]$_2$

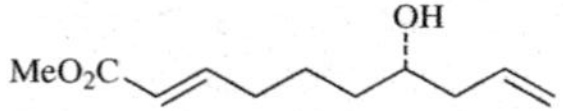

B[(+)-Ipc]$_2$

B[(+)-Ipc]$_2$

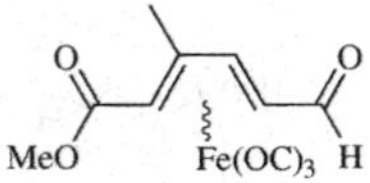

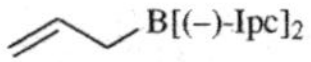

B(4-d-Icr)$_2$

B(2-d-Icr)$_2$

B[(−)-Ipc]$_2$

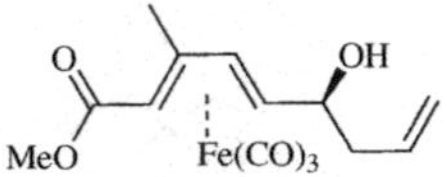

R = Me$_2$CHCH$_2$C(O)

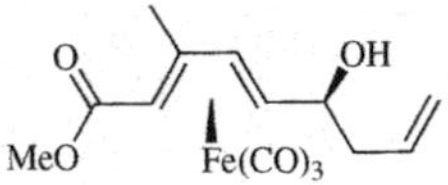

TABLE 1. ADDITION OF UNSUBSTITUTED ALLYL REAGENTS (*Continued*)

A. NON-AROMATIC CARBONYL SUBSTRATES (*Continued*)

Carbonyl Substrate	Allylborane Reagent	Conditions	Product(s) and Yield(s) (%), % ee	Refs.
C₇				
	B(OPr-*i*)₂	Et₃N, CH₂Cl₂, −10° to rt, 13 h	(80)	105
	B(2-*d*-Icr)₂	Ether, −78°, 1 h	(56)	371
	B(4-*d*-Icr)₂	Ether, −78°, 1 h	(60), 80-88	372, 373
	B(2-*d*-Icr)₂	Ether, −78°, 1 h	(62), 80-88	372, 373
		CH₂Cl₂, −78°	(91)	364
	B[(+)-Ipc]₂	Ether, −100°	(66), 12:1 d.r.	374

144

Substrate	Reagent	Conditions	Product	Reference
(4-bromothiazol-2-yl, gem-dimethyl aldehyde)	B[(+)-Ipc]$_2$ allyl	Ether, −78°	OH, (70), 54	375
n-C$_7$H$_{15}$CHO	B[(−)-Ipc]$_2$ allyl	Ether, −100° to −48°	OH, (55), 92	376, 377
	"	THF, −100°, 1 h	I (44), 90	378
	B(OPr-i)$_2$ allyl	bmimBr, rt, 3 h	OH, (86)	353
	BF$_3^-$K$^+$ allyl	BF$_3$•OEt$_2$ (2 equiv), CH$_2$Cl$_2$, −78°, 15 min	I (82)	98
	"	BF$_3$•OEt$_2$ (5 mol%), CH$_2$Cl$_2$, rt, 3–6 h	I (84)	99
	"	n-Bu$_4$NI (10 mol%), CH$_2$Cl$_2$, H$_2$O, rt, 15 min	I (94)	99
n-C$_5$H$_{11}$ (ynal)	B[(+)-Ipc]$_2$ allyl	Ether, pentane, −100°	OH, n-C$_5$H$_{11}$ I (80), 99	379
	(pinanediol Ph boronate)	Sc(OTf)$_3$ (10 mol%), CH$_2$Cl$_2$, −78°, 12 h	I (87), 95	28
(isopropenyl aldehyde)	B[(−)-Ipc]$_2$ allyl	Ether, pentane	OH, (85)	380

C$_8$

145

TABLE 1. ADDITION OF UNSUBSTITUTED ALLYL REAGENTS (*Continued*)

A. NON-AROMATIC CARBONYL SUBSTRATES (*Continued*)

Carbonyl Substrate	Allylborane Reagent	Conditions	Product(s) and Yield(s) (%), % ee	Refs.
C₈				
(PhCH₂CHO)	B[(+)-Ipc]₂	CH₂Cl₂, −78°	(—)	381
(cyclohexenyl acetaldehyde)	B[(−)-Ipc]₂	S₂(OTf)₃, THF, −78, 6 h	(92), 96	101
(1-acetylcyclohexene)	(allyl pinacol boronate)	(*R*,*R*)-*i*-Pr-DuPHOS, La(OPr-*i*)₃, CuF, DMF, −40°, 1 h	(87), 90	164
(*t*-Bu / Pr-*i* ketone)	BBN	Pentane, rt, 7 d	*i*-Pr, *t*-Bu OH (74)	103
(*t*-Bu / Pr-*n* ketone)	BBN	Pentane, rt, 2 h	*n*-Pr, *t*-Bu OH (97)	103
(2,2-dimethylcyclohexanone)	(allyl pinacol boronate)	(*R*,*R*)-*i*-Pr-DuPHOS, La(OPr-*i*)₃, CuF, DMF, −40°, 1 h	(98), 84	164
(BnO aldehyde)	B[(+)-Ipc]₂	Ether, −80°, 3 h	(65), 93:7 d.r.	382

Substrate	Reagent	Conditions	Product	Ref.
TBDMSO, n-Bu (aldehyde)	B(2-d-Icr)$_2$	—	TBDMSO, HO, n-Bu (79)	383
TBDMSO-CH$_2$ (cyclohexane aldehyde)	B[(−)-Ipc]$_2$	4 Å MS, toluene, −78° to rt	OH, TBDMSO-CH$_2$ (70), 80	384
PivO (aldehyde)	allyl boronate	—	OH, PivO (60)	385
OH, i-Pr (aldehyde)	MeO–B, OMe, i-Pr, MgBr$^+$	1. $\begin{array}{c}\text{OMe}\\\text{B}\\\text{O O}\end{array}$ i-Pr, OMe, THF, −78° to rt, 1.5 h; 2. allyl MgBr, −78° to rt, 3 h	OH, OH, i-Pr (77), 70:30 d.r.	299
OH, n-Bu (ketone)	B(OH)$_2$	CH$_2$Cl$_2$, rt	OH, OH, n-Bu (87), 88:12 d.r.	110
Ph, OTBDMS (aldehyde)	BF$_3^-$K$^+$	n-Bu$_4$NI (10 mol%), CH$_2$Cl$_2$/H$_2$O, rt, 30 min	OH, Ph, OTBDMS (94), 65:35 d.r. **I**	276
	"	BF$_3$•OEt$_2$ (5 mol%), CH$_2$Cl$_2$, rt, 6 h	**I** (91), 65:35 d.r.	99

147

TABLE 1. ADDITION OF UNSUBSTITUTED ALLYL REAGENTS (*Continued*)

A. NON-AROMATIC CARBONYL SUBSTRATES (*Continued*)

Carbonyl Substrate	Allylborane Reagent	Conditions	Product(s) and Yield(s) (%), % ee	Refs.
C$_8$				
		4 Å MS, toluene, −78° to rt	(77), 91	328, 329
		4 Å MS, toluene, −78° to rt	(77), 91	331
	B[(−)-Ipc]$_2$	Toluene, −78°	(64), >90, 14:1 d.r.	330
	B[(−)-Ipc]$_2$	Ether	(56), 3:1 d.r.	386, 387
	B[(+)-Ipc]$_2$	—	(65), 87	388
	B(4-*d*-Icr)$_2$	Ether, −78°, 1 h[a]	(52)	389, 390

Ether, −78° B[(−)-Ipc]$_2$ (85), 85:15 d.r. 391

Ether, −78° B[(+)-Ipc]$_2$ (79), >19:1 d.r. 392, 391

Toluene, −78°, 5 h B[(−)-Ipc]$_2$ (72), 96:4 d.r. 393

Ether, −78° to rt, 2 h B[(+)-Ipc]$_2$ (78) 394

Ether, pentane, −100° B[(+)-Ipc]$_2$ (85), 8:1 d.r. 395

Ether, −78° B[(+)-Ipc]$_2$ (85), 88:12 d.r. 396, 397

Ether, −78° B[(−)-Ipc]$_2$ (84), 96:4 d.r. 396

R = TBDMS

TABLE 1. ADDITION OF UNSUBSTITUTED ALLYL REAGENTS (*Continued*)

A. NON-AROMATIC CARBONYL SUBSTRATES (*Continued*)

Carbonyl Substrate	Allylborane Reagent	Conditions	Product(s) and Yield(s) (%), % ee	Refs.
C$_8$ (R = TBDMS)	B[(+)-Ipc]$_2$	Ether, −78°	(80), 96:4 d.r.	396
	B[(−)-Ipc]$_2$	Ether, −78°	(88), >19:1 d.r.	396
(TBDPSO, OTBDMS, MOMO, OMOM)	B[(+)-Ipc]$_2$	Ether, −78°, 3 h	(74), 88:12 d.r.	398
(i-Pr)	B[(−)-Ipc]$_2$	CH$_2$Cl$_2$, −78° to rt	(83)	399
(thiazole)	B[(+)-Ipc]$_2$	Ether, −100°, 0.5 h	(89-96), >97	188, 400

C9

Substrate	Reagent	Conditions	Product	Refs.
	"	—	I (96), >91	401
	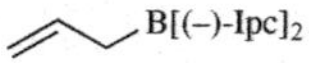 B[(–)-Ipc]$_2$	Ether, –100°	I (83), >95	402
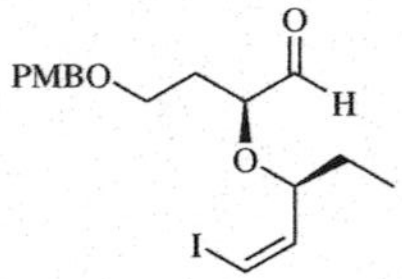	 B[(–)-Ipc]$_2$	Ether, –95° to –100°, 2 h^a	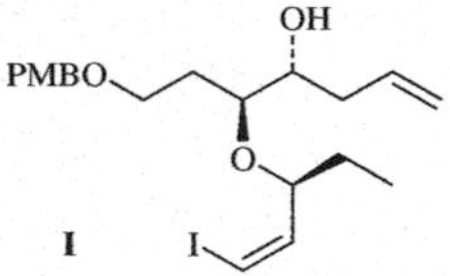 (89), 97:3 d.r.	403, 192
	"	Ether, –95° to –100°, 2 h	I (72), 93:7 d.r.	192
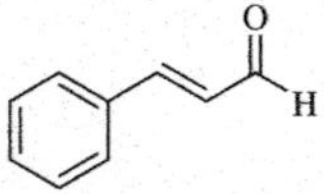	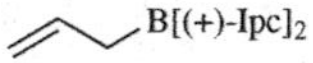 B[(+)-Ipc]$_2$	Pentane, ether, –100°	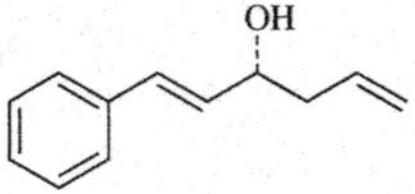 (72), 92	204
	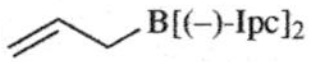 B[(–)-Ipc]$_2$	Ether, –100° to –78°, 3 h	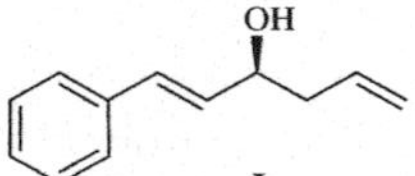 (73), 94	292
	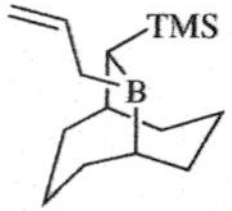	Ether, –78°, 3 h	I (92), 97	143
	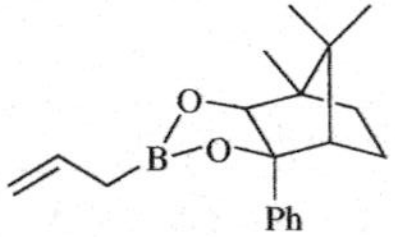	Hexane, –40°	I (81), 24	122

151

TABLE 1. ADDITION OF UNSUBSTITUTED ALLYL REAGENTS (*Continued*)

A. NON-AROMATIC CARBONYL SUBSTRATES (*Continued*)

Carbonyl Substrate	Allylborane Reagent	Conditions	Product(s) and Yield(s) (%), % ee	Refs.
C₉	B(OPr-*i*)₂	bmimBr, rt, 3 h	(84)	353
		L.A. (10 mol%), toluene, −78°, 24 h	**I** $\quad$ L.A. $\quad$ AlCl₃ (69) $\quad$ Sc(OTf)₃ (74)	25
		THF, −78°, 1 h	(98), 76	135
		−78°, 2 h	**I** $\quad$ Solvent $\quad$ toluene (>90), 97 $\quad$ CH₂Cl₂ (>90), 98	50
	BF₃⁻K⁺	BF₃•OEt₂ (5 mol%), CH₂Cl₂, rt, 3-6 h	(91)	99
	"	*n*-Bu₄NI (10 mol%), CH₂Cl₂, H₂O, rt, 15 min	**I** (95)	100

Substrate	Reagent	Conditions	Product (yield)	Refs.
(MeO-aryl acrolein)	B[(+)-Ipc]₂-allyl	Ether, −78°, 2 h	(77), 93	404
(C₆F₅ enal)	B[(−)-Ipc]₂-allyl	Ether, −100° to −78°, 3 h	(74), 93	292
(n-C₅H₁₁ enal)	B[(−)-Ipc]₂-allyl	n-C₅H₁₁ epoxide, Sc(OTf)₃, THF, −78°, 6 h	(80), 97	101
(BnO-aryl propanal)	B[(+)-Ipc]₂-allyl	—	(75), 94	405
(phenyl propanal)	B[(−)-Ipc]₂-allyl	Pentane, ether, −100°, 1 h	(76), 96	251
(phenyl propanal)	allyl-B(pinanyl, Ph)	Sc(OTf)₃ (10 mol%), CH₂Cl₂, −78°, 12 h	(64), 97	27, 28
(2-phenylpropanal)	B(OPr-i)₂-allyl	bmimBr, rt, 3 h	(85), 82:18 d.r	353

153

TABLE 1. ADDITION OF UNSUBSTITUTED ALLYL REAGENTS (*Continued*)

A. NON-AROMATIC CARBONYL SUBSTRATES (*Continued*)

Carbonyl Substrate	Allylborane Reagent	Conditions	Product(s) and Yield(s) (%), % ee	Refs.
C$_9$				
t-Bu–CO–Bu-*t*	allyl–BBN	Octane, 125°, 5 d	(25)	103
bicyclo[3.3.1]nonan-9-one	allyl–BBN	Pentane, rt, 2 h	(85)	103
resin-bound aldehyde (Ph, N–CO–Ph)	allyl–Bpin	CH$_2$Cl$_2$, rt, 16 h	(70)	235
3-hydroxy-3-phenylpropanal	allyl–B(OH)$_2$	CH$_2$Cl$_2$, rt	(85), 72:28 d.r.	110
TBDMSO 3-phenylpropanal	allyl–BF$_3^-$K$^+$	*n*-Bu$_4$NI (1 mol%), CH$_2$Cl$_2$, H$_2$O, rt, 30 min	(94), 65:35 d.r.	276
"	"	BF$_3$•OEt$_2$ (5 mol%), CH$_2$Cl$_2$, rt, 3-6 h	**I** (86), 70:30 d.r.	347
OH, Cy propanal	allyl–B(OH)$_2$	CH$_2$Cl$_2$, rt	(85), 67:33 d.r.	110

154

Substrate	Reagent	Conditions	Product (yield)	Refs.
(aldehyde/ketone structure, Ph)	B(OPr-i)$_2$ allyl	Ether, −10°, 18 h	(89)	108
(ketone, C$_6$H$_{13}$-n)	B(OPr-i)$_2$ allyl	Ether, −10°, 18 h	(80)	108
(ketone, Cy)	B(OPr-i)$_2$ allyl	Ether, −10°, 18 h	(70)	108
TBDMSO, i-Pr	B[(−)-Ipc]$_2$ allyl	Ether, −78° to rt, 14 h	(84)	347
PMBO, i-Pr	B[(+)-Ipc]$_2$ allyl	Ether, −90° to rt	(71), 10:1 d.r.	406
TBDMSO, OMEM, Ph	B[(+)-Icr]$_2$ allyl	Ether, pentane, −100°	(—), 96:4 d.r.	407
PMPO, OPMP, Ph	B[(+)-Icr]$_2$ allyl	Ether, pentane, −100°	(75), 97.5:2.5 d.r.	407
(dioxolane, Ph)	B[(+)-Ipc]$_2$ allyl	Ether, pentane, −100°	(—)	407

155

TABLE 1. ADDITION OF UNSUBSTITUTED ALLYL REAGENTS (*Continued*)

A. NON-AROMATIC CARBONYL SUBSTRATES (*Continued*)

Carbonyl Substrate	Allylborane Reagent	Conditions	Product(s) and Yield(s) (%), % ee	Refs.
C₉	B[(−)-Ipc]₂	Ether, 18°	(50), 1:1 d.r.	408
	(allyl diisopropyl tartrate boronate)	Ether, −78°	(89), 11:1 d.r.	408
	B[(+)-Ipc]₂	Ether, pentane, −100°	(—)	407
	B[(−)-Ipc]₂	Ether, −78°	(71), >95:5 d.r.	310
	B(4-*d*-Icr)₂	Ether, −78°, 1 h [a]	(56)	389, 373, 390
	B(2-*d*-Icr)₂	Ether, −78°, 1 h [a]	(52)	389

156

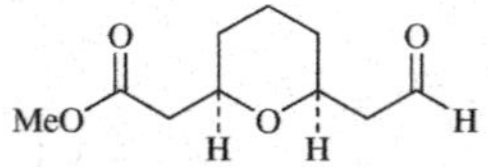		Ether, –78°	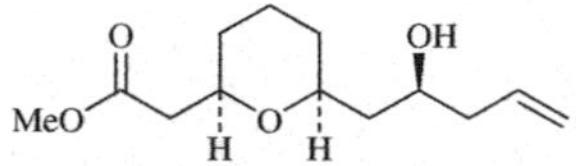 (71), >95:5 d.r.	310
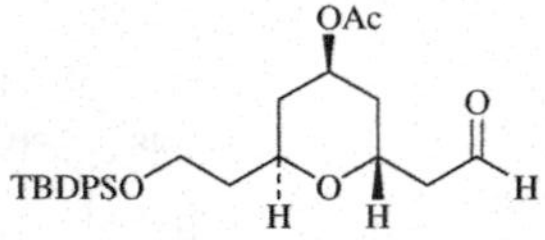 racemic	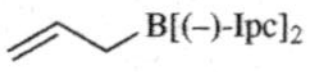	Toluene, –78° [a]	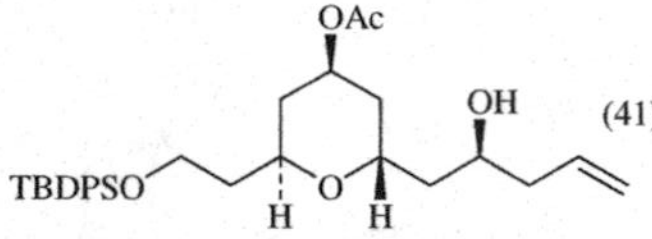 (41) + 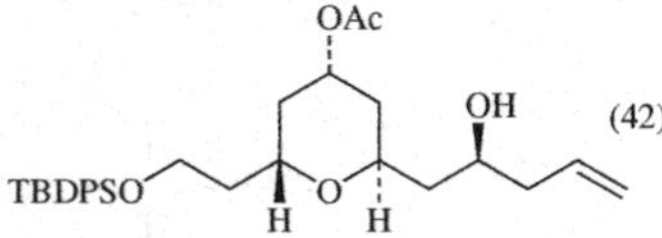 (42)	409, 410
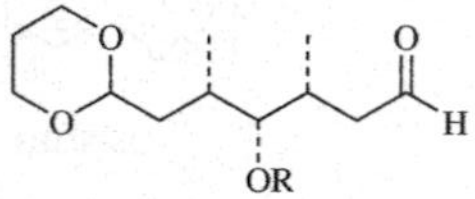		Ether, –78°	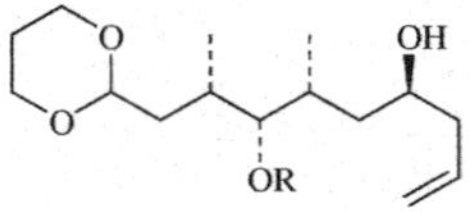 R: TBDMS (72); TES (95)	411
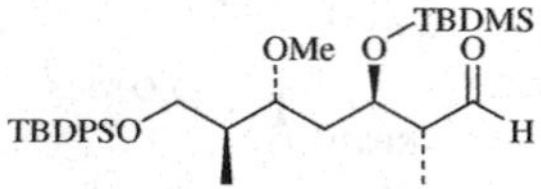		THF, –78° to rt, 3.5 h	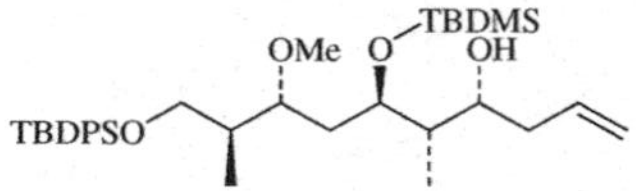 (65-70), 95:5 d.r.	412
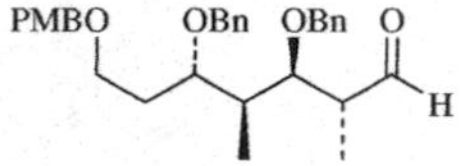		THF, –78°	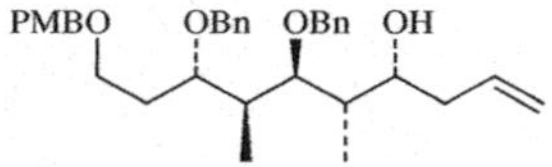 (73), >15:1 d.r.	413
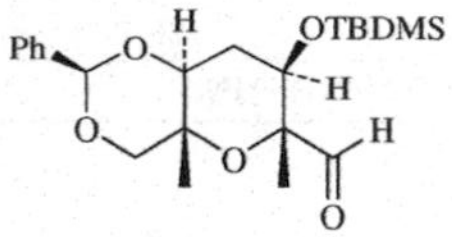	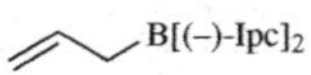	Ether, –78°	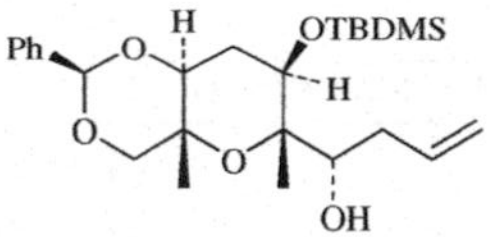 (92)	414, 415

TABLE 1. ADDITION OF UNSUBSTITUTED ALLYL REAGENTS (*Continued*)

A. NON-AROMATIC CARBONYL SUBSTRATES (*Continued*)

Carbonyl Substrate	Allylborane Reagent	Conditions	Product(s) and Yield(s) (%), % ee	Refs.
C$_9$	B[(+)-Ipc]$_2$	Ether, –100°, 2 h	(70), 95	416, 417
	TMS-pyrrolidine allylborane	Ether, –100°, 3 h	(95), 9.5:1 d.r.	418
	B[(–)-Ipc]$_2$	Ether, –78°	(80), 89	419
	B[(+)-Ipc]$_2$	THF, –78° to rt, 3.5 h	(64), 95:5 d.r.	420, 421, 422
C$_{10}$	B[(+)-Ipc]$_2$	Pentane, ether, –100°	(74), 92	204
	B(2-*d*-Icr)$_2$	Ether, –50°, 17 h	**I** (22), 76	423

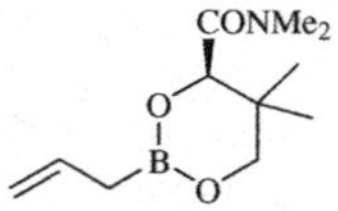

n-C$_9$H$_{19}$ (36), 47   424

I (29), 28   424

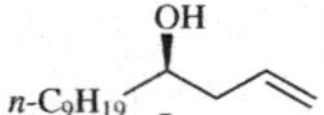

n-C$_6$H$_{13}$ (55-70), 70   130

n-C$_6$H$_{13}$ (55-70), 31   130

n-C$_7$H$_{15}$ (85-95), 92   128

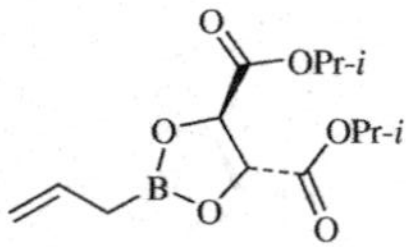

(96), 96   101

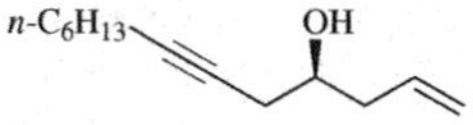

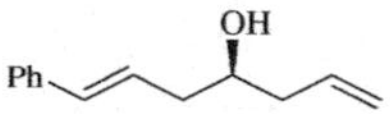

Work-up	
NH$_4$Cl	(48), 93
NaOH, H$_2$O$_2$	(80), 96

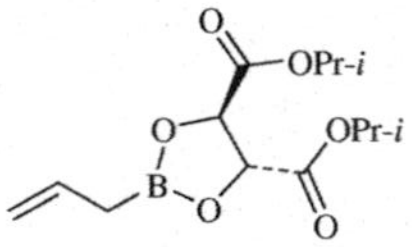

101

CH$_2$Cl$_2$, −78°, 24 h

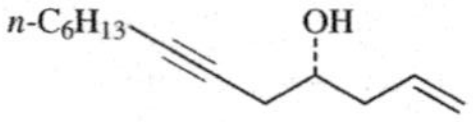

Toluene, −78°, 24 h

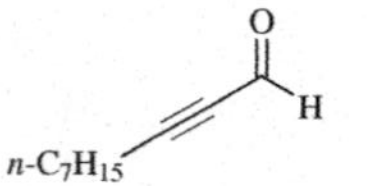

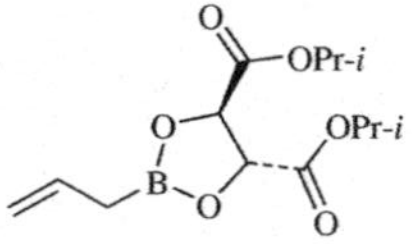

4 Å MS, toluene, −78°

Cr(CO)$_3$, 4 Å MS, toluene, −78°

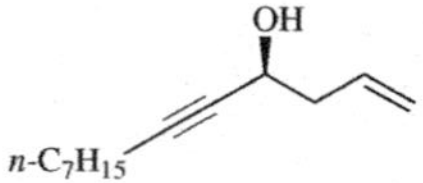

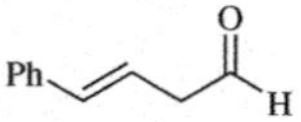

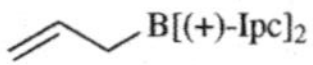

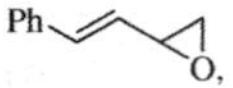

Cr(CO)$_3$, 4 Å MS, toluene, −78°

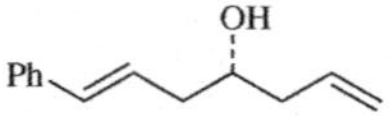

Sc(OTf)$_3$, THF, −78°, 6 h

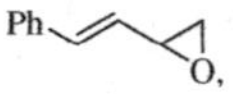

BF$_3$•OEt$_2$, THF, −78°, 6 h; work-up

CONMe$_2$

OPr-i

BI(+)-Ipc]$_2$

BI(−)-Ipc]$_2$

n-C$_6$H$_{13}$

n-C$_7$H$_{15}$

Ph

159

TABLE 1. ADDITION OF UNSUBSTITUTED ALLYL REAGENTS (*Continued*)

A. NON-AROMATIC CARBONYL SUBSTRATES (*Continued*)

Carbonyl Substrate	Allylborane Reagent	Conditions	Product(s) and Yield(s) (%), % ee	Refs.
C$_{10}$	B[(−)-Ipc]$_2$	Sc(OTf)$_3$, THF, 6 h	**I** <table><tr><td>Temp</td><td></td></tr><tr><td>−96°</td><td>(94), >98</td></tr><tr><td>−78°</td><td>(96), 97</td></tr><tr><td>−78°</td><td>(100), 95</td></tr><tr><td>−78°</td><td>(100), 89</td></tr><tr><td>−78°</td><td>(100), 86</td></tr></table>	101
	"	Sc(OTf)$_3$, THF, −78°, 6 h	**I** (88), 96	101
	"	L.A., THF, −78°, 6 h	**I** <table><tr><td>L.A.</td><td></td></tr><tr><td>Sm(OTf)$_3$</td><td>(28), 96</td></tr><tr><td>SnCl$_4$</td><td>(18), 83</td></tr><tr><td>Ag(OTf)</td><td>(33), 94</td></tr><tr><td>Y(OTf)$_3$</td><td>(65), 94</td></tr><tr><td>Sc(OTf)$_3$</td><td>(92), 96</td></tr><tr><td>BF$_3$•OEt$_2$</td><td>(59),96</td></tr></table>	101
		L.A., THF, −78°, 6 h	**I** <table><tr><td>L.A.</td><td></td></tr><tr><td>none</td><td>(40), 27</td></tr><tr><td>BF$_3$•OEt$_2$</td><td>(70), 47</td></tr></table>	101
		BF$_3$•OEt$_2$, THF, −78°, 6 h	**I** (33), 37	101

160

Substrate	Reagent	Conditions	Product	Yield	Refs.
(o-NO₂-cinnamaldehyde)	B[(+)-Ipc]₂ (butenyl)	Sc(OTf)₃, THF, −78°, 4 h; −70°, 36 h	(o-NO₂-phenyl allylic alcohol)	(88), 96	101
(o-R-cinnamaldehyde)	B[(−)-Ipc]₂ (butenyl)	Sc(OTf)₃, THF, −78°, 6 h	(o-R-phenyl allylic alcohol)	R: OMe (68), 96 / NO₂ (35), 96	101
(Ph-methyl enal)	B[(−)-Ipc]₂ (butenyl)	Ether, −100° to −78°, 3 h	(Ph-methyl allylic alcohol)	(73), 91	292
(MeO-phenyl butanal)	(pinacol allyl boronate)	n-Bu₃P, PhSeCl, AIBN, n-Bu₃SnH, THF, reflux	(MeO-phenyl alcohol)	(80)	352
	(tartrate diethyl ester allyl boronate)	n-Bu₃P, PhSeCl, Et₃B, n-Bu₃SnH, THF, −78°	(MeO-phenyl alcohol) **I**	(61), 73	352
	"	n-Bu₃P, PhSeCl, AIBN, n-Bu₃SnH, THF, reflux	**I** (85), 46		352

Carbonyl Substrate	Allylborane Reagent	Conditions	Product(s) and Yield(s) (%), % ee	Refs.
C₁₀				
	B[(–)-Ipc]₂	Ether, –78°	(72), 67:33 d.r.	166, 167
	B[(+)-Ipc]₂	Ether, –78°	(72), 98:2 d.r.	166, 167
	B[(–)-Ipc]₂	Ether, –78°	(74), 97:3 d.r.	166, 167
	B[(+)-Ipc]₂	Ether, –78°	(74), 74:26 d.r.	166, 167
	B(2-d-Icr)₂	Ether, –78°, 4 h	(63), 98.5:1.5 d.r.	425
	B[(–)-Ipc]₂	—	(—)	426
		Toluene, –78°	(94), 81:19 d.r.	427

162

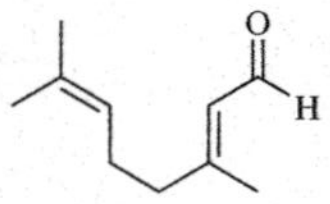

			238
			238
			238
			238
			135
			135
			164

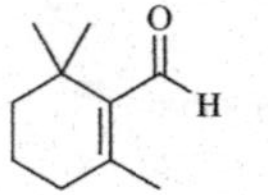

(73)

(80)

(84)

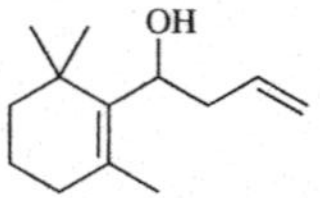

(97)

OH

(91), 76

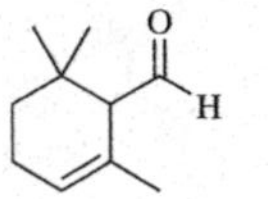

I

I (98), 50

(96), 67

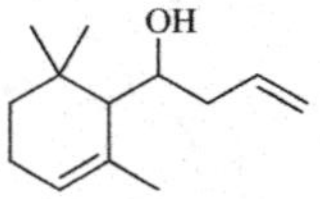

HO

HO

Ph

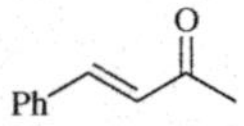

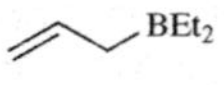

Ph

Ph

Neat, 0-10°

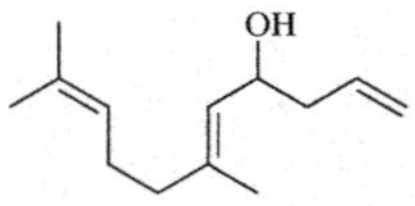

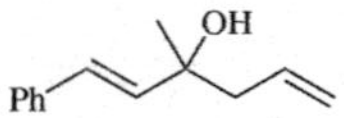

Neat, 0-10°

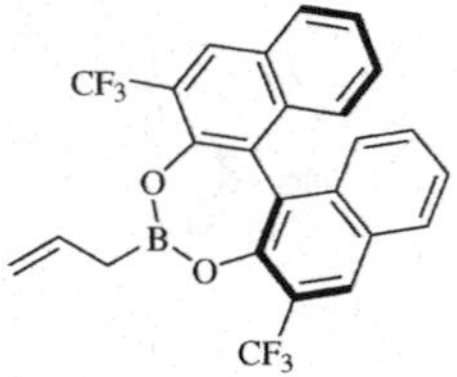

Neat, 0-10°

Neat, 0-10°

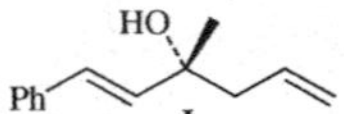

THF, −78° to −40°, 48 h

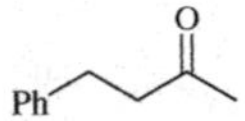

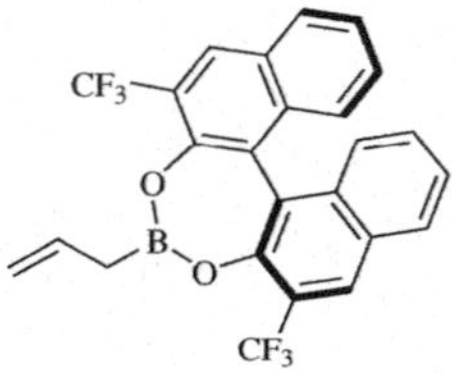

THF, −78° to −40°, 48 h

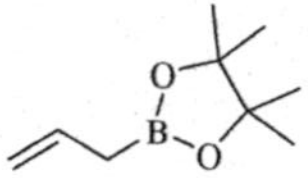

(R,R)-i-Pr-DuPHOS, La(OPr-i)$_3$, CuF, DMF, −40°, 1 h

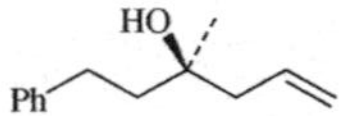

BEt$_2$

BEt$_2$

BEt$_2$

BEt$_2$

CF$_3$

CF$_3$

CF$_3$

CF$_3$

O—B

O

Ph

Carbonyl Substrate	Allylborane Reagent	Conditions	Product(s) and Yield(s) (%), % ee	Refs.
C_{10} 4-*t*-Bu cyclohexanone	allyl–BBN	Pentane, rt, 2 h	**I** + **II** **I + II** (100), **I:II** = 55:45	103
4-MeO phenyl butanoic acid (OH)	allyl pinacol boronate	*n*-Bu$_3$P, PhSeCl, AIBN, *n*-Bu$_3$SnH, THF, reflux	(80)	352
	allyl tartrate boronate (OEt)	*n*-Bu$_3$P, PhSeCl, AIBN, *n*-Bu$_3$SnH, THF, reflux	**I** (85), 46	352
	"	*n*-Bu$_3$P, PhSeCl, Et$_3$B, *n*-Bu$_3$SnH, THF, –78°	**I** (61), 73	352
Ph amino ketone (HN–R)	allyl–BBN	CH$_2$Cl$_2$, rt	**I** (see below)	107

R		d.r.
Boc	(80)	80:20
Ts	(94)	79:21
Ac	(80)	80:20
Bn	(80)	85:15
p-Tol	(85)	85:15
4-(*t*-Bu)Bn	(90)	99:1
DMPM	(90)	88:12

R = Boc, Ts, Ac, Bn, *p*-Tol, 4-(*t*-Bu)Bn, DMPM

R = Bz	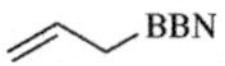	rt	**I** (—)	Solvent / d.r. THF 52:48 ether 73:27 CH_2Cl_2 85:15 toluene 91:9	107
	"	—	**I** (—)	Solvent / Temp / d.r. THF 0° 52:48 THF –78° 53:47 ether 0° 84:16 ether –78° 83:17 CH_2Cl_2 0° 92:8 CH_2Cl_2 –78° 94:6 toluene 0° 91:9 toluene –78° 93:7	107
	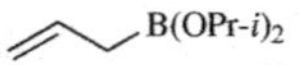	Base, CH_2Cl_2, rt	**I**	Base / Time / d.r. — 15 h (85) 74:26 DBU 45 h (26) 85:15 Et_3N 45 h (29) 88:12	107
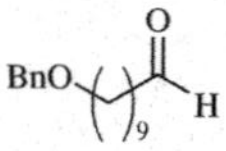	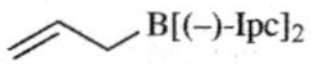	Ether, –78°, 1.5 h	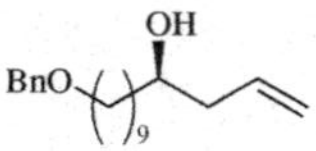	(66), 92	316
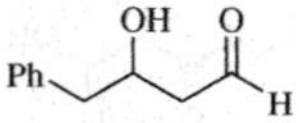		CH_2Cl_2, rt	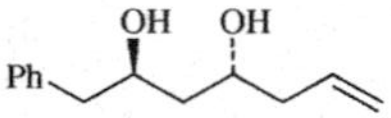	(83), 73:27 d.r.	110
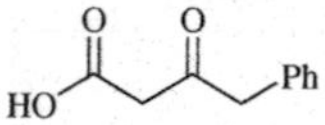	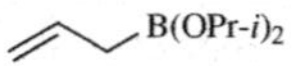	Ether, –10°, 18 h	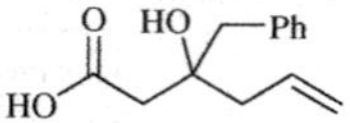	(65)	108

TABLE 1. ADDITION OF UNSUBSTITUTED ALLYL REAGENTS (*Continued*)

A. NON-AROMATIC CARBONYL SUBSTRATES (*Continued*)

Carbonyl Substrate	Allylborane Reagent	Conditions	Product(s) and Yield(s) (%), % ee	Refs.
C₁₀				
	B(OPr-*i*)₂	Ether, −10°, 18 h	(83)	108
	B[(−)-Ipc]₂	Ether, −78° to rt	(82), >10:1 d.r.	428
	B[(−)-Ipc]₂	Ether, −78°	(79)	429
	(tartrate allylboronate, OPr-*i*)	Toluene, −78°	(78), 8:1 d.r.	430
	B[(−)-Ipc]₂	THF, −78°	(73), 85:15 d.r.	431, 432
	B(4-*d*-Icr)₂	Ether, −78°, 1 h [a]	(56)	390

![allyl] B[(+)-Ipc]₂	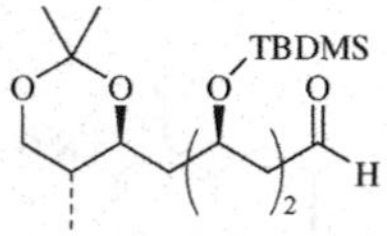 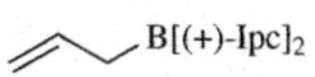 Ether, −78°	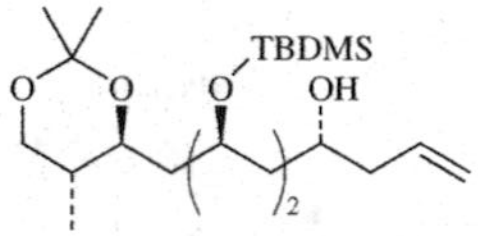 (71), >20:1 d.r.	433
B[(+)-Ipc]₂	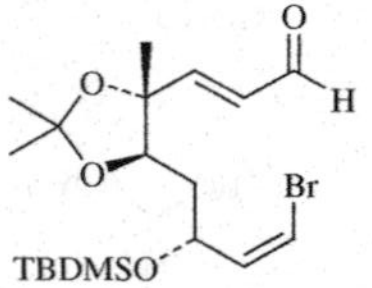  Ether, −100° to rt, 1.5 h	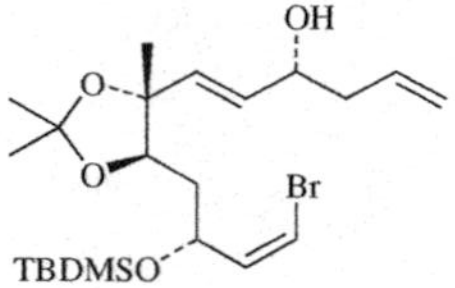 (71), 98	261
B(2-*d*-Icr)₂	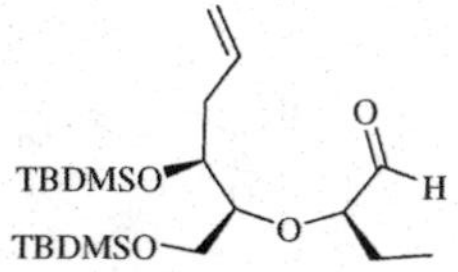 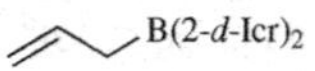 Ether, −78°, 3 h	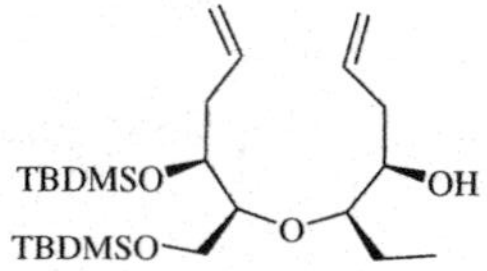 (52), 4:1 d.r.	434
B[(−)-Ipc]₂	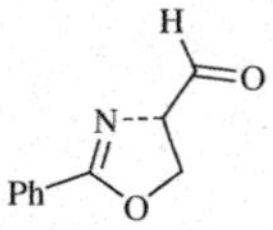  Ether, −78° to rt, 2 h	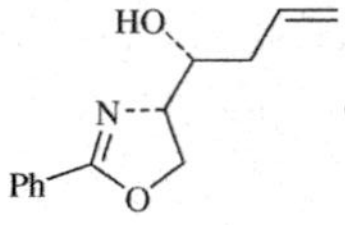 (89–91), 96:4–92:8 d.r.	435
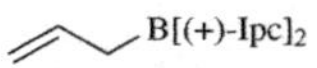 B[(+)-Ipc]₂	Ether, −78° to rt, 2 h	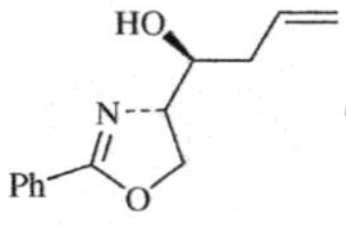 (89–91), 96:4–92:8 d.r.	435

C_{11}

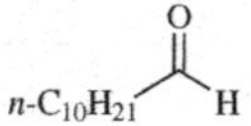

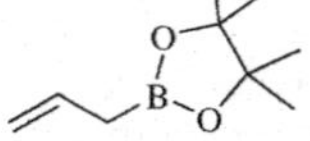

n-$C_{10}H_{21}$CHO

L.A. (10 mol%), toluene, −78°, 16 h

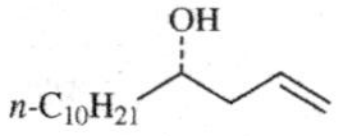

L.A.	
AlCl₃	(54)
Sc(OTf)₃	(73)

25

167

TABLE 1. ADDITION OF UNSUBSTITUTED ALLYL REAGENTS (Continued)
A. NON-AROMATIC CARBONYL SUBSTRATES (Continued)

Carbonyl Substrate	Allylborane Reagent	Conditions	Product(s) and Yield(s) (%), % ee	Refs.
		4 Å MS, toluene, −78°	(86), 86	125
	$B[(-)-Ipc]_2$	$Sc(OTf)_3$, THF, −78°, 6 h	(83), 96	101
	$B[(-)-Ipc]_2$	$Sc(OTf)_3$, THF, −78°, 6 h	(92), >95. 1:1 d.r.	101
		$n\text{-}Bu_3P$, PhSeCl, AIBN, $n\text{-}Bu_3SnH$, THF, reflux	(54)	352
	$B[(-)-Ipc]_2$	—	(82), 9:1 d.r.	436

C_{11}

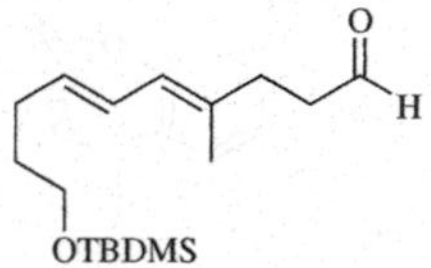

			437
			438
			439
			110
			110
			440
			364, 441

Ether, −100°, 1 h[a] **I** (70), >20:1 d.r.

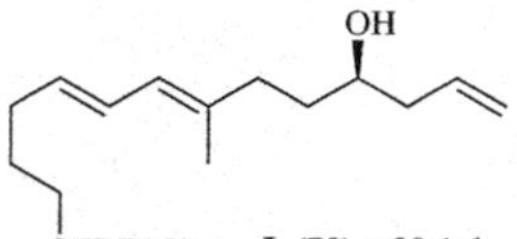

I (79), >96

Ether, −78°, 4 h

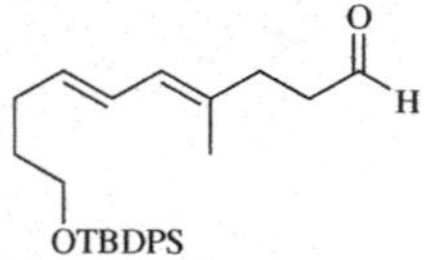

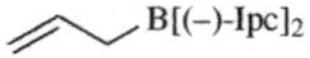

Ether, −78°, 4 h (63), 93 OTBDPS

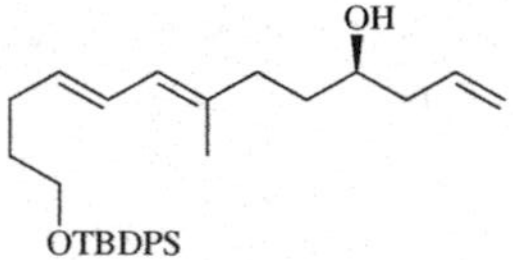

CH₂Cl₂, rt (90), 60:40 d.r.

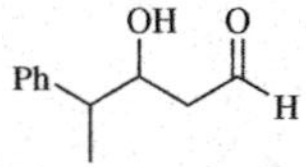

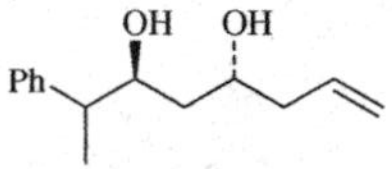

CH₂Cl₂, rt (84), 90:10 d.r.

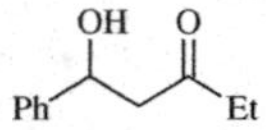

— (—)

Ether, −78° (92), 3:1 d.r.

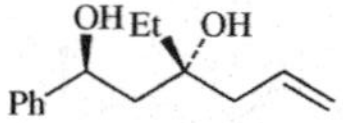

B[(−)-Ipc]₂

"

B[(−)-Ipc]₂

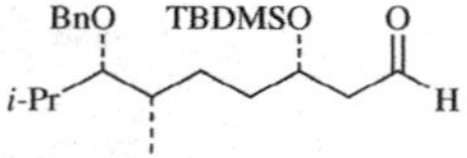

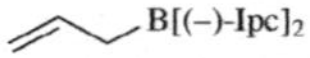

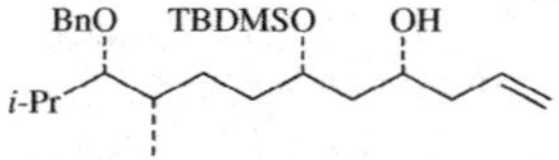

B(OH)₂

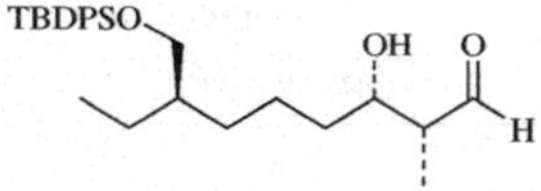

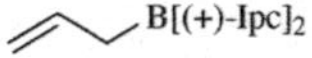

B(OH)₂

B[(−)-Ipc]₂

B[(+)-Ipc]₂

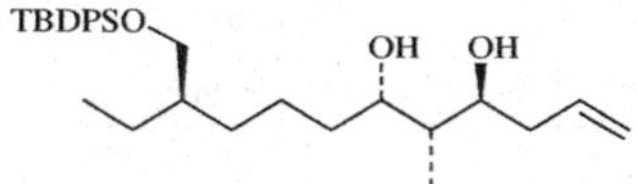

OTBDMS

OTBDPS

TABLE 1. ADDITION OF UNSUBSTITUTED ALLYL REAGENTS (*Continued*)

A. NON-AROMATIC CARBONYL SUBSTRATES (*Continued*)

Carbonyl Substrate	Allylborane Reagent	Conditions	Product(s) and Yield(s) (%), % ee	Refs.
C₁₁				
(structure)	B[(+)-Ipc]₂	Ether, –78°	(85), 95:5 d.r.	442
(structure)	B[(+)-Ipc]₂	Ether, –78°, 1 h	(74), 81:19 d.r.	443
(structure)	B[(+)-Ipc]₂	Ether, –100°	(80), >10:1 d.r.	444
(structure)	B[(–)-Ipc]₂	Ether, –78°	(—), 2:1 d.r.	444
(structure)	B[(+)-Ipc]₂	THF, –78° to rt, 2 h	(87), 17.6:1 d.r.	445
(structure)	B[(–)-Ipc]₂	Ether, –100°, 2 h	(62)	416, 417

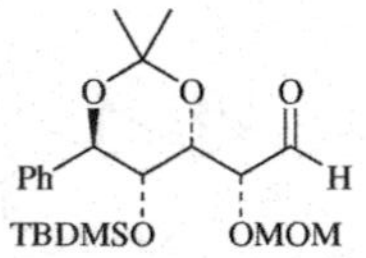	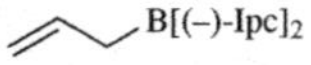 B[(–)-Ipc]_2	Ether, –90°	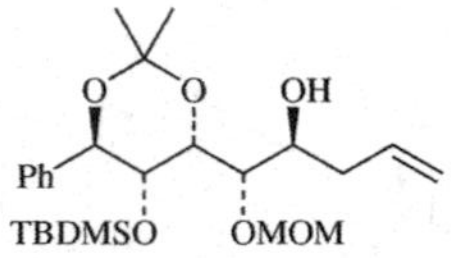 (69)	446
	B[(+)-Ipc]_2	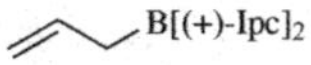 Ether, –90°	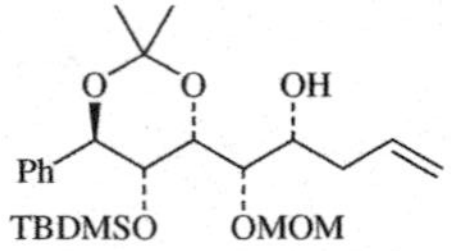 (94)	446
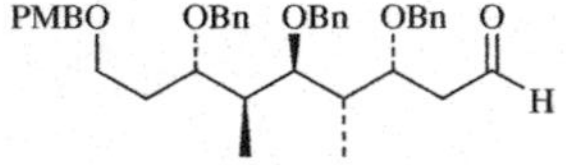	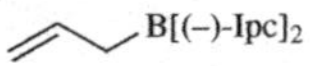 B[(–)-Ipc]_2	THF, –78°	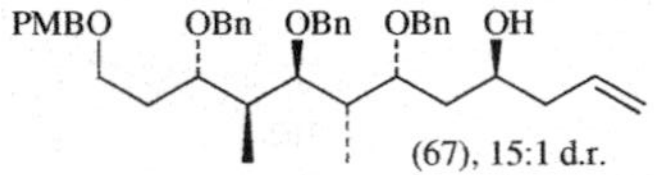 (67), 15:1 d.r.	413
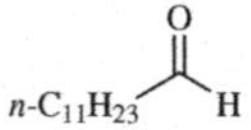	 B[(–)-Ipc]_2	Ether, –100°	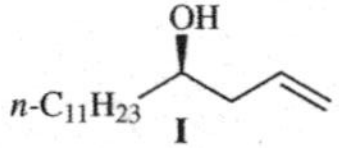 (84), 91	447
	"	Ether, –78° to 0°, 6 h	(75), 99	448
	"	Ether, –78°	(60), 90	449
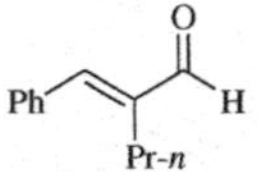	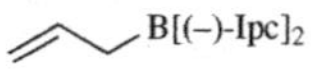 B[(–)-Ipc]_2	Ether, –100° to –78°, 3 h	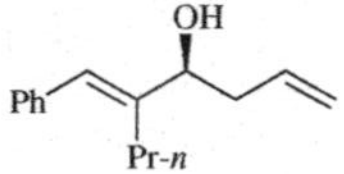 (71), 91	292
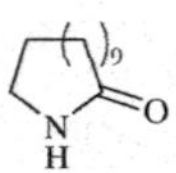	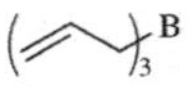 B_3	THF, reflux, 2 h	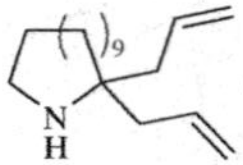 (83)	293, 294
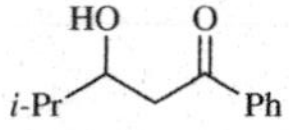	 B(OH)_2	CH_2Cl_2, rt	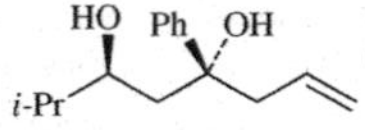 (89), 92:8 d.r.	110

C_{12}

$n\text{-}C_{11}H_{23}$

TABLE I. ADDITION OF UNSUBSTITUTED ALLYL REAGENTS (*Continued*)

A. NON-AROMATIC CARBONYL SUBSTRATES (*Continued*)

Carbonyl Substrate	Allylborane Reagent	Conditions	Product(s) and Yield(s) (%), % ee	Refs.
C$_{12}$				
	B[(−)-Ipc]$_2$	CH$_2$Cl$_2$, −78° to rt	(84)	399
	"	Ether, −78°	I (67)	326
	B(2-d-Icr)$_2$	Ether, −78°	(92), >98:2 d.r.	450
	BEt$_2$	Neat, 0–10°	(90)	238
	BEt$_2$	Neat, 0–10°	(86)	238
C$_{13}$				
	B[(−)-Ipc]$_2$	Ether, −78°, 4 h	(48)	451

172

Aldehyde	Allylborane	Conditions	Product (yield), d.r.	Ref.
(structure)	B[(+)-Ipc]₂	Ether, THF, −78° to rt, 6 h	(77), >19:1 d.r.	452, 453, 454
R = TBDMS	B[(+)-Ipc]₂	Ether, −90°, 1 h	(55), >99:1 d.r.	455
(structure)	(tartrate allylboronate, OPr-i)	4 Å MS, toluene, −78°, 5 h	(>70), 91:1 d.r.	456, 297
R = TBDMS	B[(−)-Ipc]₂	Ether, −100°	(80), >98, >15:1 d.r.	457
R = TBDMS	B[(+)-Ipc]₂	Ether, −100°	(80), >98, >15:1 d.r.	457
(structure)	BEt₂	Ether, −10°	(94)	458

TABLE 1. ADDITION OF UNSUBSTITUTED ALLYL REAGENTS (*Continued*)

A. NON-AROMATIC CARBONYL SUBSTRATES (*Continued*)

Carbonyl Substrate	Allylborane Reagent	Conditions	Product(s) and Yield(s) (%), % ee	Refs.
C13				
(structure)	⟋⟍B(4-*d*-Icr)₂	Ether, –78°, 2 h	(93), 94:6 d.r.	434
C14				
(structure)	⟋⟍B(OH)₂	CH₂Cl₂, rt	(88), 65:35 d.r.	110
(structure)	⟋⟍B[(–)-Ipc]₂	Ether, –78°	(71), 20:1 d.r.	433
(structure)	⟋⟍B[(–)-Ipc]₂	Toluene, –78°	(99), 5:1 d.r.	459
(structure)	⟋⟍B[(–)-Ipc]₂	Ether, –78° to rt	(90)	428
(structure)	⟋⟍B[(+)-Ipc]₂	Ether, –78°, 1.5 h	(60), 71	460

174

Substrate	Reagent	Conditions	Product	Ref.
C₁₅ aldehyde (BnO-substituted)	B(2-d-Icr)₂	Ether, −78°	(92), >97:3 d.r.	450
n-C₁₄H₂₉ aldehyde	B[(+)-Ipc]₂	Ether, −100°, 1 h	(75), 92	461
polyene aldehyde	B[(+)-Ipc]₂	Ether, THF, −100°, 4 h	(78)	462
dioxane aldehyde (Ph)	B[(+)-Ipc]₂	Ether, −78°	(−), 2.5:1 d.r.	444
dioxane aldehyde (Ph)	B[(−)-Ipc]₂	Ether, −78°	(−), 3.5:1 d.r.	444
OTIPS/NC aldehyde	(tartrate boronate)	4 Å MS, toluene, −78°, 1 h	(87), >19:1 d.r.	463
pyran/dioxane aldehyde	B[(−)-Ipc]₂	Ether, −78° to rt	(75), >10:1 d.r.	464

Carbonyl Substrate	Allylborane Reagent	Conditions	Product(s) and Yield(s) (%), % ee	Refs.
C$_{15}$				
	B[(–)-Ipc]$_2$	Ether, –78°, 3 h	(77), 9:1 d.r.	465
	B[(–)-Ipc]$_2$	Ether, –100°, 2 h	(74), 90:5 d.r.	416, 417
	B[(–)-Ipc]$_2$	Ether, –78°	(65), 77:23 d.r.	466
C$_{16}$				
		Toluene, 20°, 0.5 h	(98)	467

176

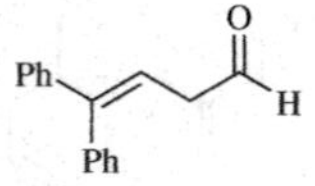	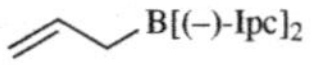	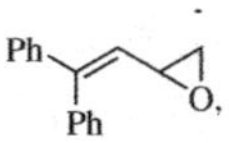 Sc(OTf)$_3$, THF, –78°, 6 h	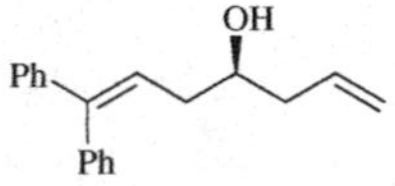 (94), 95	101
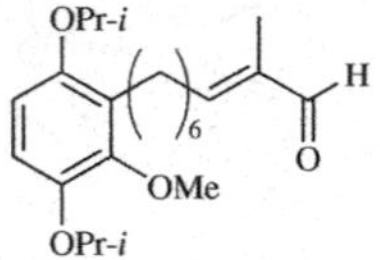	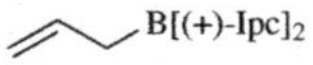	Ether, –78°, 1.5 h	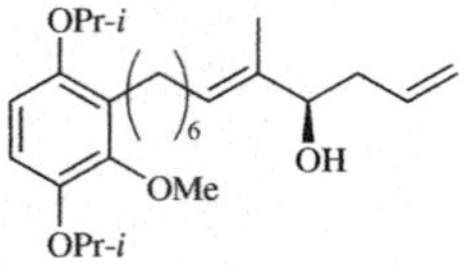 (65), 58	460
C$_{17}$				
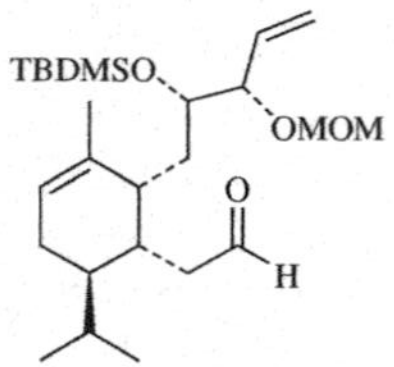	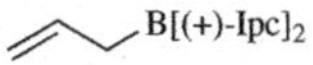	Ether, THF, –78° to –20°, 23 h	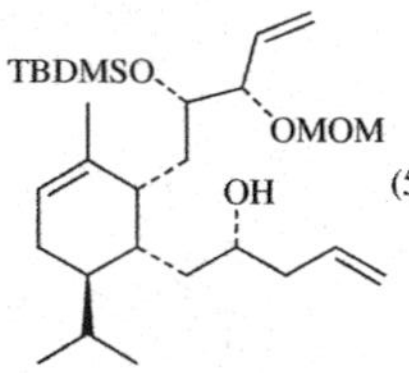 (52), >97.5:2.5 d.r.	468
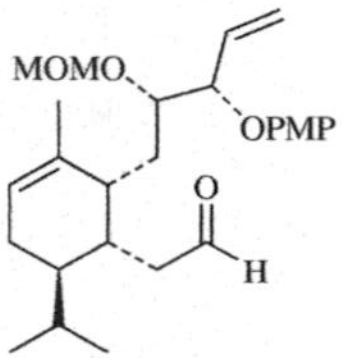	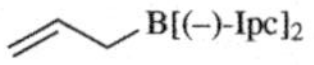	Ether, THF, –78° to –20°, 23 h	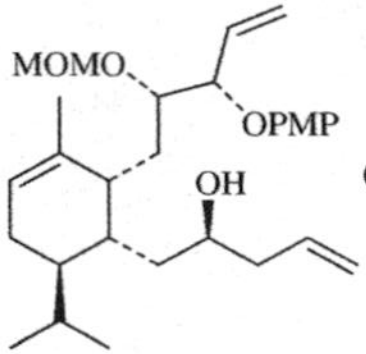 (70)	468
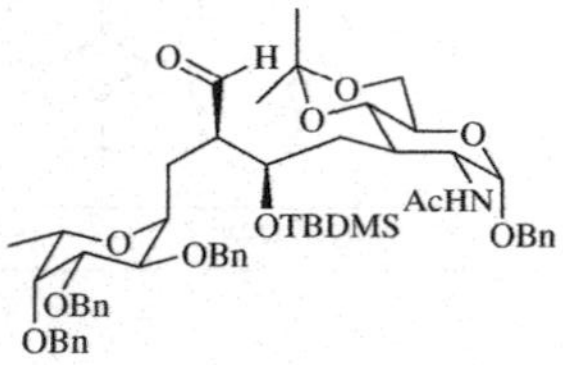		THF, –78° to rt, 18 h	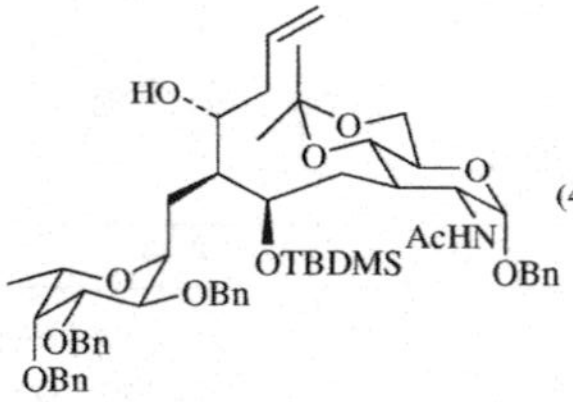 (48), 6:1 d.r.	469, 470

TABLE 1. ADDITION OF UNSUBSTITUTED ALLYL REAGENTS (*Continued*)

A. NON-AROMATIC CARBONYL SUBSTRATES (*Continued*)

Carbonyl Substrate	Allylborane Reagent	Conditions	Product(s) and Yield(s) (%), % ee	Refs.
C$_{17}$	B[(–)-Ipc]$_2$	Ether, –78°	(78)	326
C$_{18}$	B[(–)-Ipc]$_2$	Sc(OTf)$_3$, THF, –78°, 6 h	(23), 93	101
	B[(+)-Ipc]$_2$	Sc(OTf)$_3$, THF, –78°, 4 h; –70°, 36 h	(76), 93	101
	B[(–)-Ipc]$_2$	Ether, THF, –78° to –20°, 23 h	(48), 3:1 d.r.	468

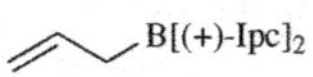

468

(54), 6:1 d.r.

Ether, THF,
−78° to −20°, 23 h

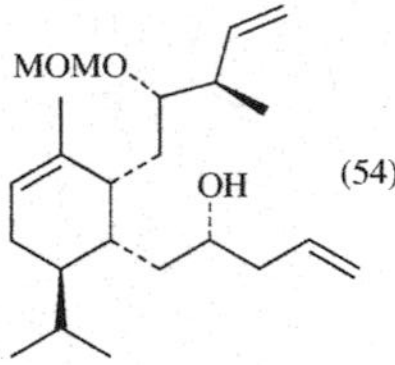

B[(+)-Ipc]₂

MOMO OH
n-C₁₄H₂₉ ... OTBDMS → O H

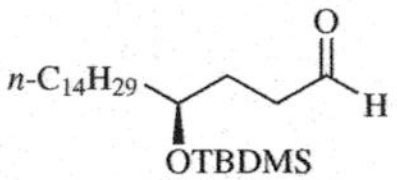

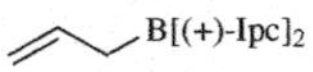

461

(55), 96:4 d.r.

Ether, −100°

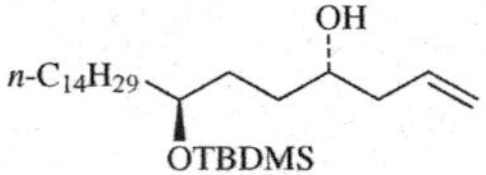

B[(+)-Ipc]₂

OH
n-C₁₄H₂₉ OTBDMS

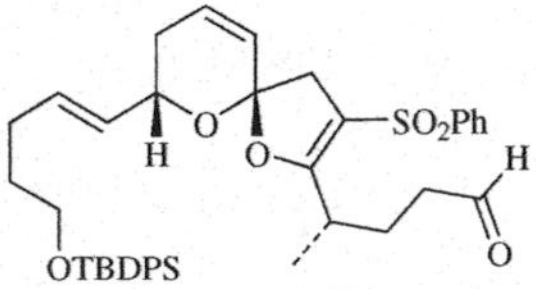

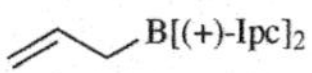

471

(70), >20:1 d.r.

Ether, pentane, −100°, 0.5 h

B[(+)-Ipc]₂

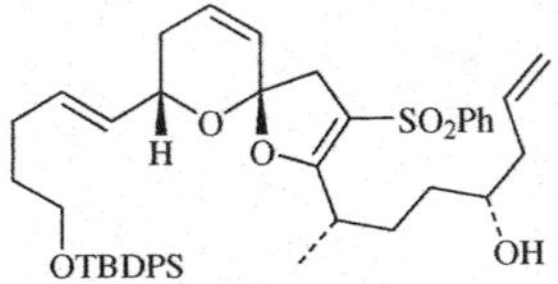

SO₂Ph OH
OTBDPS

352

(88), 72:28 d.r.

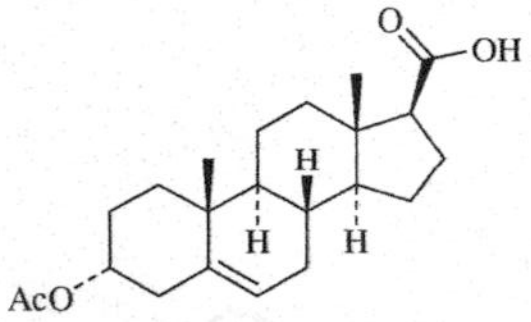

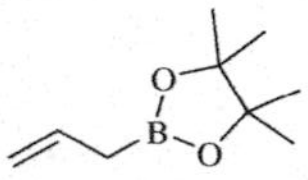

n-Bu₃P, PhSeCl, AIBN, n-Bu₃SnH, THF, reflux

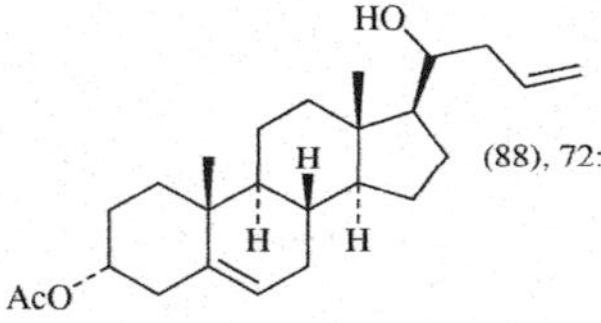

HO H H AcO

C₂₀

HOOC O H H AcO

461

(64)

Ether, −100°, 1 h

B[(+)-Ipc]₂

TBDMSO OH
n-C₁₄H₂₉ OTBDMS

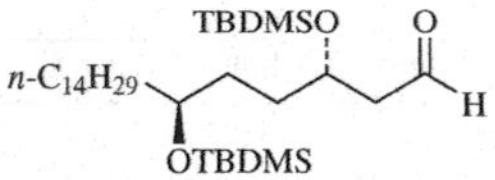

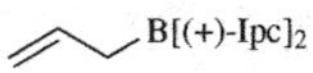

TBDMSO O H
n-C₁₄H₂₉ OTBDMS

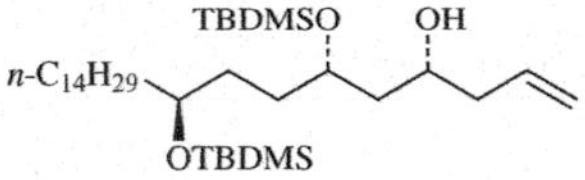

TABLE 1. ADDITION OF UNSUBSTITUTED ALLYL REAGENTS (*Continued*)

A. NON-AROMATIC CARBONYL SUBSTRATES (*Continued*)

Carbonyl Substrate	Allylborane Reagent	Conditions	Product(s) and Yield(s) (%), % ee	Refs.
C_{20}		4 Å MS, toluene, −78°, 3 h	(78), 3.8:1 d.r.	472
C_{22}	BBN	—	(88), 88:12 d.r.	473
	B[(+)-Ipc]$_2$	Ether, −100°, 1 h	(65), 95:5 d.r.	461
	B[(−)-Ipc]$_2$	Toluene, −78°	(83), 88:12 d.r.	474
C_{23}	B[(−)-Ipc]$_2$	Ether, −100°	(78), >95:5 d.r.	475

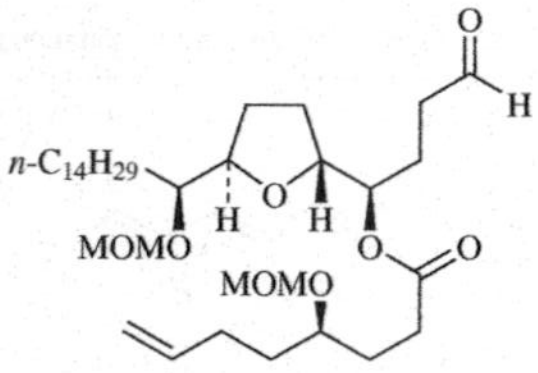

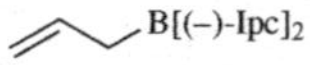

n-C$_{14}$H$_{29}$

MOMO

MOMO

OH

(50), >95:5 d.r.   475

Ether, −100°

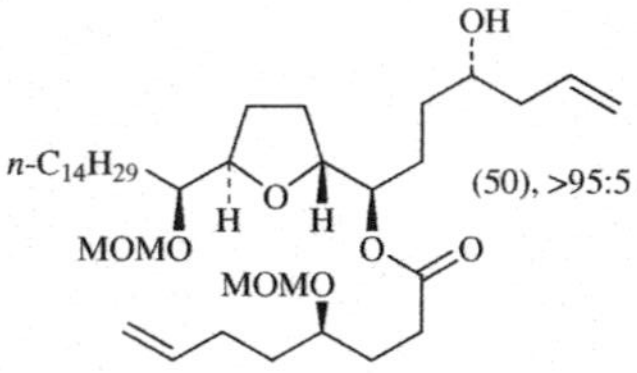

B[(−)-Ipc]$_2$

O

O

MOMO

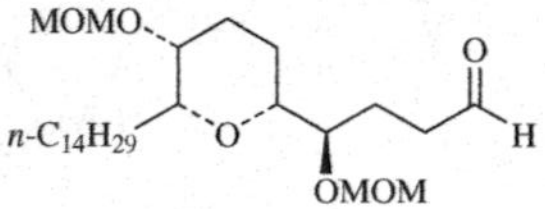

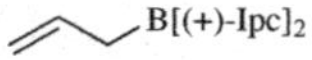

n-C$_{14}$H$_{29}$

H

O

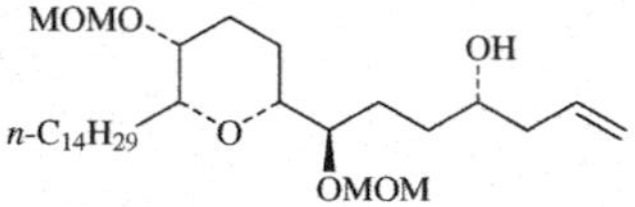

H

MOMO

n-C$_{14}$H$_{29}$

O

OMOM

OH

(91), 96.5:3.5 d.r.   476

Ether

B[(+)-Ipc]$_2$

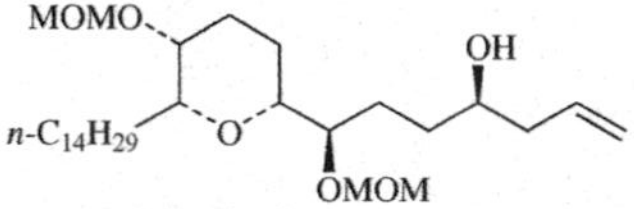

MOMO

n-C$_{14}$H$_{29}$

O

OMOM

OH

(95), 96:4 d.r.   476

Ether

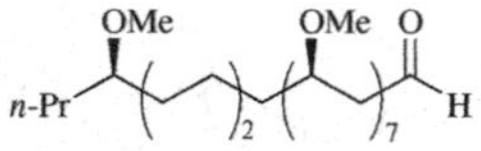

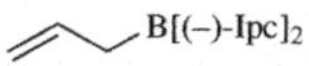

B[(−)-Ipc]$_2$

MOMO

OMOM

n-C$_{14}$H$_{29}$

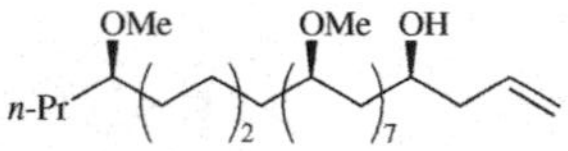

O

H

O

OMe OH

OMe

n-Pr

]$_7$

]$_2$

(95), 86:14 d.r.   474

Toluene, −78°

B[(−)-Ipc]$_2$

C$_{24}$

OMe O

OMe

n-Pr

]$_7$

]$_2$

H

AcO

H

H

H

OH

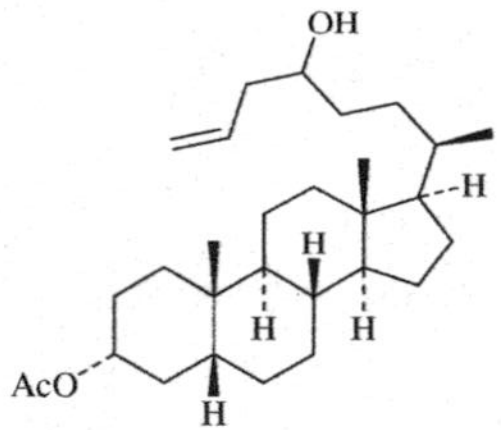

(79), 52:48 d.r.   352

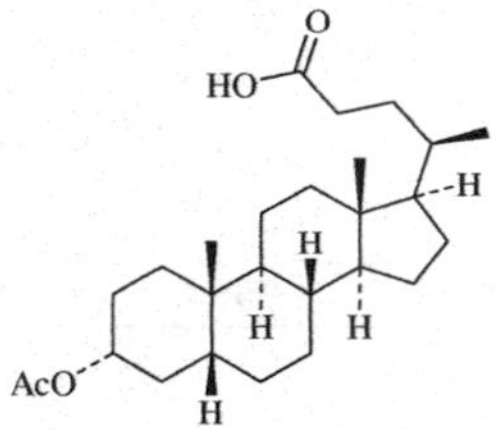

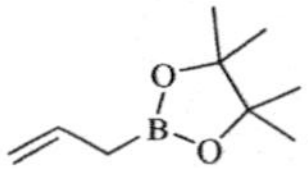

n-Bu$_3$P, PhSeCl, AIBN,
n-Bu$_3$SnH, THF, reflux

O

O

B

O

HO

AcO

H

H

H

181

TABLE 1. ADDITION OF UNSUBSTITUTED ALLYL REAGENTS (*Continued*)

A. NON-AROMATIC CARBONYL SUBSTRATES (*Continued*)

Carbonyl Substrate	Allylborane Reagent	Conditions	Product(s) and Yield(s) (%), % ee	Refs.
C$_{24}$		n-Bu$_3$P, PhSeCl, Et$_3$B, n-Bu$_3$SnH, THF, $-78°$	(60), 85:15 d.r.	352
C$_{32}$	B(OH)$_2$	CH$_2$Cl$_2$, ether, 0°, 0.33 h	(95)	477
	B(OH)$_2$	CH$_2$Cl$_2$, ether, 0°, 0.33 h	(95)	477

[a] The reaction mixture was free of Mg^{2+} ions.

TABLE 1. ADDITION OF UNSUBSTITUTED ALLYL REAGENTS (*Continued*)

B. AROMATIC AND HETEROAROMATIC CARBONYL SUBSTRATES

Carbonyl Substrate	Allylborane Reagent	Conditions	Product(s) and Yield(s) (%), % ee	Refs.
C₅				
	B[(–)-Ipc]₂	Ether, –100° to rt, 18 h	(44), 92	378
	Ts–N,N–Ts B with Ph, Ph	CH₂Cl₂, –78°	(—), >95	280
	B[(+)-Ipc]₂	Ether, –100°	(84), >95	374
	B[(–)-Ipc]₂	Ether, –100°, 1 h	(78), >99	478
	BEt₂	Neat, 0–10°	(92)	238

183

TABLE 1. ADDITION OF UNSUBSTITUTED ALLYL REAGENTS (*Continued*)

B. AROMATIC AND HETEROAROMATIC CARBONYL SUBSTRATES (*Continued*)

Carbonyl Substrate	Allylborane Reagent	Conditions	Product(s) and Yield(s) (%), % ee	Refs.
C₅		Neat, 0–10°	(71)	238
	B[(−)-Ipc]₂	Ether, −100°, 1 h	(91), >99	478
		CH₂Cl₂, H₂O, rt, 6 h	(88) **I**	234
	BF₃⁻K⁺	n-Bu₄NI (10 mol%), CH₂Cl₂, H₂O, rt, 15 min	**I** (95)	100
	B[(−)-Ipc]₂	Ether, −100°, 1 h	(80), >91	478
	B(2-d-Icr)₂	Ether, −100°, 1 h	(88), >99	478
	B[(−)-Ipc]₂	Ether, −100°, 1 h	(80), 90 **I**	478

Aldehyde	Borane reagent	Conditions	Product (yield), % ee	Refs.
	TMS	Ether, −78°, 3 h	I (86), 97 (95)	143
		CH$_2$Cl$_2$, H$_2$O, rt, 14 h	(75), 82	234
	B[(−)-Ipc]$_2$	Ether, −100°, 1 h	(90), 79	478
	B[(+)-Ipc]$_2$	Ether, −100°, 1 h	I	479
	B(2-d-Icr)$_2$	Ether, −100°, 1 h	I (83), >99	478
C$_6$	B[(+)-Ipc]$_2$	Ether, −78° to rt, 3 h	(81), 17 PhMe$_2$Si	111
	B[(+)-Ipc]$_2$	THF, −100°, 1 h	(94), 94	480, 481
	B[(−)-Ipc]$_2$	Ether, −100°, 1 h	(84), >99	478

185

TABLE 1. ADDITION OF UNSUBSTITUTED ALLYL REAGENTS (*Continued*)

B. AROMATIC AND HETEROAROMATIC CARBONYL SUBSTRATES (*Continued*)

Carbonyl Substrate	Allylborane Reagent	Conditions	Product(s) and Yield(s) (%), % ee	Refs.
C$_6$				
(pyridine-3-carbaldehyde, R^1, R^2)	B[(+)-Ipc]$_2$	Ether, –100°, 1 h	R^1 H H Cl Cl H MeO / R^2 H Br H Cl MeO H — (94), 94; (75), 66; (74), 85.5; (73), 90; (75), 91.5; (78), 88	479
(pyridine-2-carbaldehyde)	B[(–)-Ipc]$_2$	Ether, –100°, 1 h	(85), >99	478
	(allylboronate, NHBn/NHBn tartaramide)	Toluene, –78°, 18 h	**I** (92), 94	290
	(allylboronate, camphor diol)	CH$_2$Cl$_2$, H$_2$O, rt, 3 h	(79)	234
(pyridine-4-carbaldehyde)	B(4-d-Icr)$_2$	Ether, –100°, 1 h	(78), >99	478

186

Substrate	Boron reagent	Conditions	Product(s) (%)	Refs.
3-acetylthiophene (C7)	allyl-B(pinacolato)	(R,R)-i-Pr-DuPHOS, CuF, La(OPr-i)$_3$, DMF, –40°, 1 h	(84), 85	164
thiophene-2,5-dicarbaldehyde	B[(–)-Ipc]$_2$	THF, –100°, 4 h	(93), >98, 98:2 d.r.	482
benzaldehyde	(3-butenyl)$_3$B	–70° to 130°	(89)	262, 1
	B(Bu-n)$_2$	Neat, 0–10°	I (95)	238
	BBN	Pentane, rt, 2 h	I (96)	103
	B(R^1)$_2$	BR$^1{}_3$, Pd(PPh$_3$)$_4$, THF, allyl-OR^2	I	483
	B[(+)-Ipc]$_2$	—	I (76), 98	250
	"	Ether, –100°, 1 h	I (83), 95	479, 484

R^1	R^2	Temp	Time	
Et	COPh	rt	4 h	(68)
Bu	COPh	rt; 40°	16 h; 4 h	(74)
Et	Ph	rt	6 h	(59)
Et	Bn	rt; 50°	45 h; 17 h	(47)

TABLE 1. ADDITION OF UNSUBSTITUTED ALLYL REAGENTS (Continued)
B. AROMATIC AND HETEROAROMATIC CARBONYL SUBSTRATES (Continued)

Carbonyl Substrate	Allylborane Reagent	Conditions	Product(s) and Yield(s) (%), % ee	Refs.
C$_7$	$\sim$B(2-d-Icr)$_2$	THF, −78°, 3 h	I (—), 95	240
	"	Ether, −100°^{a}	I (—), >99	139, 241
	TMS B reagent	Ether, −78°, 3 h	I (80), >98	143
	"	Ether, 3 h	I (—) Temp %ee: rt 90, 0° 93, −25° 94	143
	Ph B reagent	Ether, −78°	I (82), 90	145
	$\sim$B((−)-Ipc)$_2$	Ether, −78° to rt	(81), 96	136, 136a
	"	Ether, −78°, 3 h	I (—), 94	240
	"	Ether, −100°^{a}	I (—), 96	139, 289
	$\sim$B(4-d-Icr)$_2$	THF, −78°, 3 h	I (—), 87	240
	"	Ether, −100°^{a}	I (—), 98	139

Reagent column (B(OP-i)$_2$ and various boronate structures with conditions, products, and references)

	Conditions	Product	Ref.
B(OP-i)$_2$	Ionic liquid, rt, 3 h	OH **I** Ionic liquid bnimBr (87) bnimBF$_4$ (89) emimBr (88) TEAN (85)	353
CONMe$_2$ boronate	L.A. (10 mol%), toluene, −78°, 16 h	**I** L.A. AlCl$_3$ (88) Sc(OTf)$_3$ (80)	25
"	CH$_2$Cl$_2$, −78°, 24 h	OH **I** (76), 59	424
biscyclopentyl boronate	Toluene, −78°, 24 h	**I** (82), 37	424
bis(MeO)cyclopentyl boronate	4 Å MS, CH$_2$Cl$_2$, −78°, 12 h	**I** (62), 19	485
bis(NHBn) boronate	4 Å MS, CH$_2$Cl$_2$, −78°, 12 h	**I** (72), 23	485
	See table	**I** | Solvent	Temp	Time	
toluene	0°	14 h	(77.7), 34
toluene	−20°	14 h	(80.5), 46
ether	−78°	18 h	(87), 83
toluene	−78°	14 h	(87.8), 90		290

189

TABLE 1. ADDITION OF UNSUBSTITUTED ALLYL REAGENTS (*Continued*)

B. AROMATIC AND HETEROAROMATIC CARBONYL SUBSTRATES (*Continued*)

Carbonyl Substrate	Allylborane Reagent	Conditions	Product(s) and Yield(s) (%), % ee	Refs.
C₇		4 Å MS, THF, −78°	(78), 72	125
	"	Cr(CO)₃, 4 Å MS, toluene, −78°	I (90), 83	128
	"	4 Å MS, toluene, −78°	I (—), 60	131
	"	Sc(OTf)₃ (10 mol%), CH₂Cl₂, −78°, 2 h	I (100), 7	27, 28
		Hexane, −40°	I (90), 36	122
		L.A. (10 mol%)	(see table below)	27, 28

L.A.	Solvent	Temp	Time	% Conv.	% ee
I none	CH₂Cl₂	rt	72 h	50	11
AlCl₃	CH₂Cl₂	−78°	2 h	14	63
TiCl₄	CH₂Cl₂	−78°	2 h	22	78
TfOH	CH₂Cl₂	−78°	2 h	72	84
Cu(OTf)₂	CH₂Cl₂	−40°	2 h	4	52
Yb(OTf)₃	CH₂Cl₂	−40°	2 h	4	38
Sc(OTf)₃	CH₂Cl₂	−78°	2 h	90	92
Sc(OTf)₃	toluene	−78°	2 h	30	46
Sc(OTf)₃	hexane	−78°	2 h	20	8

	4 Å MS, toluene, −78°, 47 h	**I** (—), 85	131
	Sc(OTf)₃ (10 mol%), CH₂Cl₂, −78°, 2 h	**I** (100), 9	27, 28
	Hexane, −40°	**I** (76), 46	122
	Ether, −78° to rt	**I** (92), 88	486
	Sc(OTf)₃ (10 mol%), CH₂Cl₂, −78°, 12 h	**I** (85), 92	27, 28
	Sc(OTf)₃ (10 mol%), CH₂Cl₂, −78°, 2 h	**I** (100), 84	27, 28

TABLE 1. ADDITION OF UNSUBSTITUTED ALLYL REAGENTS (*Continued*)

B. AROMATIC AND HETEROAROMATIC CARBONYL SUBSTRATES (*Continued*)

Carbonyl Substrate	Allylborane Reagent	Conditions	Product(s) and Yield(s) (%), % ee	Refs.
C$_7$				
		Pd$_2$(dba)$_3$, DMSO, toluene, 20°, 44 h	**I** (63), 42	349
	"	4 Å MS, −78°	**I** (—) Solvent / % ee: toluene 60; ether 59; THF 72; CH$_2$Cl$_2$ 59	125
		Pd$_2$(dba)$_3$, DMSO, toluene, 20°, 63 h	**I** (83), 45	349
		See table	**I** R / Solvent / Temp / Time: Ph / ether / rt / 18 h (62.5), 6.4 Bn / THF / −40° / 16 h (79.6), 61 Bn / toluene / −60° / 16 h (83.5), 85 Ph / THF / −78° / 14 h (76.2), 72 Ph / toluene / −78° / 18 h (82.3), 80	290

Reagent	Conditions	R	Product (%), ee	Refs.
	THF, −78°, 1 h	H	**I** (50), 42	135
		I	(91), 74	
		Me	(91), 86	
		Ph	(90), 46	
		4-MeOC₆H₄	(89), 62	
		3,5-Me₂C₆H₃	(94), 60	
		3,5-(t-Bu)₂C₆H₃	(92), 66	
		3,5-(CF₃)₂C₆H₃	(92), 80	
		2-naphthyl	(90), 50	
		CF₃	(90), 96	
	CH₂Cl₂, −78°, 2 h		**I** (>90), 94	50
	Toluene, −78°, 2 h		**I** (>90), 95	50
	Toluene, −78°, 2 h		**I** (92), 96	351
	Ether, −78° to rt, 2 h		**I** (91), 88	133
	−78°	Solvent THF	(88), 74	245, 246
		ether	(90), 81	

TABLE 1. ADDITION OF UNSUBSTITUTED ALLYL REAGENTS (*Continued*)
B. AROMATIC AND HETEROAROMATIC CARBONYL SUBSTRATES (*Continued*)

Carbonyl Substrate	Allylborane Reagent	Conditions	Product(s) and Yield(s) (%), % ee	Refs.
C₇		Ether, −78°, 7 h	(89), 71	245, 246
		Ether, −78°, 7 h	**I** (93), 75	245, 246
		Ether, −78°	**I** (90), 66	246
		Ether, −78°, 7 h	**I** (93), 75	245, 246
		Ether, −78°, 7 h	(93), 75	245, 246

194

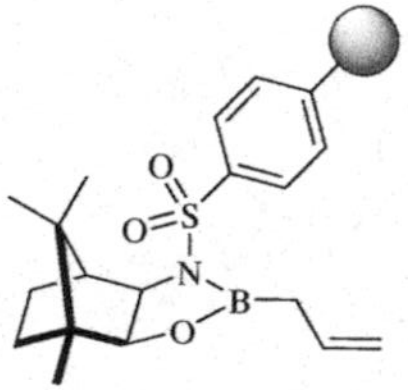

Solvent	Temp	
THF	–40°	(91), 65
THF	–78°	(94), 78
THF	–100°	(96), 88
toluene	–78°	(91), 72
ether	–78°	(93), 85

See table

245, 246

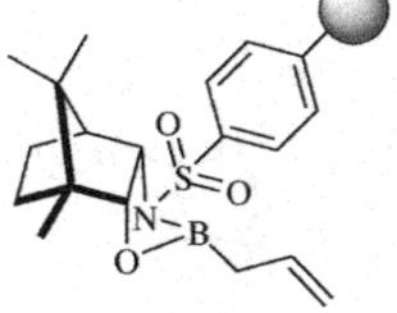

Ether, –78°, 7 h

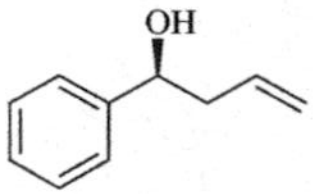

(92), 74

245, 246

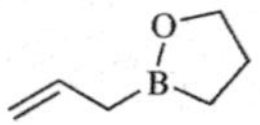

Neat, 80-100°

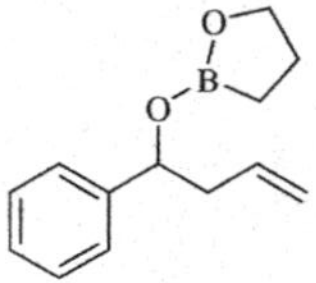

(81)

263

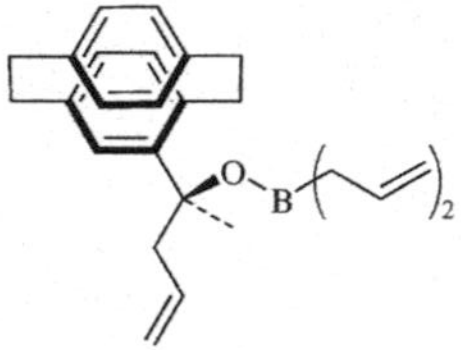

THF, –70°, 17 h

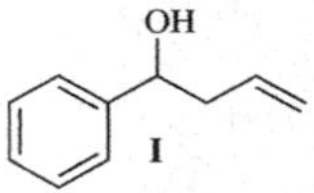

(—), 18

487

CH₂Cl₂, –40° to rt

X	
Cl	(57)
Br	(53)

BX_2

97

CH₂Cl₂, –40° to rt

X	
Cl	(67)
Br	(38)

$BX_2 \cdot OEt_2$

97

195

TABLE 1. ADDITION OF UNSUBSTITUTED ALLYL REAGENTS (Continued)

B. AROMATIC AND HETEROAROMATIC CARBONYL SUBSTRATES (Continued)

Carbonyl Substrate	Allylborane Reagent	Conditions	Product(s) and Yield(s) (%), % ee	Refs.
C7	BBN	See table	 	103

Sub-table for the first entry:

R	Solvent	Temp	Time	
Cl	pentane	rt	2 h	(89)
OC(O)Ph	pentane	rt	2 h	(82)
OEt	hexane	68°	50 h	(59)
NMe$_2$	pentane	rt	3 h	(91)

Carbonyl Substrate	Allylborane Reagent	Conditions	Product(s) and Yield(s) (%), % ee	Refs.
		n-Bu$_3$P, PhSeCl, AIBN, n-Bu$_3$SnH, THF, reflux	(79)	352
		OAc, Pd$_2$(dba)$_3$, DMSO, toluene, 20°, 21 h	(73)	349
		OAc, Pd$_2$(dba)$_3$, DMSO, toluene, 20°, 22 h	(92), 49	349

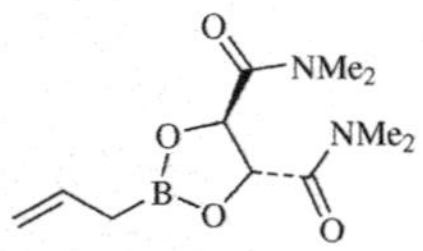

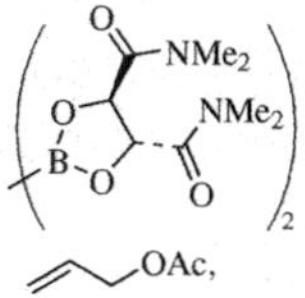

349

349

349

349

(83), 53

I

O₂N ... OH (allyl)

I (73), 3

I (76), 45

I (62), 1

NMe₂ ... NMe₂ ... OAc, Pd₂(dba)₃, DMSO, toluene, 20°, 21 h

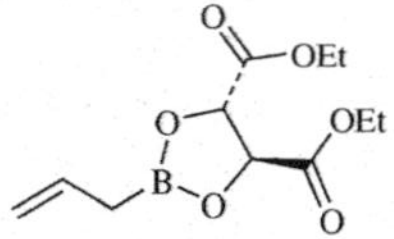

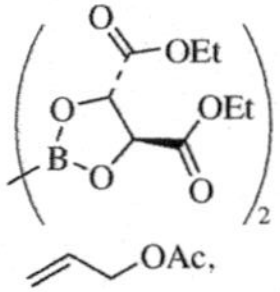

OEt ... OEt ... OAc, Pd₂(dba)₃, DMSO, toluene, 20°, 19 h

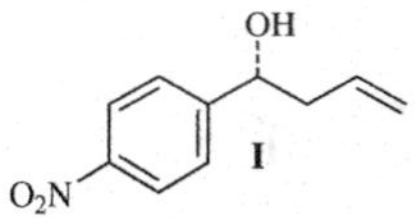

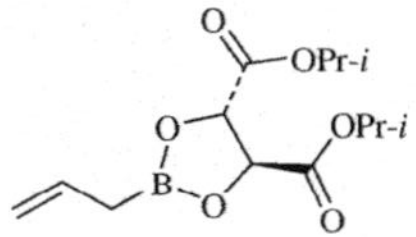

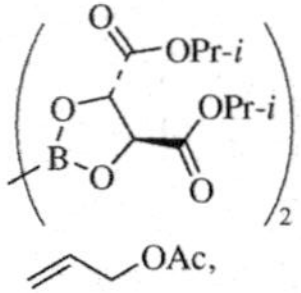

OPr-i ... OPr-i ... OAc, Pd₂(dba)₃, DMSO, toluene, 20°, 20 h

OAc, Pd₂(dba)₃, DMSO, toluene, 20°, 21 h

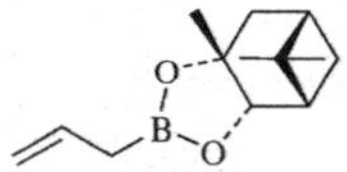

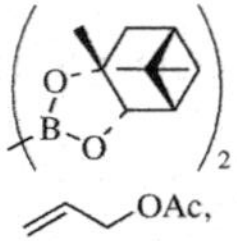

NMe₂ ... NMe₂ ... B, O

OEt ... OEt ... B, O

OPr-i ... OPr-i ... B, O

B, O

TABLE 1. ADDITION OF UNSUBSTITUTED ALLYL REAGENTS (*Continued*)

B. AROMATIC AND HETEROAROMATIC CARBONYL SUBSTRATES (*Continued*)

Carbonyl Substrate	Allylborane Reagent	Conditions	Product(s) and Yield(s) (%), % ee	Refs.
C7				
4-O_2N-benzaldehyde	allyl–$BF_3^- K^+$	L.A., CH_2Cl_2, −78°, 15 min	L.A. B(OMe)$_3$ (0) AlCl$_3$ (16) Ti(OPr-i)$_4$ (20) TMSCl (28) SnCl$_4$ (41) BF$_3$•OEt$_2$ (93)	99
2-NO_2-benzaldehyde	allyl–$BF_3^- K^+$	n-Bu$_4$NI (10 mol%), CH_2Cl_2, H_2O, rt, 15 min	(99)	100
4-MeO-benzaldehyde	allyl–B[(−)-Ipc]$_2$	THF, −78° to rt, 2 h	(72), 89	488
4-MeO-benzaldehyde	allyl-Bpin	L.A. (10 mol %), toluene, −78°, 36 h	L.A. AlCl$_3$ (82) Sc(OTf)$_3$ (84)	25
4-MeO-benzaldehyde	allyl–$BF_3^- K^+$	n-Bu$_4$NI (10 mol%), CH_2Cl_2/H_2O, rt, 15 min	(98)	100

Aldehyde	Reagent	Conditions	Product	Refs.
4-bromobenzaldehyde	allyl–B[(−)-Ipc]$_2$	Ether, −78° to rt, 3 h	(58-73)	489
4-bromobenzaldehyde	allyl–BF$_3^-$K$^+$	PTC (10 mol%), CH$_2$Cl$_2$/H$_2$O, rt, 15 min	PTC none (15) n-Bu$_4$NI (99) n-Bu$_4$NBr (98) Et$_3$NBnCl (97) Ph$_3$PMeCl (97)	100
3-bromobenzaldehyde	allyl–B[(−)-Ipc]$_2$	Ether, −78° to rt, 3 h	(58-73)	489
3-bromobenzaldehyde	allyl–B(OPr-i)$_2$	bmimBr, rt, 3 h	(86)	353
2-chlorobenzaldehyde	allyl–B[(−)-Ipc]$_2$	THF, −78° to rt, 2 h	(77), 94	488
3-hydroxybenzaldehyde	allyl–BF$_3^-$K$^+$	n-Bu$_4$NI (10 mol%), CH$_2$Cl$_2$/H$_2$O, rt, 15 min	(99)	100

TABLE 1. ADDITION OF UNSUBSTITUTED ALLYL REAGENTS (*Continued*)

B. AROMATIC AND HETEROAROMATIC CARBONYL SUBSTRATES (*Continued*)

Carbonyl Substrate	Allylborane Reagent	Conditions	Product(s) and Yield(s) (%), % ee	Refs.
C7 R = MeO, O₂N		Ether, −78°, 3 h	R MeO (90), 96 O₂N (87), 97	143
R = H, MeO, Cl, O₂N		THF, −78°, 1 h	I R H (90), 96 MeO (93), 94 Cl (93), 94 O₂N (96), 92	135
		CH₂Cl₂, H₂O, rt	R Time H 6 h (87) MeO 7 h (93) Cl 4 h (85) O₂N 3 h (90)	234
R = H, MeO, MeS, O₂N		BF₃•OEt₂ (2 equiv), CH₂Cl₂, −78°, 15 min	I R H (93) MeO (95) MeS (90) O₂N (96)	98, 99
	"	BF₃•OEt₂ (5 mol%), CH₂Cl₂, rt, 3–6 h	I R H (91) MeO (89) MeS (93) O₂N (95)	99

Substrate	Reagent	Conditions	Product	Yield (% ee)	Refs.
3,4-(AcO)(OAc)-benzaldehyde	allyl–B[(−)-Ipc]$_2$	−78°	aryl–CH(OH)CH$_2$CH=CH$_2$	(76), 90	490
aryl (R^1, R^2)-benzaldehyde	allyl–BF$_3^-$K$^+$	BF$_3$•OEt$_2$ (5 mol%), CH$_2$Cl$_2$, rt, 3–6 h	**I** R^1 R^2: Cl Cl (86); HO MeO (85)		99
"	"	BF$_3$•OEt$_2$ (2 equiv), CH$_2$Cl$_2$, −78°, 15 min	**I** R^1 R^2: Cl Cl (85); HO MeO (84)		98
pentafluorobenzaldehyde	allyl–B[(+)-Ipc]$_2$	—	C$_6$F$_5$–CH(OH)CH$_2$CH=CH$_2$	(86), 97	250
pentafluorobenzaldehyde	allyl–B[(−)-Ipc]$_2$	Ether/pentane, −100°, 2 h	C$_6$F$_5$–CH(OH)CH$_2$CH=CH$_2$	(89), 99	491
pyridine-2,6-dicarbaldehyde	allyl–B[(−)-Ipc]$_2$	THF, −100°, 4 h	2,6-bis[CH(OH)CH$_2$CH=CH$_2$]pyridine	(95), >98, 96:4 d.r.	254, 482

TABLE 1. ADDITION OF UNSUBSTITUTED ALLYL REAGENTS (*Continued*)
B. AROMATIC AND HETEROAROMATIC CARBONYL SUBSTRATES (*Continued*)

Carbonyl Substrate	Allylborane Reagent	Conditions	Product(s) and Yield(s) (%), % ee	Refs.
C₇	B[(+)-Ipc]₂	Pentane, ether, −100°		248

R¹	R²	R³	R⁴	R⁵	
F	F	F	F	F	(86), 99.8
F	H	H	H	F	(88), 99.6
H	F	H	H	H	(90), 95.2
F	H	F	H	H	(86), 95.3
F	H	H	H	H	(80), 98.1
H	F	H	H	H	(85), 95.0
H	H	F	H	H	(90), 98.0

	B(2-*d*-Icr)₂	Pentane, ether, −100°	**I** (—)	248, 249

R¹	R²	R³	R⁴	R⁵	% ee
F	F	F	F	F	96
F	H	F	H	F	97
H	F	H	F	H	98
F	H	F	H	H	96
F	H	H	H	H	96
H	F	H	H	H	97
H	H	F	H	H	98

C₈	B[(−)-Ipc]₂	−78°	(98)	492

Substrate	Reagent	Conditions	Product (Yield)	Refs.
NC–C₆H₄–CHO	BF₃⁻K⁺ (allyl)	BF₃•OEt₂ (2 equiv), CH₂Cl₂, −78°, 15 min	**I** (95)	98, 99
	"	BF₃•OEt₂ (5 mol%), CH₂Cl₂, rt, 3-6 h	**I** (95)	99
MeO₂C–C₆H₄–CHO	B[(−)-Ipc]₂ (allyl)	Ether, −78° to rt, 3 h	(58-73)	489
MeO₂C–C₆H₄–CHO	allyl pinacol boronate	CH₂Cl₂, H₂O, rt, 4 h	(95)	234
n-Bu–N(resin)–CO–C₆H₄–CHO	allyl pinacol boronate	CH₂Cl₂, rt, 16 h	(100)	235
CF₃–C₆H₄–CHO	chiral binaphthyl allyl boronate	THF, −78°, 1 h	(94), 94	135

TABLE 1. ADDITION OF UNSUBSTITUTED ALLYL REAGENTS (*Continued*)

B. AROMATIC AND HETEROAROMATIC CARBONYL SUBSTRATES (*Continued*)

Carbonyl Substrate	Allylborane Reagent	Conditions	Product(s) and Yield(s) (%), % ee			Refs.

C₈ row 1:

Conditions: L.A. (10 mol%), toluene, −78°

Product I (OH, R):

R	L.A.	Time	
CF₃	AlCl₃	16 h	(80)
CF₃	Sc(OTf)₃	16 h	(69)
Me	AlCl₃	24 h	(47)
Me	Sc(OTf)₃	24 h	(62)

R = CF₃, Me; Reagent: allyl (pinacol boronate); Refs. 25

Row 2:

Reagent: allyl BF₃⁻K⁺; Conditions: n-Bu₄NI (10 mol%), CH₂Cl₂/H₂O, rt, 15 min

R = MeO₂C, NC

Product I:

R	
MeO₂C	(97)
NC	(98)

Refs. 100

Row 3:

Substrate: SiMe₂Ph ketone, R; Reagent: allyl B[(+)-Ipc]₂; Conditions: Ether, −78° to rt, 3 h

Product: HO SiMe₂Ph

R	
CF₃	(81), 36
Me	(65), 26

Refs. 111

Row 4:

Substrate: CF₃ benzaldehyde; Reagent: allyl B[(+)-Ipc]₂; Conditions: Pentane, ether, −100°

Product I (OH, CF₃):

Isomer	
ortho	(96), 96.8
meta	(80), 99.9
para	(81), 95.2

Refs. 248

Row 5:

Reagent: allyl B(2-d-Icr)₂; Conditions: Pentane, ether, −100°

Product: I (—)

Isomer	% ee
ortho	96
meta	97
para	97

Refs. 248

Row 6:

Substrate: acetophenone; Reagent: allyl BBN; Conditions: Pentane, rt, 2 h

Product: HO (100)

Refs. 103

Reagent	Conditions	Product		Refs.
B[(−)-Ipc]₂ (allyl)	THF, −78° to rt	HO—(CH₂CH=CH₂)(Ph) **I**	(63), 5	166

x	% Cat.	La-additive	Time	
3	15	none	20 h	(42), 79
3	15	La(OPr-i)₃	3.5 h	(95), 77
1.2	3	La(OPr-i)₃	1 h	(94), 82
3	15	(R)-BINOL–La(OPr-i)₃	20 h	(52), 80
3	15	(S)-BINOL–La(OPr-i)₃	20 h	(48), 79

pinacol allyl boronate, x equiv — (R,R)-i-Pr-DuPHOS, La-additive, CuF, DMF, −40° — **I** — 164

Reagent	Conditions	Product	Refs.
B(OPr-i)₂ (allyl)	bmimBr, rt, 3 h	HO—(CH₂CH=CH₂)(Ph) (83)	353

Solvent	
THF	(88), 92
toluene	(60), 96

BINOL(CF₃) allyl boron — −78° to −40°, 48 h — 135

Reagent	Conditions	Product	Refs.
dioxaborolane (allyl)	Neat, 80-100°	(90)	263

TABLE 1. ADDITION OF UNSUBSTITUTED ALLYL REAGENTS (*Continued*)

B. AROMATIC AND HETEROAROMATIC CARBONYL SUBSTRATES (*Continued*)

Carbonyl Substrate	Allylborane Reagent	Conditions	Product(s) and Yield(s) (%), % ee	Refs.
C8				
R = H, O2N		Ether, −78°	R: H (92), 96; O2N (90), >98	145
R = MeO, Br		Ether, −78°	R: MeO (89), 94; Br (96), 98	145
R = MeO, Cl		THF, −78° to −40°, 48 h	R: MeO (95), 98; Cl (94), >98	135
		THF, −78° to −40°, 48 h	(87), 92	135
	B(OPr-*i*)2	Et3N, CH2Cl2, −10° to rt, 13 h	(94)	105

206

Substrate	Reagent	Conditions	Product	Refs.
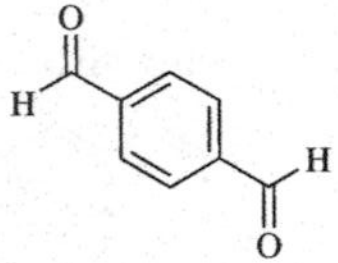 terephthalaldehyde	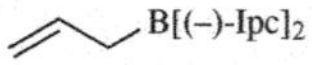 allyl–B((−)-Ipc)₂	THF, −100°, 4 h	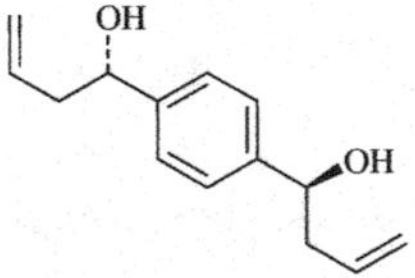 (94), >98, 99:1 d.r.	254
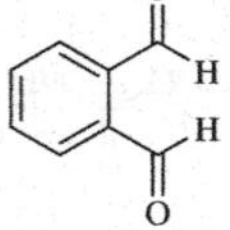 phthalaldehyde	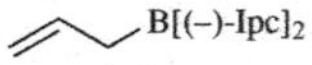 allyl–B((−)-Ipc)₂	THF, −100°, 4 h	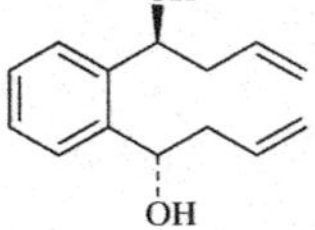 (94), >98, 99:1 d.r.	254
phthalaldehyde	allyl–B((+)-Ipc)₂	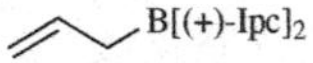 THF, −100°, 4 h	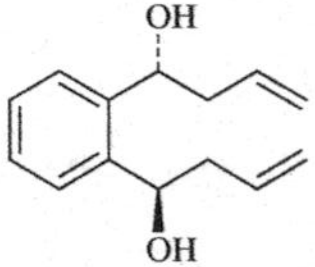 (91), >98, 99:1 d.r.	254
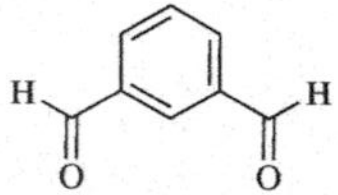 isophthalaldehyde	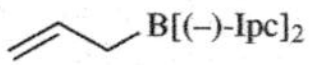 allyl–B((−)-Ipc)₂	THF, −100°, 4 h	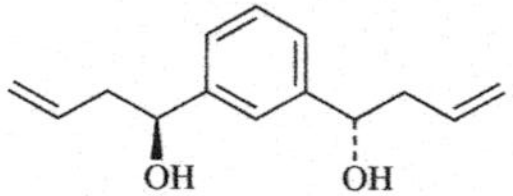 (89), >98, 94:6 d.r.	254
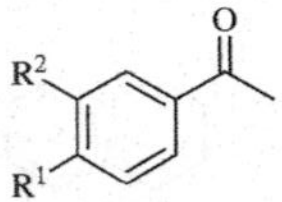 aryl methyl ketone (R¹, R²)	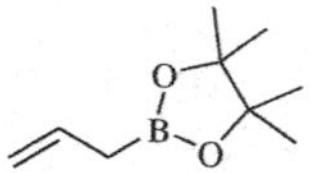 allyl-Bpin	(R,R)-i-Pr-DuPHOS, La(OP-i)₃, CuF, DMF, −40°, 1 h	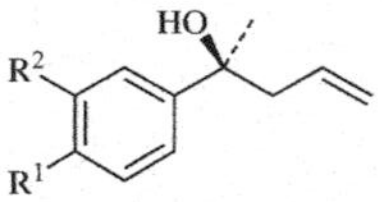 R^1 = Me, R^2 = H (89), 84; R^1 = H, R^2 = Me (83), 83	164
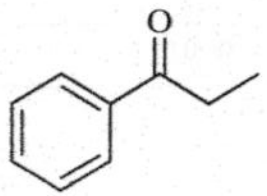 propiophenone	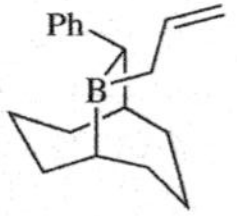 allyl-B-9-BBN (Ph)	Ether, −78°	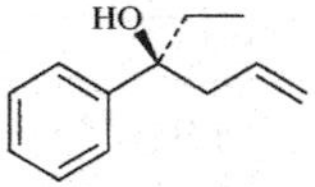 (70), 94	145

C₉

207

Carbonyl Substrate	Allylborane Reagent	Conditions	Product(s) and Yield(s) (%), % ee	Refs.
C_{10}				
		(*R,R*)-*i*-Pr-DuPHOS, La(OPr-*i*)$_3$, CuF, DMF, –40°, 1 h	(88), 85	164
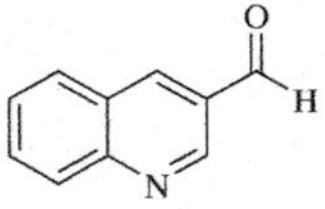	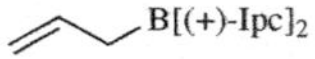 B[(+)-Ipc]$_2$	Ether, –100°, 1 h	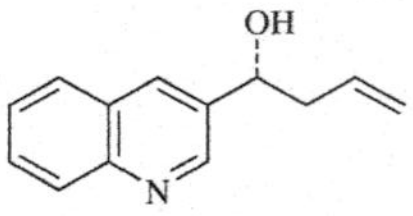 (75), 92	479
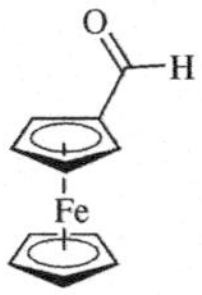	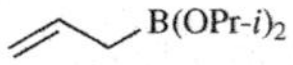 B(OPr-*i*)$_2$	bmimBr, rt, 3 h	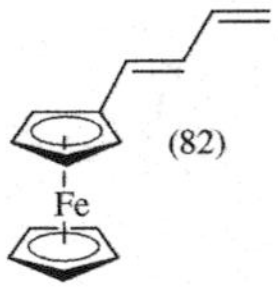 (82)	353
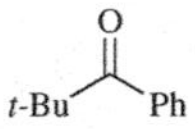 C_{11}	 BBN	Pentane, rt, 4 h	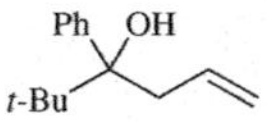 (82)	103
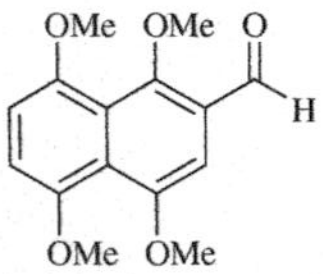	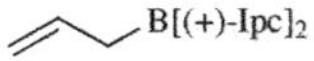 B[(+)-Ipc]$_2$	Ether, –78°, 30 min	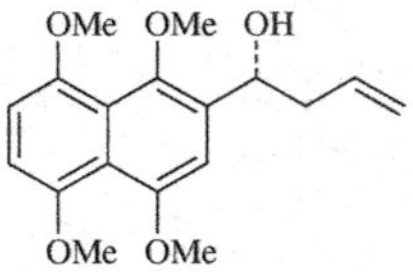 (78), 82	493

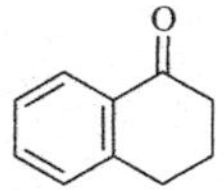

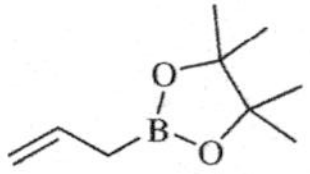

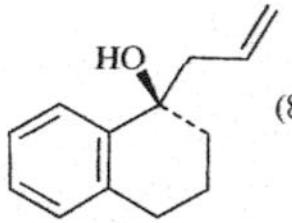

208

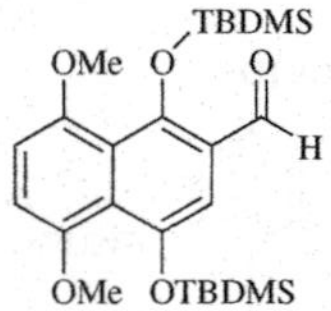

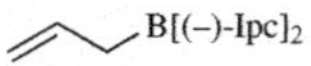

	494		(84), 72 (—)
	353		(86)
	238		(79) I (100)
	103		
	109		(94), >99:1 d.r.
	110		(90), 91:9 d.r.
	110		(93), 86:14 d.r.

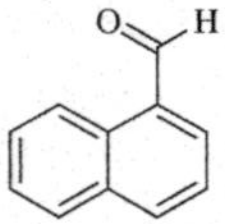

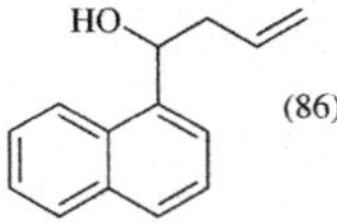

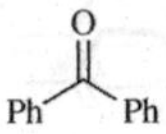

Ether, −78° to rt, 2-3 h

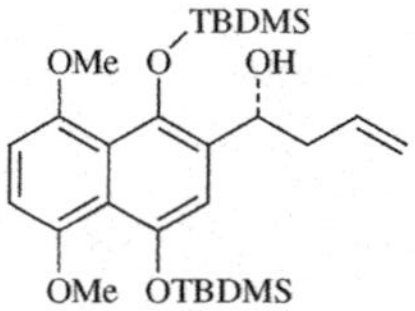

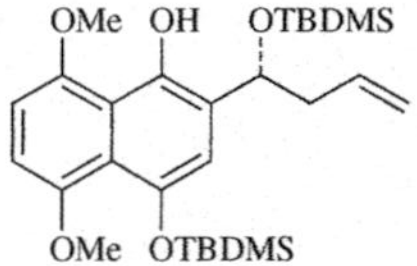

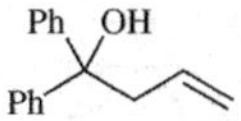

bmimBr, rt, 3 h

Neat, 0-10°

Pentane, rt, 2 h

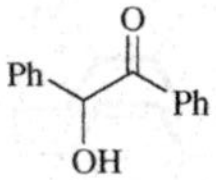

Et₃N, CH₂Cl₂, −10° to 40°, 15 h

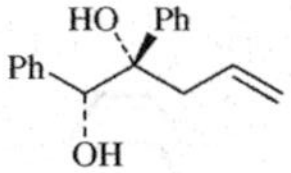

CH₂Cl₂, rt

CH₂Cl₂, rt

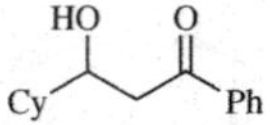

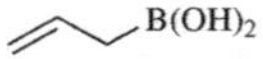

$B[(-)\text{-Ipc}]_2$

$B(OPr\text{-}i)_2$

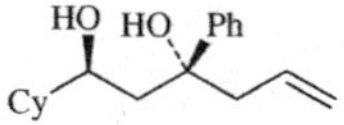

BEt_2

BBN

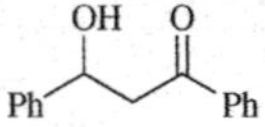

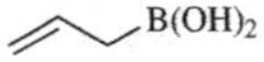

$B(OPr\text{-}i)_2$

$B(OH)_2$

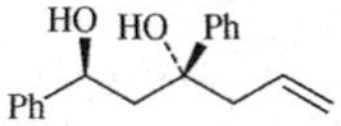

$B(OH)_2$

C₁₃

C₁₄

C₁₅

TABLE 1. ADDITION OF UNSUBSTITUTED ALLYL REAGENTS (*Continued*)

B. AROMATIC AND HETEROAROMATIC CARBONYL SUBSTRATES (*Continued*)

Carbonyl Substrate	Allylborane Reagent	Conditions	Product(s) and Yield(s) (%), % ee	Refs.
C_{16} 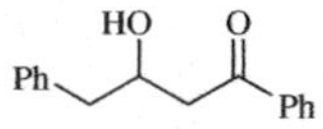		CH_2Cl_2, rt	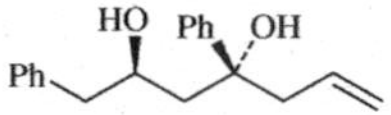 (93), 92:8 d.r.	110
C_{17} 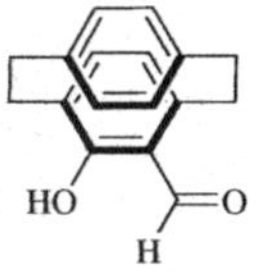	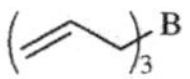	CH_2Cl_2, –78° to rt, 0.5 h	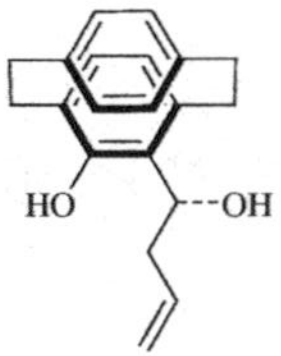 All$_3$B/Ald — d.r. 1:1 (85) >99:1 2:1 (87) >99:1	487, 464
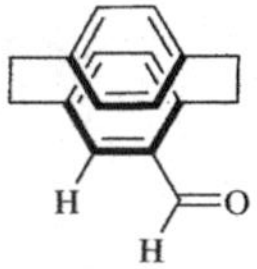	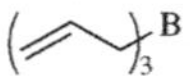	CH_2Cl_2, –78° to rt, 0.5 h	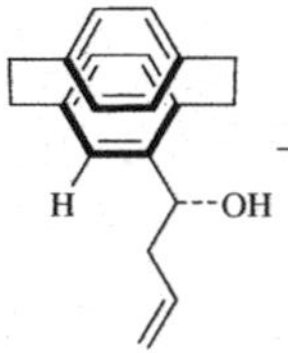 All$_3$B/Ald — Conc. — d.r. 1:1 0.08 M (98) 1:1 1:2 0.08 M (93) 1.5:1 1:3 0.08 M (85) 1.8:1 1:3 0.01 M (60) 7:1	487, 464
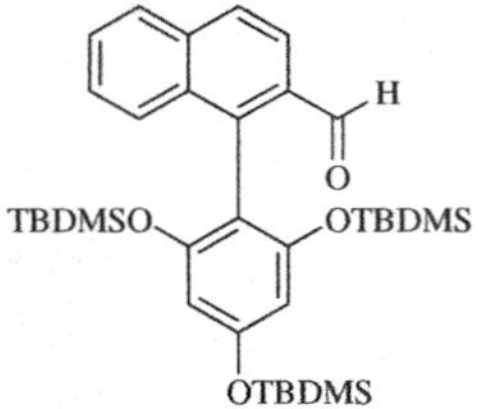		THF	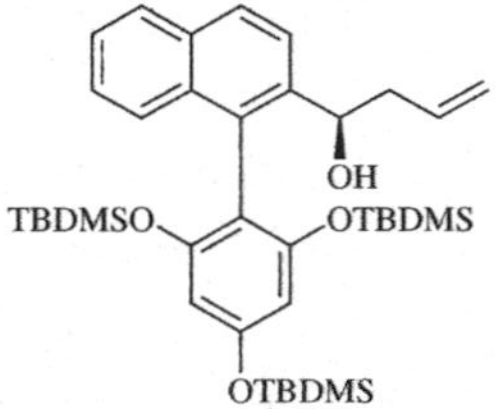 (95), 86	495

C$_{18}$

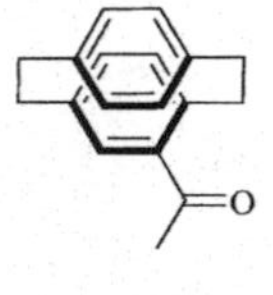

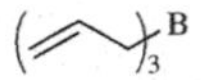

C$_{19}$

All$_3$B/Ket	Conc.		d.r.	
1:2	0.08 M	(75)	>99:1	487, 464
1:1	0.08 M	(93)	>99:1	
1:1	0.04 M	(90)	>99:1	
2:1	0.04 M	(95)	>99:1	

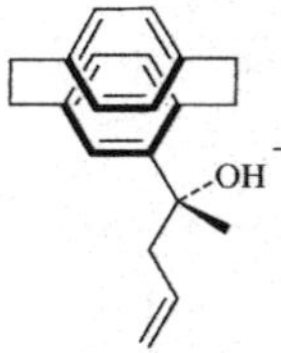

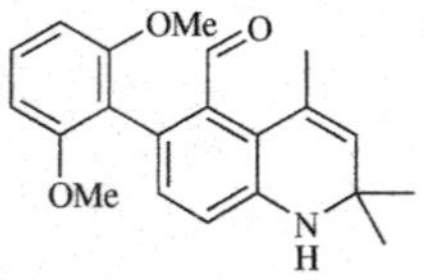

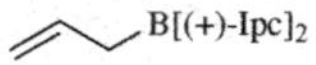

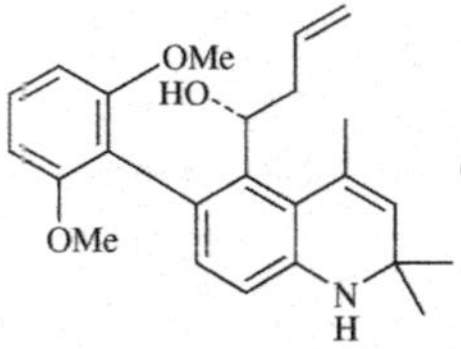

CH$_2$Cl$_2$, 40°, 5 h

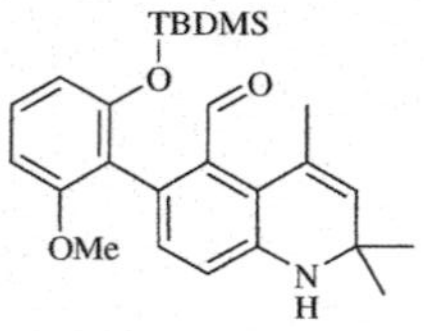

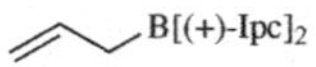

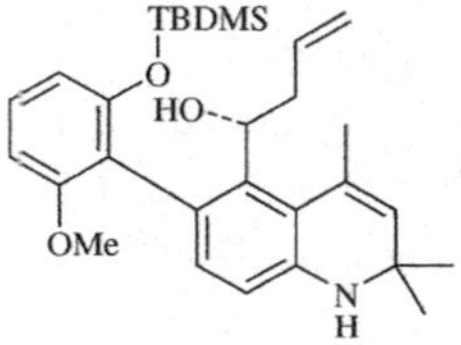

B[(+)-Ipc]$_2$ THF   (95), 69   496

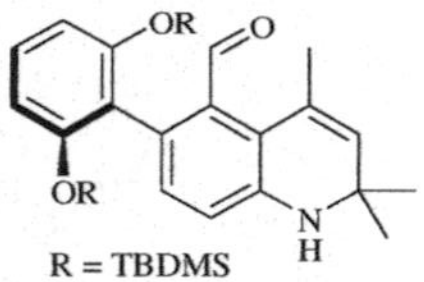

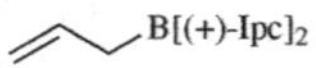

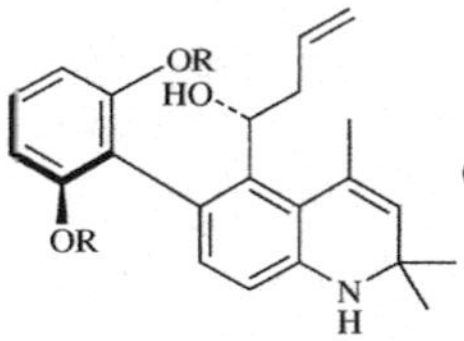

B[(+)-Ipc]$_2$ THF   (95), 90-95, 1:1 d.r.   495, 496

B[(+)-Ipc]$_2$ THF   (95), 90   495, 496

R = TBDMS

[a] The reaction mixture was free of Mg^{2+} ions.

TABLE 2. ADDITION OF Z-CROTYL REAGENTS

Carbonyl Substrate	Allylborane Reagent	Conditions	Product(s) and Yield(s) (%), % ee	Refs.
C₁	BBN	Pentane, rt, 2 h	(88)	103
C₂	B(Bu-n)₂	Neat, 0–5°	(92)	238
	B[(+)-Ipc]₂	THF, –78°, 3 h	(72), 96, >99:1 d.r.	137
	"	Ether, –78°, 3 h	I (72), 92, >99:1 d.r.	497
	B(2-d-Icr)₂	Ether, –78°, 3 h	I (75), 94, >99:1 d.r.	497
	B-TMS	Ether, –78°, 3 h	I (68), 95, >98:2 d.r.	143
	B[(–)-Ipc]₂	Ether, –78°, 3 h	(75), 90, >99:1	497
	"	THF, –78°, 3 h	I (75), 95, >99:1 d.r.	137
	B(4-d-Icr)₂	Ether, –78°, 3 h	I (—), 94, >99:1 d.r.	497

212

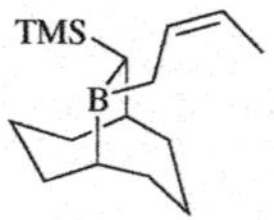

143

3

22

498

499

103

273, 500,
295

502

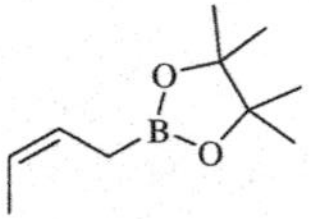

503, 504,
505

R	
Cl	(93)
OAc	(97)

(70), >90, >20:1 d.r.

(80)

I (68), 95, >98:2 d.r.

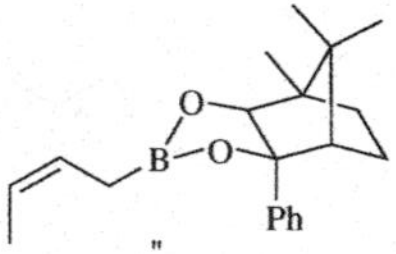

I (92), 97:3 d.r.

I (20), 97:3 d.r.

I (93), 60-70, 98:2 d.r.

I (93), 95:5 d.r.

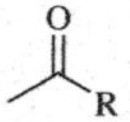

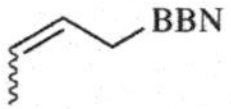

I (86), 94

Ether, –78°, 3 h

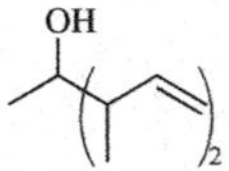

—

Ether, –78° to rt

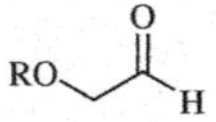

—

Hexanes, –55° to –20°

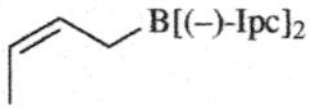

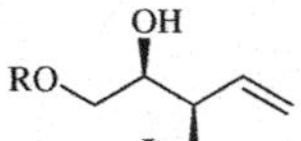

Pentane, rt, 2 h

THF, –78°, 3 h

THF, –78°

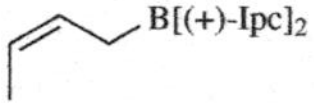

THF, –78°, 12 h

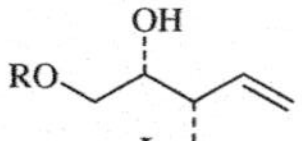

"

"

BBN

B[(–)-Ipc]$_2$

B[(+)-Ipc]$_2$

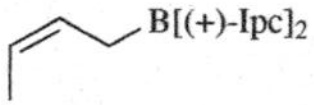

B[(+)-Ipc]$_2$

R = Bn

R = Bn

R = TBDMS

213

TABLE 2. ADDITION OF Z-CROTYL REAGENTS (*Continued*)

Carbonyl Substrate	Allylborane Reagent	Conditions	Product(s) and Yield(s) (%), % ee	Refs.
C₂				
RO⟋⟍CHO R= TBDMS		Sc(OTf)₃ (10 mol%), CH₂Cl₂, –78°, 24 h	(57), 96, >49:1 d.r.	27, 28
C₃				
CH₂=CH–CHO	B[(–)-Ipc]₂	THF, –78°, 3 h	(63), 90, >99:1 d.r.	137
CH₃CH₂–CHO		THF, –78°, 4 h	(73), 86, 93:7 d.r.	142
	B[(–)-Ipc]₂	THF, –78°, 3 h	I (70), 90, 99:1 d.r.	137, 497
	"	THF, ether, –78°, 3 h	I (70), 95-96, 98.5:1.5 d.r.	506
	B(4-d-Icr)₂	THF, –78°, 3 h	I (75), 94, >99:1 d.r.	137, 497
	B[(+)-Ipc]₂	THF, ether, –78°, 3 h	(70), 95-96, 98.5:1.5 d.r.	506
	"	THF, –78°, 3 h	I (78), 92, 99:1 d.r.	137, 497
	B(2-d-Icr)₂	Ether, –78°, 3 h	I (78), 96, >99:1 d.r.	497

B(OMe)$_2$		Ether, −78° to rt	OH — (62), 96:4 d.r.	43
pinacol boronate (2-butenyl)		Ether, −78° to rt	I (20), 97:3 d.r.	22
B(Bu-n)$_2$		Neat, 0-5°	OH — (85), I	238
BBN		Pentane, rt, 2 h	I (100)	103
B(Bu-n)$_2$		Neat, 0-5°	CF$_3$ OH, CF$_3$ — (88)	238
tartrate boronate, OPr-i, OPr-i	CF$_3$COCF$_3$	4 Å MS, toluene, −78°, 4 h	OH — (43), 80	507
B[(+)-Ipc]$_2$	TES-C≡C-CHO	THF, −78°, 4 h	OH, TES — (82)	354, 355, 356

215

TABLE 2. ADDITION OF Z-CROTYL REAGENTS (Continued)

Carbonyl Substrate	Allylborane Reagent	Conditions	Product(s) and Yield(s) (%), % ee	Refs.
C₃				
RO~CHO, R = TBDPS	B[(−)-Ipc]₂	THF, −78°	(82), 90	508, 509, 510
R = 3,4-DMB	B[(−)-Ipc]₂	THF, ether, −78°	I (72), >97.5:2.5 d.r.	511
R = TBDMS	B[(−)-Ipc]₂	THF, −78°, 5 h	I (82)	512, 513, 501
R = PMB	B[(−)-Ipc]₂	THF, −78°	I (72), 97, >95:5 d.r.	514
R = Bn	B[(−)-Ipc]₂	THF, −78°	I (78), 90	515
TBDPSO~CHO	B[(−)-Ipc]₂	BF₃•Et₂O, THF, −78°, 3 h	(81)	533
R = TBDMS	B[(+)-Ipc]₂	—	I (—)	517
R = TBDMS	B[(−)-Ipc]₂	THF, −78°, 3 h	I (82), >98, >99:1 d.r.	501
R = TBDMS	B(2-d-Icr)₂	THF, −78°	I (62)	518
		4 Å MS, toluene, −78°, 4 h	(68), 72, >98:2 d.r.	519, 37

	Reagent	Conditions	Product	Refs.
(aldehyde, R = TBDPS)	pinanediol allylboronate, Ph	Sc(OTf)$_3$ (10 mol%), CH$_2$Cl$_2$, −78°, 24 h	(57), 96, >49:1 d.r.	27, 28
	B[(+)-Ipc]$_2$	—	(—), >97:3 d.r. **I**	520
	B[(−)-Ipc]$_2$	—	**I** (—), 1:1 d.r.	520
(aldehyde, NBn$_2$)	pinacol allylboronate	Neat, rt, 4 d	(50), 51:49 d.r.	521
	B[(+)-Ipc]$_2$	THF, −78°, 3 h	(78), 92, 99:1 d.r.	167
	B[(−)-Ipc]$_2$	THF, −78°, 3 h	(74), 90, 73:27 d.r.	167
(aldehyde, OBn)	B[(−)-Ipc]$_2$	THF, ether, −78°, 1.5 h	(75), >99:1 d.r.	522
(aldehyde, OR; R = TES)	BIpc$_2$(+)	−78°	(83), 98.5:1.5 d.r.	523

R = TBDPS

R = TES

R = Bn

TABLE 2. ADDITION OF Z-CROTYL REAGENTS (Continued)

Carbonyl Substrate	Allylborane Reagent	Conditions	Product(s) and Yield(s) (%), % ee	Refs.
C₃				
R = Bn		Neat, rt, 4 d	(98), 88:12 d.r.	521
R = TBDPS		4 Å MS, toluene, −78°, 1.5 h	**I** (98), >99:1 d.r.	524
TBDMSO		BF₃•OEt₂ (5 mol%), CH₂Cl₂, rt, 6 h	(73), 95:5 d.r.	99
	"	n-Bu₄NI (10 mol%), CH₂Cl₂, H₂O, rt, 30 min	**I** (95), 95:5 d.r.	276
		THF, rt, 72 h	(46), 75:25 d.r.	525
		THF, −78°, 4 h	(65), 5.3:1 d.r., 97:3 syn:anti	142
		THF, −78°, 4 h	(66), 57:1 d.r., 93:7 syn:anti	142

218

Reagent	Conditions	Product (yield), selectivity	Ref.
(2,5-dimethylborolane with crotyl group)	THF, −78°, 4 h	I (57), 4.2:1 d.r., 96:4 syn:anti	142
(pinacol crotylboronate)	CH₂Cl₂, −78° to rt, 24–48 h	I (75–85), 96:4 d.r., 94:6 syn:anti	118
(bis-CF₃ diamide dioxaborolane, crotyl)	4 Å MS, toluene, −78° to rt, 10 h	I (95), >99:1 d.r.	132
(bis-CF₃ diamide dioxaborolane, crotyl)	4 Å MS, toluene, −78°, 12 h	I (58), 61:39 d.r.	132
(pinanediol phenyl boronate, crotyl)	Petroleum ether, −78° to rt, 18 h	I (86), >99:1 d.r.	288
(pinanediol phenyl boronate, crotyl) racemic	Petroleum ether, −78° to rt, 18 h	I (—), 97:3 d.r.	288
B(OPr-i)₂ (crotyl)	Et₃N, CH₂Cl₂, −10° to rt, 36 h	(65), 94:6 d.r.	106

Substrate (bottom): $HOCH_2C(O)CH_3$

TABLE 2. ADDITION OF Z-CROTYL REAGENTS (Continued)

Carbonyl Substrate	Allylborane Reagent	Conditions	Product(s) and Yield(s) (%), % ee	Refs.
C₃				
HO_2C–C(=O)CH₃	(Z)-crotyl–B(OPr-i)₂	Et₃N, CH₂Cl₂, –10° to rt	HO, HO_2C (96), 99:1 d.r.	106
EtO_2C–C(=O)CH₃	(Z)-crotyl tartrate diisopropyl ester borane	4 Å MS, toluene, –78° to rt, 6 h	HO, EtO_2C (87), 9, 5.4:1 d.r.	134
EtO_2C–C(=O)CH₃	(Z)-crotyl–$BF_3^-$$K^+$	BF_3•OEt_2, CH₂Cl₂, –78°, 2 h	HO, EtO_2C (90), >98:2 d.r.	98
C₄				
n-Pr–CHO	(E/Z)-crotyl–B(OC₆H₁₃-n)₂	Ether, rt	OH, n-Pr (79)	31
n-Pr–CHO	(Z)-crotyl-9-BBN (TMS)	Ether, –78°, 3 h	OH, n-Pr (91), 94, >98:2 d.r.	143
i-Pr–CHO	(Z)-crotyl pyrrolidine borane	THF, –78°, 4 h	OH, i-Pr **I** (70), 93, 96:4 d.r.	142
i-Pr–CHO	(Z)-crotyl pinacol borane	Ether, –78°	**I** (51), 94:6 d.r.	22

Neat, rt, 4 d

OH

(71), 90:10 d.r.

521

THF, −78°, 3 h

B[(+)-Ipc]₂

OH   I

(78), 95:5 d.r.

167, 295

R = Bn

Ether, THF, −100°
−78°, 4 h

"

I (73)

527, 528

"

I (70), 87.5:12.5 d.r.

529

THF, −90°, 16 h

I (70)

530

THF, −78°, 3 h

B[(−)-Ipc]₂

OH   I

RO

(83), 92:8 d.r.

167

R = TBDMS, TBDPS, Bn

Toluene, rt

B[(+)-Ipc]₂

R	d.r.	Syn:Anti
TBDMS	62:38	84:16
TBDPS	53:47	81:19
Bn	61:39	86:14

I (—)

169, 297

R = TBDMS

THF, −78°, 3 h

OH

RO

(75), >95:5 d.r.

250

R = TBDPS, TBDMS

4 Å MS, toluene,
−78° to rt, 10 h

R		d.r.	Syn:Anti
TBDPS	(74)	>99:1	—
TBDMS	(83)	95:5	96:4

132

R = PMB

4 Å MS, toluene, −78°,
18 h

I (—)

531

TABLE 2. ADDITION OF Z-CROTYL REAGENTS (*Continued*)

Carbonyl Substrate	Allylborane Reagent	Conditions	Product(s) and Yield(s) (%), % ee	Refs.
C₄				

C₄

Carbonyl Substrate: structure with OR, R = TBDMS, TBDPS, Bn

Allylborane Reagent: Z-crotyl dioxaborolane with (OPr-i) diester

Conditions: 4 Å MS, toluene, −78°

Product(s) and Yield(s) (%), % ee:

R		d.r.	Syn:Anti	
TBDMS	(71)	99:1	96:4	
TBDPS	(—)	97:3	89:11	
Bn	(—)	91:9	97:3	

Refs. 169, 297

Carbonyl Substrate: R = TBDPS, TBDMS

Allylborane Reagent: Z-crotyl dioxaborolane with (OPr-i) diester

Conditions: 4 Å MS, toluene, −78°

Product(s): I (—)

R		d.r.	Syn:Anti
TBDMS		53:47	86:14
TBDPS		73:27	88:12
Bn		53:47	90:10

Refs. 169

Allylborane Reagent: Z-crotyl dioxaborolane with bis-CF₃ diamide

Conditions: 4 Å MS, toluene, −78° to rt

Product(s): I

	Time		d.r.	Syn:Anti
TBDPS	12 h	(90)	88:12	95:5
TBDMS	10 h	(82)	63:37	99:1

Refs. 132

Carbonyl Substrate: structure, R = Bn

Allylborane Reagent: B[(+)-Ipc]₂ Z-crotyl

Conditions: THF, −78°, 3 h

Product(s): (—), 99:1 d.r.

Refs. 167

Allylborane Reagent: B[(−)-Ipc]₂ Z-crotyl

Conditions: THF, −78°, 3 h

Product(s): I (—), 94:6 d.r.

Refs. 167

Allylborane Reagent: "

Conditions: THF, −90°, 16 h

Product(s): I (72), 9:1 d.r.

Refs. 269

Carbonyl Substrate: structure OR, R = TES

Allylborane Reagent: B[(+)-Ipc]₂ Z-crotyl

Conditions: THF, ether, −78°, 3 h

Product(s): I (41), 3:1 d.r.

Refs. 522

222

Substrate	Reagent	Conditions	Product	Refs.
R = MOM	B[(+)-Ipc]₂	Ether, THF, −78°, 3 h	**I** (—), 3:1 d.r.	532
	B[(+)-Ipc]₂	THF, −78°, 2 h	(76)	534, 535
	B[(+)-Ipc]₂	THF, −78°, 6 h	(98)	354, 355
	B[(+)-Ipc]₂	THF, −78°, 4 h	(70), >97.5:2.5 d.r.	536
	B[(+)-Ipc]₂	THF, −78°, 4 h	(72), >97.5:2.5 d.r.	536
		CH₂Cl₂, −78° to rt, 24–48 h	(75–85), 98:2 d.r., 97:3 syn:anti	537, 118

223

TABLE 2. ADDITION OF Z-CROTYL REAGENTS (*Continued*)

Carbonyl Substrate	Allylborane Reagent	Conditions	Product(s) and Yield(s) (%), % ee	Refs.
C₄				
(HO₂C, Et ketone)	B(OPr-i)₂ (Z-crotyl)	Et₃N, CH₂Cl₂, –10° to rt	(92), 99:1 d.r.	106
C₅				
(methyl propenyl ketone)	B(OPr-i)₂ (crotyl)	Neat, 0–5°	(81)	238
(t-Bu aldehyde)	(chiral 2,5-dimethylborolane crotyl)	THF, –78°, 4 h	(75), 97, 95:5 d.r.	142
	(TMS-borabicyclononane crotyl)	Ether, –78°, 3 h	I (92), 94, >98:2 d.r.	143
	(tartrate-derived dioxaborolane crotyl)	4 Å MS, toluene, –78°, 7 d	I (66), 70, 99:1 d.r.	519, 37
	BF₃⁻K⁺ (crotyl)	n-Bu₄NI (10 mol%), CH₂Cl₂, H₂O, rt, 15 min	I (96), >98:2 d.r.	100
(Et aldehyde)	B[(–)-Ipc]₂ (crotyl)	Ether, –78°, 3 h	(73), 82:18 d.r.	538, 167

224

Reagent	Conditions	Product	Refs.
B[(+)-Ipc]$_2$	Ether, −78°, 3 h	(79), 96:4 d.r.	538, 167
(pinacol boronate)	Neat, rt, 3 d	I (49), 70:30 d.r.	539, 113
"	Petroleum ether, −78° to rt	I (98), 70:30 d.r.	114
"	Neat	I (—) (see table below)	539
(bicyclic boronate, Ph)	Petroleum ether, −78° to rt	(74), 55:45 d.r., >99:1 syn:anti	114, 540
(bicyclic boronate, Ph, racemic)	Petroleum ether, −78° to rt	(90), 70:30 d.r., >99:1 syn:anti	114, 540
(bicyclic boronate, Ph)	Petroleum ether, −78° to rt	(95), 70:30 d.r., >99:1 syn:anti	114, 540
(bicyclic boronate, Ph, racemic)	Petroleum ether, −78° to rt	(94), 66:34 d.r.	114

Temp	d.r.
rt	72:28
42°	70:30
61°	69:31
81°	68:32

225

TABLE 2. ADDITION OF Z-CROTYL REAGENTS (*Continued*)

	Carbonyl Substrate	Allylborane Reagent	Conditions	Product(s) and Yield(s) (%), % ee	Refs.
C_5	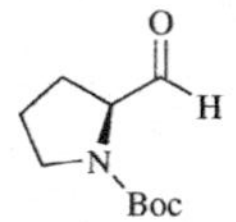	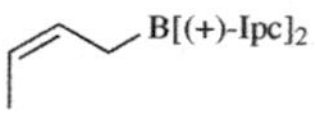	—	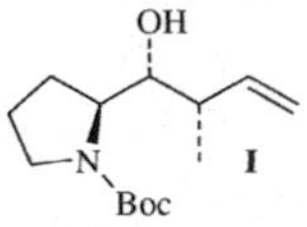 (50), >95:5 d.r.	278, 525
		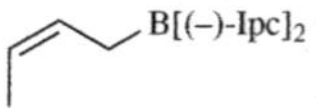	—	**I** (50), 80:20 d.r.	278, 525
		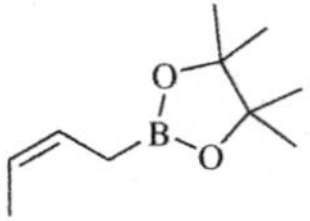	THF, rt, 72 h	**I** (70), 93:7 d.r., 89:11 syn:anti	525, 541
		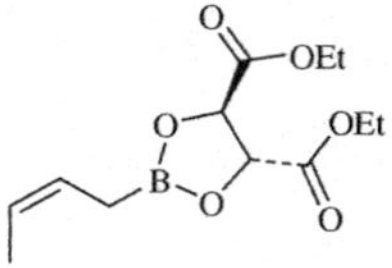	—	**I** (72), 87:13 d.r.	278, 525
		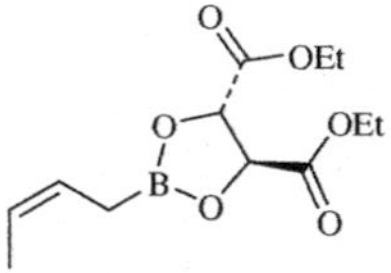	—	**I** (68), 96:4 d.r.	278, 525
	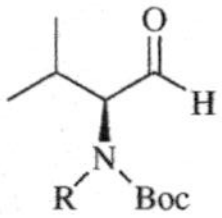	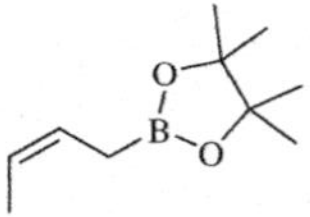	THF	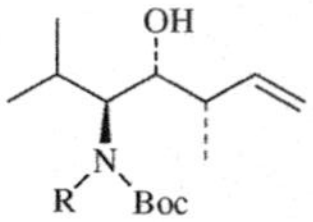	525

R	Temp	Time		d.r.
H	rt	24 h	(78)	72:28
Me	0°	96 h	(85)	86:14

					Ref.
		THF, −78°		(65), 90, >20:1 d.r.	330
		THF, −78°		(68), 90	542
		THF, −78°		(—), 93	543
		Neat, rt, 3 d		R d.r. TBDMS 91:1 MOM 78:22 (—)	115, 113
		Neat, rt, 3 d		R d.r. TBDMS 60:40 MOM 59:41 (—)	115
		THF, −78°		(72)	411
		THF, −78°, 7 h		(39), 85, >99:1 d.r.	544

TABLE 2. ADDITION OF Z-CROTYL REAGENTS (Continued)

Carbonyl Substrate	Allylborane Reagent	Conditions	Product(s) and Yield(s) (%), % ee	Refs.
C₅				
	B[(−)-Ipc]₂	THF/ether, −78°	(—), 97, 15:1 d.r.	522, 532
	B[(+)-Ipc]₂	THF/ether, −78°	I (—), 1:1 d.r.	522
	B[(+)-Ipc]₂	—	(49–82)	341
	B[(−)-Ipc]₂	—	(49–82)	341
	(pinacol boronate)	Neat, rt, 3 d	(—), 68:32 d.r.	115
	(pinacol boronate)	—	(—), >99:1 d.r.[a]	545
C₆	BBN	Pentane, rt, 2 h	(89), 60:40 d.r.	103

228

	Reagent	Conditions	Product	Reference
	$B[(+)\text{-Ipc}]_2$	—	$n\text{-}C_5H_{11}$, OH (product)	(56), 95, >99:1 d.r.	250
	$B(OMe)_2$	$-120°$ to $-75°$	$n\text{-}C_5H_{11}$, OH, **I**	(51), 98:2 d.r.	546
	(bis-CF₃ diamide boronate)	4 Å MS, toluene, $-78°$, 9 h	**I**	(76), 92	132
	$B(OC_6H_{13}\text{-}n)_2$	Ether, rt	(cyclohexanol allyl)	(78)	31
TIPSO / BnO aldehyde	(tartrate boronate, OPr-i)	4 Å MS, toluene	TIPSO, OH, BnO	(64), >20:1 d.r.	547
OBn / dioxolane aldehyde	(tartrate boronate, OPr-i)	4 Å MS, toluene, $-78°$	OBn, OH	(73), 94:6 d.r., 99:1 syn:anti	548, 549
OBn / dioxolane aldehyde	(tartrate boronate, OPr-i)	4 Å MS, toluene, $-78°$	OBn, OH	(—), 82:18 d.r., 97:3 syn:anti	549

TABLE 2. ADDITION OF Z-CROTYL REAGENTS (Continued)

Carbonyl Substrate	Allylborane Reagent	Conditions	Product(s) and Yield(s) (%), % ee	Refs.
C₇ Ph–CHO	⌁⌁ BBN	Pentane, rt, 2 h	OH / Ph structure (87)	103
	B[(+)-Ipc]₂	—	(60), 96, >99:1 d.r.	250
	B[(−)-Ipc]₂	THF, −78°, 3 h	(72), 88, 99:1 d.r.	137
	TMS–B (9-BBN type)	Ether, −78°, 3 h	I (72), 98, >98:2 d.r.	143
	B(OMe)₂	Ether, −78°	I (40), 96:4 d.r.	43
	pinacol boronate	Ether, −78°	I (22), 96:4 d.r.	22
	"	—	I (97), 20:1 d.r.	52
	"	L.A. (10 mol%), toluene, −78°, 4 h	I L.A. (d.r.): AlCl₃ (87) 99:1; Sc(OTf)₃ (89) 99:1	25
	catechol boronate	Pd(PPh₃)₄	I (81)	550

	Conditions	Product	Ref.
(boronate, Cy-substituted dioxaborolane, crotyl)	Petroleum ether, rt, 12 h	**I** (62)	64
(boronate, tartrate-derived dioxaborolane OPr-i, crotyl)	4 Å MS, THF, −78°, 6 h	**I** (90), 55, 98:2 d.r.	37
(pinacol crotylboronate)	Et₂AlCl/(S)-BINOL (10 mol%), toluene, −78°, 6 h	(19), 8, 99:1 d.r.	25
(pinanediol crotylboronate, Ph)	Sc(OTf)₃ (10 mol%), CH₂Cl₂, −78°, 24 h	**I** (53), 59, >49:1 d.r.	27, 28
B[(+)-Ipc]₂ (crotyl)	—	(77), 97, >99:1 d.r.	250
"	Pentane/ether (1:1), −100°	**I** (75), 94, >19:1 d.r.	249
B[(−)-Ipc]₂ (crotyl)	THF, −78°, 3 h	(75), 94, >95:5 d.r.	491

(pentafluorobenzaldehyde structure)

TABLE 2. ADDITION OF Z-CROTYL REAGENTS (*Continued*)

Carbonyl Substrate	Allylborane Reagent	Conditions	Product(s) and Yield(s) (%), % ee	Refs.
C₇ R = H, MeO, O₂N		CH₂Cl₂, −78° to rt	 R Me (100) TBDMS (99)	551
  R = H, MeO, O₂N	BF₃⁻K⁺	BF₃•OEt₂ (2 equiv), CH₂Cl₂, −78°, 15 min	I d.r. R H (91) >98:2 MeO (91) 96:4 O₂N (95) >98:2	98, 99
	"	BF₃•OEt₂ (5 mol%), CH₂Cl₂, rt, 3–6 h	I d.r. R H (92) >98:2 MeO (93) 96:4 O₂N (94) >98:2	99
R = H, MeO	BF₃⁻K⁺	n-Bu₄NI (10 mol%), CH₂Cl₂, H₂O, rt, 15 min	I d.r. R H (96) >98:2 MeO (98) >98:2	100
		4 Å MS, toluene, −78°, 6 h	 (90), 83, 98:2 d.r.	519, 37
		4 Å MS, toluene, −78°, 9 h	I (80), 92	132

232

Aldehyde	Reagent	Conditions	Product	Refs.
t-Bu		CDCl₃, 4 kbar, rt	t-Bu, OH (50), 60:40 d.r.	521
p-MeOC₆H₄	B[(−)-Ipc]₂	THF, ether, −78°, 2 h	p-MeOC₆H₄, OH (60)	552
TBDPSO	B(OPr-i)₂	Toluene, −23°	TBDPSO, OH, OH (78), 1.6:1 d.r.	553
PMBO, MeO, TBDMS	B[(−)-Ipc]₂	THF, −78°, 6 h	PMBO, MeO, TBDMS, OH (—), 6:1 d.r.	354
i-Pr	B[(+)-Ipc]₂	THF, −78°, 6 h	I (—), 1:1 d.r.	354
i-Pr	B[(−)-Ipc]₂	THF, −78°, 4 h	i-Pr, OH (56)	554
i-Pr	B[(−)-Ipc]₂	THF, −78°, 4 h	i-Pr, OH (62)	554
TBDPSO	B, OPr-i, OPr-i	4 Å MS, toluene, −78°	TBDPSO, OH (85), 5:1 d.r.	364, 441

233

TABLE 2. ADDITION OF Z-CROTYL REAGENTS (*Continued*)

Carbonyl Substrate	Allylborane Reagent	Conditions	Product(s) and Yield(s) (%), % ee	Refs.
C₇				
		4 Å MS, toluene, −78°, 6 h	(99)	555
C₈				
n-C₇H₁₅ CHO	B[(−)-Ipc]₂	THF, −78° to rt	(90), 95:5 d.r. **I**	420
	"	BF₃·OEt₂, THF, −78° to rt	**I** (72), 95:5 d.r.	422
	"	BF₃·OEt₂	**I** (90)	421
	"	THF, −78° to rt, 3.5 h	**I** (90), 95:5 d.r.	420
	BF₃⁻K⁺	BF₃·OEt₂ (2 equiv), CH₂Cl₂, −78°, 15 min	**I** (74), >98:2 d.r.	98, 99
	"	BF₃·OEt₂ (5 mol%), CH₂Cl₂, rt, 3–6 h	**I** (76), >98:2 d.r.	99
	"	n-Bu₄NI (10 mol%), CH₂Cl₂·H₂O, rt, 15 min	**I** (94), >98:2 d.r.	100
Ph CHO	B[(+)-Ipc]₂	THF, −78°, 18 h	(75), >95, >95:5 d.r. **I**	556
	"	THF, −78° to rt, 2 h	**I** (80)	557
	"	—	**I** (85–90), >95	558

234

Substrate	Reagent	Conditions	Product (yield)	Ref.
PhCOCH₃ (methyl phenyl ketone)	allyl-BBN (with internal alkene)	Pentane, rt, 2 h	OH, Ph, methyl (97)	103
PhCOCH₃	pinacol crotylboronate	(R,R)-i-Pr-DuPHOS, La(OPr-i)₃, CuF, DMF, –40°, 5 h	HO, Ph (94), 87, 84:16 d.r.	164
1-acetylcyclohexene	pinacol crotylboronate	(R,R)-i-Pr-DuPHOS, La(OPr-i)₃, CuF, DMF, –40°, 4 h	HO, cyclohexenyl (90), 92, 62:38 d.r.	164
Ph, TBDMSO aldehyde	BF₃⁻K⁺	n-Bu₄NI (10 mol%), CH₂Cl₂/H₂O, rt, 30 min	**I** Ph, OH, TBDMSO (97), 90:10 d.r.	276
"	"	BF₃•OEt₂ (5 mol%), CH₂Cl₂, rt, 6 h	**I** (85), 90:10 d.r.	99
PMBO, BnO aldehyde	B[(+)-Ipc]₂	THF, –78°	PMBO, BnO, OH (—)	543
PMBO, BnO aldehyde	B[(−)-Ipc]₂	THF, –78°	PMBO, BnO, OH (—), 2:1 d.r.	543

TABLE 2. ADDITION OF Z-CROTYL REAGENTS (*Continued*)

Carbonyl Substrate	Allylborane Reagent	Conditions	Product(s) and Yield(s) (%), % ee	Refs.
C₈	B[(+)-Ipc]₂	THF, –78°	(—), 2:1 d.r.	543
	B[(–)-Ipc]₂	THF, –78°	(—)	543
		4 Å MS, toluene, –78°	(—), >99:1 d.r. **I**	559
		4 Å MS, toluene, –78°	**I** (—), 1.5:1 d.r.	559
	B(OPr-*i*)₂	Et₃N, CH₂Cl₂, –10° to rt	(95), 99:1 d.r.	106
		Sc(OTf)₃ (10 mol%), CH₂Cl₂, –78°, 24 h	(61), 95, >49:1 d.r.	28
C₉	BF₃[–] K⁺	*n*-Bu₄NI (10 mol%), CH₂Cl₂, H₂O, rt, 15 min	(99), >98:2 d.r.	100

236

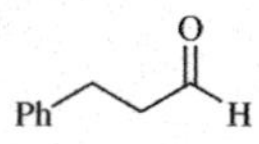

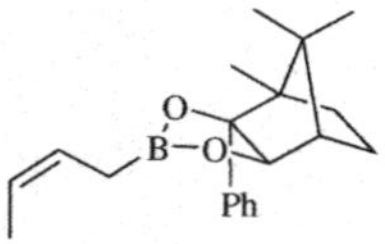

L.A.		d.r.	
Sc(OTf)₃	(52), 96	>49:1	27, 28
TiCl₄	(48), —	—	
TfOH	(<10), —	—	
TFA	(65), —	—	

L.A. (10 mol%), CH₂Cl₂, −78°, 24 h

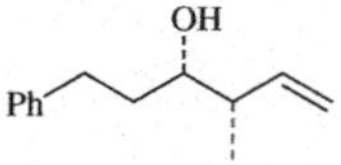

(92), 77:23 d.r. — 521

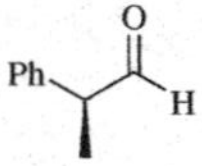

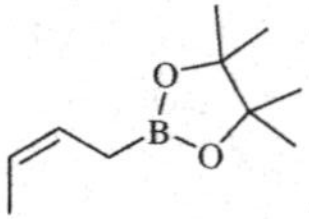

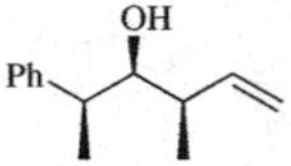

Neat, rt, 4 d

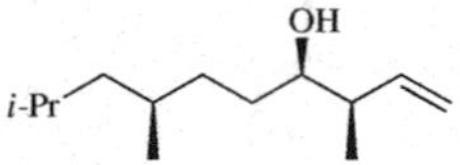

(70), 96.4:3.6 d.r. — 526

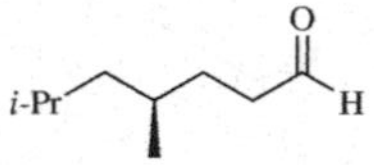

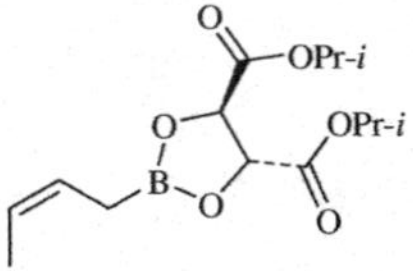

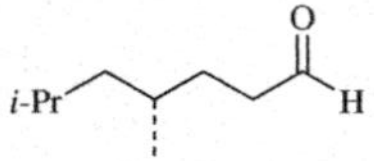

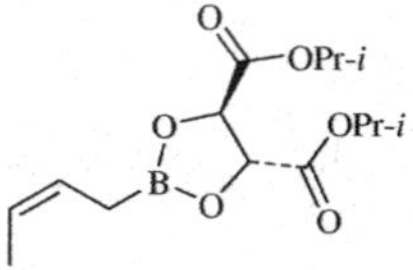

4 Å MS, toluene, −78°, 2 h

(70), 96:4 d.r. — 526

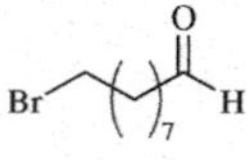

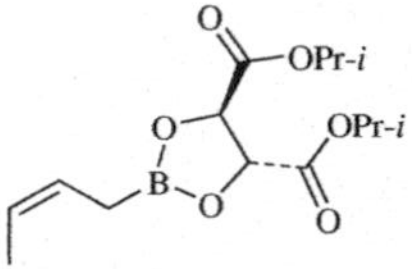

4 Å MS, toluene, −78°, 2 h

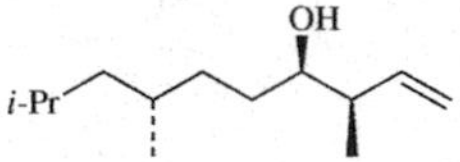

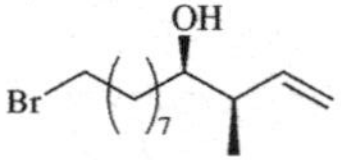

(91), 74 — 560

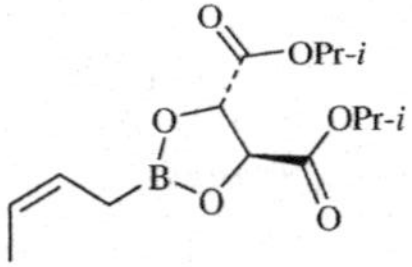

Toluene, −78°

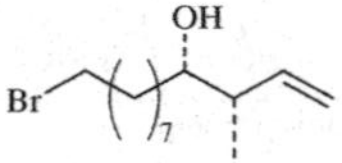

(91), 76 — 560

Toluene, −78°

237

TABLE 2. ADDITION OF Z-CROTYL REAGENTS (*Continued*)

Carbonyl Substrate	Allylborane Reagent	Conditions	Product(s) and Yield(s) (%), % ee	Refs.
C₉				
TBDMSO / Ph aldehyde	$BF_3^-K^+$ (Z-crotyl trifluoroborate)	$BF_3 \cdot OEt_2$ (5 mol%), CH_2Cl_2, rt, 3-6 h	TBDMSO, OH / Ph; (84), 75:25 d.r.	99
"	"	$n\text{-}Bu_4NI$ (10 mol%), CH_2Cl_2, H_2O, rt, 30 min	TBDMSO, OH / Ph; (96), 75:25 d.r.	276
C₁₀				
$n\text{-}C_6H_{13}$ (alkynal)	tartrate-derived Z-crotylboronate (OPr-i)	4 Å MS, toluene, −78°	OH, $n\text{-}C_6H_{13}$; (55-70), 61	130
"	"	$Cr(CO)_3$, 4 Å MS, toluene	OH, $n\text{-}C_6H_{13}$; Temp: −78° (55-70), 83; −90° (55-70), 88	130
$n\text{-}C_7H_{15}$ (alkynal)	tartrate-derived Z-crotylboronate (OPr-i)	$Cr(CO)_3$, 4 Å MS, toluene, −78°	OH, $n\text{-}C_7H_{15}$; (85-95), 86, 97:3 d.r.	128
$n\text{-}C_7H_{15}$ (enal)	tartrate-derived Z-crotylboronate (OPr-i)	4 Å MS, toluene, −78°, 6 h	OH, $n\text{-}C_7H_{15}$; (83), 62, 97:3 d.r.	519, 37

	Conditions	Product	Refs.
	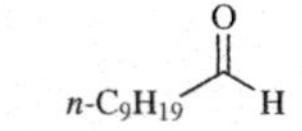 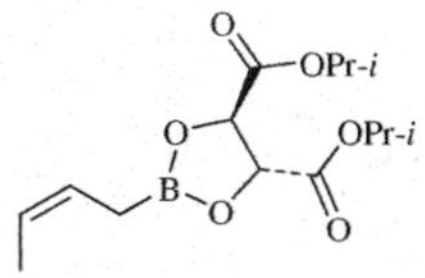 4 Å MS, toluene, −78°, 6 h	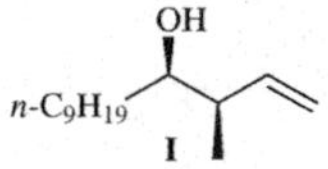 (80), 82, 99:1 d.r.	519, 37
	4 Å MS, toluene, −95°, 6 h	I (69), 86	519
	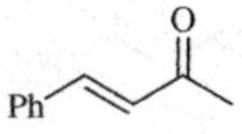 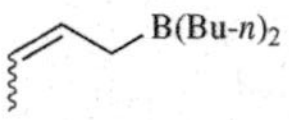 Neat, 0-5°	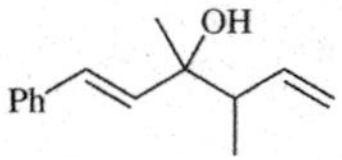 (77)	238
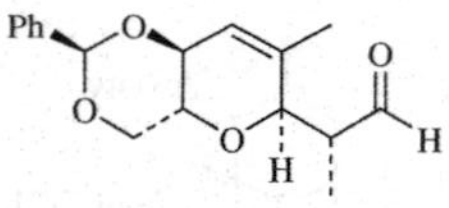	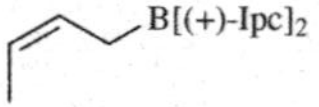 —	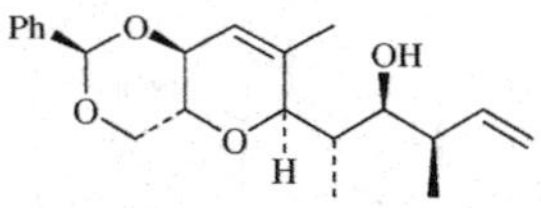 (78)	561
	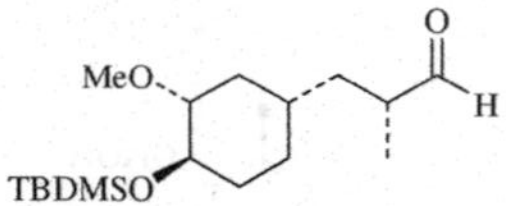 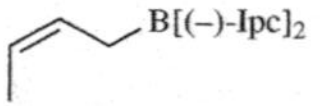 THF, −78°	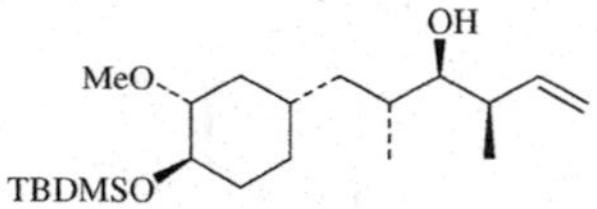 (90), 19:1 d.r.	562
	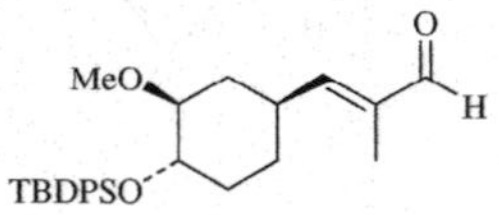 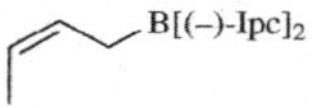 Ether, −78°, 1.5 h	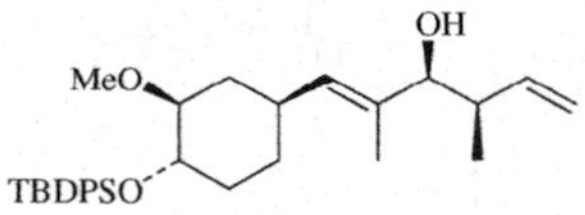 (51), 6.2:1 d.r.	563, 430
	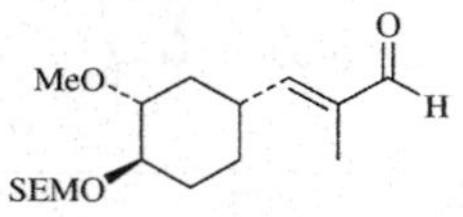 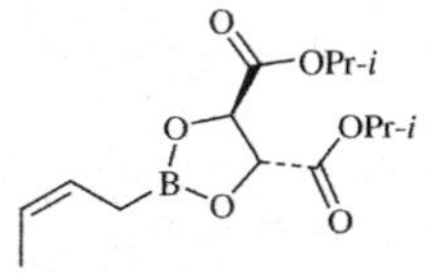 4 Å MS, toluene, −78°, 91 h	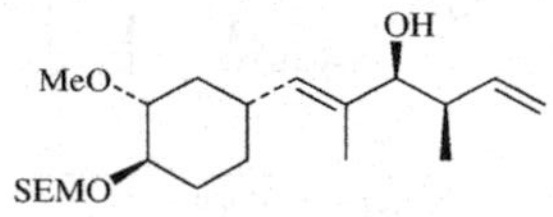 (98), 2.4:1 d.r.	564

B(Bu-n)₂

B[(+)-Ipc]₂

B[(−)-Ipc]₂

B[(−)-Ipc]₂

B[(−)-Ipc]₂

n-C₉H₁₉

Ph

MeO TBDMSO

MeO TBDPSO

MeO SEMO

TABLE 2. ADDITION OF Z-CROTYL REAGENTS (*Continued*)

Carbonyl Substrate	Allylborane Reagent	Conditions	Product(s) and Yield(s) (%), % ee	Refs.
C₁₀				
		THF, −78°	(59), >95	565
C₁₁				
		Toluene, −80° to −70°	(52)	566
		4 Å MS, toluene, −78°	(79)	566, 567, 568, 569, 570
		4 Å MS, toluene, −78°	(—), 1:1.3 d.r.	559
		4 Å MS, toluene, −78°	**I** (—), 1:1 d.r.	559

C_{13}	4 Å MS, toluene, −78°	(—), >99:1 d.r.	559
	4 Å MS, toluene, −78°	**I** (—), >99:1 d.r.	559
	Pentane, rt, 2 h	(96)	103
C_{14}	Et$_3$N, CH$_2$Cl$_2$, −10° to 40°, 32 h	(80), >99:1 d.r., 5:95 syn:anti	109
	4 Å MS, toluene, −78°	(80), 20:1 d.r.	571
	−78°	(61), >97.5:2.5 d.r.	572

241

TABLE 2. ADDITION OF Z-CROTYL REAGENTS (*Continued*)

Carbonyl Substrate	Allylborane Reagent	Conditions	Product(s) and Yield(s) (%), % ee	Refs.
C₁₄	B[(+)-Ipc]₂	–78°	(65), 11:1 d.r.	572
C₁₆	B[(+)-Ipc]₂	THF, –78°, 2 h; rt	R / H (67°), 7:1 d.r. / TES (67°), 3:1 d.r.	573, 574
	(boronate with OPr-i groups)	4 Å MS, toluene, –90° to rt, 18 h	(80), 67:33 d.r.	575, 576
	B[(+)-Ipc]₂	THF, –78°, 2 h	I (60), >95:5 d.r.	575, 576

[a] The reference reports uncertainty over the stereochemical assignment of the product.

TABLE 3. ADDITION OF E-CROTYL REAGENTS

Carbonyl Substrate	Allylborane Reagent	Conditions	Product(s) and Yield(s) (%), % ee	Refs.
C₂				
(acetaldehyde)	B[(+)-Ipc]₂	THF, −78°, 3 h	**I** (76), 96, 99:1 d.r.	44, 497
	B(2-d-Icr)₂	Ether, −78°, 3 h	**I** (75), 96, >99:1 d.r.	497
	B[(−)-Ipc]₂	THF, −78°, 3 h	**I** (78), 95, 99:1 d.r.	44, 497
	B(4-d-Icr)₂	Ether, −78°, 3 h	**I** (−), 94, >99:1 d.r.	497
	(TMS-allylborabicyclononane)	Ether, −78°, 3 h	**I** (73), 97, >98:2 d.r.	143
	(allyl pinacol boronate)	Ether, −78° to rt	**I** (40), 93:7 d.r.	22
	B(Bu-n)₃⁻Li⁺	(crotyl)Li⁺, B(Bu-n)₃, ether, −70°	**I** (−), 80:20–92.5:7.5 d.r.	577
(Boc-NH-CH₂-CHO)	B[(+)-Ipc]₂	—	(−), 97	517

TABLE 3. ADDITION OF E-CROTYL REAGENTS (Continued)

Carbonyl Substrate	Allylborane Reagent	Conditions	Product(s) and Yield(s) (%), % ee	Refs.
C₂ R = TBDPS	B[(−)-Ipc]₂	THF, ether, −78°	(61), 92, >99:1 d.r. **I**	578, 579, 580
R = PMB		4 Å MS, toluene, −78°	**I** (79)	559
R = TIPS		Toluene, −78°	**I** (80), 92, 20:1 d.r.	330
R = Bn		Ether, rt	**I** (86), 52, >95:5 d.r.	581
R = TBDMS		Sc(OTf)₃ (10 mol%), CH₂Cl₂, −78°, 24 h	(74), 95, >49:1 d.r.	27, 28
C₃	B[(−)-Ipc]₂	THF, −78°, 3 h	(65), 90, 99:1 d.r.	137

244

B(2-d-Icr)$_2$	Ether, −78°, 3 h	(76), 98, >99:1 d.r.	497
B[(+)-Ipc]$_2$	THF, −78°, 3 h	I (78), 92, 99:1 d.r.	137, 497
"	THF, ether, −78°, 3 h	I (70), 95-96, 98:2 d.r.	506
(2,5-dimethylborolane)	THF, −78°, 4 h	(81), 96, 93:7 d.r.	142
B[(−)-Ipc]$_2$	Ether, −78°, 3 h	I (70), 90, 99:1 d.r.	137, 497
"	THF/ether, −78°, 3 h	I (70), 95-96, 98:2 d.r.	506
B(4-d-Icr)$_2$	THF, −78°, 3 h	I (75), 94, >99:1 d.r.	137
"	Ether, −78°, 3 h	I (—), 94, >99:1 d.r.	497
B(OMe)$_2$	Ether, −78° to rt	I (61), 97:3 d.r.	43
(pinacol boronate)	Ether, −78°	I (62), 97:3 d.r.	22
B(Bu-n)$_3^-$Li$^+$	Li$^+$ B(Bu-n)$_3$, ether, −70°	I (—), 80:20-92.5:7.5 d.r.	577

TABLE 3. ADDITION OF E-CROTYL REAGENTS (*Continued*)

Carbonyl Substrate	Allylborane Reagent	Conditions	Product(s) and Yield(s) (%), % ee	Refs.
C₃				
(aldehyde, HN–Boc)	B[(+)-Ipc]₂	—	(—), >97:3 d.r.	520
	B[(−)-Ipc]₂	—	(—), >97:3 d.r. **I**	520, 582
R = Boc	"	−78°, 4 h	**I** (81)	583
	B[(−)-Ipc]₂	4 Å MS, toluene, −78°, 15 h	(85), >95:5 d.r.	278
	(dioxaborolane)	4 Å MS, toluene, −78°, 15 h	(53), 75:25	278
R = Ac	(tartrate borolane, OPr-*i*)	4 Å MS, toluene, acetone, −78°	**I** (67), 87:13 d.r.	584, 278
R = Boc	(tartrate borolane, OPr-*i*)	4 Å MS, toluene, −78°, 15 h	**I** (68), 75:25 d.r.	278

246

Substrate	Reagent	Conditions	Product	Refs.
aldehyde (NBn₂)	allyl(crotyl)boronate (OPr-i ester)	4 Å MS, toluene, −78°, 15 h	(68), >95:5 d.r.	278
aldehyde (NBn₂)	allyl(crotyl)boronate (OPr-i ester)	4 Å MS, toluene, acetone, −78°	(94), >95:5 d.r.	278
aldehyde (NBn₂)	allyl pinacol boronate	Neat, rt, 4 d	(92), 9:1 d.r.	521
RO-CHO	B[(+)-Ipc]₂	THF, −78° to −30°	(55)	406
"	"	−78° to rt, 2 h	**I** (71), >85, >50:1 d.r.	585
"	"	"	**I** (48)	586
"	"	−78°	**I** (67), >99, >99:1 d.r.	587
R = TBDMS	"	—	(70)	341
"	B[(+)-Ipc]₂	THF, −78°	(62), 92	413, 339
R = PMB	"	Ether, THF, −78°, 3 h	**I** (76), 91	588, 589

247

TABLE 3. ADDITION OF E-CROTYL REAGENTS (*Continued*)

Carbonyl Substrate	Allylborane Reagent	Conditions	Product(s) and Yield(s) (%), % ee	Refs.
C_3 (RO—CH₂CH₂—CHO, O=CH, R = TBDMS)	B[(−)-Ipc]₂ (E-crotyl)	THF, −78°	(72), 95, 20:1 d.r. **I** (RO—...—OH, vinyl)	590, 591
	"	THF, −78°	**I** (81), 95, 20:1 d.r.	592, 593
	"	THF, −78°, 2 h	**I** (68)	594
	(tartrate crotylboronate, OPr-*i* / OPr-*i*)	THF, −78° to rt, 4 h	(70), >95, >20:1 d.r.	595
	B[(+)-Ipc]₂	THF, −78°	(71), 85, >98:2 d.r. **I**	519, 37
R = TBDPS	(pinanyl crotylborane, Ph)	Sc(OTf)₃ (10 mol%), CH₂Cl₂, −78°, 24 h	(63), 94, >49:1 d.r.	27, 28
(O=CH—CH(OR)CH₃, R = Bn)	B[(+)-Ipc]₂	THF, −78°, 3 h	(85), 97:3 d.r.	167, 538

Substrate	Reagent	Conditions	Product	Refs.
R = PMB	B[(−)-Ipc]₂	THF, −78°, 3 h	(80), 95:5 d.r. **I**	167, 538
R = Bn	B[(−)-Ipc]₂	Ether, −78°	**I** (64)	597
(TBDMSO aldehyde)	pinacolboronate	Neat, rt, 4 d	**I** (46), 77:23 d.r.	521
	BF₃⁻K⁺	BF₃•OEt₂ (5 mol%), CH₂Cl₂, rt, 6 h	(72), 75:25 d.r. **I**	99
	"	n-Bu₄NI (10 mol%), CH₂Cl₂, H₂O, rt, 30 min	**I** (95), 75:25 d.r.	276
(HO ketone)	B(OPr-i)₂	Et₃N, CH₂Cl₂, −10° to rt, 15 h	(72), 97:3 d.r.	109
(TMS ynal)	B[(−)-Ipc]₂	THF, −78°	(66), 90	266
(TES ynal)	B[(+)-Ipc]₂	THF, −78°	(67)	573

TABLE 3. ADDITION OF E-CROTYL REAGENTS (Continued)

Carbonyl Substrate	Allylborane Reagent	Conditions	Product(s) and Yield(s) (%), % ee	Refs.
C₃ RO₂C–C(=O)CH₃	crotyl–BBN	Ether, −76° to 0°, 30 min	 d.r. R Me (96) 73:27 t-BuCH₂ (90) 85:15 PhCH₂ (94) 80:20 Ph (89) 80:20 2,6-Me₂C₆H₃ (79) 75:25 3,5-Me₂C₆H₃ (88) 65:35 2,3,6-Me₃C₆H₂ (80) 70:30 2,6-(t-Bu)₂-3-MeC₆H₂ (90) 100:0	598
HO₂C–C(=O)CH₃	crotyl–B(OPr-i)₂	Et₃N, CH₂Cl₂, −10° to rt	(95), 99:1 d.r.	106
EtO₂C–C(=O)CH₃		4 Å MS, toluene, −78° to rt, 6 h	**I** (86), 6, 53:47 d.r.	134
		4 Å MS, −78° to rt, 6 h	**I** Solvent — d.r. toluene (84), 73 — 7:1 THF (82), 53 — 3.2:1 ether (80), 62 — 5:1 CH₂Cl₂ (83), 33 — 2:1	134
	crotyl–BF₃⁻K⁺	BF₃•OEt, CH₂Cl₂, −78°, 2 h	**I** (92), >98:2 d.r.	98

THF, −78°, 4 h		**I** (71), 34:1 d.r., 99:1 anti:syn		142
THF, −78°, 4 h		**I** (62), 2:1 d.r., 89:11 anti:syn		142
—		**I** (72), >20:1 d.r.		599
THF, −78°, 4 h		(74), 6.9:1 d.r., 98:2 anti:syn		142
CH₂Cl₂, −78° to rt, 24-48 h		(75-85), 52:42 d.r., 94:6 anti:syn		537, 118

		Temp	d.r.	Anti:Syn	
CH₂Cl₂, 24 h	**I** (—)	−78° to rt	52:42	94:6	118
		rt	51:44	95:5	
		−20°	51:46	97:3	
		−78°	53:47	>99:1	

TABLE 3. ADDITION OF E-CROTYL REAGENTS (*Continued*)

Carbonyl Substrate	Allylborane Reagent	Conditions	Product(s) and Yield(s) (%), % ee	Refs.
C₃		—	(37)	600
	"	4 Å MS, toluene, −78°	**I** (85), 98:2 d.r., 98:2 anti:syn	601, 602, 170
	"	4 Å MS, toluene, −78° to rt, 18 h	**I** (65)	603
		4 Å MS, toluene, −78°, 7 h	**I** (84), >99:1 d.r.	132
		4 Å MS, toluene, −78°	(87), 91:9 d.r., 96:4 anti:syn	126, 548, 602, 170
		4 Å MS, toluene, −78°, 9 h	**I** (77), 95:5 d.r.	132
		Petroleum ether, −78° to rt, 18 h	**I** (—), 72:28 d.r.	288

C₄

Petroleum ether, −78° to rt, 18 h	**I** (—), 2:1 d.r.	288	
4 Å MS, toluene, −78°, 1.5 h	(77), 88:12 d.r.	524	
4 Å MS, toluene, −78°	(85), 98:2 d.r., 98:2 anti:syn	549	
4 Å MS, toluene, −78°, 1.5 h	(84), 91:9 d.r.	524	
Ether, −78°, 2 h	(58), 99, 14:1 d.r.	604	
Ether, THF, −78°, 3 h	(—), 84, 9:1 d.r.	605, 606	

TABLE 3. ADDITION OF E-CROTYL REAGENTS (Continued)

Carbonyl Substrate	Allylborane Reagent	Conditions	Product(s) and Yield(s) (%), % ee	Refs.
C₄				
_i_-PrCHO		THF, −78°, 4 h	(76), 97, 96:4 d.r.	142
		Ether, −78°, 3 h	I (80), 88:12 d.r.	305, 607, 608
		Ether, −78°, 3 h	(94), 96, >98:2 d.r.	143
		Ether, −78°	(59), 96:4 d.r.	22
	B(Bu-_n_)₃·Li⁺	Li⁺ / B(Bu-_n_)₃, ether, −70°	I (—), 80:20–92.5:7.5 d.r.	577
		Neat, rt, 4 d	(71), 53:47 d.r.	521
		4 Å MS, toluene, −78°	(87), 80	609

Substrate	Reagent	Conditions	Product	Refs.
(aldehyde) RO—CH₂—CH(CH₃)—CHO R = Bn	CH₂=CH—CH₂—CH=CH—B[(+)-Ipc]₂	THF, −78°, 4 h	**I** (48), 95:5 d.r.	167, 610, 611
R = TES	B[(+)-Ipc]₂ "	THF/ether, −78° THF, −78°	**I** (76), >98:2 d.r. **I** (68), >98:2 d.r.	167 513
R = TBDPS	B[(+)-Ipc]₂	THF/ether, −85°, 20 h	**I** (74)	530
R = Bn	B[(−)-Ipc]₂	THF, −78°, 3 h	**I** (74)	167
R = TBDPS	B[(−)-Ipc]₂	Ether/THF, −100°	**I** (87), 98:2 d.r. **I** (74)	527
R = TBDMS, TBDPS, Bn	(allyl pinacol boronate)	Toluene, rt	**I** (—), >99:1 anti:syn	169, 297

Product structures: allylation adducts drawn as RO–CH₂–CH(CH₃)–CH(OH)–CH(CH₃)–CH=CH₂ (syn/anti isomers).

R	d.r.
TBDMS	61:39
TBDPS	62:38
Bn	68:32

TABLE 3. ADDITION OF E-CROTYL REAGENTS (Continued)

Carbonyl Substrate	Allylborane Reagent	Conditions	Product(s) and Yield(s) (%), % ee	Refs.
C₄ R = TBDMS, TBDPS, Bn		4 Å MS, toluene, −78°	 R — d.r. — Anti:Syn TBDMS (80) — 97:3 — >99:1 TBDPS (—) — 82:16 — 98:2 Bn (—) — 93:5 — 98:2	297, 456, 169, 612, 613, 614
		4 Å MS, toluene, −78°	 R — d.r. — Anti:Syn TBDMS (—) — 81:16 — 97:3 TBDPS (85) — 88:11 — 99:1 Bn (—) — 85:14 — 99:1	297, 169, 615
R = TBDPS, TBDMS		4 Å MS, toluene, −78°, 18 h	 R — d.r. TBDMS (77), 60 — — TBDMS (81), — — >97:3 d.r.	596
R = TBDPS, TBDMS		4 Å MS, toluene, −55° to 0°	 R — Time — d.r. TBDPS — 12 h (87) — 95:5 TBDMS — 10 h (85) — >97:3	132
R = TBDPS, TBDMS		4 Å MS, toluene, −55° to 0°	 R — Time — d.r. TBDPS — 12 h (80) — 96:4 TBDMS — 10 h (78) — 90:10	132

RO–CH(...)–CHO R = Bn

Reagent	Conditions	Product	Refs.
B[(+)-Ipc]₂ (pentenyl)	THF, −78°, 3 h	**I** (—), 98:2 d.r.	167
B[(+)-Ipc]₂	THF, −90°, 16 h	**I** (50), 93:7 d.r.	616
B[(−)-Ipc]₂	THF, −78°, 3 h	**I** (—), 94:6 d.r.	167
tartrate crotylboronate, OPr-i	4 Å MS, toluene, −78°, 8 h	**I** (78), 85:15 d.r.	195, 617
tartrate crotylboronate, OPr-i	4 Å MS, toluene, −78°, 4 h	**I** (71), 82:18 d.r.	618
tartrate crotylboronate, OPr-i	4 Å MS, toluene, −78°	**I** (86), 89:11 d.r.	619, 266, 615
tartrate crotylboronate, OPr-i	4 Å MS, toluene, −78°	see below	364, 271

R = Bn
R = TIPS
R = Bn
R = PMB
R = t-Bu
R = TBDMS
R = TBDMS, t-Bu

R	Time		d.r.
TBDMS	—	(—)	9:1
t-Bu	4 h	(74)	18:1

I

TABLE 3. ADDITION OF E-CROTYL REAGENTS (Continued)

Carbonyl Substrate	Allylborane Reagent	Conditions	Product(s) and Yield(s) (%), % ee	Refs.
C$_4$				
(BnO, epoxide aldehyde)	(crotyl pinacol boronate)	—	(95), 9:1 d.r.	52
(TBDMSO, BnO aldehyde)	B[(+)-Ipc]$_2$	THF, −78°, 3 h	(82)	250
(OMe, TBDPSO aldehyde)	B[(+)-Ipc]$_2$	THF, −78°, 4 h	(73), >97.5:2.5 d.r.	536
	B[(−)-Ipc]$_2$	THF, −78°, 4 h	(75), >97.5:2.5 d.r.	536
(acetonide aldehyde)	(tartrate crotyl borate, OPr-i)	4 Å MS, toluene, −78°	(63), 87:13 d.r.	620
(3,3-diethyl dioxolane aldehyde)	(tartrate crotyl borate, OPr-i)	4 Å MS, toluene, −78°, 1 h	(88), 95:5 d.r.	621
(3,3-diethyl dioxolane aldehyde)	(tartrate crotyl borate, OPr-i)	Toluene, −78°	(88), 95:5 d.r.	622

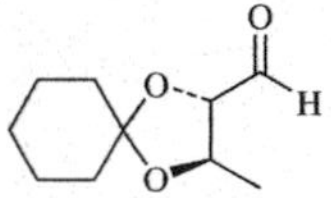	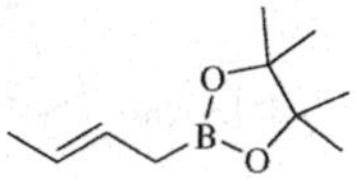	CH_2Cl_2, –78° to rt, 24-48 h	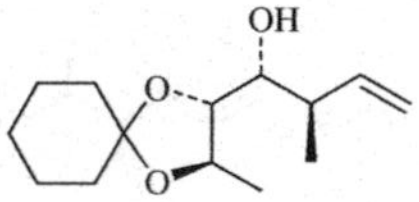	(75-85), 52:48 d.r., 98:2 anti:syn	537, 118
	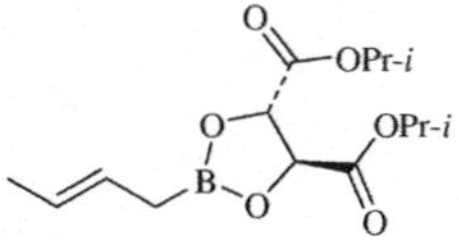	4 Å MS, toluene, –78°	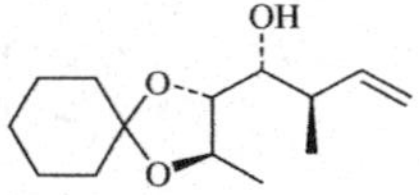	(80), 96:4 d.r., 92:8 anti:syn	126
	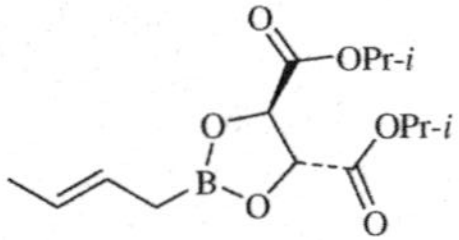	4 Å MS, toluene, –78°	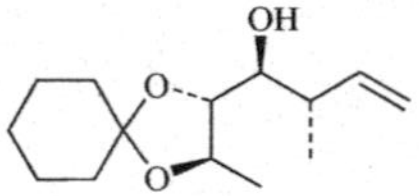	(88), 93:4 d.r., 93:7 anti:syn	126
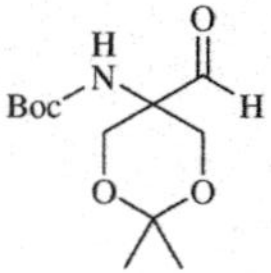		THF, ether, –78°	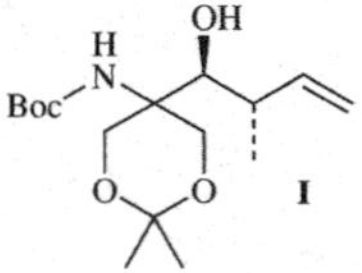 **I**	(68), 90, >99:1 d.r.	623
	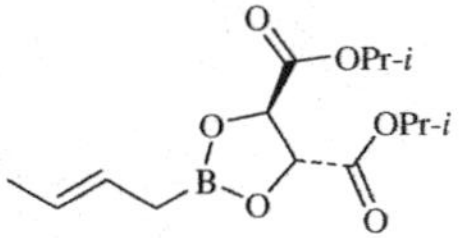	Toluene, 0°	**I** (—), 16, 99:1 d.r.		623
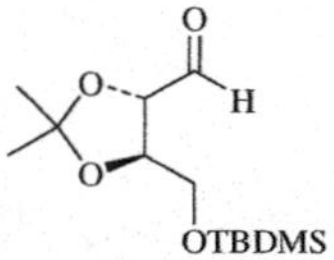	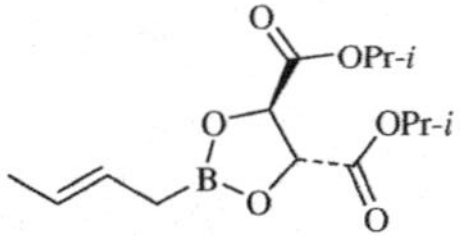	Toluene, 20°	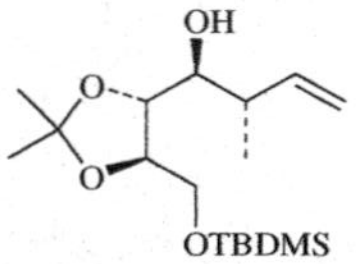	(80), 7:1 d.r.	624
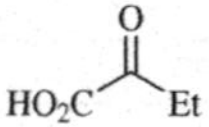		Et_3N, CH_2Cl_2, –10° to rt	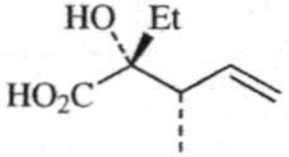	(93), 99:1 d.r.	106

TABLE 3. ADDITION OF E-CROTYL REAGENTS (Continued)

Carbonyl Substrate	Allylborane Reagent	Conditions	Product(s) and Yield(s) (%), % ee	Refs.
C₅ (t-Bu—CHO)	(2,5-dimethylborolane crotyl reagent)	THF, −78°, 4 h	(72), 95, 96:4 d.r.	142
	(TMS-substituted borabicyclononane crotyl reagent)	Ether, −78°, 3 h	(69), 99, >98:2 d.r.	143
	(diisopropyl tartrate boronate crotyl reagent, OPr-i)	4 Å MS, toluene, −78°, 6 d	(41), 73, 95:5 d.r. **I**	519, 37
	BF₃⁻K⁺ (crotyltrifluoroborate)	n-Bu₄NI (10 mol%), CH₂Cl₂, H₂O, rt, 15 min	**I** (95), >98:2 d.r.	100
	B[(−)-Ipc]₂	THF, −78°, 3 h	(75), 96:4 d.r. **I**	167
C₅ (Et—CH(CH₃)—CHO)	"	Ether, −78°	**I** (75), 96:4 d.r.	538
	"	Ether, −78°, 3 h	**I** (65), 92, 78:22 d.r., 85:15 anti:syn	607
	B[(+)-Ipc]₂	Ether, −70°, 3 h	(65), >90, >3:1 d.r.	306

Reagent	Conditions	Product	Refs.
B[(+)-Ipc]$_2$	THF, −78°, 3 h	OH / Et, (70), 91:9 d.r.	167
"	Ether, −78°	**I** (70), 91:9 d.r.	538
	Neat, rt, 4 d	OH / Et, (46), 77:23 d.r.	539, 115, 113
	Petroleum ether, −78° to rt	**I** (95), 76:24 d.r., >99:1 anti:syn	114
	Petroleum ether, −78° to rt	**I** (99), 92:8 d.r., >99:1 anti:syn	114, 540
racemic	Petroleum ether, −78° to rt	**I** (—), 76:24 d.r., >99:1 anti:syn	540
B[(+)-Ipc]$_2$	Ether, −78°, 3 h	OH / Et, (65), 54, 78:22 d.r.	607
	Petroleum ether, −78° to rt	OH / Et, (98), 75:25 d.r., >99:1 anti:syn	114, 540
racemic	Petroleum ether, −78° to rt	**I** (96), 76:24 d.r.	114

261

TABLE 3. ADDITION OF E-CROTYL REAGENTS (*Continued*)

Carbonyl Substrate	Allylborane Reagent	Conditions	Product(s) and Yield(s) (%), % ee	Refs.
C₅				
(pyrrolidine-N-Boc aldehyde)	(pinacol E-crotylboronate)	THF, rt, 60 h	(81), 95:5 d.r., 74:26 anti:syn	525
(BnO, 4, aldehyde)	B[(–)-Ipc]₂	THF, –78°	(—), 75	543
TBDMSO, BnO aldehyde	B[(+)-Ipc]₂	—	(49–82)	341
TBDMSO, BnO aldehyde	B[(–)-Ipc]₂	—	(49–82)	341
OR aldehyde	(pinacol E-crotylboronate)	Neat, rt, 3 d		115, 113
OR aldehyde	(pinacol E-crotylboronate)	Neat, rt, 3 d		115
OR aldehyde, R = DMPM	(tartrate E-crotylboronate)	4 Å MS, toluene, –78°	(85), >98:2 d.r.	625, 626

For rows 115, 113:

R	d.r.
TBDMS	98:2
MOM	89:11

(—)

For row 115:

R	d.r.
TBDMS	98:2
MOM	94:6

(—)

R = TES

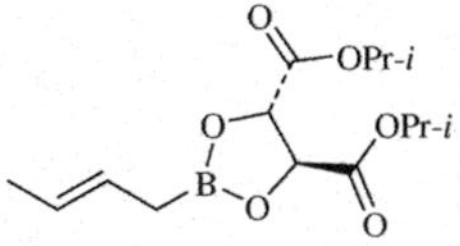

4 Å MS, toluene, −78°,
4 h

I (70), >98:2 d.r.

617

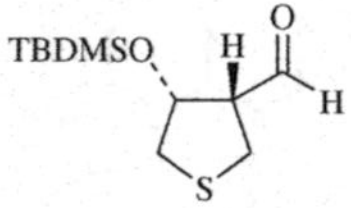

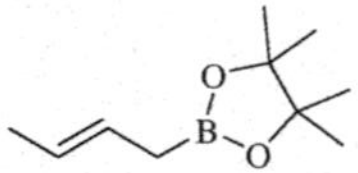

Neat, rt, 3 d

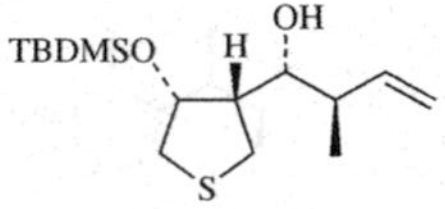

(—), 90:10 d.r.

115

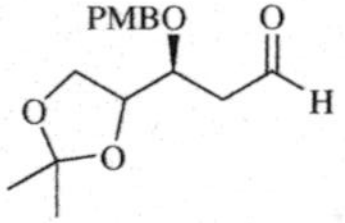

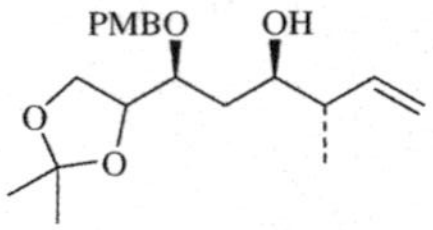

—

(58)

627

—

(67), 95, >99:1 d.r. **I**

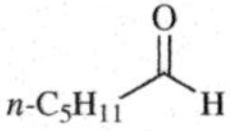

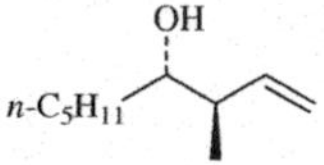

250

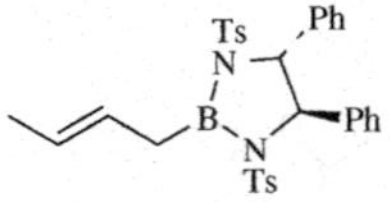

THF, −78°

I (74-82), 91-95

50

4 Å MS, toluene, −78°
4 h

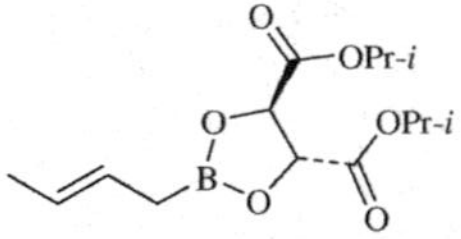

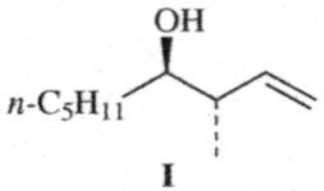

(84), 86, 99:1 d.r. **I**

519, 37

4 Å MS, toluene, −78°,
6 h

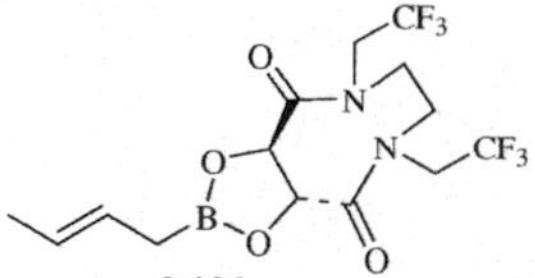

I (83), 92

132

0.4 M

C$_6$

n-C$_5$H$_{11}$

TABLE 3. ADDITION OF E-CROTYL REAGENTS (*Continued*)

Carbonyl Substrate	Allylborane Reagent	Conditions	Product(s) and Yield(s) (%), % ee	Refs.
C$_6$				
		4 Å MS, toluene, −78°	(90), >98, >50:1 d.r.	342, 190
		Petroleum ether, 0° to rt, 14 h	(99), 95:5 d.r., >99:1 anti:syn	149, 628, 197
		4 Å MS, toluene, −78°	(87), 94:6 d.r.	629, 630
		4 Å MS, toluene, −78°	(76), 80:20 d.r.	584, 278

Ref.	Product (yield), d.r.	Conditions	Reagent
278	(76), 80:20 d.r.	4 Å MS, toluene, −78°	tartrate allylboronate, OPr-i
583	(80)	−78°, 4 h	B[(+)-Ipc]$_2$
597	(83), 6:1 d.r.	Ether, −78°	B[(+)-Ipc]$_2$
413	(60), 98, 10:1 d.r.	THF, −78°	B[(−)-Ipc]$_2$
631	(62), 95:5 d.r.	THF, −78°	B[(−)-Ipc]$_2$
632	(—), 3.5:1 d.r.	4 Å MS, toluene, −78°, 2 h	tartrate allylboronate, OPr-i
633	(75), 4:1 d.r.	4 Å MS, toluene, −78° to rt, 18 h	tartrate allylboronate, OPr-i

TABLE 3. ADDITION OF E-CROTYL REAGENTS (*Continued*)

Carbonyl Substrate	Allylborane Reagent	Conditions	Product(s) and Yield(s) (%), % ee	Refs.
C₆				
		4 Å MS, toluene, −78° to rt, 18 h	(60)	603
		4 Å MS, toluene, −78°	(—), 86:14 d.r., 97:3 anti:syn	549
	B[(−)-Ipc]₂	THF, ether, −78°, 3 h	(31), 83, >96:4 d.r.	634
	BEt₂	⟍⟍⟍OCOPh, BEt₃, Pd(PPh₃)₄, THF, rt, 24 h	(64), 2.6:1 d.r.	483
	"	⟍OCOPh, BEt₃, Pd(PPh₃)₄, THF, rt, 24 h; 40°, 2 h	**I** (60), 2.2:1 d.r.	483
C₇	B[(+)-Ipc]₂	—	**I** (59), 96, >99:1 d.r.	250
	B[(−)-Ipc]₂	THF, −78°, 3 h	(79), 88, 99:1 d.r.	137

266

<image: TMS-substituted allylborane>	Ether, −78°, 3 h	**I** (82), 97, >98:2 d.r.	143
<image: B(OMe)₂ crotyl> B(OMe)$_2$	Ether, −75°	**I** (50), 99:1 d.r.	43
<image: pinacol crotylboronate>	Ether, −78°	**I** (80), 94:6 d.r.	22
"	—	**I** (97), 10:1 d.r.	52
"	L.A. (10 mol%), toluene, −78°, 4 h	**I** L.A. d.r. AlCl$_3$ (92) >99:1 Sc(OTf)$_3$ (94) >99:1	25
"	AlCl$_3$/(S)-BINOL (10 mol%), toluene, −78°, 4 h	<image: OH product> (92), 39, >99:1	25
"	Et$_2$AlCl/(S)-BINOL (10 mol%), toluene, −78°, 6 h	**I** (40), 51, 99:1	25
<image: tartrate-derived crotylboronate>	<image: (EtO₂C)₂ / OAc fragment>, Pd$_2$(dba)$_3$, DMSO, toluene, 20°, 21 h	**I** (76), 33	349

TABLE 3. ADDITION OF E-CROTYL REAGENTS (*Continued*)

Carbonyl Substrate	Allylborane Reagent	Conditions	Product(s) and Yield(s) (%), % ee	Refs.
C7 (benzaldehyde)	(pinanediol crotylboronate w/ Ph)	Sc(OTf)₃ (10 mol%), CH₂Cl₂, −78°, 24 h	(60), 97, >49:1 d.r.	27, 28
	(tartrate crotylboronate, OPr-*i*)	4 Å MS, THF, −78°, 3 h	(91), 67, >99:1 d.r.	519, 37
	"	Co(CO)₃, 4 Å MS, toluene, −78°	**I** (90), 92, 98:2 d.r.	128
	(bis-Ts diamine boron, Ph)	CH₂Cl₂, −78°	**I** (74-82), 91-95	50
	BR₃⁻Li⁺	Li⁺ BR₃, ether, −70°	**I** R — d.r. Et (—) 90:10-91.5:8.5 *n*-Bu (—) 90:10-91.5:8.5	577
	B[(−)-Ipc]₂	THF, ether, −78°, 4.5 h	(82), >99:1 d.r.	635
R = BocO	BF₃⁻K⁺	BF₃•OEt₂ (2 equiv), CH₂Cl₂, −78°, 15 min	**I** R — d.r. H (94) >98:2 MeO (91) 97:3 O₂N (96) >98:2	98, 99
R = H, MeO, O₂N				

	"	$BF_3{\bullet}OEt_2$ (5 mol%), CH_2Cl_2, rt, 3-6 h	**I** R = H (93) d.r. >98:2; MeO (95) 97:3; O_2N (94) >98:2	99
R = H, MeO		*n*-Bu_4NI (10 mol%), CH_2Cl_2, H_2O, rt, 15 min	**I** R = H (95) d.r. >98:2; MeO (98) >98:2	100
R = O_2N	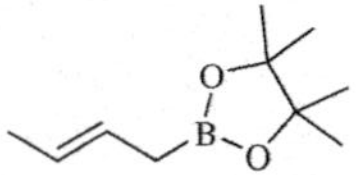	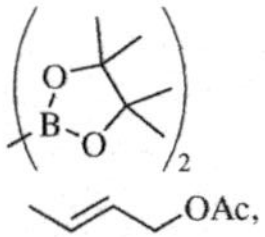 $Pd_2(dba)_3$, DMSO, 60°, 21 h	**I** (81), >99:1 d.r.	636, 349
	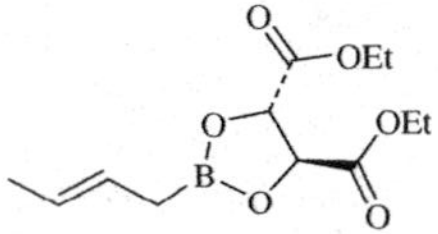	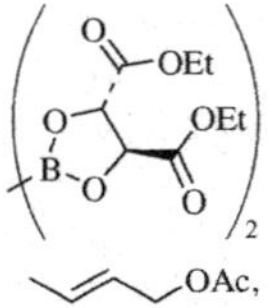 $Pd_2(dba)_3$, DMSO, toluene, 20°, 21 h	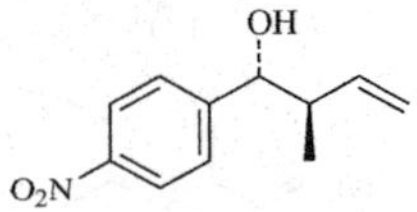 (67), 50	349
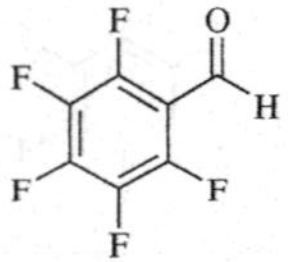		Pentane/ether, –100°	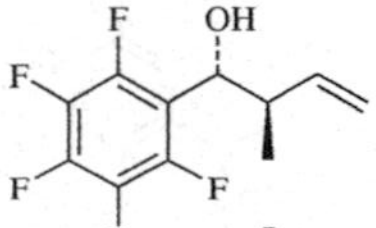 **I** (72), 92, >19:1 d.r.	249
	"	—	**I** (84), 97, >99:1 d.r.	250

TABLE 3. ADDITION OF E-CROTYL REAGENTS (*Continued*)

Carbonyl Substrate	Allylborane Reagent	Conditions	Product(s) and Yield(s) (%), % ee	Refs.
C₇	B[(−)-Ipc]₂	THF, −78°, 3 h	(72), 92, >95:5 d.r.	491
	B[(+)-Ipc]₂	THF, −85°, 18 h	(68)	348
		4 Å MS, toluene, −95°, 3 h	**I** (100), 91, >99:1 d.r.	519, 37
	"	4 Å MS, toluene, rt, 1 h	**I** (96), 46, >99:1 d.r.	37
	"	4 Å MS, toluene, −78°, 3 h	**I** Solvent / d.r. toluene (85), 87 / 96:4 ether (87), 77 / >99:1 THF (88), 70 / >99:1 CH₂Cl₂ (87), 62 / >99:1	37
	0.4 M	4 Å MS, toluene, −78°, 6 h	**I** (85), 94	132
		THF, −78°	(74-82), 91-95	50

270

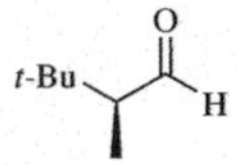

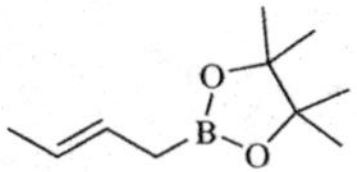

521   (65), >97:3 d.r.   Neat, rt, 4 d

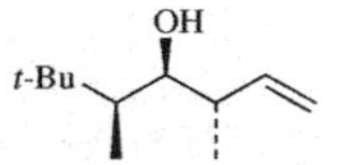

554   (72), 98:2 d.r.   THF, −78°, 4 h

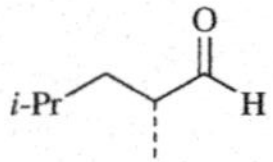

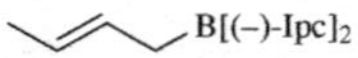

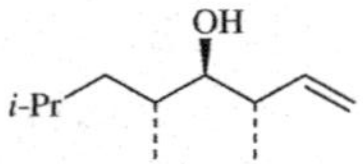

554   (56)   THF, −78°, 4 h

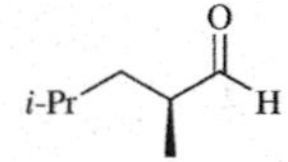

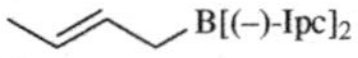

637   (70)   4 Å MS, toluene, −78°

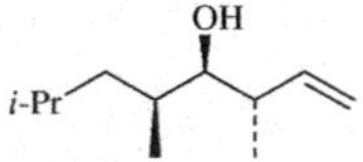

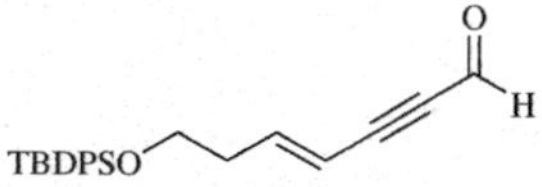

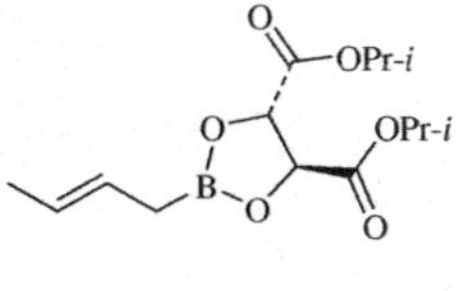

638   (54)   THF, −78°, 6 h

639   (75), 94:6 d.r.   4 Å MS, toluene, −78°

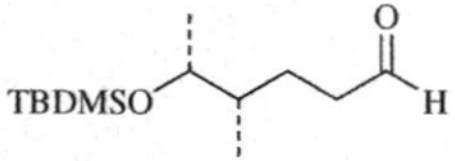

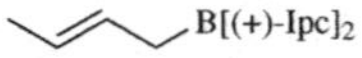

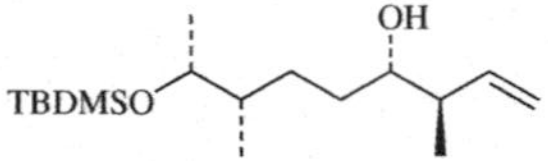

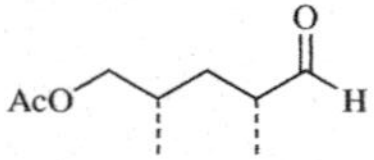

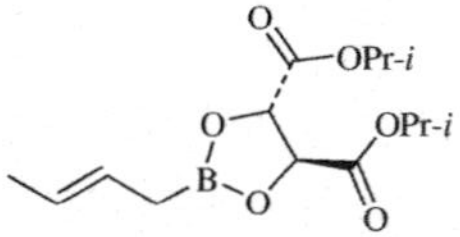

640   4 Å MS, toluene, −78°

R		d.r.
Bz	(71)	50:1
Ac	(—)	2:1
H	(—)	1:1

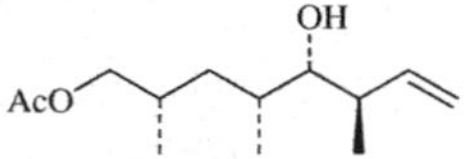

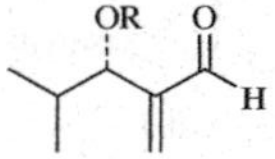

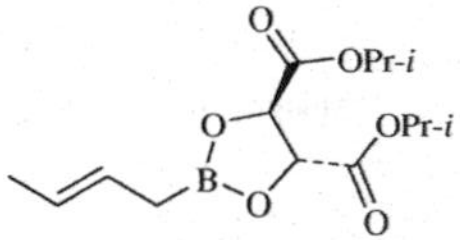

t-Bu, i-Pr, TBDPSO, TBDMSO, AcO, OH, OR

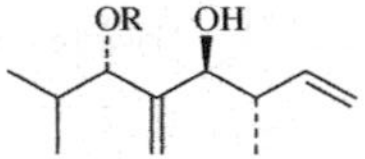

B[(−)-Ipc]₂   B[(+)-Ipc]₂   OPr-i

TABLE 3. ADDITION OF E-CROTYL REAGENTS (*Continued*)

Carbonyl Substrate	Allylborane Reagent	Conditions	Product(s) and Yield(s) (%), % ee	Refs.
C$_7$	B[(–)-Ipc]$_2$	BF$_3$•OEt$_2$, THF, –78°, 12 h	**I** (70), 70:30 d.r.	641
	B[(+)-Ipc]$_2$	BF$_3$•OEt$_2$, THF, –78°, 12 h	**I** (75), 98:2 d.r.	641
		4 Å MS, toluene, –78°, 18 h	**I** (70), 98:2 d.r.	641
		4 Å MS, toluene, –78°, 18 h	(85), 90:10 d.r.	641
			I (45), 85:15 d.r.	641
	B[(–)-Ipc]$_2$	BF$_3$•OEt$_2$, THF, –78°, 12 h	**I** (72), 98:2 d.r.	641
	B[(+)-Ipc]$_2$	BF$_3$•OEt$_2$, THF, –78°, 12 h	**I** (78), 98:2 d.r.	641
		4 Å MS, toluene, –78°, 18 h		641

272

Reagent substrate	Reagent (boronate)	Conditions	Product (yield)	Ref.
O=CH, OTBDMS, OMe (aldehyde)	allyl boronate, OPr-*i*, OPr-*i*	4 Å MS, toluene, −78°, 18 h	OTBDMS, OH, OMe **I** (52), 98:2 d.r.	641
(same aldehyde)	allyl boronate, OPr-*i*, OPr-*i*	Toluene, −75°	OH (78)	642
TESO, TBDPSO, O=CH	allyl boronate, N(CH₂CF₃)₂ diamide	4 Å MS, toluene, −78°	TESO, OH, TBDPSO (—), >9:1 d.r.	553
TBDMSO, TBDPSO, O=CH	allyl boronate, OPr-*i*, OPr-*i*	4 Å MS, toluene, −78°	TBDMSO, OH, TBDPSO (72), >98:2 d.r.	364
OBn, TBDPSO, O=CH	allyl boronate, OPr-*i*, OPr-*i*	toluene, −78°, 1.5 h	OBn, OH, TBDPSO (90)	649
acetonide, TBDMSO aldehyde	B[(−)-Ipc]₂	BF₃•OEt₂, THF, −78°	OH, acetonide, TBDMSO (67), >10:1 d.r.	464
OTBDMS, HO, OH (pyranose)	pinacol allyl boronate	4 Å MS, 40°, 25 h; 100°, 3.5 d	OH, OH, O-TBDMS (53)	643

273

TABLE 3. ADDITION OF E-CROTYL REAGENTS (*Continued*)

Carbonyl Substrate	Allylborane Reagent	Conditions	Product(s) and Yield(s) (%), % ee	Refs.
C8				
$n\text{-}C_5H_{11}$		Sc(OTf)$_3$ (10 mol%), CH$_2$Cl$_2$, −78°, 24 h	(60), 97, >49:1 d.r.	28
$n\text{-}C_7H_{15}$	$B[(+)\text{-Ipc}]_2$	THF, −78° to rt	(86), >95, >97.5:2.5 d.r.	322
	$BF_3^-K^+$	BF$_3$•OEt$_2$ (2 equiv), CH$_2$Cl$_2$, −78°, 15 min	**I** (84), >98:2 d.r.	98, 99
	"	BF$_3$•OEt$_2$ (5 mol%), CH$_2$Cl$_2$, rt, 3-6 h	**I** (85), >98:2 d.r.	99
	"	n-Bu$_4$NI (10 mol%), CH$_2$Cl$_2$:H$_2$O, rt, 15 min	**I** (96), >98:2 d.r.	100
		4 Å MS, toluene, −78°, 4 h	(72), 95:5 d.r.	644, 645
		(R,R)-i-Pr-DuPHOS, CuF, La(OPr-i)$_3$, DMF, −40°, 1 h	(73), 90, 70:30 d.r.	164
		(R,R)-i-Pr-DuPHOS, CuF, La(OPr-i)$_3$, DMF, −40°, 3 h	(80), 93, 73:27 d.r.	164

Substrate	Reagent	Conditions	Product	Refs.
(aldehyde with Ph, TBDMSO)	(allyl) BF₃⁻K⁺	n-Bu₄NI (10 mol%), CH₂Cl₂, H₂O, rt, 30 min	OH (99), 90:10 d.r.; TBDMSO **I** (82), 90:10 d.r.	276
"	"	BF₃•OEt₂ (5 mol%), CH₂Cl₂, rt, 3–6 h	**I** (82), 90:10 d.r.	99
PMBO, BnO aldehyde	B[(+)-Ipc]₂	THF, −78°	PMBO, OH (—)	543
PMBO, BnO aldehyde	B[(−)-Ipc]₂	THF, −78°	PMBO, OH (—), 1:1 d.r.	543
PMBO, BnO aldehyde	B[(+)-Ipc]₂	THF, −78°	PMBO, OH (—)	543
I	B[(−)-Ipc]₂	THF, −78°	**I** (—)	543
TIPSO aldehyde	B[(+)-Ipc]₂	Ether, −78°, 3 h	TIPSO, OH (51), 9:1 d.r.	646
TIPSO aldehyde	B[(−)-Ipc]₂	Ether, −78°, 3 h	TIPSO, OH (56), 9:1 d.r.	646
OTBDMS, TBDMSO aldehyde	B[(+)-Ipc]₂	THF, −78°	OTBDMS, OH (85), >97.5:2.5 d.r.	590

275

TABLE 3. ADDITION OF E-CROTYL REAGENTS (*Continued*)

Carbonyl Substrate	Allylborane Reagent	Conditions	Product(s) and Yield(s) (%), % ee	Refs.
C₈				
		4 Å MS, toluene, −78°	(78), >99:1 d.r.	619
		4 Å MS, toluene, −78°	(—), 2.5:1	647
		4 Å MS, toluene, −78° to rt, 15 h	(93), >99:1 d.r.	648
		Et₃N, CH₂Cl₂, −10° to rt	(96), 99:1 d.r.	106
		4 Å MS, toluene, −78°	(—), 9:1 d.r.	559

C₉ replaced with LaTeX: C_9

Reference	Conditions	Product (yield), d.r.
559	4 Å MS, toluene, −78°	(—), 1.2:1 d.r.
374	THF, −70°, 6 h	(67), >92, >96:4 d.r.
292	Ether, −78°, 5 h	(69), 89, >99:1 d.r.
50	THF, −78°	(74–82), 91–95
100	n-Bu$_4$NI (10 mol%), CH$_2$Cl$_2$, H$_2$O, rt, 15 min	(99), >98:2 d.r.
650, 651	THF, ether, −78°, 15 h	(34), 4:1 d.r.
197	—	(83), 80:20 d.r.
197	—	(90), 68:32 d.r.
519, 37	4 Å MS, toluene, −78°, 3 h	(89), 86, 99:1 d.r.
27, 28	Sc(OTf)$_3$ (10 mol%), CH$_2$Cl$_2$, −78°, 24 h	(71), 96, >49:1 d.r.

TABLE 3. ADDITION OF E-CROTYL REAGENTS (*Continued*)

Carbonyl Substrate	Allylborane Reagent	Conditions	Product(s) and Yield(s) (%), % ee	Refs.
C₉				
		Neat, rt, 4 d	(53), 75:25 d.r.	521
		4 Å MS, toluene, −78°, 2 h	(70), 97:3 d.r.	526
		4 Å MS, toluene, −78°, 2 h	(70), 95.5:4.5 d.r.	526
	BF₃⁻K⁺	BF₃•OEt₂ (5 mol%), CH₂Cl₂, rt, 3–6 h	(87), 75:25 d.r.	99
	"	n-Bu₄NI (0.1 eq), CH₂Cl₂/H₂O, rt, 30 min	(96), 75:25 d.r.	276
C₁₀				
		Cr(CO)₃, 4 Å MS, toluene, −78°	(85-95), 96, 97:3	128
		4 Å MS, toluene, −78°	(55-70), 69	130

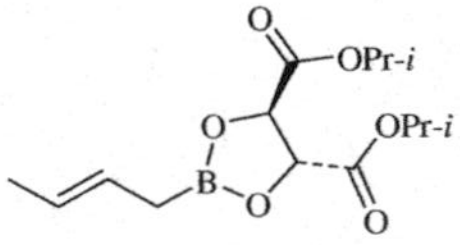

n-C$_6$H$_{13}$ Cr(CO)$_3$, 4 Å MS, toluene, −78°   (55-70), 28   130

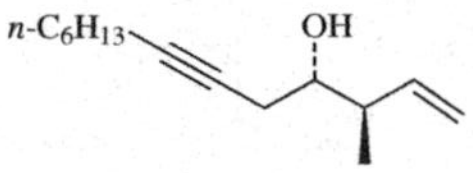

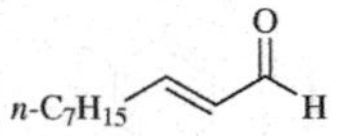

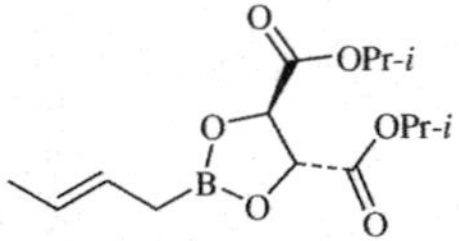

n-C$_7$H$_{15}$ 4 Å MS, toluene, −78°, 4 h   (91), 74, >99:1 d.r.   519, 37

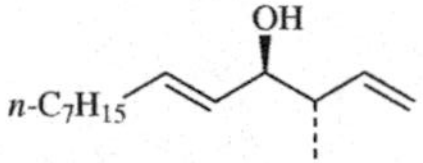

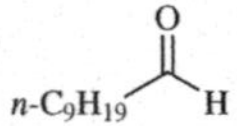

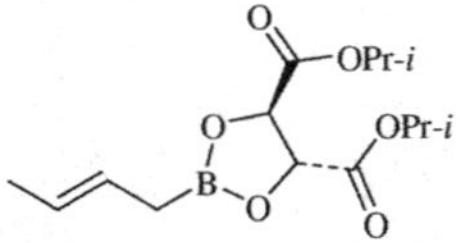

n-C$_9$H$_{19}$ 4 Å MS, toluene, −78°, 3 h   (90), 88, >99:1 d.r.   519, 37

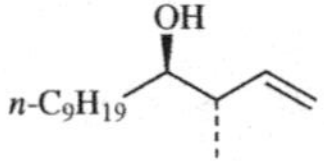

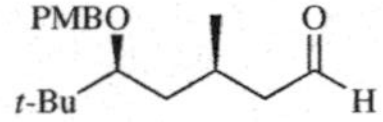

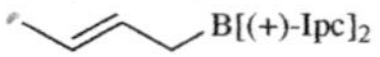

PMBO, t-Bu   THF, −100° to −78°, 12 h   (53), >97.5:2.5 d.r.   313

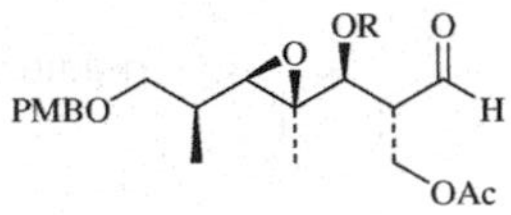

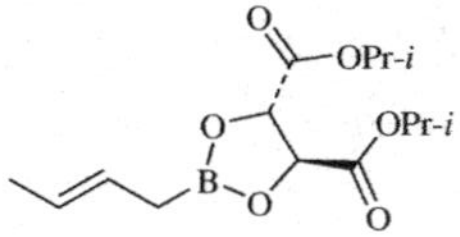

PMBO (OR epoxide)   4 Å MS, toluene, −78°   (67)   R / TBDMS   652

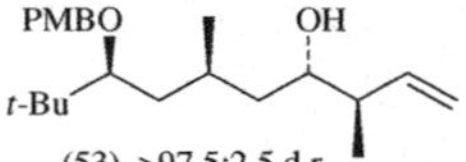

DEIPSO, OAc, PMBO   4 Å MS, toluene, −78°   (97)   653, 196

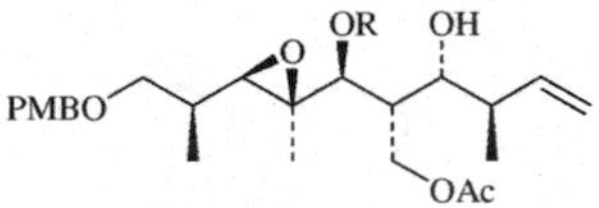

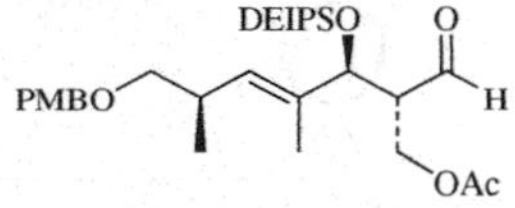

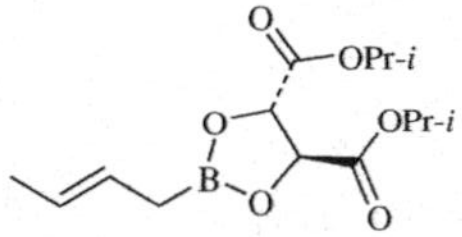

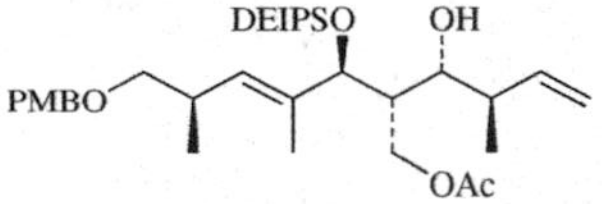

OPr-i
OPr-i
B[(+)-Ipc]$_2$

279

TABLE 3. ADDITION OF E-CROTYL REAGENTS (Continued)

Carbonyl Substrate	Allylborane Reagent	Conditions	Product(s) and Yield(s) (%), % ee	Refs.
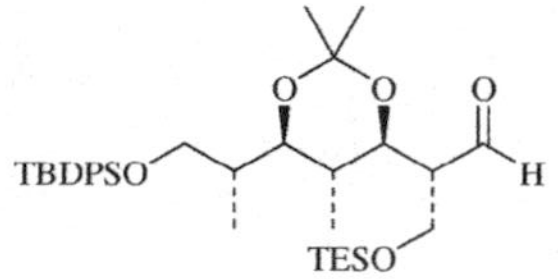	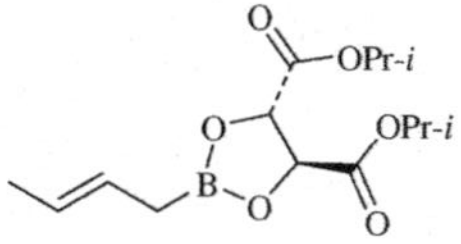	4 Å MS, toluene, −78°	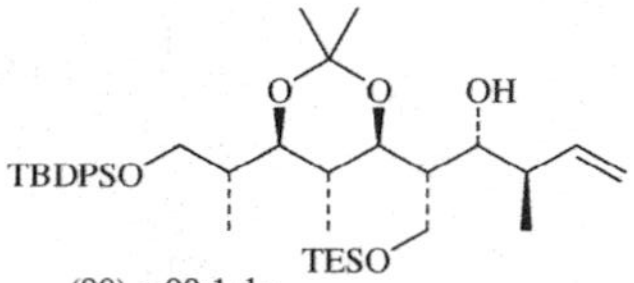 (90), >99:1 d.r.	654
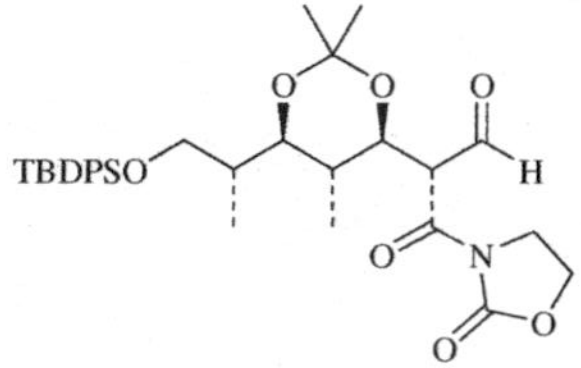	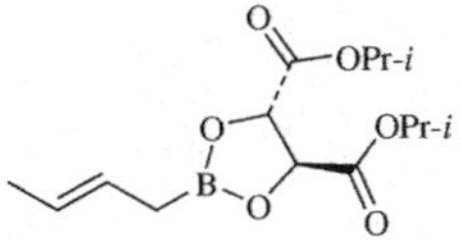	4 Å MS, toluene, −78°	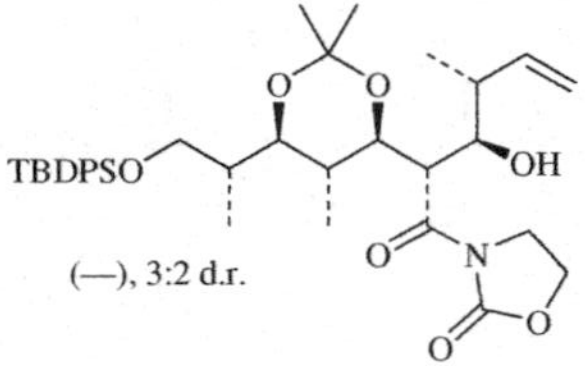 (—), 3:2 d.r.	654
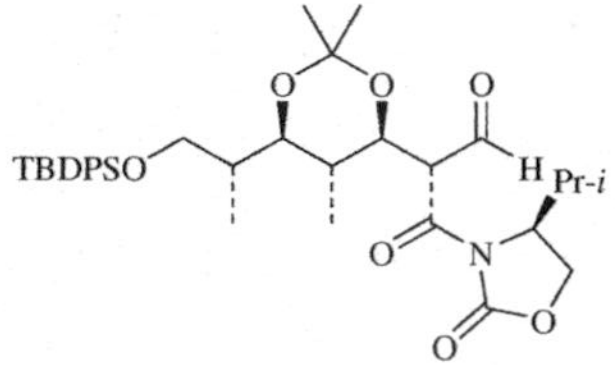	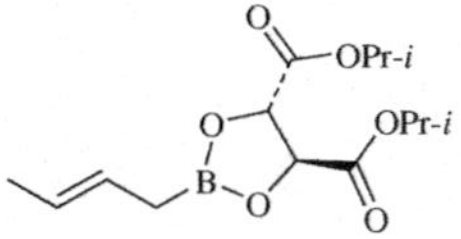	4 Å MS, toluene, −78°	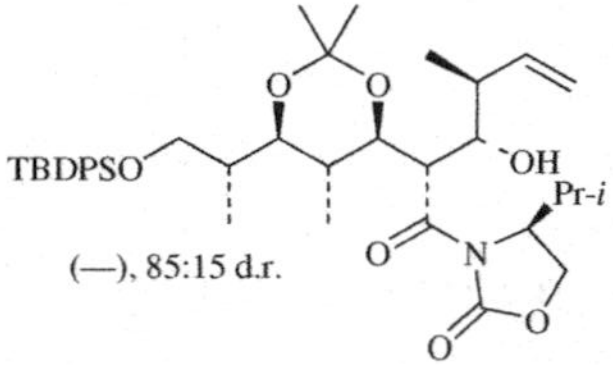 (—), 85:15 d.r.	654
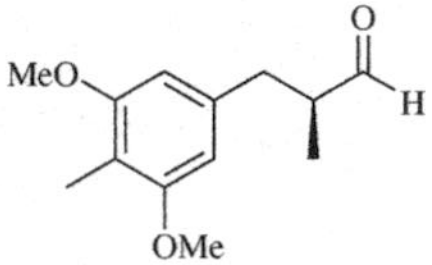	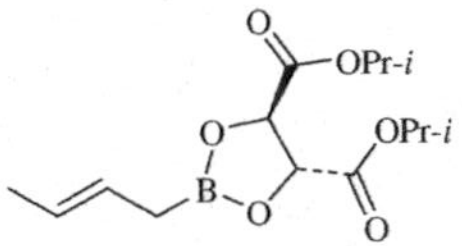	4 Å MS, toluene, −78°	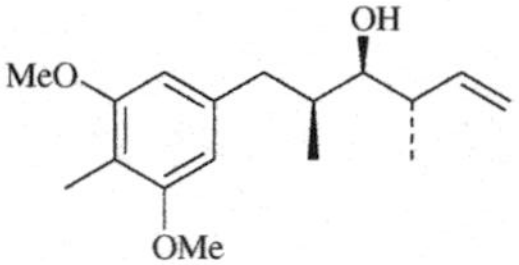 (71)	655
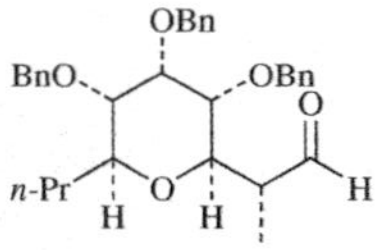	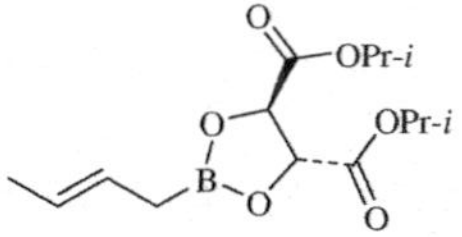	4 Å MS, toluene, −78°	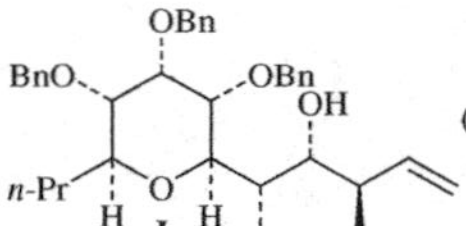 (80), 2.2:1 d.r.	559

C_{10}

C_{11}

280

C_{12}

Substrate	Reagent	Conditions	Product	Refs.
(OBn, OBn, BnO, n-Pr sugar aldehyde)	$B[(-)\text{-}Ipc]_2$ (tartrate-derived boronate, OPr-i)	4 Å MS, toluene, –78°	**I** (80), >99:1 d.r.	559
	(tartrate-derived boronate, OPr-i)	4 Å MS, toluene, –78°	(80), >99:1 d.r.	559
	(tartrate-derived boronate, OPr-i)	4 Å MS, toluene, –78°	**I** (80), >99:1 d.r.	559
(PMBO, oxazole aldehyde)	$B[(-)\text{-}Ipc]_2$	THF	(53), 6:1 d.r.	374
(Ph aldehyde)	$B[(+)\text{-}Ipc]_2$	THF/ether, –78°, 15 h	(79), 7:1 d.r.	650, 651
	"	—	**I** (80)	656
(Cy aldehyde)	$B[(+)\text{-}Ipc]_2$	THF/ether, –78°, 15 h	(65), 7:1 d.r.	650, 651
(MeO, MeO aldehyde)	$B[(+)\text{-}Ipc]_2$	THF/ether, –78°, 15 h	(52), 5:1 d.r.	650, 651

TABLE 3. ADDITION OF E-CROTYL REAGENTS (*Continued*)

Carbonyl Substrate	Allylborane Reagent	Conditions	Product(s) and Yield(s) (%), % ee	Refs.
C$_{12}$				
		Toluene, −78°	(82)	334
	B[(−)-Ipc]$_2$	—	(71)	561
C$_{13}$				
		4 Å MS, toluene, −78°	(82), >98.5:1.5 d.r.	654
	B[(+)-Ipc]$_2$	THF, −78° to rt, 1 h	(77), 14.1:1 d.r.	445
		—	(77)	657, 658

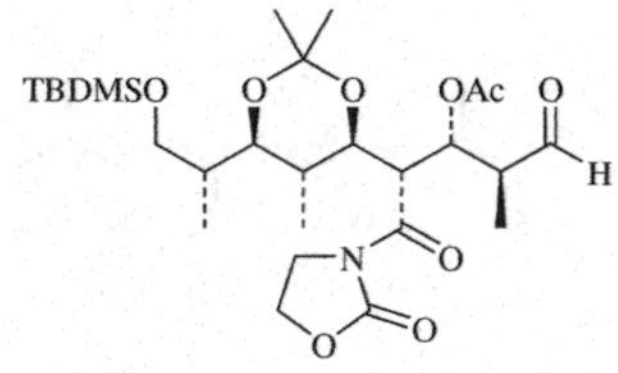

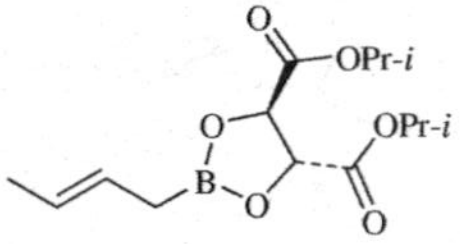

C₁₄ → C_{14}

260

(90), 88, >99:1 d.r.

4 Å MS, toluene, −78°, 3 h

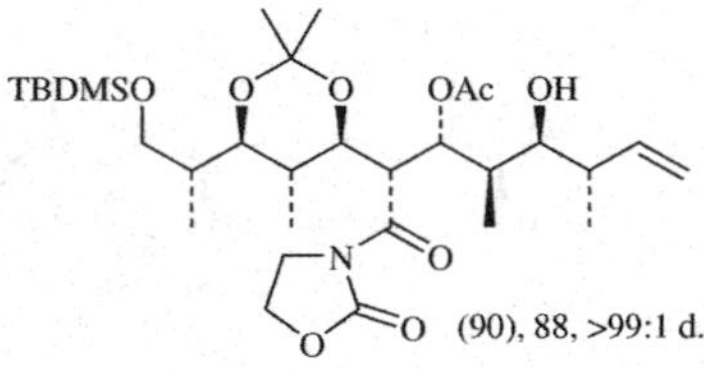

(69), 7:1 d.r. 656

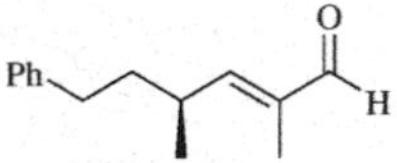

THF, −78°, 15 h

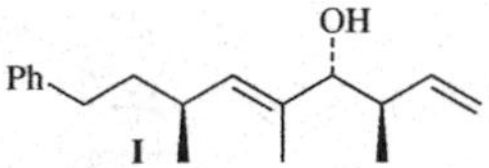

I (80), >99:1 d.r.

659

THF, ether, −78°, 15 h

(60), 7:1 d.r. 660

—

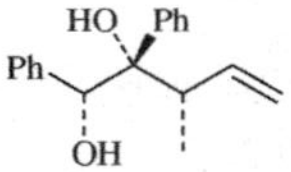

(83), 99:1 d.r., >99:1 syn:anti

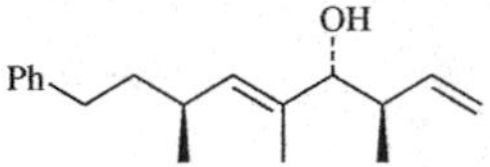

109

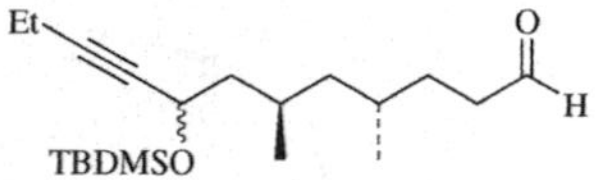

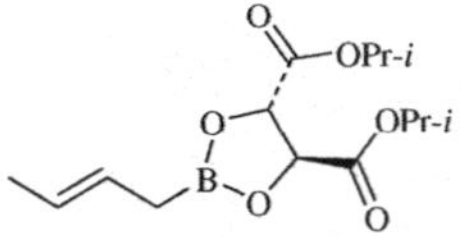

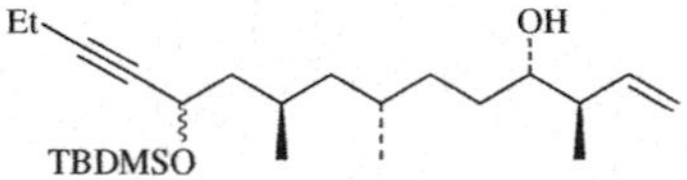

Et₃N, CH₂Cl₂, 0-40°, 24 h

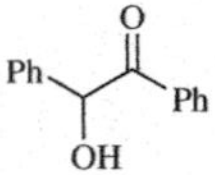

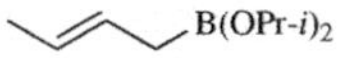

(—), 5:1 d.r. 629

—

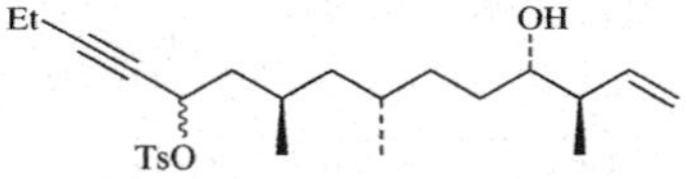

(61) 629, 630

Ether, THF, −78°

OPr-i OPr-i

B[(+)-Ipc]₂

"

B[(−)-Ipc]₂

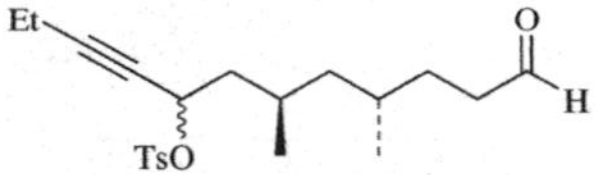

B(OPr-i)₂

OPr-i OPr-i

B[(−)-Ipc]₂

TABLE 3. ADDITION OF E-CROTYL REAGENTS (Continued)

Carbonyl Substrate	Allylborane Reagent	Conditions	Product(s) and Yield(s) (%), % ee	Refs.
C_{14}				
	n-C_6H_{13} BF_3^- K^+	n-Bu_4NI (10 mol%), CH_2Cl_2, H_2O, rt, 30 min	(68), 75:25 d.r.	276
	B[(+)-Ipc]_2	—	I (—), 4:1 d.r.	661
	(pinacol boronate)	—	I (—), 2:1 d.r.	661
	(DIPT boronate)	—	I (—), 3:1 d.r.	661
	B[(+)-Ipc]_2	THF, –78°, 18 h	(70), 4:1 d.r.	661, 662
C_{15}				
	(DIPT boronate)	—	(65), 1.5:1 d.r.	334

				d.r.	
			R		650, 651
			1-Adm (70)	5:1	
			2-Nph (72)	8:1	

C_{16} ... B[(+)-Ipc]$_2$ THF, ether, $-78°$, 15 h ... (OTBDPS, BnO) (60) ... 533

BF$_3$•Et$_2$O, THF, $-78°$, 1 h

THF, $-78°$, 4 h ... OMe OH, OMEM, OR^2 ... 168

	R^1	R^2	
	TBDMS	TBDMS	(66)
	TBDMS	Bz	(55)

C_{17} ... OPr-i, OPr-i ... $-78°$ ... (82) ... NH$_2$, n-C$_{13}$H$_{27}$, BnO ... 663

C_{18} ... B[(+)-Ipc]$_2$ ... THF, $-78°$, 3 h ... OH, OTBDMS, i-Pr, OTBDMS ... (73) ... 607

TABLE 3. ADDITION OF E-CROTYL REAGENTS (Continued)

Carbonyl Substrate	Allylborane Reagent	Conditions	Product(s) and Yield(s) (%), % ee	Refs.
C₁₉				
	B[(−)-Ipc]₂	−78°	(83)	361
		4 Å MS, toluene, −78°, 1.5 h	(92), 4:1 d.r.	664
	B[(+)-Ipc]₂	BF₃•Et₂O, THF, −78°, 1 h	(40), 1.3:1 d.r.	533
		4 Å MS, toluene, −78°	(92), 4:1 d.r.	665

286

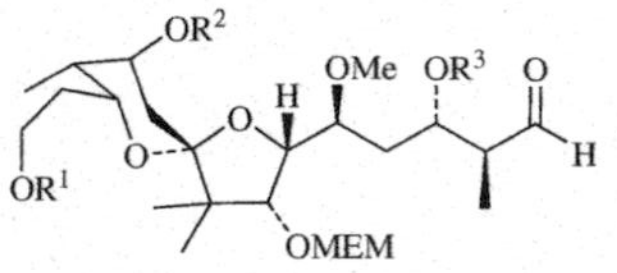

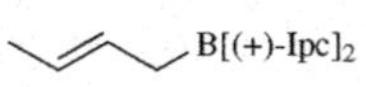

168

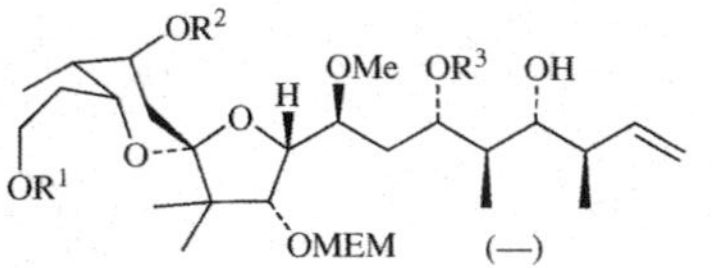

R^1	R^2	R^3	d.r.
TBDMS	TBDMS	TBDMS	48:0
Bz	TBDMS	TBDMS	31:22
Bz	Bz	TBDMS	39:35
Bz	Bz	TBDPS	30:40

THF, −78°, 4 h

(—)

630, 578, 579

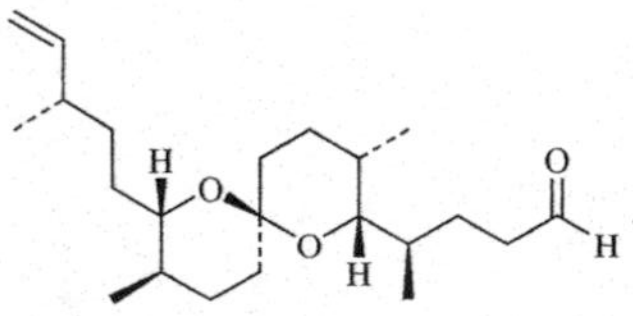

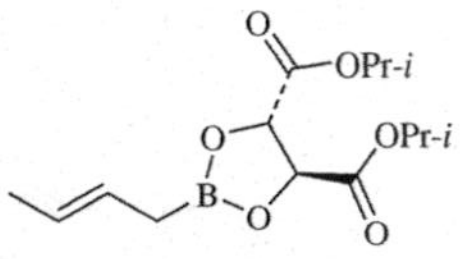

4 Å MS, toluene, −78°

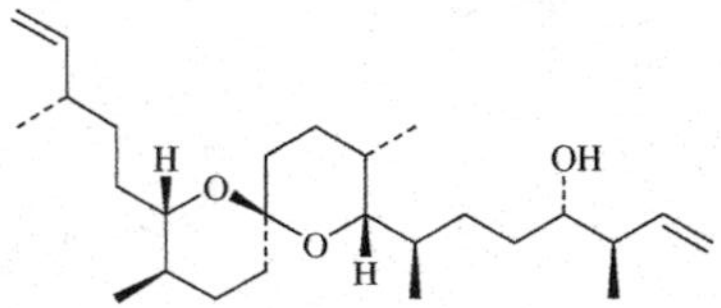

(97), 9:1 d.r.

C$_{22}$

666, 667

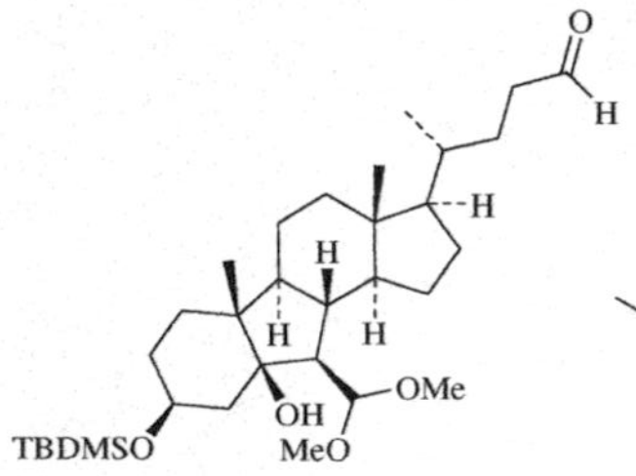

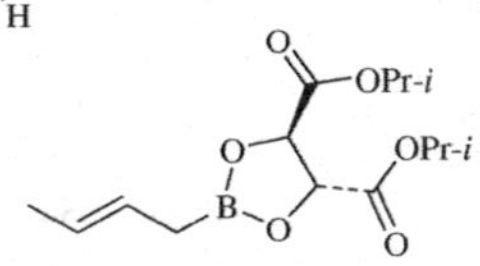

4 Å MS, toluene, −78°

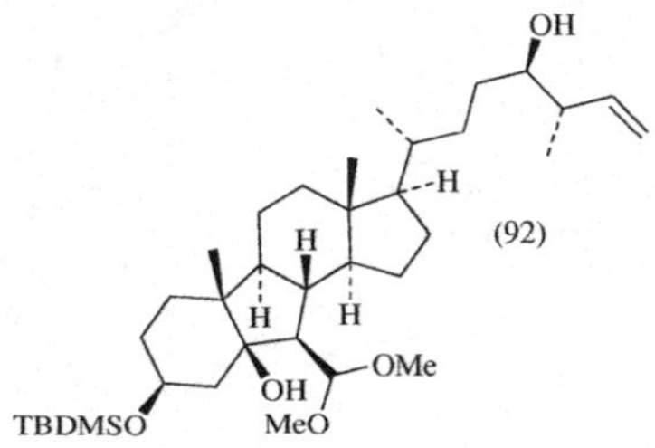

(92)

C$_{24}$

TABLE 4. ADDITION OF α-SUBSTITUTED REAGENTS

Carbonyl Substrate	Allylborane Reagent	Conditions	Product(s) and Yield(s) (%), % ee	Refs.
C₁		Toluene, 80°		236
C₂		Neat, 0°, 15 h	(63), 7:93 E:Z	146, 668
		Neat, 0°, 15 h	(—), 92, 5:95 E:Z	66, 669
		THF, −78° to rt, 18 h	(78) >99, >78:1 Z:E	65
		Neat, 0°, 15 h	(78) 7:93 E:Z	146, 668
		−78°, 2 h; rt, 24 h	(95), >95	670

For the C₁ product:

R		d.r.
i-Pr	(79)	4:1
n-Pr	(78)	6:1

Reagent	Conditions	Product	Yield	Ref.
(pinacol boronate, but-3-en-2-yl)	Ether, −70° to rt, 16 h	(OH, CH$_3$, alkene)	(71), 21:79 E:Z	148
(pinacol boronate, OMe allyl)	20°, 60 h	(OH, OMe alkene)	(64), 94, >95:5 Z:E	147
	Petroleum ether, −78° to rt, 2 d	**I** (64), 94		68
(pinacol boronate, Me–N–Ts)	4 kbar, rt, 3 d	(OH, Me–N–Ts)	(69), 97:3 d.r.	672
(pinacol boronate, Me–N–SO$_2$Me)	1. 4 kbar, rt, 3 d 2. H$^+$, MeOH	(MeO tetrahydrofuran)	(48), 96:4 d.r.	672
(pinacol boronate, OPh)	Petroleum ether, rt, 24 h	(OH, OPh alkene)	(54), 5:95 E:Z	68
(pinacol boronate, Cl)	CH$_2$Cl$_2$, 0° to rt, 15 h	(OH, Cl alkene)	(53-64), 95:5 d.r., 5:95 E:Z	67

TABLE 4. ADDITION OF α-SUBSTITUTED REAGENTS (*Continued*)

Carbonyl Substrate	Allylborane Reagent	Conditions	Product(s) and Yield(s) (%), % ee	Refs.
C$_2$		Petroleum ether, rt, 18 h	(53-64), 95, 95:5 d.r.	67
	"	Petroleum ether, rt, 12 h	(53-60), 92, 95:5 d.r.	83
		CH$_2$Cl$_2$, 0° to rt, 15 h	(70) 97:3 d.r., 85:12 E:Z **I**	67
		CH$_2$Cl$_2$, 0° to rt, 15 h	**I** (70), 73, 97:3 d.r.	67
		Neat, rt, 5-8 d	(79), 96:4 d.r. **I**	673
		Neat, rt, 5-8 d	**I** (73), 97:3 d.r	673
		Neat, rt, 5-8 d	(70), 87, 97:3 d.r.	673

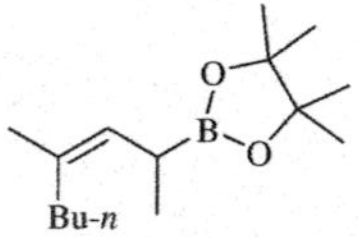 Neat, rt, 5-8 d	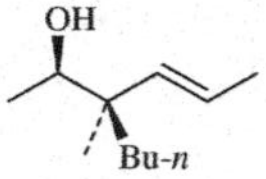 (69), 96:4 d.r	673
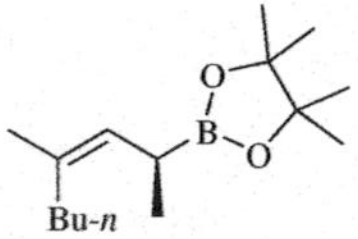 Neat, rt, 5-8 d	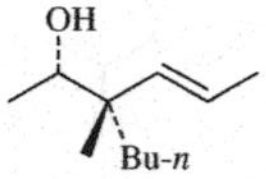 (70), 93, 96:4 d.r.	673
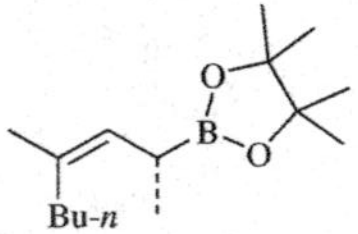 Neat, rt, 5-8 d	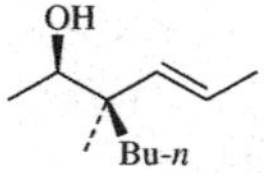 (53), 91, 97:3 d.r.	673
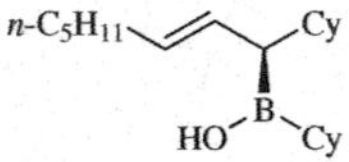 THF, −15° to rt	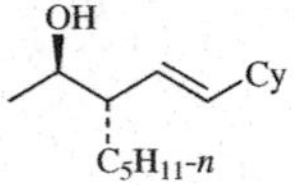 (76), 50, 96:4 d.r.	674
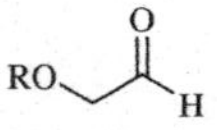 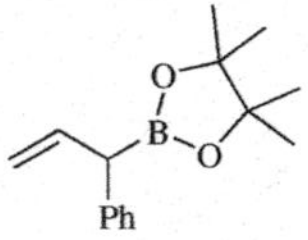 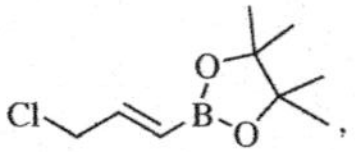 PhMgCl, −15°, 45 min; rt, 24 h	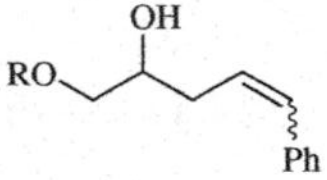 (35), 2:98 E:Z	87
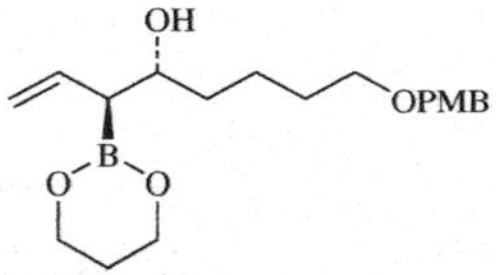 —	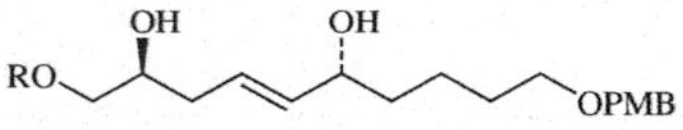 (73), 94	202

R = Bn

R = TBDMS

291

TABLE 4. ADDITION OF α-SUBSTITUTED REAGENTS (Continued)

Carbonyl Substrate	Allylborane Reagent	Conditions	Product(s) and Yield(s) (%), % ee	Refs.
C₃		Neat, 0°, 15 h	(86), 6:94 E:Z	146, 668
		Neat, 0°, 15 h	I (—), 89, 4:96 E:Z	66, 669
		THF, −78° to rt, 18 h	(81), >99, >81:1 Z:E	65
		Neat, 0°, 15 h	(82), 6:94 E:Z	146, 668
		THF, 100°, 24 h	(74), 15:85 E:Z	675
		Petroleum ether, −78° to rt	(81), 35:65 E:Z	676
		CH₂Cl₂, 0° to rt, 15 h	(47-65), 95:5 d.r., 5:95 E:Z	67

Reactant	Conditions	Product(s)	Refs.
(pinacol boronate, Cl)	Petroleum ether, rt, 12 h	**I** (47–65), 96, 95:5 d.r.	83, 67
(pinacol boronate, OMe)	Petroleum ether, −78° to rt, 2 d	(81), 90, 97.5:1.5 d.r.	68
	20°, 60 h	**I** (81), 90, >95:5 Z:E	147
(pinacol boronate, OPh)	Petroleum ether, rt, 24 h	(57), 5:95 E:Z	68
(pinacol boronate, Me–N–Ts)	4 kbar, rt, 3 d	(62), 97:3 d.r.	672
(pinacol boronate, Me–N–SO$_2$Me)	1. 4 kbar, rt, 3 d 2. H$^+$, MeOH	(81), 97:3 d.r.	672
(pinacol boronate)	CH$_2$Cl$_2$, rt, 20 h	(92), 96:4 d.r.	676

TABLE 4. ADDITION OF α-SUBSTITUTED REAGENTS (Continued)

Carbonyl Substrate	Allylborane Reagent	Conditions	Product(s) and Yield(s) (%), % ee	Refs.
C₃ (propanal)		CH₂Cl₂, 0° to rt, 15 h	**I** (81), 97:3 d.r., 91:6 E:Z	67
		CH₂Cl₂, 0° to rt, 15 h	**I** (81), 85, 97:3 d.r.	67
		Petroleum ether	(85), 99	628, 197
		Neat, rt, 10 h	R / Bn (55) / TBDMS (62)	67
		Neat, rt, 10 h	(55)	67
		Neat, rt, 10 h	(62)	67

294

Conditions	Product	Ref.
CH₂Cl₂, rt, 18 h	OH, C₅H₁₁-n (51), 1:9 E:Z	677
Neat, rt, 5-8 d	OH, Bu-n (85), 97:3 d.r.	673
Neat, rt, 5-8 d	OH, Bu-n (80), 97:3 d.r.	673
Pentane, rt, 2 h	OH, Pr-i (85) I	103
1. HBBN, pentane, rt, 24 h; 2. Ketone, rt, 2 h	I (80)	103
Petroleum ether, 8 kbar, 3 d	OH, I (—), 98:2 E:Z	102
Petroleum ether, 8 kbar, 3 d	I (—), 33:67 E:Z	102

TABLE 4. ADDITION OF α-SUBSTITUTED REAGENTS (*Continued*)

Carbonyl Substrate	Allylborane Reagent	Conditions	Product(s) and Yield(s) (%), % ee	Refs.
C₃				
		CH₂Cl₂, rt, 18 h	(64), 1:9 E:Z	677
		–78° to rt	(36–77), 83–>95	203
		(MeO)₂CH₂, –30° to rt, 18 h	(73), 92:8 d.r.	65
		(MeO)₂CH₂, –30° to rt, 18 h	(74), 97:3 d.r.	65
		(MeO)₂CH₂, –30° to rt, 18 h	(67), 92:8 d.r.	65
		(MeO)₂CH₂, –30° to rt, 18 h	(72), 96:4 d.r.	65

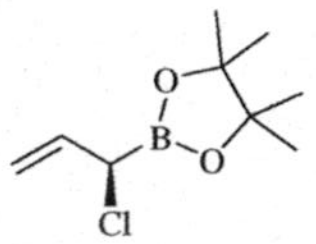

CH₂Cl₂, 15 h

I (45), 96, 96:4 d.r., 6:94 E:Z

66

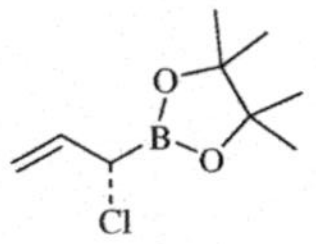

CH₂Cl₂, 15 h

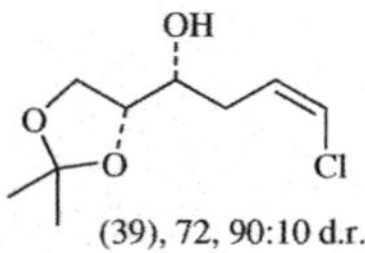

(39), 72, 90:10 d.r., 12:88 E:Z

66

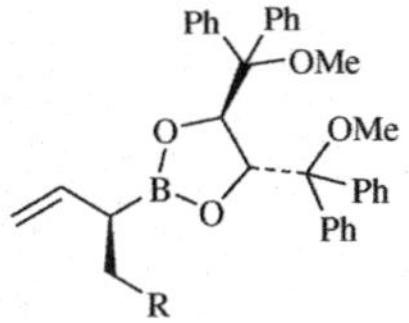

Toluene, −100° to rt

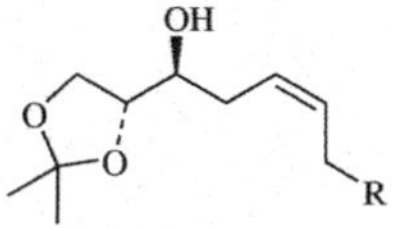

R	d.r.
CO₂Et (86)	>99:1
C(O)NMe₂ (63)	>99:1

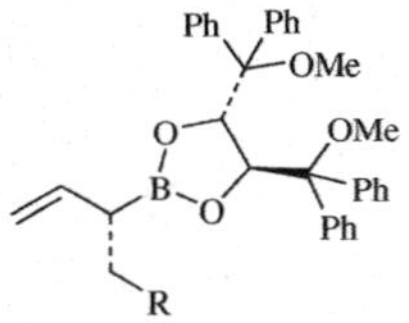

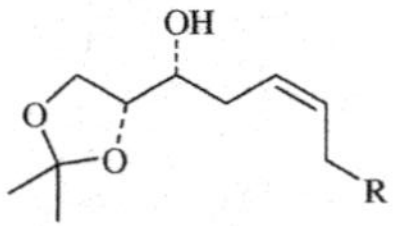

85

Toluene, −100° to rt

R	d.r.
CO₂Et (89)	69:31
C(O)NMe₂ (75)	68:32

85

TABLE 4. ADDITION OF α-SUBSTITUTED REAGENTS (*Continued*)

Carbonyl Substrate	Allylborane Reagent	Conditions	Product(s) and Yield(s) (%), % ee	Refs.

C₃

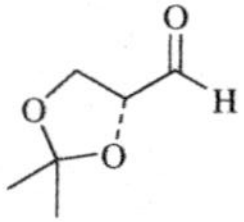

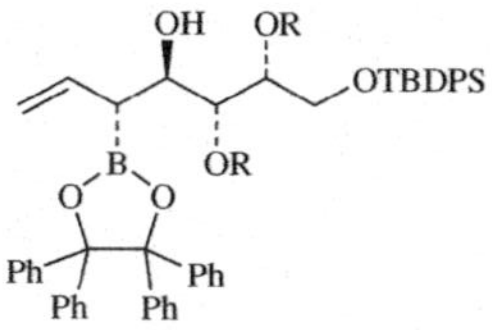

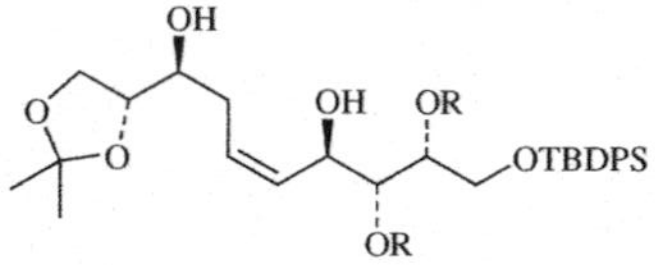

R	d.r.
PMB (57)	4:1
t-Bu (70-90)	9:1

201

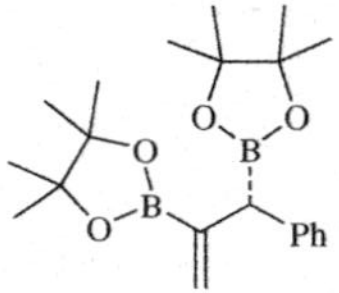

CH₂Cl₂, rt

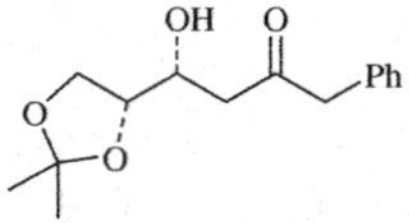

(73), 21:1 d.r.

187

1. Pd₂(dba)₃, 22°, 10 h
2. Aldehyde, 22°, 14 h
3. NaOH, H₂O₂

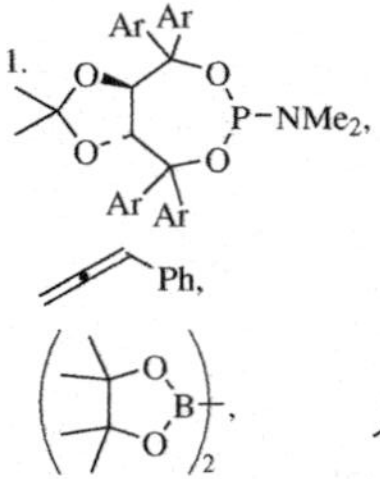

Ar = 3,5-Me₂C₆H₃

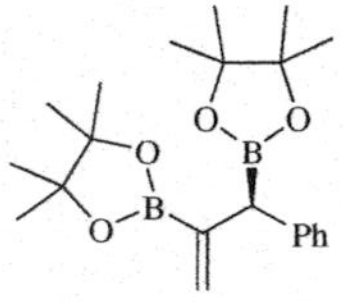

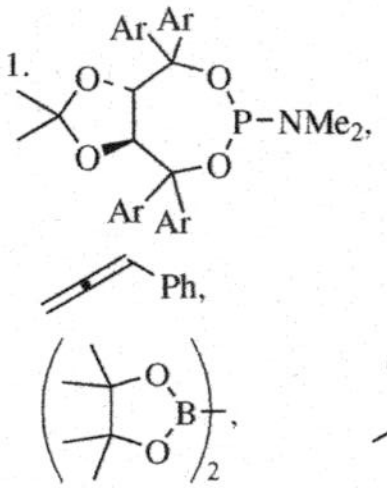

1. [P–NMe₂ phosphoramidite reagent, Ar₂, Ar = 3,5-Me₂C₆H₃]

Ph, [pinacolboron]₂

Pd₂(dba)₃, 22°, 10 h

2. Aldehyde, 22°, 14 h

3. NaOH, H₂O₂

Ar = 3,5-Me₂C₆H₃

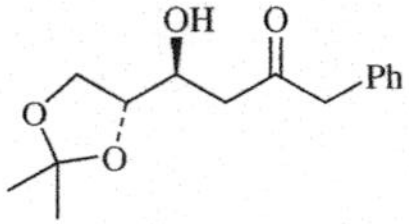

(60), >21:1 d.r.

187

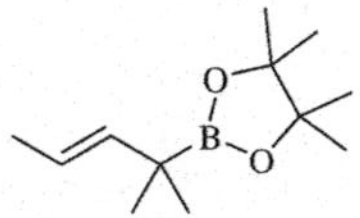

CH₂Cl₂, −78° to rt, 24-48 h

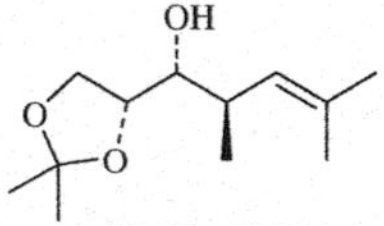

(75-85), 56:44 d.r., >99:1 anti:syn

537, 118

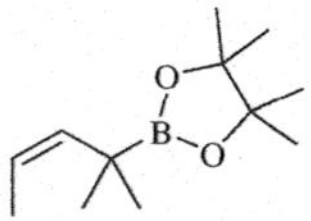

CH₂Cl₂, −78° to rt, 24-48 h

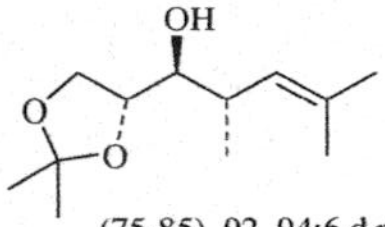

(75-85), 92, 94:6 d.r., 94:6 anti:syn

537, 118

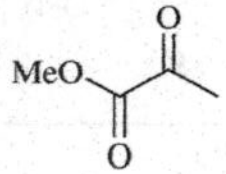

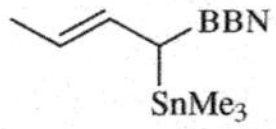

Ether, −76° to 0°, 30 min

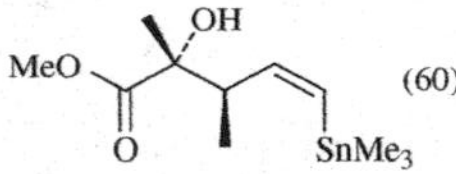

(60)

598

299

TABLE 4. ADDITION OF α-SUBSTITUTED REAGENTS (Continued)

Carbonyl Substrate	Allylborane Reagent	Conditions	Product(s) and Yield(s) (%), % ee	Refs.
C$_3$		Ether, $-76°$ to $0°$, 30 min	**I + II** (65), **I:II** = 63:37	598
C$_4$		Petroleum ether, rt, 5 d	(65), >90	678, 679
		CH$_2$Cl$_2$, rt, 18 h		677

R		E:Z
n-C$_5$H$_{11}$	(82)	1:9
Cl⌇	(80)	9:91
⌇	(75)	11:89
cyclopentyl⌇	(79)	12:88
t-Bu⌇	(62)	12:88
Ph⌇	(83)	1:9

300

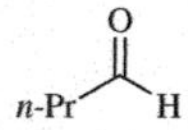

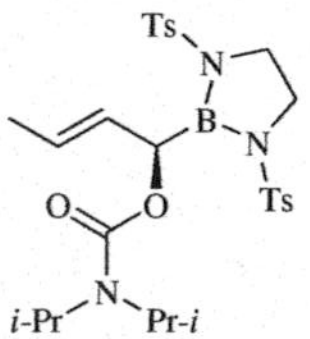

Toluene, 60°, 72 h

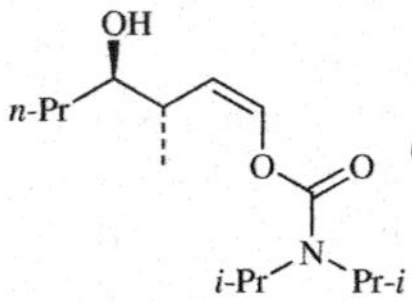

(58), 85, >95.5 Z:E 51

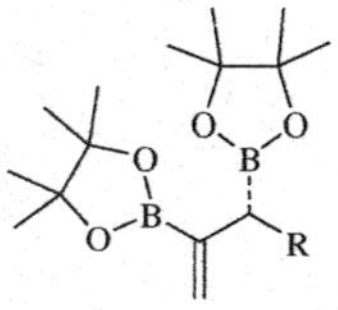

187

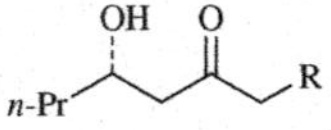

R	
Ph	(85), 94
Cy	(89), 86
$n\text{-}C_{10}H_{21}$	(88), 91

1. 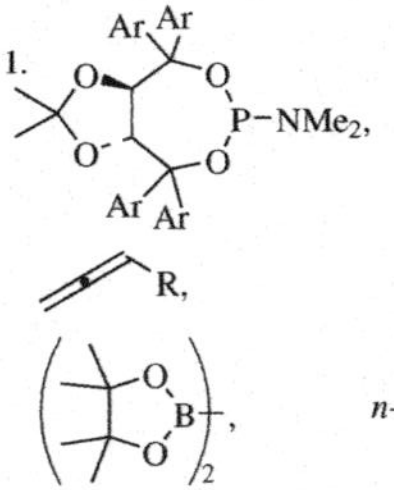 [structure] P–NMe$_2$, Ar Ar, R, (pinacolboronate)$_2$
   Pd$_2$(dba)$_3$, 22°, 10 h
2. Aldehyde, 22°, 14 h
3. NaOH, H$_2$O$_2$

Ar = 3,5-Me$_2$C$_6$H$_3$

TABLE 4. ADDITION OF α-SUBSTITUTED REAGENTS (Continued)

Carbonyl Substrate	Allylborane Reagent	Conditions	Product(s) and Yield(s) (%), % ee	Refs.
C$_4$				
i-Pr–CHO	(BCy$_2$, Br)	1. propargyl Br, Cy$_2$BH 2. Aldehyde, n-Bu$_4$NBr, 0°	OH, i-Pr, Br (46), 16:84 E:Z	680
	(pinacol boronate, Cl)	Neat, 0°, 15 h	OH, i-Pr, Cl (83), 4:96 E:Z	146, 668, 669
	(pinacol boronate, Cl)	Neat, 0°, 15 h	OH, i-Pr, Cl (—), 92, 5:95 E:Z	66
	(dioxaborolane, R, Cl)	Neat, 0°, 15 h	OH, i-Pr, Cl	146

R		E:Z
Ph	(68)	34:66
CH$_2$OMe	(78)	61:39
OCH$_2$C$_6$H$_4$Cl-p	(74)	82:18

302

Reagent	Conditions	Product		Refs.
	THF, −78° to rt, 18 h		(89), >99, >89:1 Z:E	65
	Neat, 0°, 15 h		(83), 4:96 E:Z	146, 668
	Neat, 0°, 15 h		R E:Z Et (84) 15:85 Bu-t (63) 20:80	146, 668
	Neat, 0°, 15 h		(72), 21:79 E:Z	146

TABLE 4. ADDITION OF α-SUBSTITUTED REAGENTS (*Continued*)

Carbonyl Substrate	Allylborane Reagent	Conditions	Product(s) and Yield(s) (%), % ee	Refs.
C₄				
(i-Pr-CHO)		Ether, −70° to rt, 16 h	(72), 21:79 E:Z	148
		Neat, 0°, 15 h	(55), >95:5 Z:E	146, 668
		Neat, 0°, 15 h	(94), 60:40 E:Z	146
		1. Toluene, rt, 2 d 2. I₂, 110°	(90–94)	207, 208

304

	Petroleum ether, rt, 12 h	**I** (54-84), 95	83
"	Petroleum ether, rt, 18 h	**I** (55-84), 96, 5:95 E:Z	67
	CH$_2$Cl$_2$, 0° to rt, 15 h	**I** (55-84), 95:5 d.r., 5:95 E:Z	67
	Petroleum ether, −78° to rt, 2 d	(82), 90	68
"	20°, 60 h	**I** (82), 90, >95:5 Z:E	147
	4 kbar, rt, 3 d	(75), 98:2 d.r.	672

TABLE 4. ADDITION OF α-SUBSTITUTED REAGENTS (Continued)

Carbonyl Substrate	Allylborane Reagent	Conditions	Product(s) and Yield(s) (%), % ee	Refs.
C₄				
i-Pr—CHO	(structure)	1. 4 kbar, rt, 3 d 2. H⁺, MeOH	(76), 96:4 d.r.	672
	(structure)	Toluene, 60°, 72 h	(71), 84, 6:94 E:Z	51
	(structure)	CH₂Cl₂, 0° to rt, 15 h	(58), 91, 96:4 d.r., 96:4 E:Z	67
	(structure)	CH₂Cl₂, 0° to rt, 15 h	I (58), 97:3 d.r., 91:6 E:Z	67
	(structure)	Petroleum ether, rt, 3 d	(77), 99, 80:20 d.r.	681
	(structure)	Neat, rt, 10 h	(62)	67

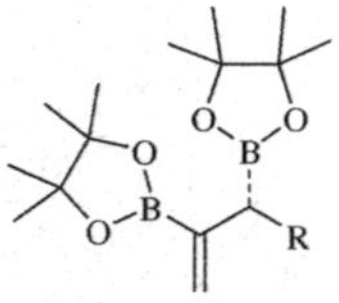

R	
Ph	(89), 95
$n\text{-}C_{10}H_{21}$	(96), 91
Cy	(83), 87

187

I

1. Ar Ar, O–O, P–NMe$_2$, Ar Ar, R, $\left(\begin{array}{c}\text{O}\\\text{O}\end{array}\text{B}\right)_2$

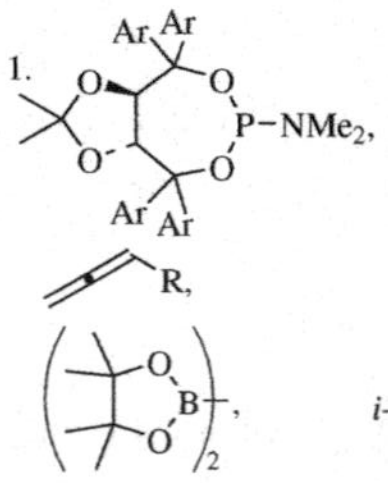

Pd$_2$(dba)$_3$, 22°, 10 h
2. Aldehyde, 22°, 14 h
3. NaOH, H$_2$O$_2$

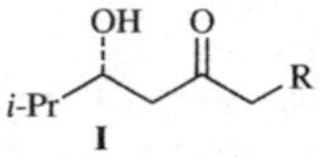

Ar = 3,5-Me$_2$C$_6$H$_3$

R	
Ph	(80), 92
$n\text{-}C_{10}H_{21}$	(68), 88

187

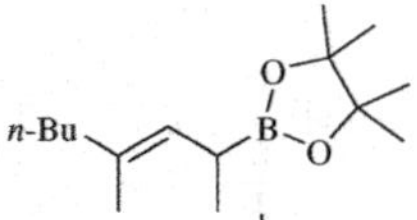

I

1. Ar Ar, O–O, P–NMe$_2$, Ar Ar, R, $\left(\begin{array}{c}\text{O}\\\text{O}\end{array}\text{B}\right)_2$

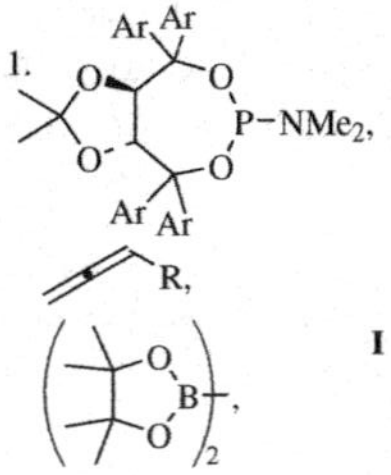

Pd$_2$(dba)$_3$, aldehyde, 22°, 14 h
2. NaOH, H$_2$O$_2$

Ar = 3,5-Me$_2$C$_6$H$_3$

Neat, rt, 5-8 d

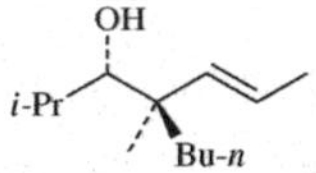

(85), 98:2 d.r.

673

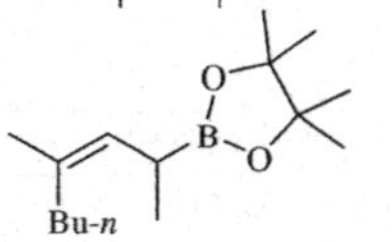

Neat, rt, 5-8 d

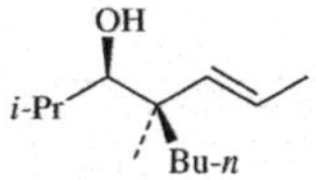

(76), >99:1 d.r.

673

TABLE 4. ADDITION OF α-SUBSTITUTED REAGENTS (*Continued*)

Carbonyl Substrate	Allylborane Reagent	Conditions	Product(s) and Yield(s) (%), % ee	Refs.
C$_4$				
		THF, −15° to 15°	Ipc-(−) (62), 85, 96:4 d.r.	674
		Petroleum ether, 8 kbar, 3 d	(—), 1:1 d.r., 98:2 E:Z	102
		Petroleum ether, 8 kbar, 3 d	(—), 72:28 d.r., 29:71 E:Z	102
		Pyr, ether, −78°, 30 min	(40)	682
C$_5$		Toluene, 60°, 48 h	(60), 85, >95:5 Z:E	51
		CH$_2$Cl$_2$, rt, 20 h	(90), 90:10 d.r.	676

308

Conditions	Product	Refs.
Petroleum ether, −78° to rt	(90), 23:77 E:Z	676
Neat, rt, 15 h	(85), 8:92 E:Z	146, 668
THF, −78° to rt, 18 h	(76), 99, 1:76 E:Z	65
Neat, rt, 10 h	(61)	67
1. 4 kbar, rt, 3 d 2. H⁺, MeOH	(87), 98:2 d.r.	672
Ether, −70° to rt, 16 h	(—), 76:24 E:Z **I**	148
Ether, −70° to rt, 16 h	**I** (—), 51:49 E:Z	148

TABLE 4. ADDITION OF α-SUBSTITUTED REAGENTS (Continued)

Carbonyl Substrate	Allylborane Reagent	Conditions	Product(s) and Yield(s) (%), % ee	Refs.
C$_5$				
		Ether, −70° to rt, 16 h	(75), 31:69 E:Z **I**	148
		Ether, −70° to rt, 16 h	**I** (—), 41:59 E:Z	148
		Toluene, 60°, 72 h	(69), 85, >95:5 Z:E	51
		Petroleum ether, 8 kbar, 3 d	(—), 49:51 d.r., 98:2 E:Z	102
		Petroleum ether, 8 kbar, 3 d	(—), >99:1 d.r., 25:75 E:Z	102
		Petroleum ether, 0° to rt, 12 h	(—), 80:20 d.r. **I**	149

Reagent	Conditions	Product	Refs.
pinacol boronate, –Cl (from 1-butenyl)	Petroleum ether, rt, 12 h	**I** (50-60), 90:10 d.r.	83
pinacol boronate, –Cl	Petroleum ether, rt, 12 h	(50-60), 98:2 d.r.	83
pinacol boronate, –OMe	Toluene, –78° to rt, 60 h	(59), 90, >99:1 d.r.	68
pinacol boronate, –OMe	Toluene, –78° to rt, 60 h	(59), 99, >99:1 d.r.	68
"	20°, 60 h	**I** (48), >95:5 d.r.	147
pinacol boronate	Toluene, 0° to rt, 18 h	(44), 90:10 d.r.	149
pinacol boronate, Me	Petroleum ether, 0° to rt, 18 h	(56), >98:2 d.r.	149

311

TABLE 4. ADDITION OF α-SUBSTITUTED REAGENTS (Continued)

Carbonyl Substrate	Allylborane Reagent	Conditions	Product(s) and Yield(s) (%), % ee	Refs.
C$_6$				
(4-methyl-2,3-pentadienal, TMS)	TMS—BBN	—	**I** (93), >98:2 d.r., >95:5 Z:E	683
	TMS—BCy$_2$	—	**I** (92), >99:1 d.r., >95:5 Z:E	683
(Et-substituted enal)	(Cy/Cy dioxaborolane allyl)	Petroleum ether, rt, 3 d	(81), 99	681
n-C$_5$H$_{11}$CHO	TMS—BCy$_2$ (R^1, R^2)	—	n-C$_5$H$_{11}$—**III**, n-C$_5$H$_{11}$—**I** + **II**, **IV**	684

R^1	R^2	Work-Up	I + II + III + IV	I:II:III:IV
Me	BBN	NaOH	(77)	94:1:4:1
Me	BBN	H$_2$SO$_4$	(80)	1:90:3:6
Me	BCy$_2$	NaOH	(78)	0:0:98:2
Me	BCy$_2$	H$_2$SO$_4$	(70)	0:0:8:92
n-Bu	BBN	NaOH	(68)	97:1:2:0
n-Bu	BBN	H$_2$SO$_4$	(65)	1:93:2:4
n-Bu	BCy$_2$	NaOH	(77)	0:0:97:3
n-Bu	BCy$_2$	H$_2$SO$_4$	(73)	0:0:9:91

	Neat, 0°, 15 h	(69), 8:92 E:Z	146, 668
	Petroleum ether, rt, 12-18 h	(60-91), 92, 95:5 d.r.	83, 67, 67
	CH$_2$Cl$_2$, 0° to rt, 15 h	**I** (60-91), 95:5 d.r., 5:95 E:Z	67
	Petroleum ether, −78° to rt, 2 d	(87), 88	68
	Petroleum ether, rt, 24 h	(49), 5:95 E:Z	68
	Neat, rt, 10 h	(61)	67
	THF, −15° to 15°	Ipc-(−) (62), 85, 96:4 d.r.	674

TABLE 4. ADDITION OF α-SUBSTITUTED REAGENTS (*Continued*)

Carbonyl Substrate	Allylborane Reagent	Conditions	Product(s) and Yield(s) (%), % ee	Refs.
C_6				
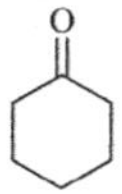	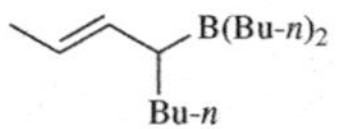	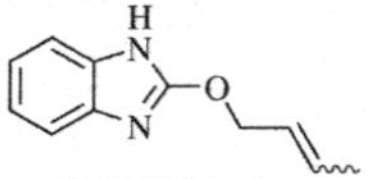 *n*-BuLi, B(Bu-*n*)$_3$, THF, –100°	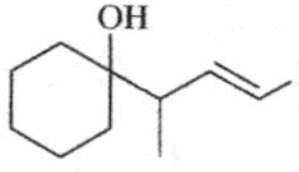 (84), >90:10 E:Z	685
	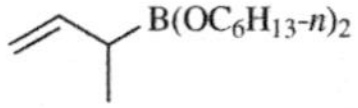	Ether, rt	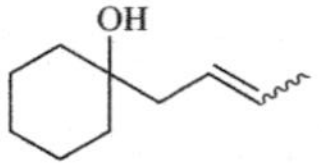 (75)	31
	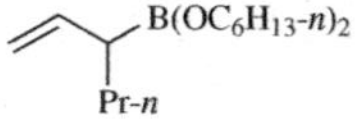	Ether, rt	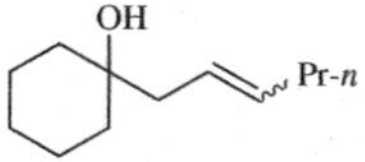 (80)	31
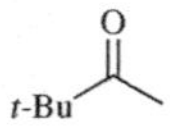	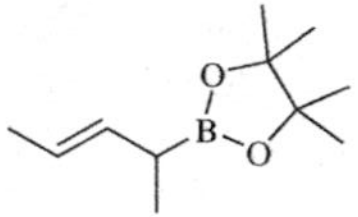	Petroleum ether, 8 kbar, 3 d	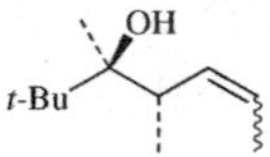 (—), >99:1 d.r., 25:75 E:Z	102
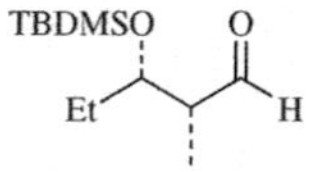	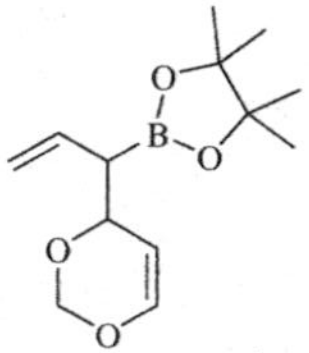	1. Toluene, rt, 2 d 2. I_2, 110°	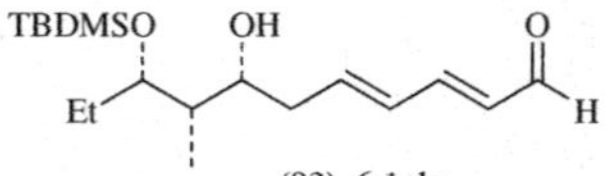 (93), 6:1 d.r.	207
	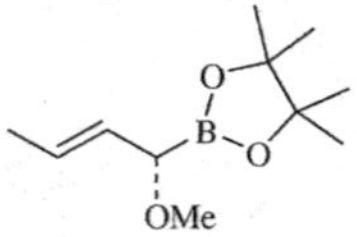	4 kbar, 60°, 12 h	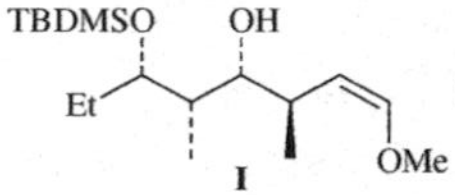 **I** (66), 88:12 d.r.	147
	"	Toluene, rt	**I** (66), >99:1 d.r.	68

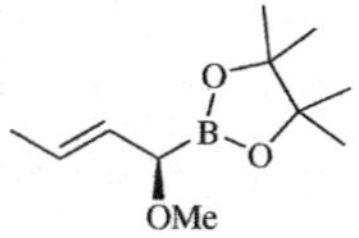

TBDMSO, OH, Et, OMe — (67), 60:6.5 d.r. — 68

Toluene, 4 kbar, 8 h

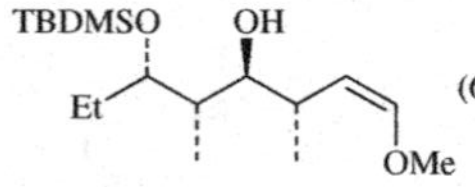

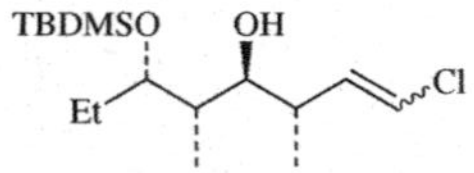

TBDMSO, OH, Et, Cl — (53), 81:19 d.r., >99:1 anti:syn — 149

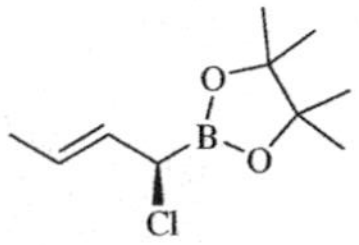

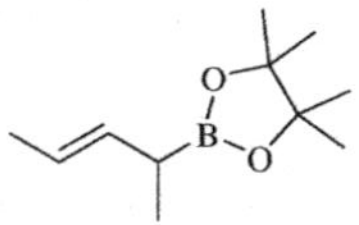

Toluene, 4 kbar, 8 h

TBDMSO, OH, Et — (94), 86:14 d.r. — 676

Petroleum ether, −78° to rt

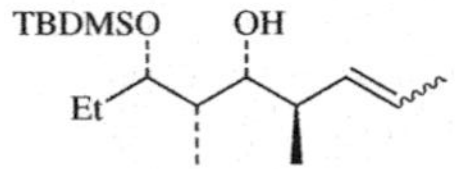

TBDMSO, OH, Et — (86) — 694, 687

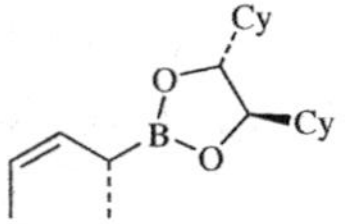

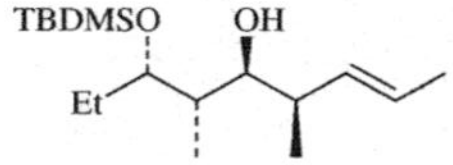

—

TBDMSO, OH, Et, Cl — (69), >99:1 d.r. — 149

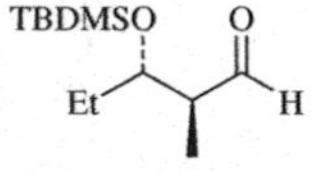

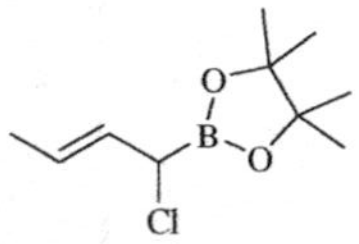

Toluene, 0° to rt, 60 h

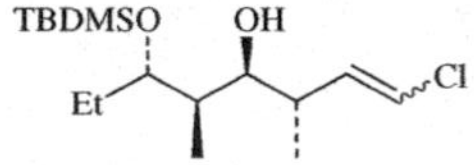

TBDMSO, OH, S — (86), 75:25 d.r. — 676

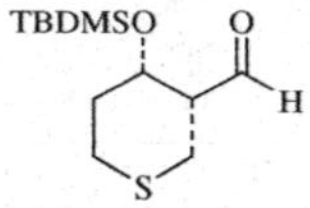

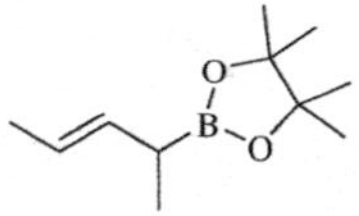

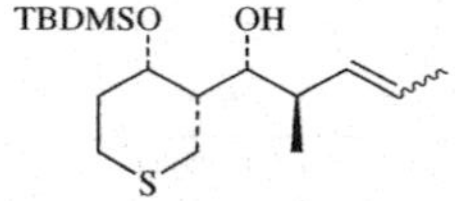

Petroleum ether, −78° to rt

TBDMSO, Et, O, H

TBDMSO, S, O, H

315

TABLE 4. ADDITION OF α-SUBSTITUTED REAGENTS (*Continued*)

Carbonyl Substrate	Allylborane Reagent	Conditions	Product(s) and Yield(s) (%), % ee	Refs.
C₆		Benzene, 80°, 3 d	(81), >96:4 d.r.	688, 681
C₇		1. propargyl Br, Cy₂BH 2. Aldehyde, n-Bu₄NBr, 0°	(56), 15:85 E:Z	680
		Base, ether, −78°, 30 min		682
		Base, ether, −78°, 30 min	(90), >98:2 d.r., >98:2 Z:E	682
		n-BuLi, B(Bu-n)₃, THF, −100°	(84), >95.5 d.r., >90:10 E:Z	685

Nested table (for BBN/TMS product):

Base	Equiv	d.r.	Z:E
pyr	1	(75)	88:5	93:5
pyr	2	(90)	>92:1	92:7
n-BuLi	1	(50)	>98:2	>98:2
s-BuLi	1	(56)	>98:2	>98:2

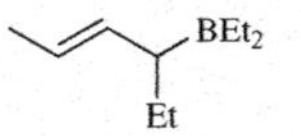

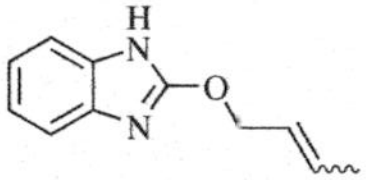

685

n-BuLi, BEt$_3$, THF, −100° 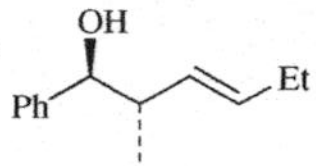   (67), >95:5 d.r., >90:10 E:Z

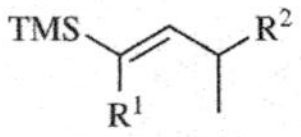

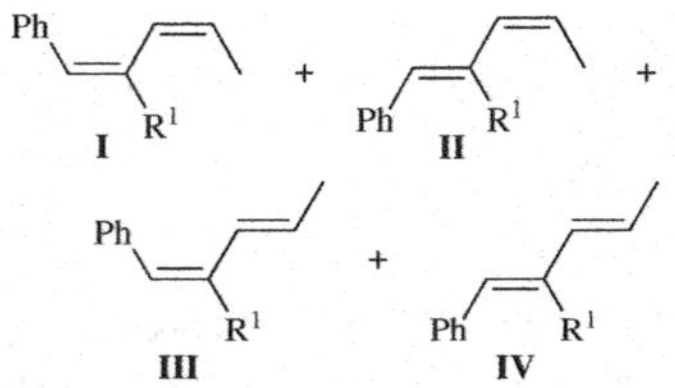

684

I + II + III + IV

R^1	R^2	Work-Up	I + II + III + IV	I:II:III:IV
Me	BBN	NaOH	(86)	92:1:5:2
Me	BBN	H$_2$SO$_4$	(87)	1:91:2:6
Me	BCy$_2$	NaOH	(83)	0:0:97:3
Me	BCy$_2$	H$_2$SO$_4$	(82)	0:0:8:92
n-Bu	BBN	NaOH	(83)	92:1:4:2
n-Bu	BBN	H$_2$SO$_4$	(86)	1:92:2:5
n-Bu	BCy$_2$	NaOH	(85)	0:0:97:3
n-Bu	BCy$_2$	H$_2$SO$_4$	(79)	0:0:3:97

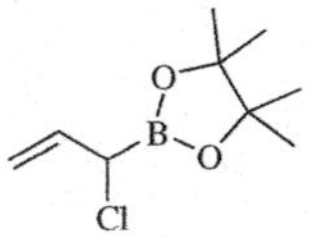

—

Neat, 0°, 15 h 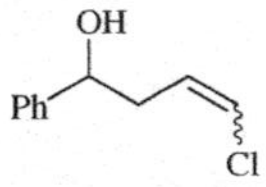   (82), 5:95 E:Z

146, 668

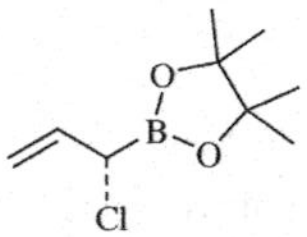

Neat, 0°, 15 h 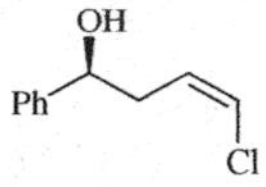   (—), 92, 6:94 E:Z

66, 669

TABLE 4. ADDITION OF α-SUBSTITUTED REAGENTS (Continued)

Carbonyl Substrate	Allylborane Reagent	Conditions	Product(s) and Yield(s) (%), % ee	Refs.
C₇		THF, −78° to rt, 18 h	(79), >99, >79:1 Z:E	65
		Ether, rt, 24 h	>14:1 d.r. R: Pr-i (88), 92; CH₂CH₂Ph (92), 94	175
		Ether, rt, 24 h	>20:1 d.r. R: Ph (65), 95; Cy (83), 92; CH₂CH₂Ph (83), 92	175
		Neat, rt, 15 h	(80), 3:97 E:Z	146, 668
		Neat, 0°, 15 h	(89), 12:88 E:Z	146

Reagent	Conditions	Product	Ref.
	Ether, reflux, 36 h	(89), 12:88 E:Z	689
	Pyr, ether to neat, 120°, 36 h	(91), 22:78 E:Z	689
	THF , 100°, 24 h	(89), 9:91 E:Z	675
	1. Toluene, rt, 2 d 2. I_2, 110°	(92)	207
	Toluene, −100° to rt		85

R		
CH_2CO_2Et	(91), 96	
$CH_2C(O)NMe_2$	(78), >94	
OMe	(—)	

Reagent	Conditions	Product	Ref.
	Toluene, −100° to rt		85

R		
CH_2CO_2Et	(89), >94	
$CH_2C(O)NMe_2$	(78), >94	

TABLE 4. ADDITION OF α-SUBSTITUTED REAGENTS (Continued)

Carbonyl Substrate	Allylborane Reagent	Conditions	Product(s) and Yield(s) (%), % ee	Refs.
C$_7$ Ph–CHO		Ether, –70° to rt, 16 h	 (73), 23:77 E:Z	148
		1. i-PrMgCl, THF, –15°, 45 min 2. Aldehyde, rt, 24 h	 (55), 30:70 E:Z	87
		1. i-PrMgCl, THF, –15°, 15 min 2. Aldehyde, rt, 24 h	I (13), >99:1 d.r. + II (52), 30:70 E:Z	87
		1. i-PrMgCl, THF, –78° to rt 2. Aldehyde, 24 h	I (13), >99:1 d.r. + II (32), 30:70 E:Z	87
		1. n-BuMgCl, THF, –15°, 45 min 2. Aldehyde, rt, 24 h	 (6), >99:1 d.r. + (59), 20:80 E:Z	87

690

Catalyst (10 mol%), additive, CH$_2$Cl$_2$

R	Catalyst	Additive	Temp	Time		E:Z
Et	—	4 Å MS	0°	15 h	(75)	2:1
Et	Sc(OTf)$_3$	4 Å MS	–78°	40 h	(63)	1:1.3
Et	TfOH	4 Å MS	–78°	40 h	(99)	1:1.7
Br	—	4 Å MS	0°	15 h	(75)	13:1
Br	Sc(OTf)$_3$	4 Å MS	–78°	40 h	(50)	10:1
Br	TfOH	4 Å MS	–78°	40 h	(24)	9:1
TMS	—	—	0°	15 h	(46)	1:6.7
TMS	Sc(OTf)$_3$	—	–78°	40 h	(40)	4:1
TMS	TfOH	—	–78°	40 h	(43)	3:1
SiMe$_2$Ph	—	—	0°	15 h	(22)	1:6.2
SiMe$_2$Ph	Sc(OTf)$_3$	—	–78°	40 h	(11)	4:1
SiMe$_2$Ph	TfOH	—	–78°	40 h	(26)	4:1

Petroleum ether, rt, 12-18 h   (53–73), 96-98, 95:5 d.r.   83, 67

CH$_2$Cl$_2$, 0° to rt, 15 h   I (53–73), 95:5 d.r., 5:95 E:Z   67

Petroleum ether, rt, 24 h   (63), 5:95 E:Z   68

TABLE 4. ADDITION OF α-SUBSTITUTED REAGENTS (Continued)

Carbonyl Substrate	Allylborane Reagent	Conditions	Product(s) and Yield(s) (%), % ee	Refs.
C₇ Ph-CHO		Neat, rt, 10 h	(55)	67
		4 kbar, rt, 3 d	(64), 99:1 d.r.	672
		1. 4 kbar, rt, 3 d 2. H⁺, MeOH	(56), 99:1 d.r.	672
		Ether to neat, 125°, 16 h	(—), 90:10 E:Z	689
		Pyr, ether to neat, 120°, 36 h	(—), 96:4 E:Z	689
		CH₂Cl₂, rt, 20 h	(95), 94:6 d.r.	676

Boronate reagent	Conditions	Product	Reference

Row 1 (boronate with OMe): Petroleum ether, −78° to rt, 2 d — product (OH, Ph, **I**, OMe) (79), 90 — 68

Row 2 (" ditto): 20°, 60 h — **I** (65), 90, >95.5 Z:E — 147

Row 3 (TBDMSO / Cl boronate): Neat, rt, 10 h — product (OH, Ph, TBDMSO, Cl) (55) — 67

Row 4 (dioxaborolane with CO₂Et): CH₂Cl₂, rt — product (OH, Ph, R, CO₂Et):

R	
Me	(95), >98
Ph	(65), >99

Reference 86

Row 5 (dioxaborolane with CO₂Et): CH₂Cl₂, rt — product (OH, Ph, R, CO₂Et):

R	
Me	(90), >98
Ph	(78), >99

Reference 86

Row 6 (pinacol boronate with R): Toluene, 60°, 12 h — product (OH, Ph, R):

R		E:Z
Me	(51), 84	1:4
n-Bu	(28), 85	1:3
Cy	(57), 83	1:2
Ph	(46), 84	9:1
t-Bu	(59), 85	1:3

Reference 691

TABLE 4. ADDITION OF α-SUBSTITUTED REAGENTS (Continued)

Carbonyl Substrate	Allylborane Reagent	Conditions	Product(s) and Yield(s) (%), % ee	Refs.
C₇				
		—	(89), 9:91 E:Z	675
		Petroleum ether, −78° to rt	(95), 94:6 d.r.	676
		Petroleum ether, 12 h	(71), 99	64, 63
		CH₂Cl₂, 0° to rt, 15 h	(56), 97, 96:4 d.r.	67
		CH₂Cl₂, 0° to rt, 15 h	(56), 96:4 d.r., 93:3 E:Z	67
		Ether to neat, 125°, 16 h	(—), 13:87 E:Z	689

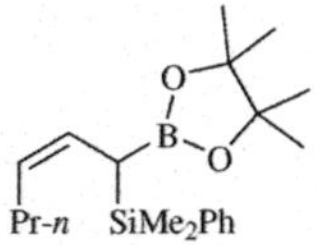

—

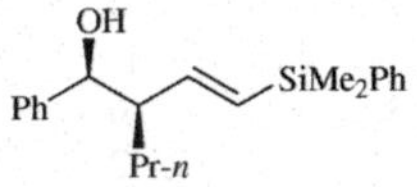

(46), 94:6 E:Z — 675

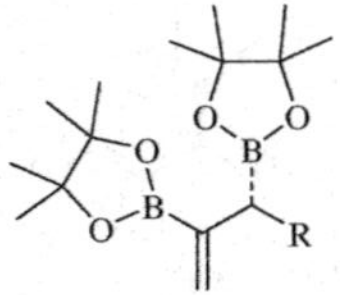

1. 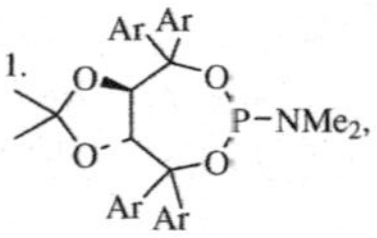

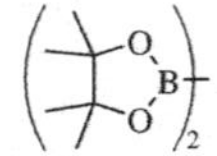

$Pd_2(dba)_3$, 22°, 10 h
2. Aldehyde, 22°, 14 h
3. NaOH, H_2O_2
Ar = 3,5-$Me_2C_6H_3$

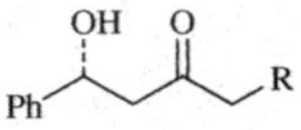

R		
Ph	(81), 93	187
n-$C_{10}H_{21}$	(96), 87	
Cy	(83), 84	

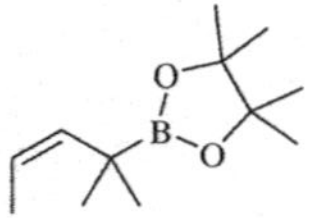

Ether, –78°

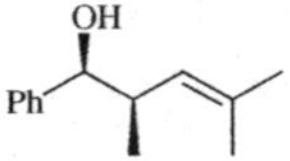

(42), 97:3 d.r. — 22

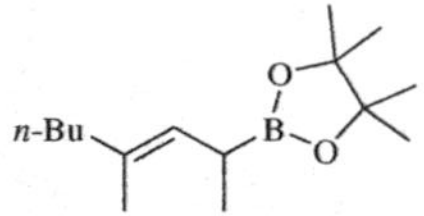

Neat, rt, 5-8 d

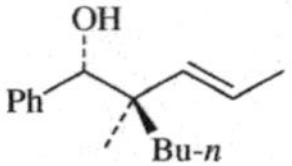

(72), >99:1 d.r. — 673

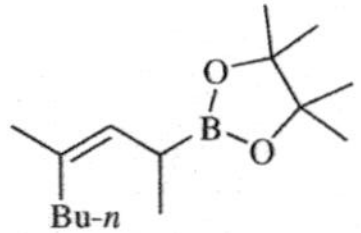

Neat, rt, 5-8 d

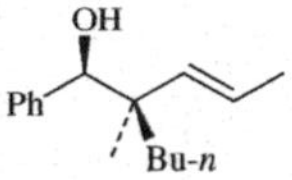

(83), >99:1 d.r. — 673

TABLE 4. ADDITION OF α-SUBSTITUTED REAGENTS (Continued)

Carbonyl Substrate	Allylborane Reagent	Conditions	Product(s) and Yield(s) (%), % ee	Refs.
C_7		CH_2Cl_2, rt, 18 h	 R n-C_5H_{11} (86) 6:94 Cl (82) 5:95 (72) 6:94 cyclopentyl (77) 5:95 t-Bu (65) 5:95 Ph (88) 4:96	677
		THF, −15° to 5°	(72), 79, 98:2 d.r.	674
		THF, −15° to 5°	(70), 60, 99:1 d.r.	674
		1. PhSe ⌇, LDA, THF, −78°, 30 min 2. Et_3B, −78° to 0°, 1 h		692

R	I + II	I:II	E:Z
H	(88)	94:6	86:14
MeO	(80)	71:29	—
O_2N	(65)	77:23	—

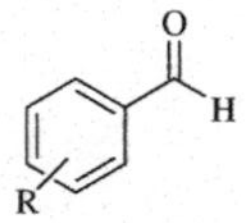

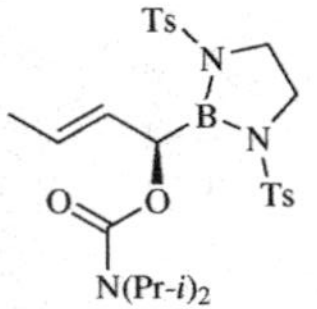

Toluene, 60°, 48 h

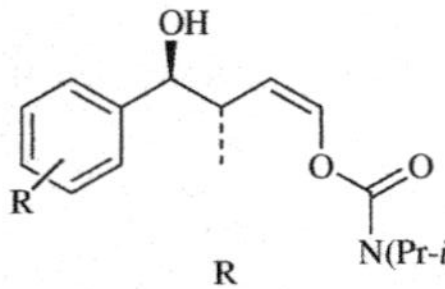

R	N(Pr-i)$_2$	Z:E
H	(87), 84	95:5
p-MeO	(66), 85	>95:5
p-Br	(74), 84	>95:5

51

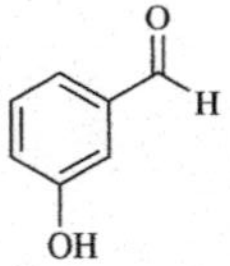

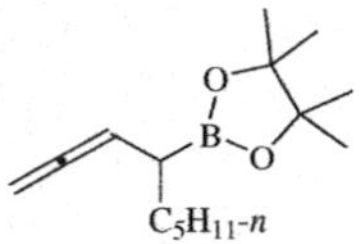

CH$_2$Cl$_2$, rt, 18 h

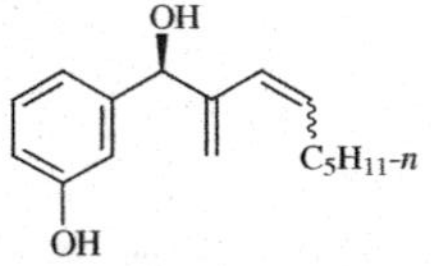

(35), 1:9 E:Z

C$_5$H$_{11}$-n

677

Toluene, 80°

(—)

Ph

Pr-n

i-Pr, Pr-i

Si

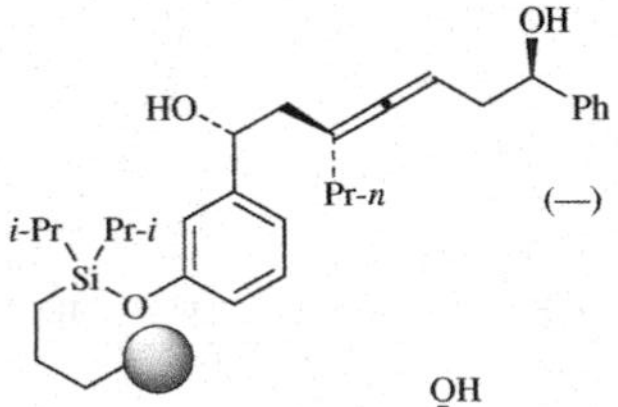

236

1. Toluene, 80°
2. HF•pyr, pyr, THF

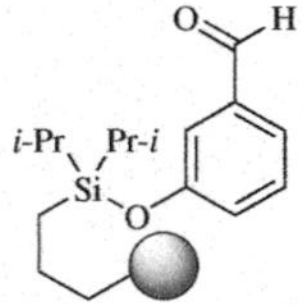

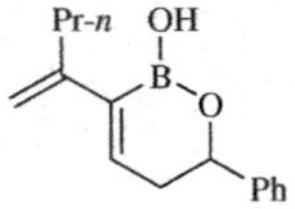

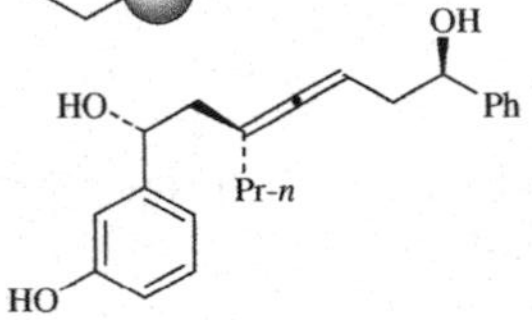

(—)

Ph

Pr-n

236

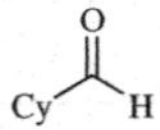

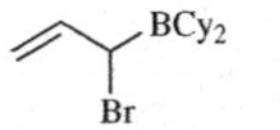

1. Br,
Cy$_2$BH
2. Aldehyde,
n-Bu$_4$NBr, 0°

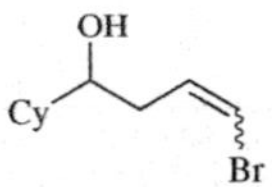

(63), 14:86 E:Z

Br

Cy

BCy$_2$

Br

680

Ts—N   N—Ts
B
N(Pr-i)$_2$

C$_5$H$_{11}$-n

Pr-n B—O
OH

Pr-i, Pr-i
Si

R

OH

OH

Cy

TABLE 4. ADDITION OF α-SUBSTITUTED REAGENTS (Continued)

Carbonyl Substrate	Allylborane Reagent	Conditions	Product(s) and Yield(s) (%), % ee	Refs.
C7 Cy-CHO	(5,5-dimethyl-1,3,2-dioxaborinane, α-Ph allyl)	1. ClCH2–B, PhMgCl, THF, –78° to rt; 2. Aldehyde, 24 h	I (9), >99:1 d.r. + II (39), 2:98 E:Z	87
	''	1. (1,3,2-dioxaborinane, Cl) PhMgCl, THF, –15°, time 1; 2. Aldehyde, rt, 24 h	I + II, Time 1 / I / d.r. / II / E:Z: 15 min (17) 98:2 (41) 2:98; 30 min (34) >99:1 (15) 98:2	87
	(1,3-dioxa, Cl, Ph)	PhMgCl, THF, –78° to rt; 2. Aldehyde, 24 h	I (8), >99:1 d.r. + II (32), 98:2 E:Z	87
	(1,3,2-dioxaborolane, Cl, Ph)	PhMgCl, THF, –15°, 45 min; 2. Aldehyde, rt, 24 h	I (21), >99:1 d.r. + II (21), 2:98 E:Z	87
	(pinacol boronate, Cl, Ph)	PhMgCl, THF, –15°, 30 min; 2. Aldehyde, rt, 24 h	II (60), 2:98 E:Z	87

Product I: Cy–CH(OH)–CH(Ph)–CH=CH2

Product II: Cy–CH(OH)–CH2–CH=CH–Ph

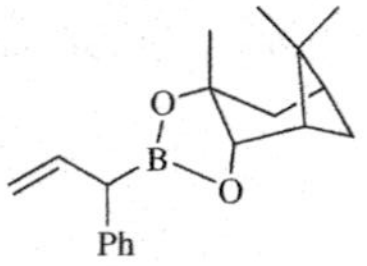

87

1. [boronate structure] Ph
PhMgCl, THF,
−15°, 30 min
2. Aldehyde, rt, 24 h

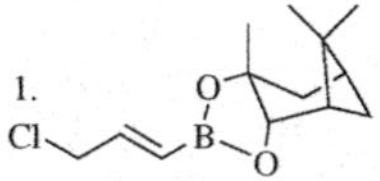

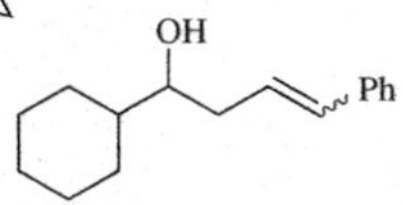

OH (67), 20
Ph

Cy

175

Ether, rt, 24 h

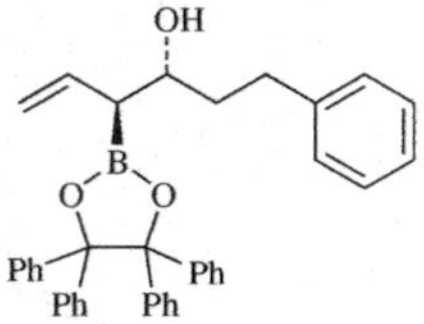

OH

OH (91), 94, >14:1 d.r.

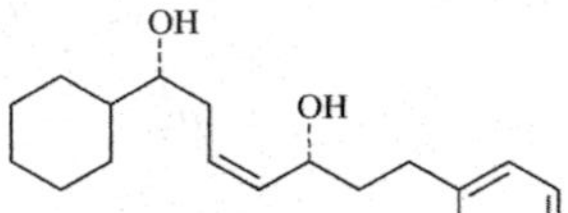

Cy

680

1. [structure] Br,
Cy$_2$BH
2. Aldehyde,
n-Bu$_4$NBr, 0°

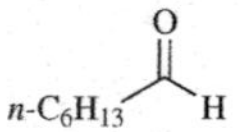

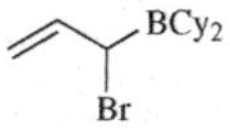

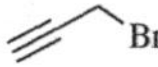

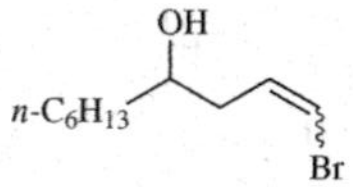

n-C$_6$H$_{13}$

OH

Br (55), 12:88 E:Z

n-C$_6$H$_{13}$

689

Ether to neat, 130°,
36 h

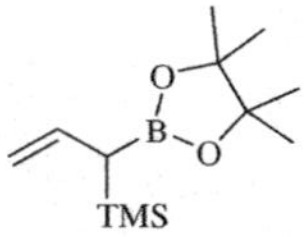

OH

TMS (80), 33:67 E:Z

n-C$_6$H$_{13}$

147

20°, 60 h

OH

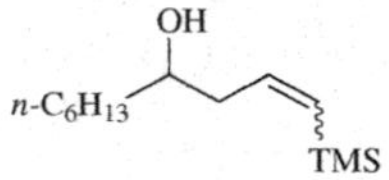

OMe (87), 88, >95:5 Z:E

n-C$_6$H$_{13}$

64

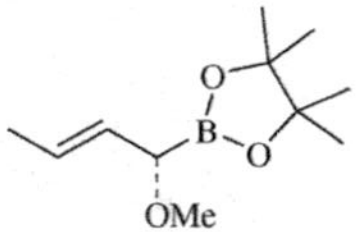

—

OH

OMe (67)

MeO

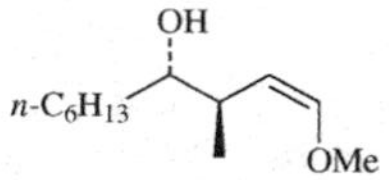

O

n-C$_6$H$_{13}$ CHO

MeO—C(O)—CH(CH$_3$)—CH$_2$—CHO

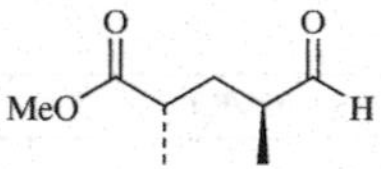

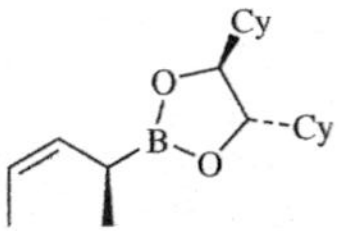

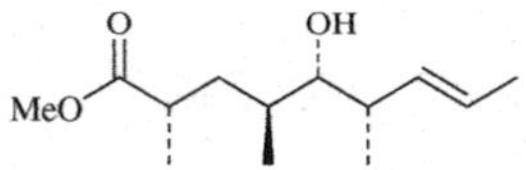

TABLE 4. ADDITION OF α-SUBSTITUTED REAGENTS (Continued)

Carbonyl Substrate	Allylborane Reagent	Conditions	Product(s) and Yield(s) (%), % ee	Refs.
C₇				
		rt, 7 d	(80), 97.5:2.5 d.r.	693
		1. PhSe , LDA, THF, −78°, 30 min 2. Et₃B, −78° to 0°, 1 h	I + II (90), I:II >99:1	692
C₈				
		CH₂Cl₂, rt, 18 h	(87), 5:95 E:Z	677
		Toluene, 60°, 48 h	(69), 85, >95:5 Z:E	51
		n-BuLi, B(Bu-n)₃, THF, −100°	(75), >95:5 d.r., >90:10 E:Z	685

330

Petroleum ether, 8 kbar, 3 d — (—), 98:2 d.r., 6:94 E:Z — 102

Petroleum ether, 8 kbar, 3 d — (—), 95:5 d.r., 30:70 E:Z — 102

Petroleum ether, rt, 3 d — (95), 4:1 d.r. — 53

1. ≡—Br, Cy$_2$BH
2. Aldehyde, n-Bu$_4$NBr, 0° — (65), 18:82 E:Z — 680

Pyr, ether, −78°, 30 min — (85) — 682

1. Cl—[boronate], i-PrMgCl, THF, −15°, 45 min
2. Aldehyde, rt, 24 h — (60), 30:70 E:Z — 87

C$_9$

TABLE 4. ADDITION OF α-SUBSTITUTED REAGENTS (Continued)

Carbonyl Substrate	Allylborane Reagent	Conditions	Product(s) and Yield(s) (%), % ee	Refs.
C9				
		CH_2Cl_2, rt, 18 h	(77), 91:9 Z:E	677
		n-BuLi, BEt_3, THF, −100°	Et (52), 97:3 d.r., >90:10 E:Z	685
		Ether, rt, 24 h	R — d.r.; Ph (95), 95 >14:1; i-Pr (95), 95 >14:1	175
		Ether, rt, 24 h	R — d.r.; Ph (87), 91 >20:1; i-Pr (69), 92 50:1	175
		CH_2Cl_2, rt, 18 h	(75), 5:95 E:Z	677

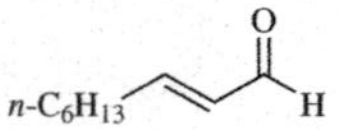

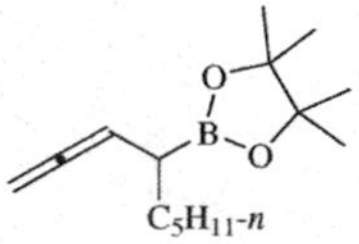

(85), 1:9 E:Z 677

(95), 80:16 d.r. 681

(60-85), 75:25 d.r. 628

197

147

149, 197

(60-85), 75:25 d.r.

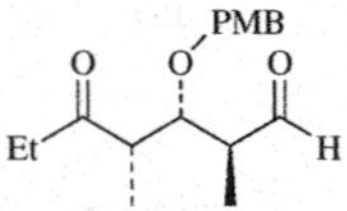

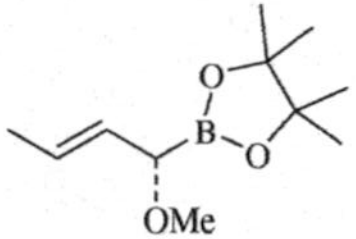

I

(56), >99:1 d.r.

(79), >95:5 d.r. 688, 681

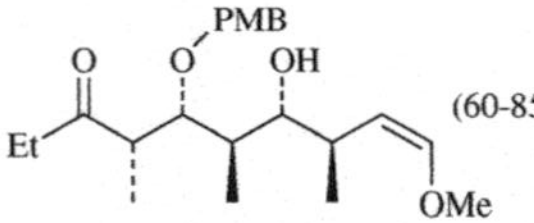

CH_2Cl_2, rt, 18 h

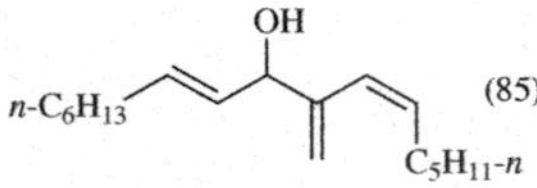

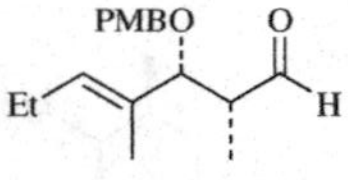

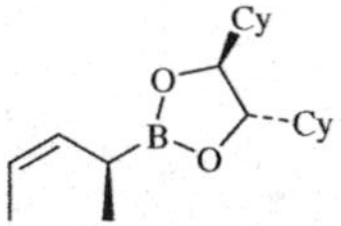

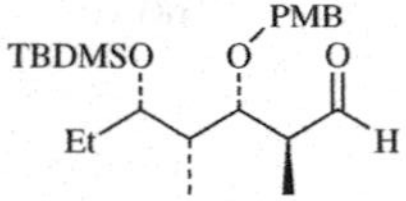

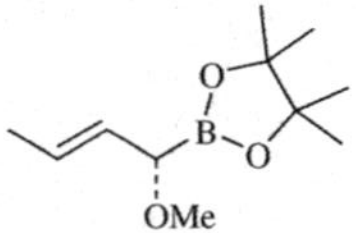

Petroleum ether, 4 kbar, rt, 3 d

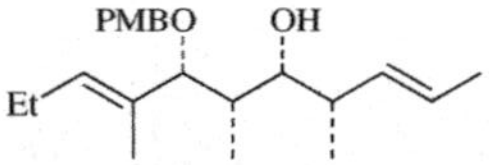

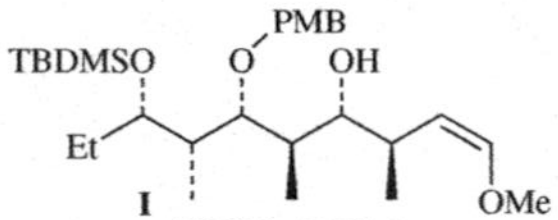

Toluene, 4 kbar, 12 h

4 Å MS, toluene, rt, 60 h

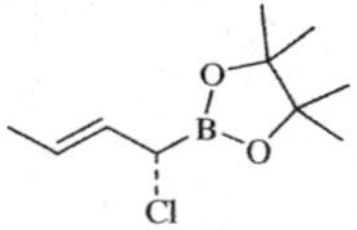

I (65), 77:23 d.r.

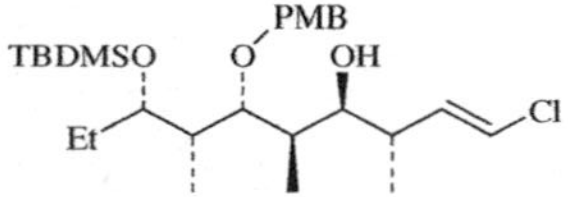

4 kbar, 20°

4 Å MS, toluene, rt, 60 h

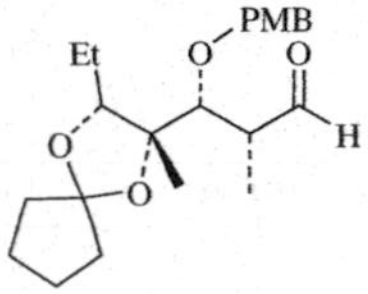

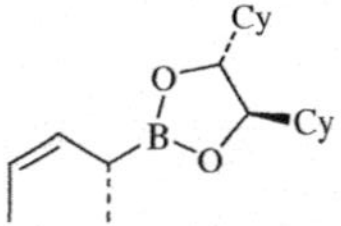

Petroleum ether, rt, 2 d

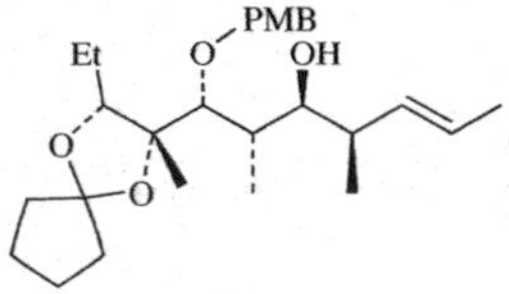

TABLE 4. ADDITION OF α-SUBSTITUTED REAGENTS (*Continued*)

Carbonyl Substrate	Allylborane Reagent	Conditions	Product(s) and Yield(s) (%), % ee	Refs.
C₁₀				
		4 kbar, 12 h	(35), 63:27 d.r.	693
		1. Toluene, rt, 2 d 2. I₂, 110°	(35)	207
		1. Toluene, rt, 2 d 2. I₂, 110°	(82)	207
		2-Hydroxypyridine (10 mol%), 10 kbar	(71)	694, 687
C₁₁		Ether, reflux, 16 h	(63), 16:84 E:Z	689

334

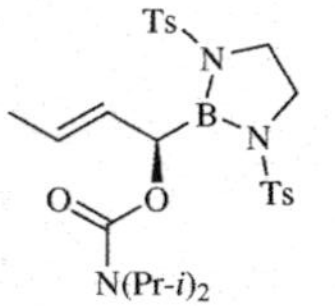

51

(67), 85, >95:5 Z:E   N(Pr-i)$_2$

Toluene, 60°, 48 h

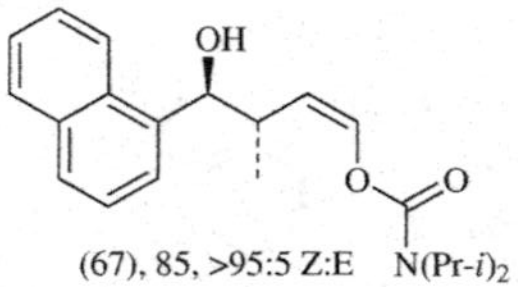

694, 687

(100), 94:6 d.r.

TBDMSO   OH   OH   Et

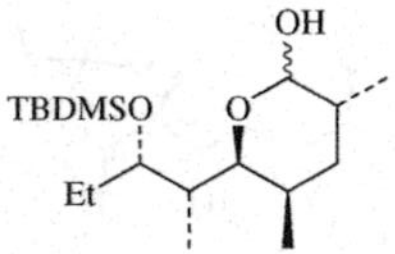

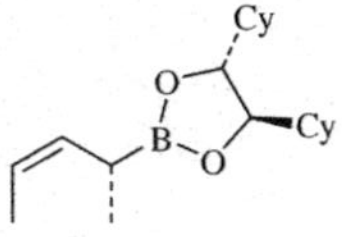

2-Hydroxypyridine
(10 mol%), 10 kbar

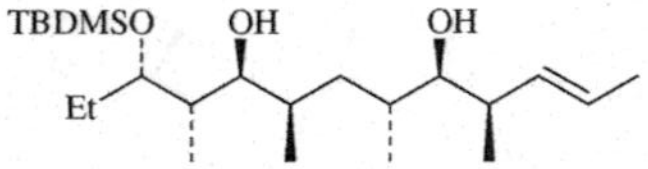

207

(40)

1. Toluene, rt, 2 d
2. I$_2$, 110°

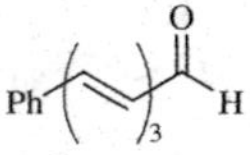

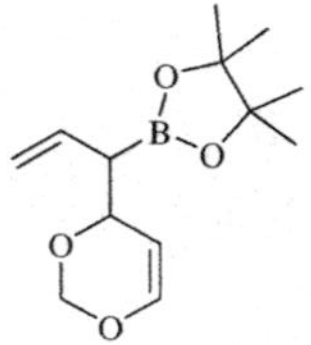

207

(80)

1. Toluene, rt, 2 d
2. I$_2$, 110°

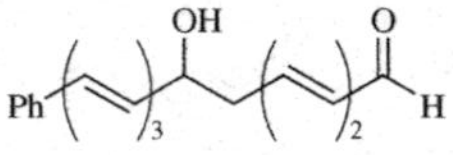

Ts—N   N—Ts   N(Pr-i)$_2$

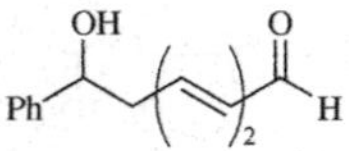

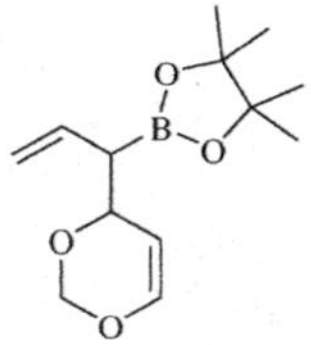

Cy   Cy

C$_{12}$ TBDMSO   Et   OH

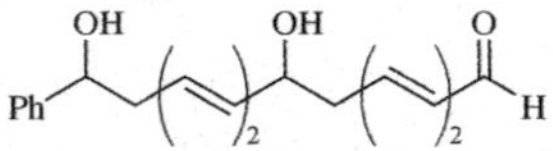

C$_{13}$ Ph

TABLE 4. ADDITION OF α-SUBSTITUTED REAGENTS (Continued)

Carbonyl Substrate	Allylborane Reagent	Conditions	Product(s) and Yield(s) (%), % ee	Refs.
C$_{13}$		CH$_2$Cl$_2$, –78° to rt, 48 h	(55)	202
C$_{14}$		Petroleum ether, rt, 4 d	(53)	695, 208
C$_{15}$		Petroleum ether, 10 kbar, 3 d	(79), >95:5 d.r.	688, 696
C$_{18}$		Petroleum ether, 10 kbar, 3 d	(70), 89:11 d.r.	688, 696

C₂₀

1. Toluene, rt, 2 d

2. I₂, 110°

208

(69), 9:1 E:Z

TABLE 5. ADDITION OF β-SUBSTITUTED REAGENTS

Carbonyl Substrate	Allylborane Reagent	Conditions	Product(s) and Yield(s) (%), % ee	Refs.
C2 (acetaldehyde)	B(Pr-n)$_2$	Neat, 0–20°	(92) **I**	238
	[allylborane]$_3$	Neat, 0–20°	**I** (93)	238
	B[(+)-Ipc]$_2$	Ether, −78° to rt	(56), 90	697, 136a
	B(Pr-n)$_2$	−70° to rt	(84) **I**	698
	BBN	THF, ether, −78°, 1 h	**I** (65)	699
	B[(−)-Ipc]$_2$	THF, ether, −78°, 1 h	(65), 90	699, 700
	N$_3$—[dioxaborinane]	Ether, 0°, 1 h	(75)	701
	[pinanediol borane, Ph]	Hexane, −40°	(82), 74	122, 244

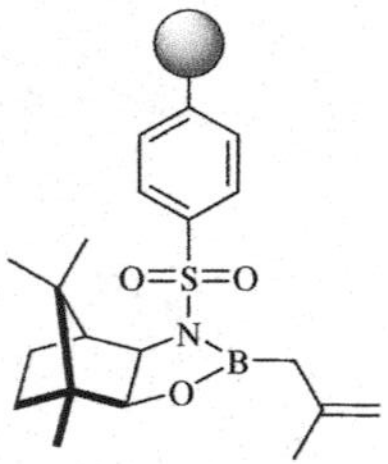

246

27

702

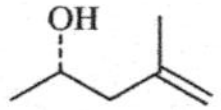

703

697, 136a

704

(69), 84

R		
TBDMS	(90), 95	
Bn	(70), 97	

(34)

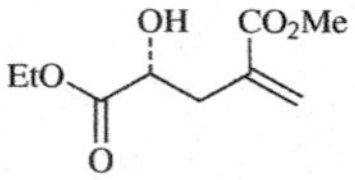

Solvent		
toluene	(67), 78
ether	(86), 77
THF	(67), 42
CH$_2$Cl$_2$	(75), 70
pentane	(47), 63

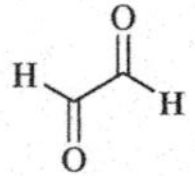

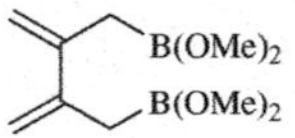

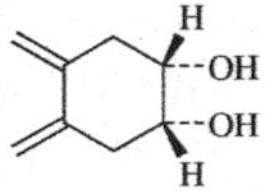

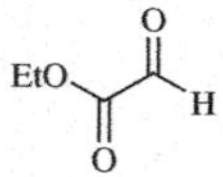

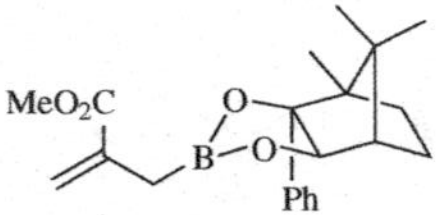

(57), 92

(60), 72

Ether, −78°

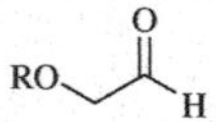

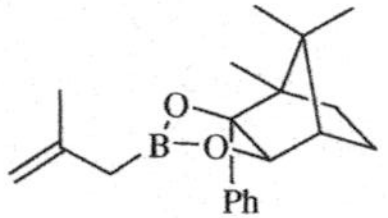

Sc(OTf)$_3$ (10 mol%), CH$_2$Cl$_2$, −78°, 12 h

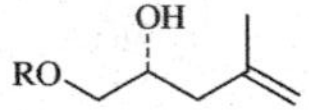

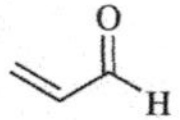

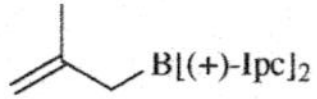

MeOH, 0° to rt, 15 h

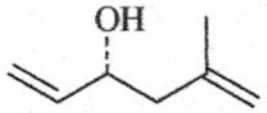

rt, 7 d

Ether, −78° to rt

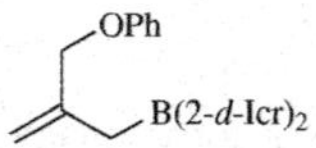

THF, −78° [a]

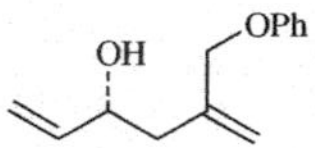

C$_3$

TABLE 5. ADDITION OF β-SUBSTITUTED REAGENTS (*Continued*)

Carbonyl Substrate	Allylborane Reagent	Conditions	Product(s) and Yield(s) (%), % ee	Refs.
C₃ (propanal)	B[(+)-Ipc]₂	Ether, −78° to rt	(54), 90	697, 136a
	B(2-*d*-Icr)₂ (OPh)	THF, −78° *a*	(67), 76	704
	[(+)-Ipc]₂B	Ether, −78° *a*	(45), >95, 93:7 d.r.	176, 705
	(pinanediol boronate, Ph)	rt, 7 d	(solvent: toluene (99), 78; ether (95), 70; pentane (100), 68; neat (99), 67)	703
	(pinanediol boronate, Ph)	Hexane, −40°	(85), 70	122, 244
	B(-)₃	Neat, 0–20°	(90) **I**	238
(acetone)	B(Pr-*n*)₂	Neat, 0–20°	**I** (94)	238

^{a}

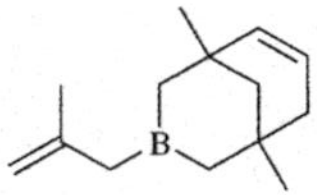	Neat, 0-20°	**I** (91)	238
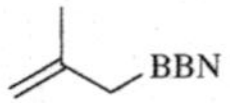 BBN	Pentane, rt, 2 h	**I** (95)	103
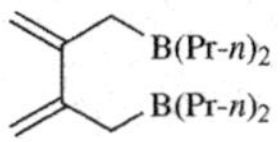 B(Pr-n)$_2$ / B(Pr-n)$_2$	Ether, −70° to rt	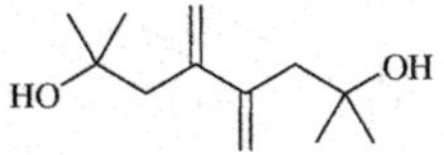 (73)	706, 702
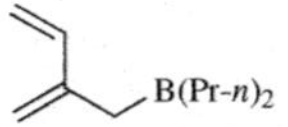 B(Pr-n)$_2$	−70° to rt	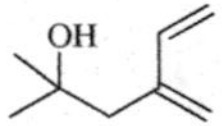 (94)	698
B(Pr-n)$_2$	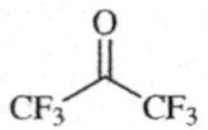 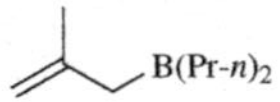 Neat, 0-20°	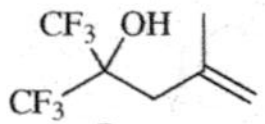 (89) **I**	238
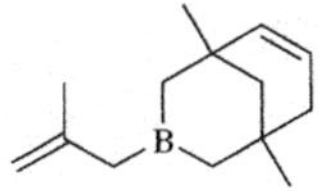	Neat, 0-20°	**I** (86)	238
[(+)-Ipc]$_2$B $\quad$ B[(+)-Ipc]$_2$	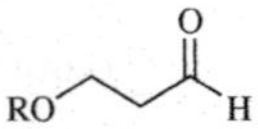 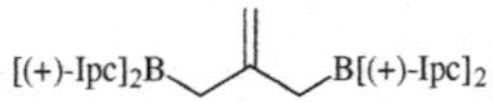 Ether, −78° ^{a}	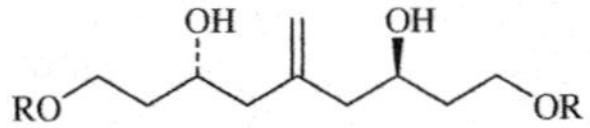 (45)	176
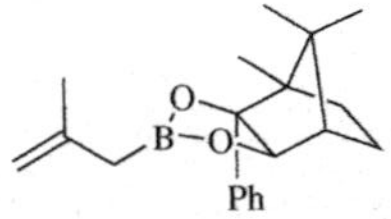	Sc(OTf)$_3$ (10 mol%), CH$_2$Cl$_2$, −78°, 12 h	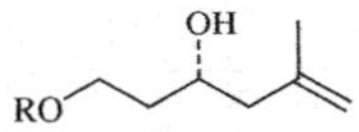 (77), 97	27

CF$_3$COCF$_3$

RO—CHO

R = THP

R = TBDPS

341

TABLE 5. ADDITION OF β-SUBSTITUTED REAGENTS (*Continued*)

Carbonyl Substrate	Allylborane Reagent	Conditions	Product(s) and Yield(s) (%), % ee	Refs.
C_3				
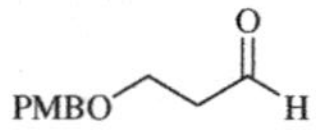	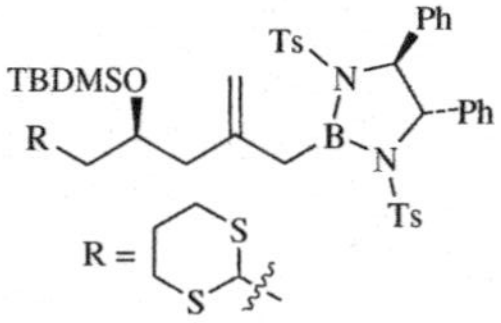	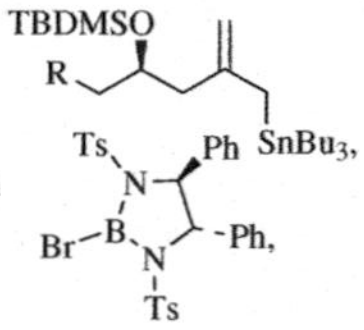 CH_2Cl_2, –78°, 1.5 h	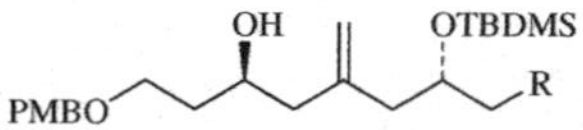 (100), 11:1 d.r.	191
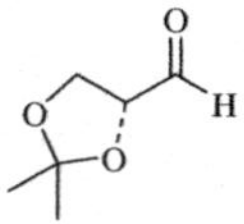	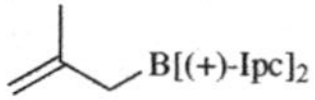	Ether, –78°	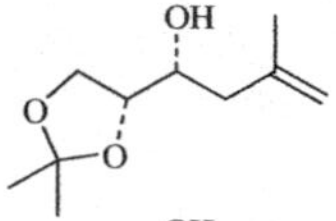 (—)	571
	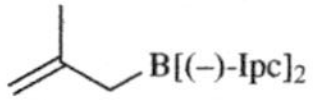	Ether, –78°	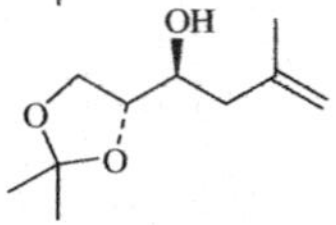 (—), 97:3 d.r.	571
	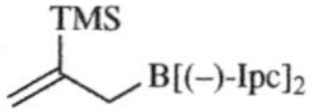	THF, –78°, 3 h	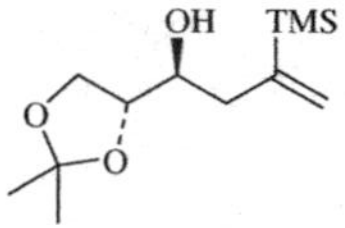 (31), 85:15 d.r.	707
	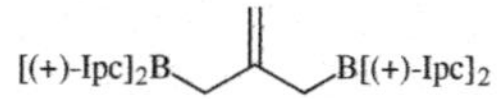	Ether, –78° [a]	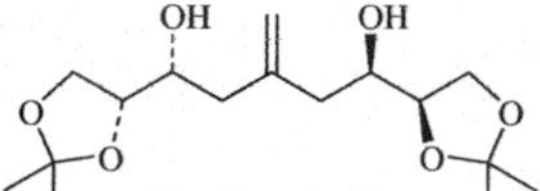 (38), 66:14.5:1 d.r.	176, 705
	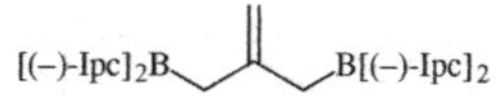	Ether, –78° [a]	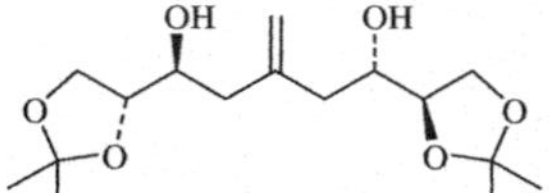 (50), 160:22:1 d.r.	176, 705

C$_4$

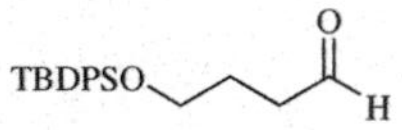

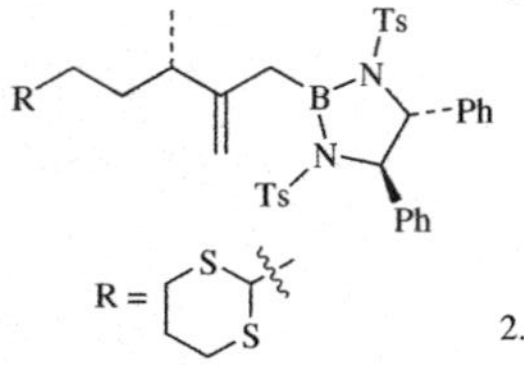

TBDPSO (aldehyde)

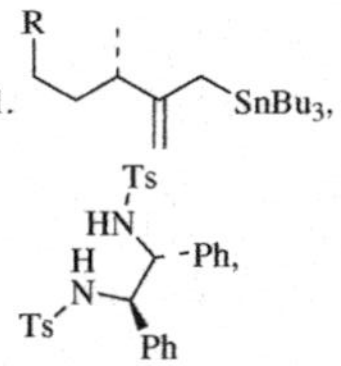

1. (reagent with SnBu$_3$), BBr$_3$, CH$_2$Cl$_2$
2. Aldehyde, −78°, 1 h

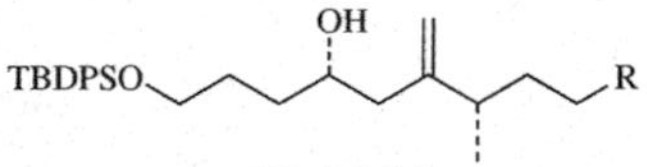

R = (dithiane)

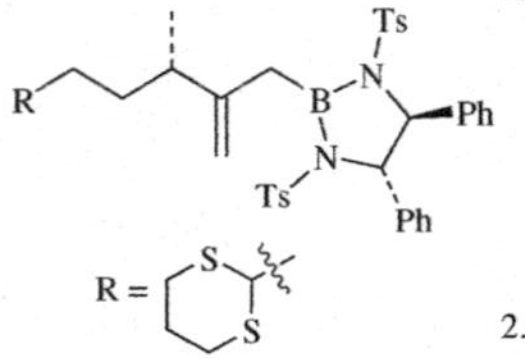

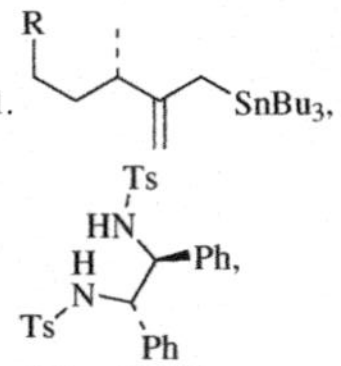

(85), 7.7:1 d.r.   708

1. (reagent with SnBu$_3$), BBr$_3$, CH$_2$Cl$_2$
2. Aldehyde, −78°, 1 h

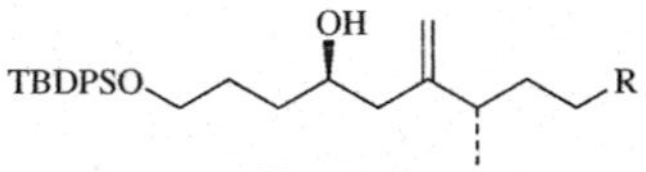

R = (dithiane)

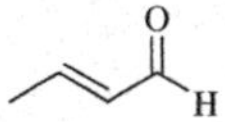

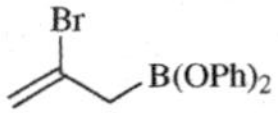

(80), 4:1 d.r.   708

Br—B(OPh)$_2$ THF, 0° to rt, 18 h   (85)   709

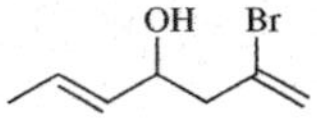

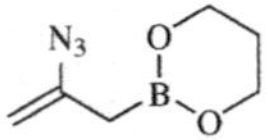

N$_3$—B(O—O)   Ether, 0°, 1 h   (64)   701

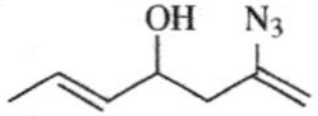

B[(+)-Ipc]$_2$ Ether, −78° to rt   (56), 91   697, 136a

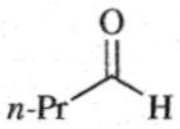

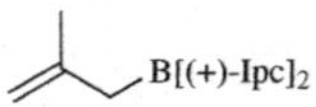

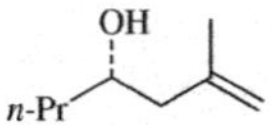

n-Pr

TABLE 5. ADDITION OF β-SUBSTITUTED REAGENTS (Continued)

Carbonyl Substrate	Allylborane Reagent	Conditions	Product(s) and Yield(s) (%), % ee	Refs.
C₄ O=CH, n-Pr	(allyl–OPh) B(2-d-Icr)₂	Ether, −78°[a]	(allyl–OPh) OH, n-Pr (78), 74	704
	(pinanyl boronate, Ph)	Hexane, −40°	OH, n-Pr (84), 65	122, 244
O=CH, i-Pr	BBN	THF, hexane, −78°, 1 h	OH, i-Pr (65)	699
	TMS, B (pyrrolidine)	Ether, −100°, 3 h	OH, i-Pr I (78), 81	141
	B (methyl pyrrolidine)	Ether, −100°, 3 h	I (80), 27	141
	B[(−)-Ipc]₂	THF, hexane, −78°, 1 h	OH, i-Pr (65), 92	699
	B[(+)-Ipc]₂	Ether, −78° to rt	OH, i-Pr (57), 96	697, 136a
	[(+)-Ipc]₂B B[(+)-Ipc]₂	Ether, −78°[a]	OH, i-Pr / Pr-i (41), >95, 93:7 d.r.	176, 705

344

TABLE 5. ADDITION OF β-SUBSTITUTED REAGENTS (*Continued*)

Carbonyl Substrate	Allylborane Reagent	Conditions	Product(s) and Yield(s) (%), % ee	Refs.
C_4				
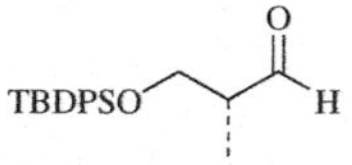	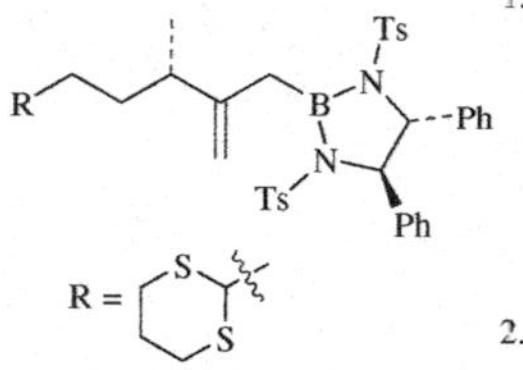	1. 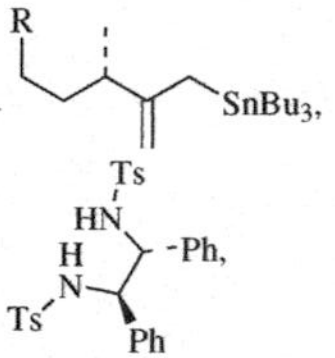 BBr_3, CH_2Cl_2 2. Aldehyde, –78°, 1 h	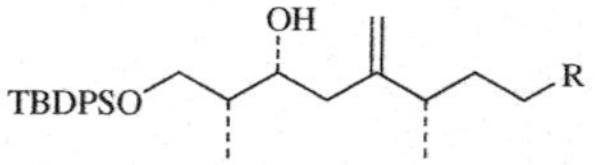 **I** (77), 15:1 d.r.	708
	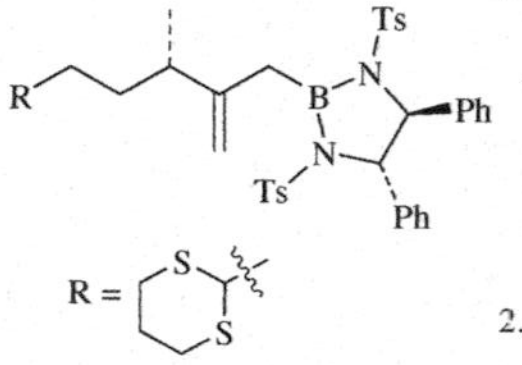	1. 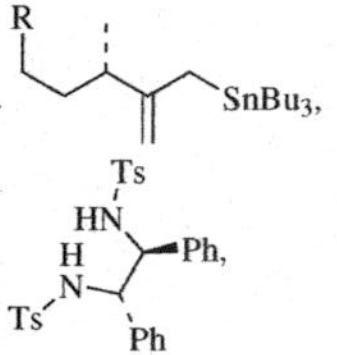 BBr_3, CH_2Cl_2 2. Aldehyde, –78°, 1 h	**I** (80), 4:1 d.r.	708
	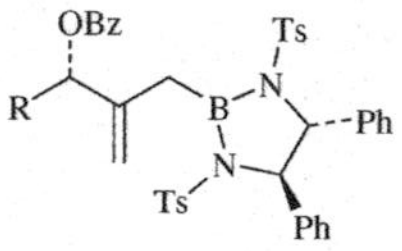 R = $(CH_2)_3OTBDPS$	1. 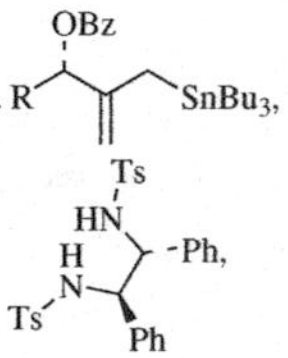 BBr_3, CH_2Cl_2 2. Aldehyde, –78°, 1 h	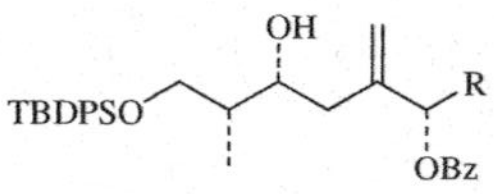 (56), >20:1 d.r.	708

708

(75), 5.4:1 d.r.

708

(76), 1.7:1 d.r.

708

(56), >25:1 d.r.

TABLE 5. ADDITION OF β-SUBSTITUTED REAGENTS (Continued)

Carbonyl Substrate	Allylborane Reagent	Conditions	Product(s) and Yield(s) (%), % ee	Refs.
C₄				
TBDPSO—CHO (with β-Me)	OBz, R / Ts–N–B–N–Ts (Ph, Ph), R = (CH₂)₃OTBDPS	1. R—OBz—SnBu₃, Ts HN, H–N Ts, Ph, Ph, BBr₃, CH₂Cl₂ 2. Aldehyde, −78°, 1 h	TBDPSO, OH, R, OBz (83), >20:1 d.r.	708
THPO—/PMBO—CHO	[(+)-Ipc]₂B—	Ether, −78° [a]	THPO, OH, OTHP, PMBO, PMBO (48)	176
	B(Pr-n)₂ / B(Pr-n)₂	Ether, −70° to rt	HO—OH (50)	702
C₅				
(diketone)	B(Pr-n)₂	−70° to rt	OH **I** (96)	698
(enal)	BBN	THF, ether, −78°, 1 h	**I** (60)	699
	B[(−)-Ipc]₂	THF, ether, −78°, 1 h	OH (60), 96	699, 700

348

Reagent	Conditions	Product	Refs.
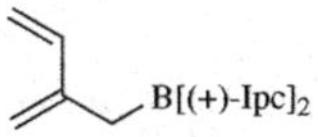 B[(+)-Ipc]₂ (3-methylenebut-3-enyl)	THF, ether, −78°, 1 h	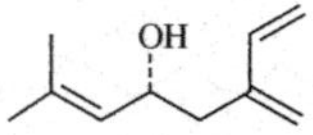 (60), 96	699, 700
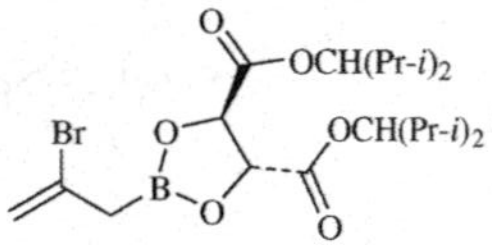 OCH(Pr-i)₂ tartrate boronate, Br-allyl	OCH(Pr-i)₂ 4 Å MS, toluene, −78°, 1 h	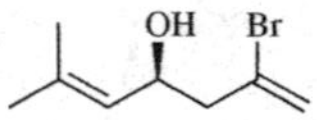 (93), 87	711, 712
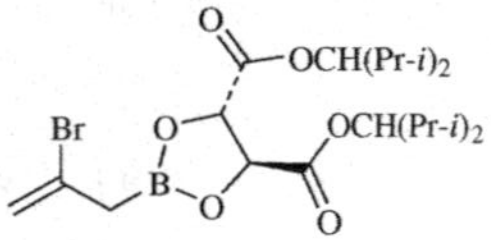 OCH(Pr-i)₂ tartrate boronate, Br-allyl	OCH(Pr-i)₂ 4 Å MS, toluene, −78°, 1 h	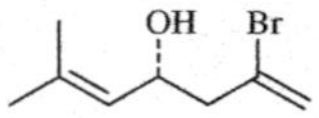 (98), 88	711, 712
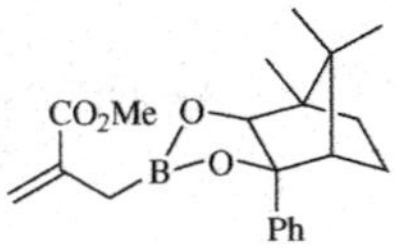 CO₂Me pinanediol boronate	rt, 8 d	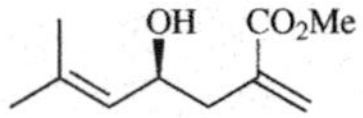 (—), 20	713
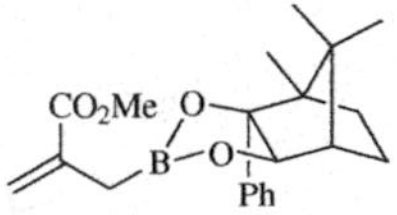 CO₂Me pinanediol boronate	rt, 8 d	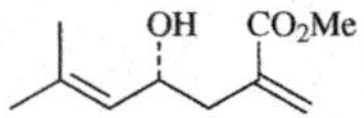 (82), 88	713
B[(+)-Ipc]₂ (3-methylbut-3-enyl)	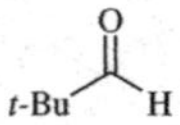 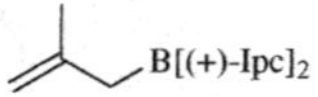 Ether, −78° to rt	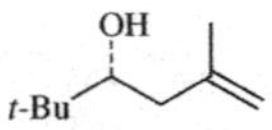 (55), 90	697, 136a
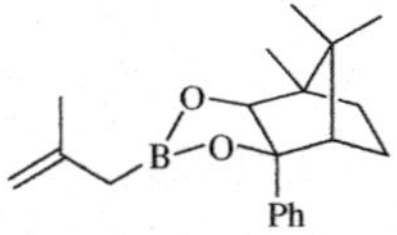 Ph pinanediol boronate	Hexane, −40°	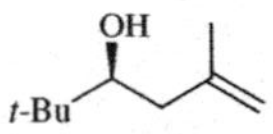 (92), 70	122, 244

t-Bu CHO

349

TABLE 5. ADDITION OF β-SUBSTITUTED REAGENTS (*Continued*)

Carbonyl Substrate	Allylborane Reagent	Conditions	Product(s) and Yield(s) (%), % ee	Refs.
C5				
t-Bu CHO	(polymer-supported sulfonamide boronate)	Ether, −78°	OH, t-Bu (89), 91	246
	N3 dioxaborinane reagent	Ether, 0°, 1 h	N3, OH, t-Bu (68)	701
i-Pr CH2CHO	B(Pr-n)2	−70° to 0°	OH, i-Pr (94) **I**	698
	BBN	THF, ether, −78°, 1 h	**I** (65)	699
	B[(+)-Ipc]2	THF, ether, −78°, 1 h	OH, i-Pr (65, 96)	699, 700
	B[(−)-Ipc]2	THF, ether, −78°, 1 h	OH, i-Pr (60, 94)	699, 700
	[(+)-Ipc]2B ... B[(+)-Ipc]2	Ether, −78° [a]	OH, Pr-i (53), >95, 93:7 d.r.	176, 705

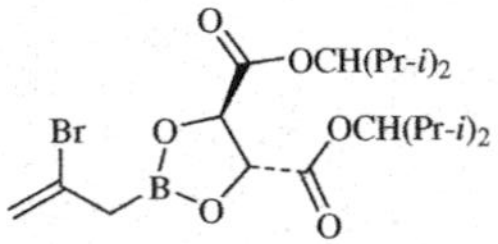

711, 712

711, 712

713

713

28

234

(83), 90

(90), 93

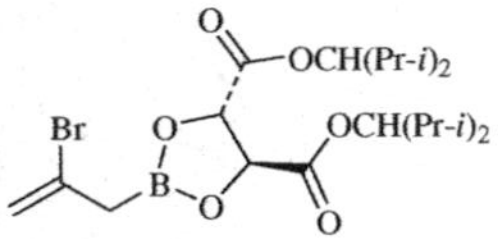

CO_2Me (—), 25

CO_2Me (—), 82

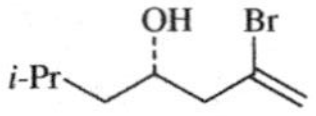

(89), 98

(68)

4 Å MS, toluene, −78°, 1 h

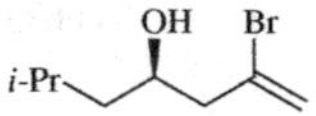

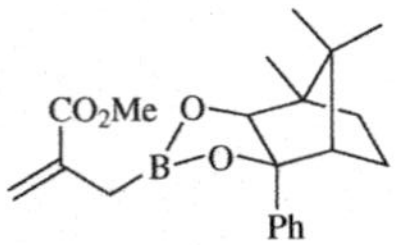

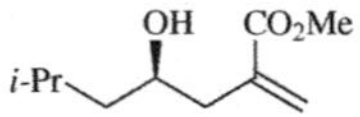

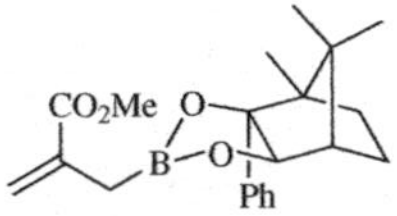

4 Å MS, toluene, −78°, 1 h

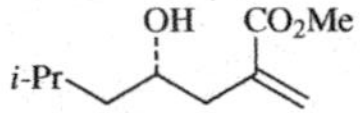

rt, 8 d

rt, 8 d

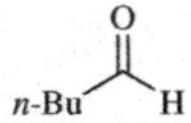

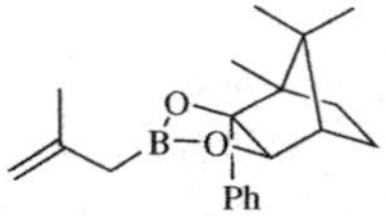

$Sc(OTf)_3$ (10 mol%), CH_2Cl_2, −78°, 16 h

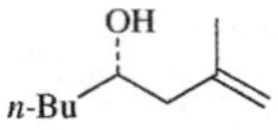

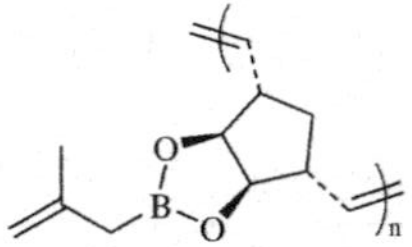

CH_2Cl_2, rt; H_2O, 1 h

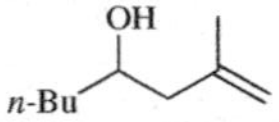

TABLE 5. ADDITION OF β-SUBSTITUTED REAGENTS (*Continued*)

Carbonyl Substrate	Allylborane Reagent	Conditions	Product(s) and Yield(s) (%), % ee	Refs.
C5 (cyclopentanone)	B(Pr-n)$_2$	Neat, 0–20°	(82)	238
	B(Pr-n)$_2$	Ether, −70° to rt	(46)	706, 702
	Br–B(OPh)$_2$	THF, 0° to rt, 18 h	(80)	709
	B(OH)$_2$	—	(64), 54:46 d.r.	714
	Br–B(OPh)$_2$	THF, 0° to rt, 18 h	(80)	709
	chiral bis-sulfonamide allylborane	CH$_2$Cl$_2$, rt; H$_2$O, 2 h	(87)	234
(oxazole carbaldehyde, PMBO)	TBDMSO / OPiv chiral allylborane	CH$_2$Cl$_2$, −78°, 2–3 h	**I** (92), >20:1 d.r.	710, 715

708

710

708

707

176, 705

I (98), >20:1 d.r.

I (98), >20:1 d.r.

OTBDMS

OH

I

(98), >20:1 d.r.

PMBO

R	
Me	(62), 28
Ph	(43), 30

OH   SiMe$_2$R

n-C$_5$H$_{11}$

OH

OH

n-C$_5$H$_{11}$

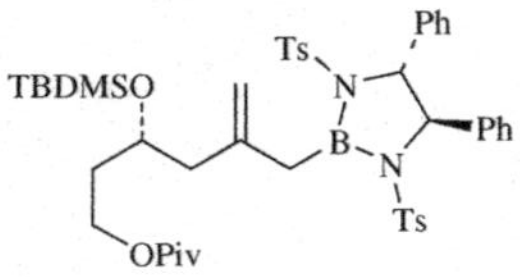

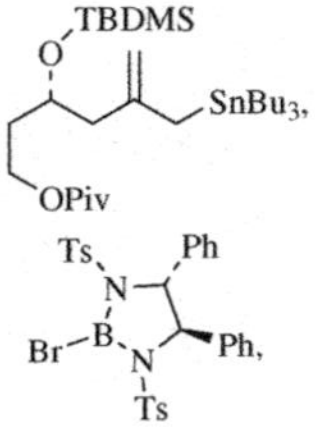

C$_5$H$_{11}$-n

(43), >95, 92:8 d.r.

TBDMS

SnBu$_3$,

Ph,

Ts N B N Ts

Ph

Br

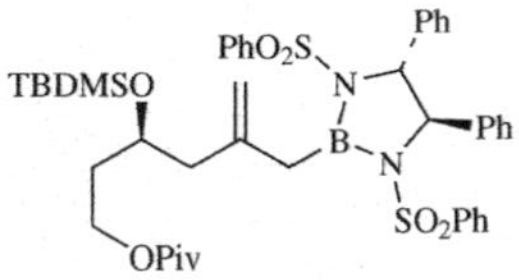

Ts N B N Ts

Ph

OPiv

CH$_2$Cl$_2$, −78°

TBDMSO

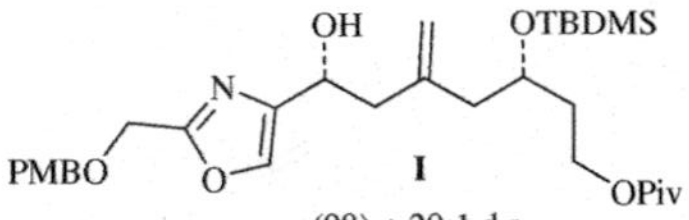

OPiv

Ph

PhO$_2$S N B N SO$_2$Ph

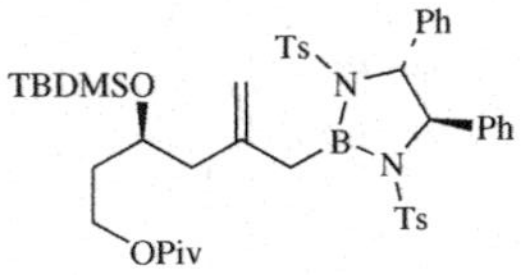

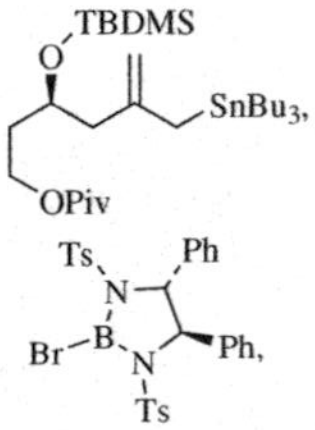

Ph

OPiv

CH$_2$Cl$_2$, −78°, 2-3 h

TBDMSO

OPiv

TBDMS

SnBu$_3$,

Ph,

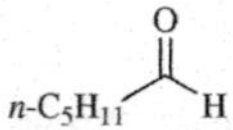

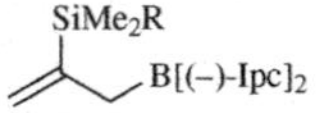

Ts N B N Ts

Ph

Br

Ph

Ts N B N Ts

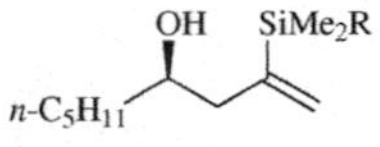

Ph

OPiv

CH$_2$Cl$_2$, −78°

TBDMSO

OPiv

SiMe$_2$R

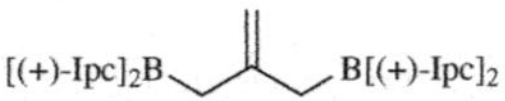

B[(−)-Ipc]$_2$

THF, −78°, 3 h

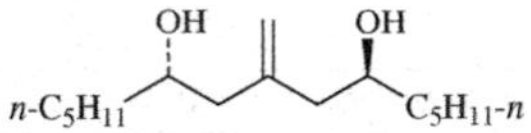

[(+)-Ipc]$_2$B

B[(+)-Ipc]$_2$

Ether, −78° ^{a}

C$_6$

O

n-C$_5$H$_{11}$ H

TABLE 5. ADDITION OF β-SUBSTITUTED REAGENTS (*Continued*)

Carbonyl Substrate	Allylborane Reagent	Conditions	Product(s) and Yield(s) (%), % ee	Refs.
C$_6$ n-C$_5$H$_{11}$CHO	(allyl)B(OPh)$_2$, β-Br	THF, 0° to rt, 18 h	(87)	709
	chiral Ts-diamine borane, β-Br	CH$_2$Cl$_2$, −78°, 2.5 h	(71), 94	50
	tartrate allylboronate, β-Br	4 Å MS, toluene, −78°, 1 h	**I** (86), 88	712
	tartrate allylboronate, β-Br	4 Å MS, toluene, −78°, 1 h	(78), 92	712
	chiral Ts-diamine borane, β-Cl	CH$_2$Cl$_2$, −78°, 2.5 h	(77), 99	50
	chiral Ts-diamine borane, β-CH$_3$	CH$_2$Cl$_2$, −78°, 2.5 h	(79), 88	50

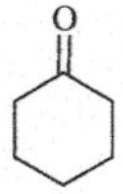

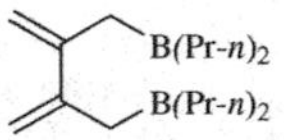

706, 702

702

714

294

716, 717

(58)

(40)

(70), 60:40 d.r.

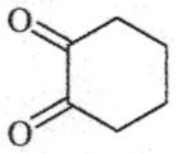

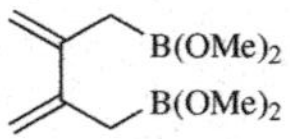

(84)

(96), 7.2:1 d.r.

Ether, −70° to rt

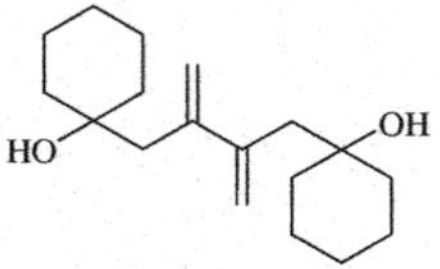

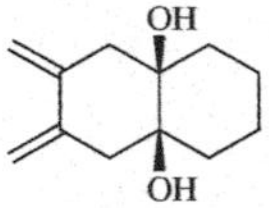

MeOH, 0° to rt, 15 h

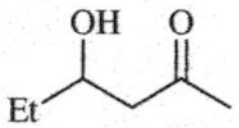

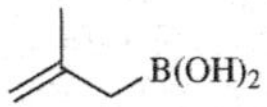

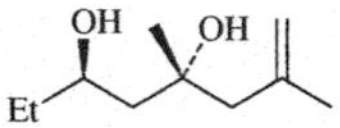

—

THF, reflux, 2 h

1. OTBDMS, SnBu₃, Ts, HN, Ph, H, N, Ts, Ph; BBr₃, CH₂Cl₂
2. Aldehyde, −78°, 1 h

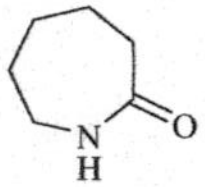

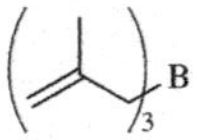

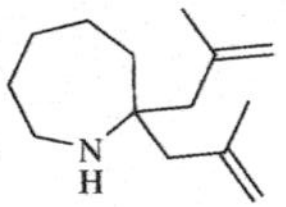

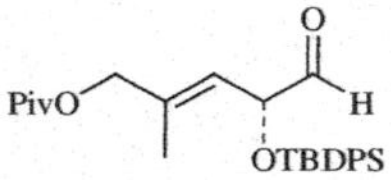

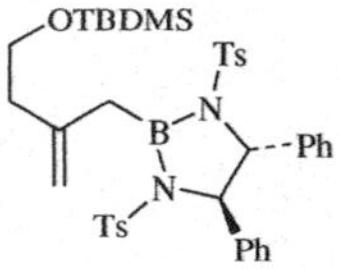

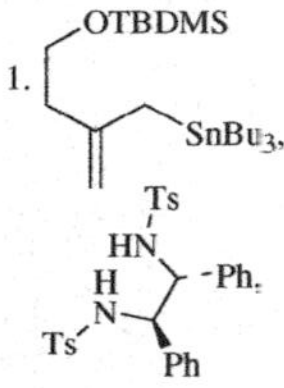

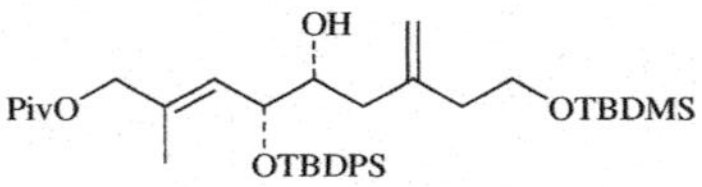

B(Pr-n)₂ / B(Pr-n)₂

B(OMe)₂ / B(OMe)₂

B(OH)₂

B

Ts, N, Ph, N, Ts, Ph, OTBDMS, B

TABLE 5. ADDITION OF β-SUBSTITUTED REAGENTS (Continued)

Carbonyl Substrate	Allylborane Reagent	Conditions	Product(s) and Yield(s) (%), % ee	Refs.
C₆		1. SnBu₃, TBDPS, HN–Ph, Ts, H, N–Ts, Ph, BBr₃, CH₂Cl₂ 2. Aldehyde, −78°, 1 h	**I** (80), 1.2:1 d.r.	708
		1. SnBu₃, TBDPS, HN–Ph, Ts, H, N–Ts, Ph, BBr₃, CH₂Cl₂ 2. Aldehyde, −78°, 1 h	**I** (70), 4:1 d.r.	708
		1. SnBu₃, TBDPS, HN–Ph, Ts, H, N–Ts, Ph, BBr₃, CH₂Cl₂ 2. Aldehyde, −78°, 1 h	**I** (60), 7:1 d.r.	708

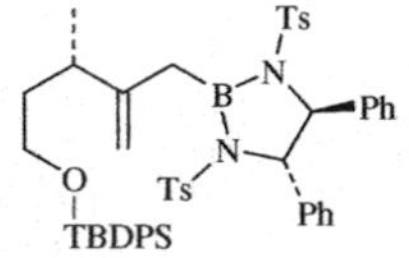

1. SnBu₃,

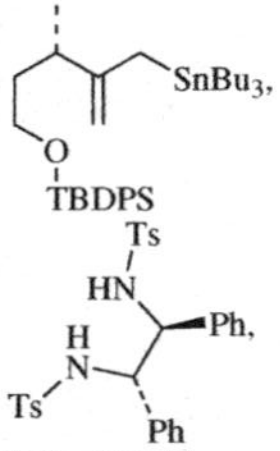

TBDPS

BBr₃, CH₂Cl₂

2. Aldehyde, −78°, 1 h

I (77), 1.4:1 d.r.

708

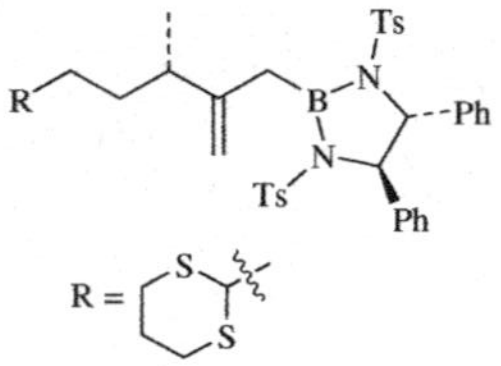

1. SnBu₃,

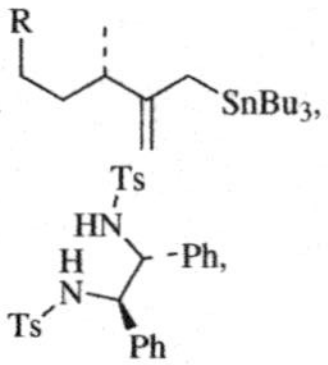

BBr₃, CH₂Cl₂

2. Aldehyde, −78°, 1 h

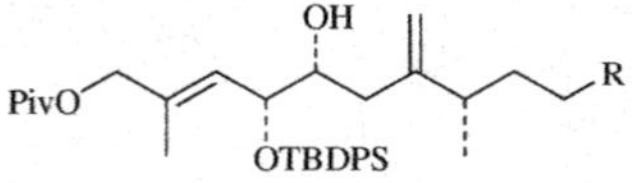

R =

I (82), 4.2:1 d.r.

708

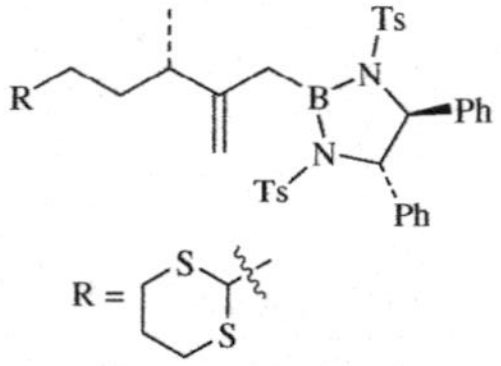

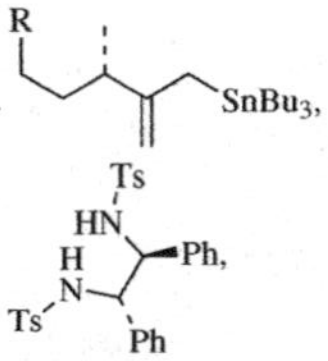

1. SnBu₃,

BBr₃, CH₂Cl₂

2. Aldehyde, −78°, 1 h

R =

I (82), 1.5:1 d.r.

708

TABLE 5. ADDITION OF β-SUBSTITUTED REAGENTS (Continued)

Carbonyl Substrate	Allylborane Reagent	Conditions	Product(s) and Yield(s) (%), % ee	Refs.
C₆				
		CH₂Cl₂, −78°	(98), >25:1 d.r.	716
		CH₂Cl₂, −78°	(98), >25:1 d.r.	716
		CH₂Cl₂, −78°, 2-3 h	(55), >20:1 d.r.	710
		THF, −78°, 3 h	(36), 54	707

358

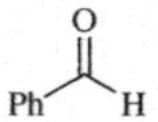

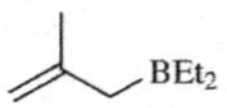

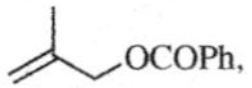

PhCHO

Reagent	Conditions	Product	Refs.
BE2 (allyl/methallyl boron)	—OCOPh, BEt3, Pd(PPh3)4, THF, rt, 9 h; 40°, 4 h	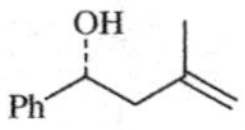 I (60)	483
"	—OCOPh, BEt3, Pd(PPh3)4, Et3N, THF, rt, 20 h	I (63)	483
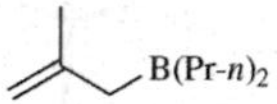 B(Pr-n)2	Neat, 0–20°	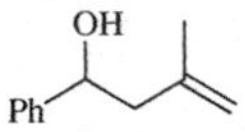 I (89)	238
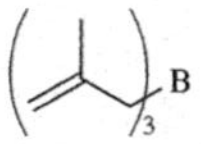 B₃	Neat, 0–20°	I (78)	238
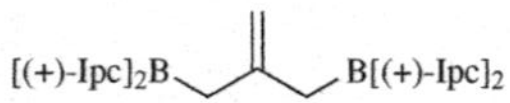 [(+)-Ipc]2B / B[(+)-Ipc]2	Ether, −78° [a]	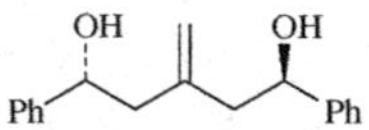 (55), >95, 84:16 d.r.	176, 705
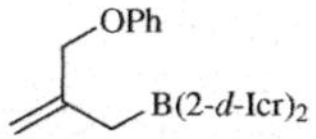 OPh / B(2-d-Icr)2	−78° [a]	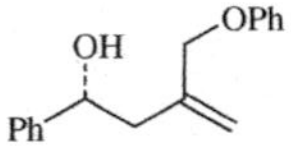 Solvent THF (72), 84 ether (74), 88	704
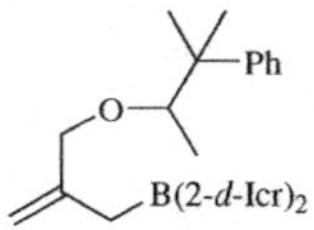 OPh / B(2-d-Icr)2	Ether, −78° [a]	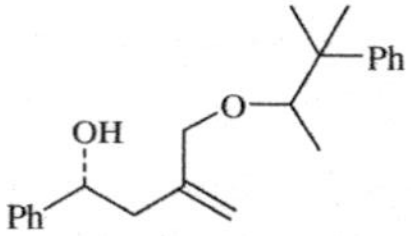 (39), 91	704

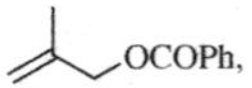

359

TABLE 5. ADDITION OF β-SUBSTITUTED REAGENTS (Continued)

Carbonyl Substrate	Allylborane Reagent	Conditions	Product(s) and Yield(s) (%)	% ee	Refs.
C₇ PhCHO	BBN	THF, ether, −78°, 1 h	(60)		699
	B[(−)-Ipc]₂	THF, ether, −78°, 1 h	(60), 93		699
	pinacol boronate, Li, Cl		(80)		52
	polymer-supported sulfonamide	Ether, −78°	(92), 89 **I**		246
	Ph-oxazaborolidine	Sc(OTf)₃ (10 mol%), CH₂Cl₂, −78°, 12 h	**I** (64), 98		27
	Ph-boronate	Hexane, −40°	(83), 40		122

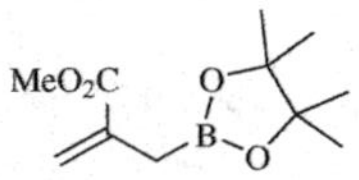

57

52

58

718

701

709

(98) OH CO$_2$Me Ph

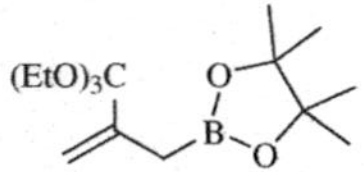

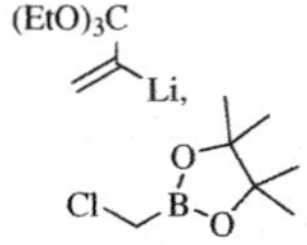

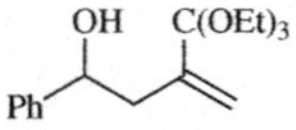

(56) OH C(OEt)$_3$ Ph

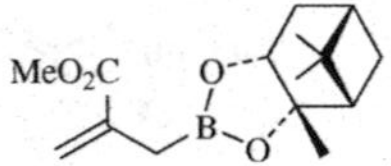

(70), 10 OH CO$_2$Me Ph

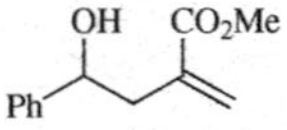

(75) OH Ph Ph

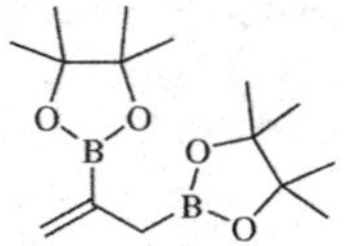

(70) OH N$_3$ Ph

(89) OH Br Ph

Neat, rt, 6 d

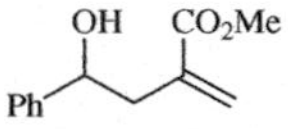

(EtO)$_3$C—Li,

Cl—

Neat, rt, 14 d

1. pinacolborane, allene, dioxane, 50°, 16 h
2. PhI, PdCl$_2$(dppf), KOH (aq), dioxane, 90°, 16 h

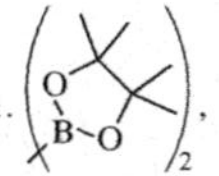

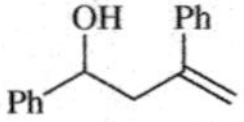

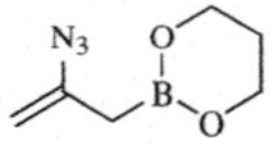

Ether, 0°, 1 h

THF, 0° to rt, 18 h

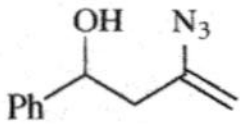

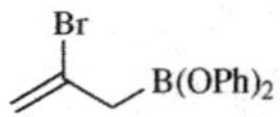

MeO$_2$C—B

(EtO)$_3$C—B

MeO$_2$C—B

B

N$_3$—B

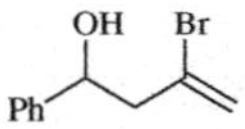

Br—B(OPh)$_2$

TABLE 5. ADDITION OF β-SUBSTITUTED REAGENTS (*Continued*)

Carbonyl Substrate	Allylborane Reagent	Conditions	Product(s) and Yield(s) (%), % ee	Refs.
C_7 PhCHO	tartrate boronate (OR/OR) with bromoallyl	4 Å MS, toluene, −78°, 1 h	R Et — (92), 47 Pr-i — (97), 47 CH(Pr-i)$_2$ — (98), 84	712
	OCH(Pr-i)$_2$ tartrate boronate with bromoallyl	4 Å MS, toluene, −78°, 1 h	(98), 84	712
	Ts–N/N–Ts diazaborolidine (Ph, Ph) with allyl–R	CH$_2$Cl$_2$, −78°, 2.5 h	R Br — (73), 79 Cl — (79), 84	50
R^1 = H, O$_2$N (ArCHO)	SiMe$_2$R^2 B[(−)-Ipc]$_2$	THF, −78°, 3 h	R^1 — R^2 H — Me — (46), 75 H — Ph — (63), 22 O$_2$N — Me — (72), 33	707
R^1 = H, MeO, O$_2$N (ArCHO)	bicyclic boronate with methallyl	CH$_2$Cl$_2$, H$_2$O, rt	R^1 — Time H — 2 h — (87) MeO — 5 h — (95) O$_2$N — 3 h — (96)	234

362

Substrate	Reagent	Conditions	Product	Yield (%), ee	Refs.
2-Cl-benzaldehyde	boronate (NHBn)	Toluene, −78°, 18 h	2-Cl product	(87), 92	290
4-Cl-benzaldehyde	boronate (CO$_2$Et)	Neat, rt, 14 d	(30), 20; I (90), 20		58
	boronate (cyclopentyl OMe)	Neat, rt, 14 d	I (90), 20		58
N-Ts-2-aminobenzaldehyde	B[(+)-Ipc]$_2$	THF, −78°[a]		(80), 84	719
N-Ts-2-aminobenzaldehyde	B[(−)-Ipc]$_2$	THF, −78°[a]		(84), 82	719
3-NO$_2$-benzaldehyde	[(+)-Ipc]$_2$B / B[(+)-Ipc]$_2$	Ether, −78°[a]		(47), >95, 94:6 d.r.	176, 705

363

TABLE 5. ADDITION OF β-SUBSTITUTED REAGENTS (*Continued*)

Carbonyl Substrate	Allylborane Reagent	Conditions	Product(s) and Yield(s) (%), % ee	Refs.
C₇				
	$[(+)\text{-Ipc}]_2\text{B}$ ... $\text{B}[(+)\text{-Ipc}]_2$	Ether, −78° [a]	R / d.r. — O₂N (51), >95, 95:5; MeO (55), >95, 95:5	176, 705
	$\text{B}[(+)\text{-Ipc}]_2$	THF, −78° [a]	(79), 80	719
	$\text{B}[(−)\text{-Ipc}]_2$	THF, −78° [a]	(74), 80	719
	MeO₂C— (pinacol boronate)	Neat, rt, 14 d	(92)	57
	MeO₂C— (pinacol boronate)	Neat, rt, 14 d	(98) **I**	57
	MeO₂C— (pinanediol boronate)	Neat, rt, 14 d	**I** (95), 6	58

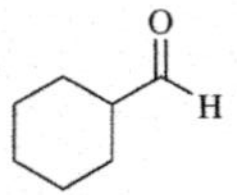

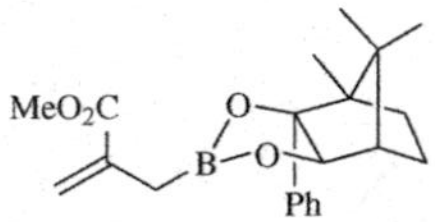

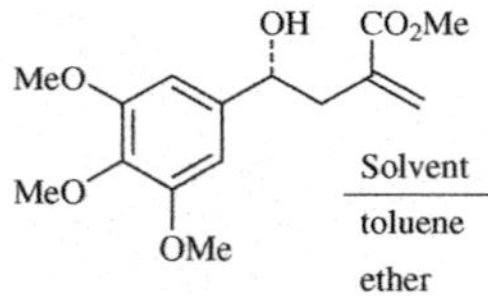

Solvent	
toluene	(88), 82
ether	(100), 76
THF	(91), 80
CH$_2$Cl$_2$	(80), 82
pentane	(89), 71

703

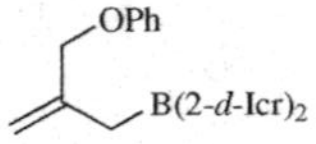

704

28

701

50

712

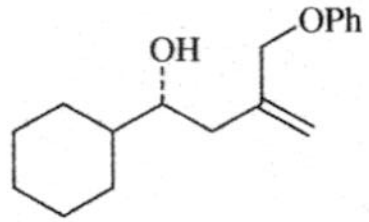

(81), 71

(63), 92

(64)

(75), 94

I (98), 88

rt, 7 d

THF, −78° [a]

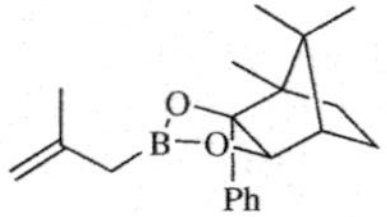

Sc(OTf)$_3$ (10 mol%), CH$_2$Cl$_2$, −78°, 12 h

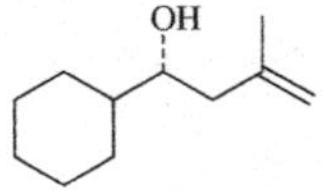

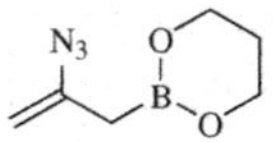

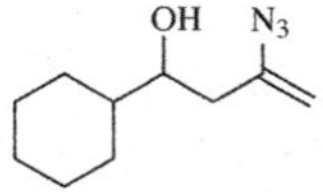

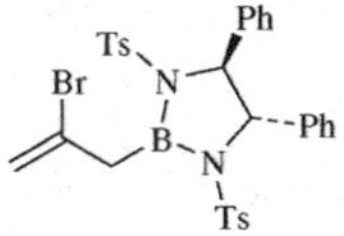

Ether, 0°, 1 h

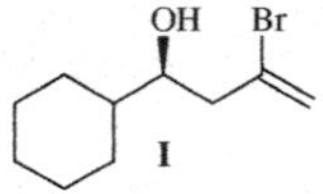

CH$_2$Cl$_2$, −78°

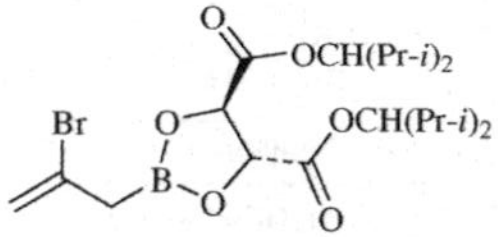

4 Å MS, toluene, −78°, 1 h

365

TABLE 5. ADDITION OF β-SUBSTITUTED REAGENTS (*Continued*)

Carbonyl Substrate	Allylborane Reagent	Conditions	Product(s) and Yield(s) (%), % ee	Refs.
C₇				
		4 Å MS, toluene, −78°, 1 h	(95), 85	712
	3,5-(CF₃)₂C₆H₃O₂S ... 3,5-(CF₃)₂C₆H₃O₂S	—	(76), 88	351
		CH₂Cl₂, −78°	(81), 99	50
		(Ph₃P)₂Pt(CH₂CH₂) (5 mol%), THF, 80°, 14 h 2. H₂O₂	(75)	186

C_7

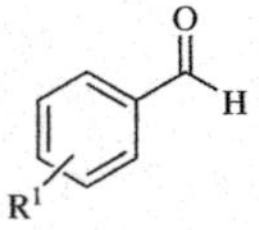

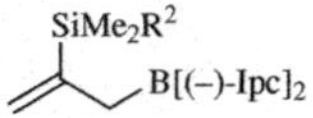

707

SiMe$_2$R^2 · B[(−)-Ipc]$_2$

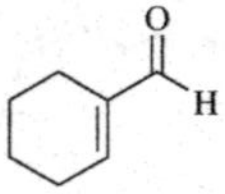

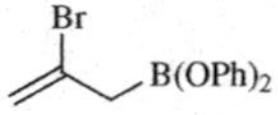

THF, −78°, 3 h

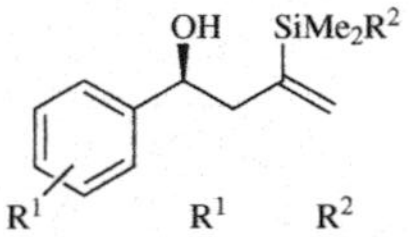

R^1	R^2	
4-CF$_3$	Me	(76), 51
4-CF$_3$	Ph	(69), 48
2-CF$_3$	Me	(51), 34

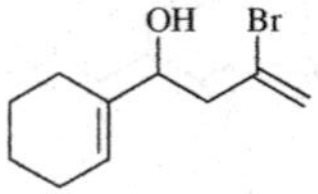

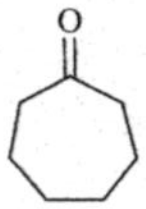

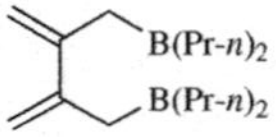

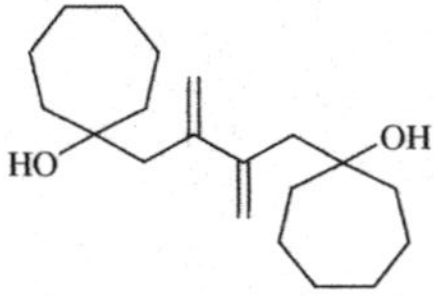

709

B(OPh)$_2$

THF, 0° to rt, 18 h

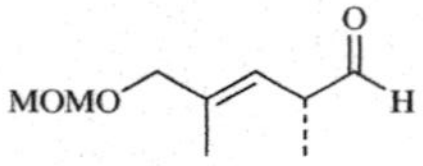

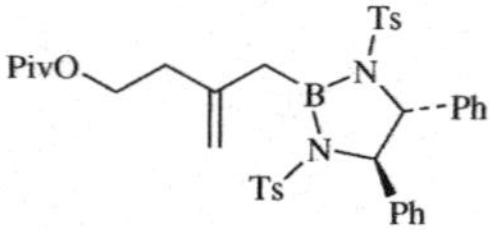

(91)

706, 702

B(Pr-n)$_2$
B(Pr-n)$_2$

Ether, −70° to rt

OH (58)

720

1. SnBu$_3$, OPiv
Ts, HN Ph, H, Ts Ph
BBr$_3$, CH$_2$Cl$_2$
2. Aldehyde, −78°

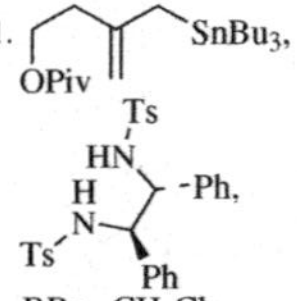

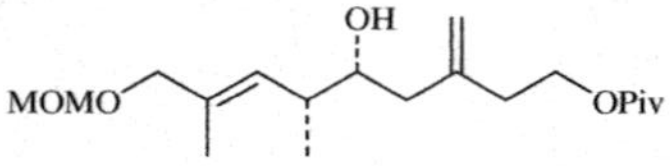

(98), 10:1 d.r.

TABLE 5. ADDITION OF β-SUBSTITUTED REAGENTS (*Continued*)

Carbonyl Substrate	Allylborane Reagent	Conditions	Product(s) and Yield(s) (%), % ee	Refs.

C₇ row:

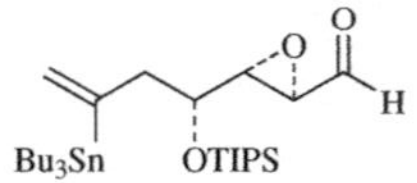

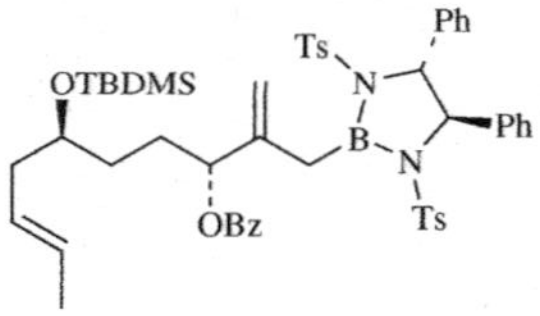

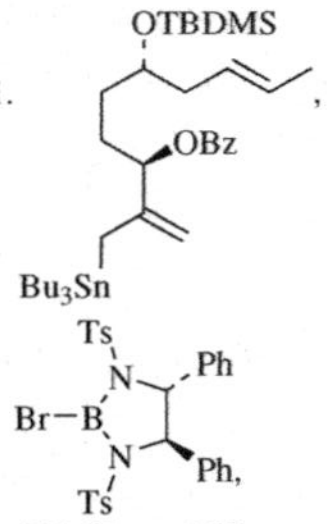

(75), 17:1 d.r. — 710

CH₂Cl₂, rt, 16 h
2. Aldehyde, −78°, 3 h

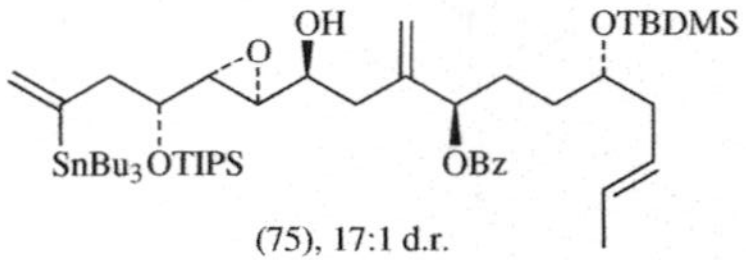

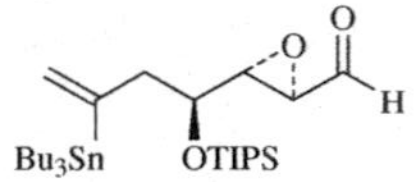

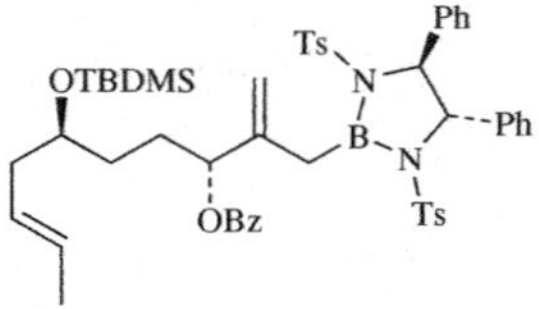

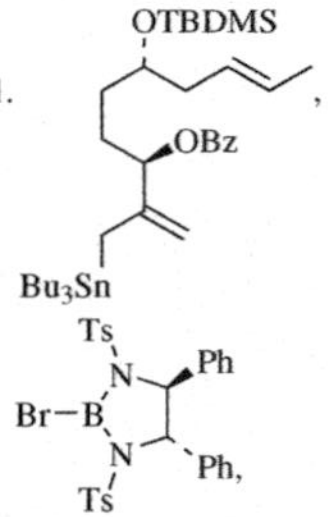

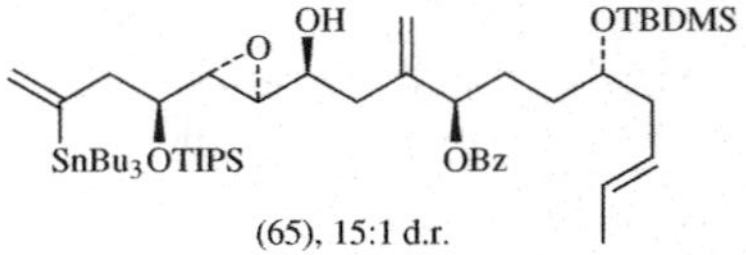

(65), 15:1 d.r. — 708

CH₂Cl₂, rt, 16 h
2. Aldehyde, −78°, 3 h

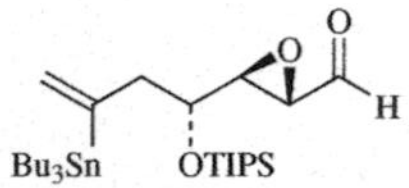

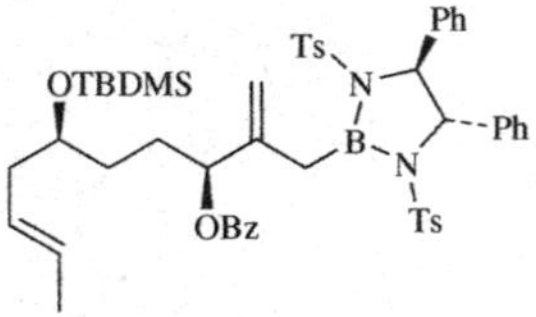

C$_8$

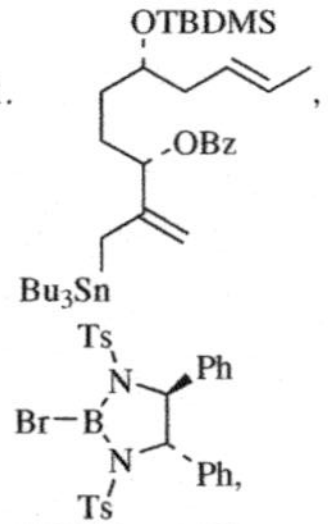

1.

2. Aldehyde, −78°, 3 h

CH$_2$Cl$_2$, rt, 16 h

(72), 17:1 d.r.

708, 721

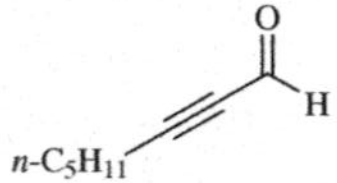

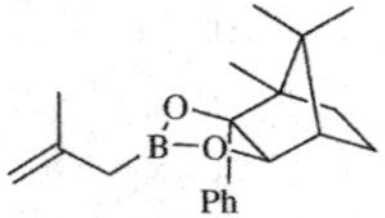

Sc(OTf)$_3$ (10 mol%), CH$_2$Cl$_2$, −78°, 12 h

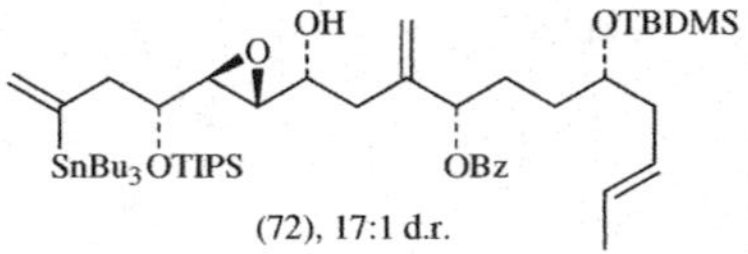

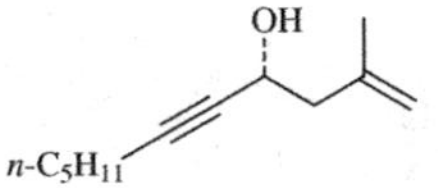

(95), 97

28

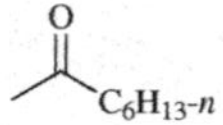

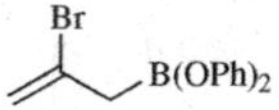

THF, 0° to rt, 18 h

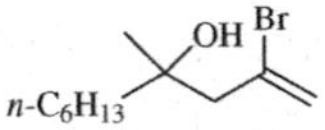

(95)

709

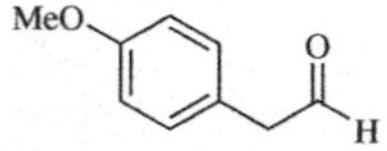

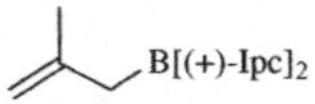

Ether, −78° to rt, 4 h

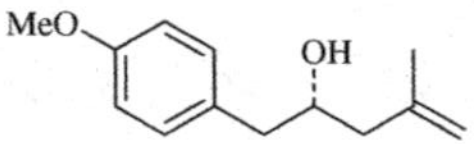

(85), 95.4

722

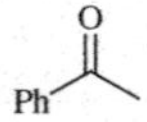

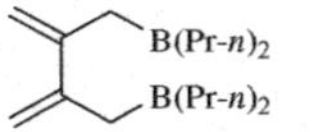

Ether, −70° to rt

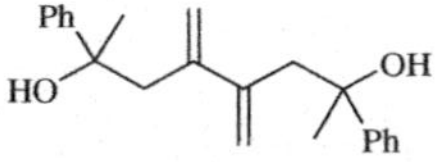

(38)

702

n-C$_5$H$_{11}$

n-C$_6$H$_{13}$

n-C$_6$H$_{13}$-n

MeO

Ph

OTBDMS

OBz

OTIPS

SnBu$_3$

OH

OBz

Bu$_3$Sn

Ts

Ph

N

B

N

Ts

Br

B(OPh)$_2$

B[(+)-Ipc]$_2$

B(Pr-n)$_2$

Ph

OH

TABLE 5. ADDITION OF β-SUBSTITUTED REAGENTS (*Continued*)

Carbonyl Substrate	Allylborane Reagent	Conditions	Product(s) and Yield(s) (%), % ee	Refs.
C₈				
		CH₂Cl₂, −78°, 2-3 h	(95), 10.5:1	710
C₉				
		THF, 0° to rt, 18 h	(95)	709
		Toluene, −78°	R Cl (84), 92 Br (79), 87	50
		Sc(OTf)₃ (10 mol%), CH₂Cl₂, −78°, 12 h	(76), 97	27, 28
		CH₂Cl₂, −78°, 2-3 h	(94), 1:1 d.r.	710, 708

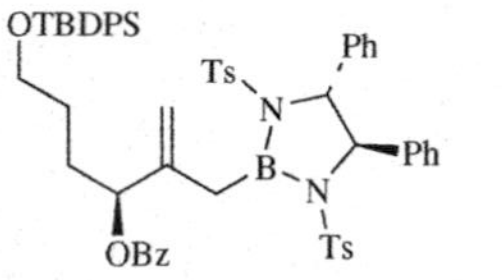

C$_{10}$

	Reagent conditions	Product	Ref.
	CH$_2$Cl$_2$, −78°, 2–3 h	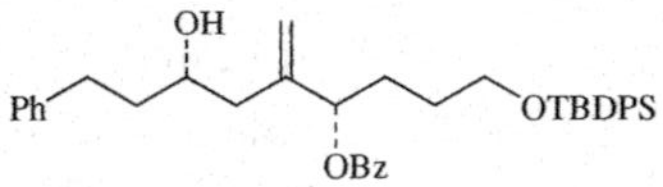 (96), >20:1 d.r.	710, 708
	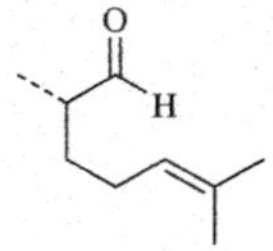 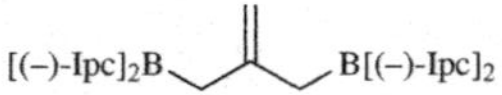 Ether, −78° ^{a}	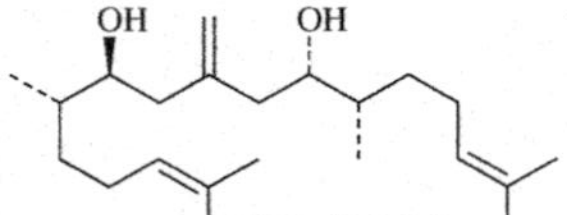 (57), 46:6:1 d.r.	176, 705
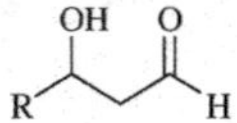	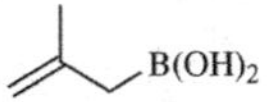 —	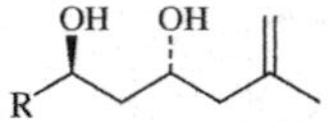 R / Ph (83) 58:42 / Cy (90) 59:41 (d.r.)	714
	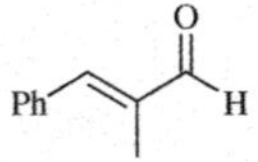 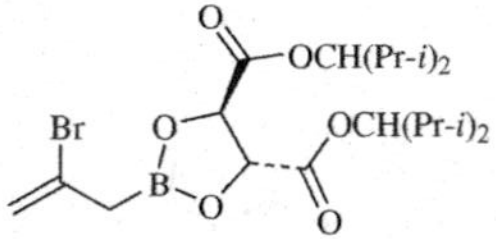 4 Å MS, toluene, −78°, 1 h	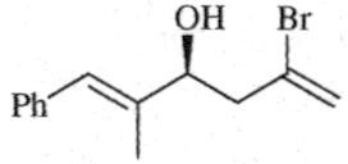 (98), 85	712
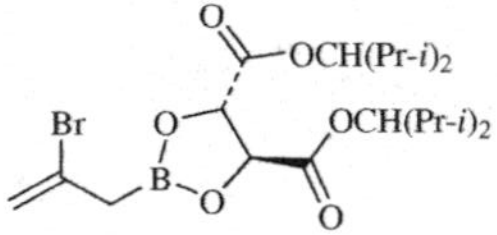	4 Å MS, toluene, −78°, 1 h	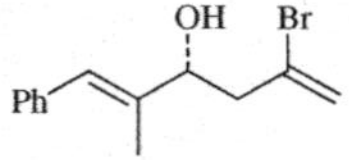 (98), 89	712
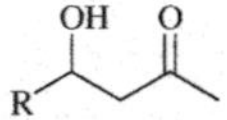	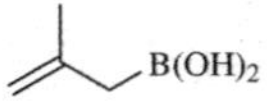 —	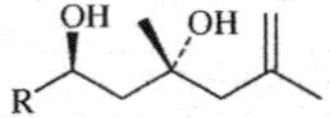 R / Ph (60) 80:20 / Cy (75) 95:5 (d.r.)	714

371

TABLE 5. ADDITION OF β-SUBSTITUTED REAGENTS (*Continued*)

Carbonyl Substrate	Allylborane Reagent	Conditions	Product(s) and Yield(s) (%), % ee	Refs.
			(81), 55:45 d.r.	714
			(100), 91:9 d.r.	723, 708
		CH$_2$Cl$_2$, −78°, 2 h	(78), 2.5:1 d.r.	708
		CH$_2$Cl$_2$, −78°, 2 h	(93), 7:1 d.r.	708

372

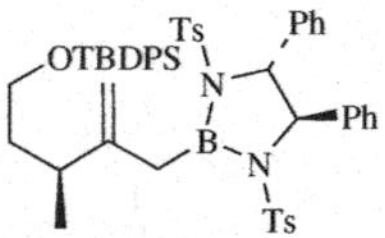

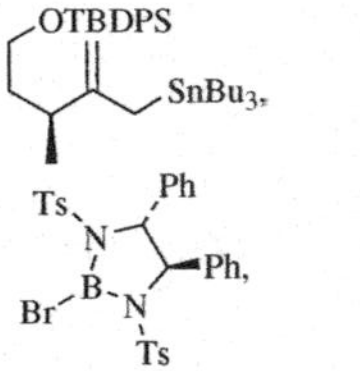

708

708

708

191

OH

OH

OH

OH

TBDPSO

TBDPSO

TBDPSO

OTBDMS

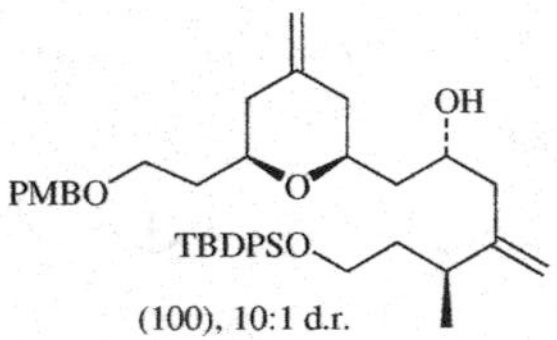

PMBO

PMBO

PMBO

PMBO

TBDMSO

(100), 10:1 d.r.

(98), 2:1 d.r.

(97), 2:1 d.r.

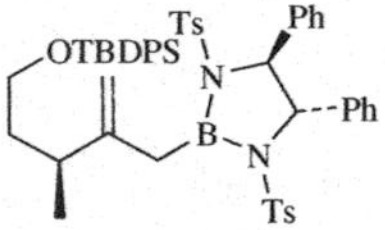

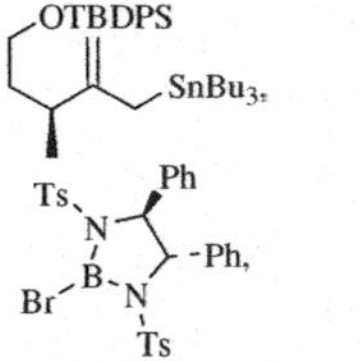

(96), 8.5:1 d.r.

OTBDPS SnBu₃

OTBDPS SnBu₃

OTBDPS SnBu₃

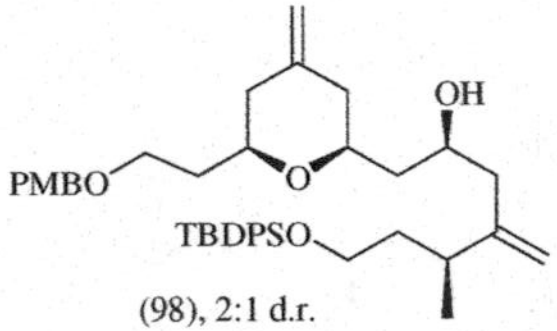

SnBu₃

Ph Ph

Ph Ph

Ph Ph

Ph

Ts–N B N–Ts

Ts–N B N–Ts

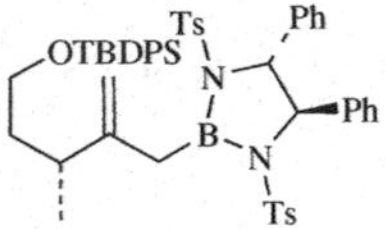

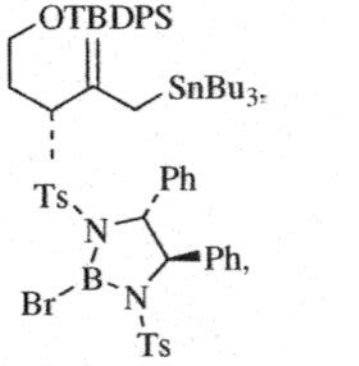

Ts–N B N–Ts

Ts–N B N–Ts

Br

Br

Br

Br

CH₂Cl₂, –78°, 2 h

CH₂Cl₂, –78°, 2 h

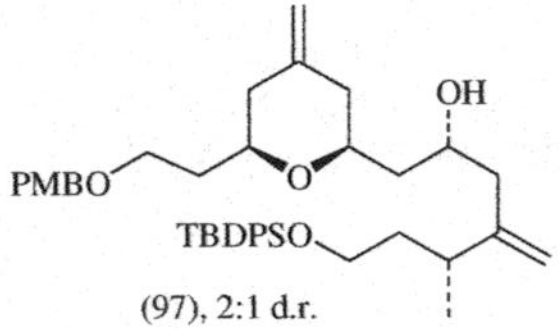

CH₂Cl₂, –78°, 2 h

CH₂Cl₂, –78°, 1.5 h

OTBDMS

OTBDMS

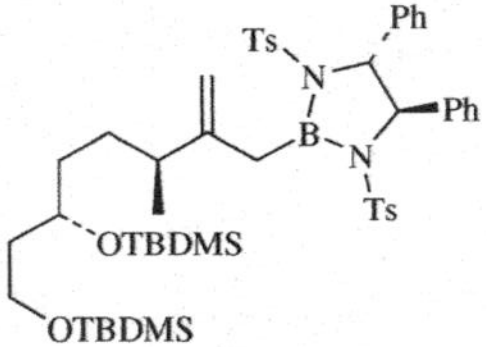

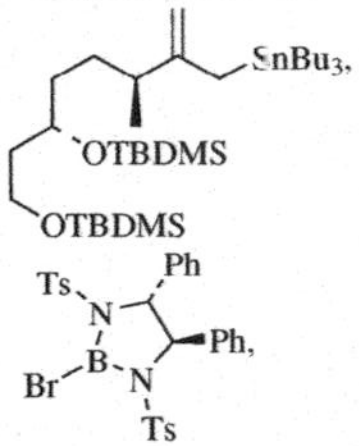

Ph Ph

Ph Ph

Ph Ph

Ph Ph

Ts–N B N–Ts

Ts–N B N–Ts

Ts–N B N–Ts

Ts–N B N–Ts

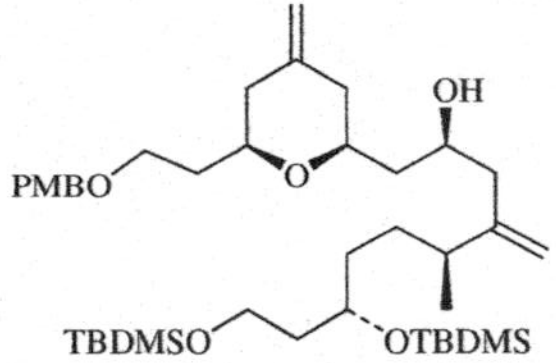

OTBDPS

OTBDPS

OTBDPS

OTBDMS

OTBDMS

373

TABLE 5. ADDITION OF β-SUBSTITUTED REAGENTS (*Continued*)

Carbonyl Substrate	Allylborane Reagent	Conditions	Product(s) and Yield(s) (%), % ee	Refs.
C_{10}				
		CH_2Cl_2, $-78°$	(95), 10.5:1 d.r.	724
C_{11}				
		—	(60), 80:20 d.r.	714
		CH_2Cl_2, $-78°$, 2-3 h	(99), 11.4:1 d.r.	710
		CH_2Cl_2, $-78°$, 2-3 h	(88), 4:1 d.r.	710

374

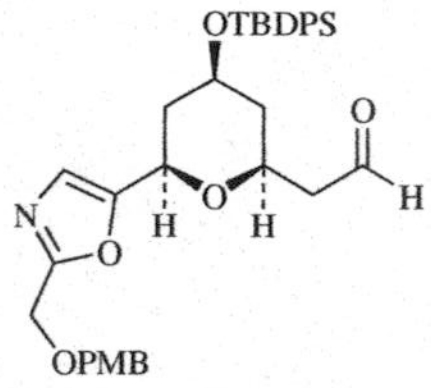

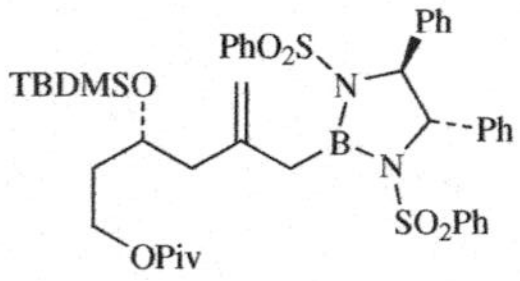

715, 716

280

725

726, 727

(96), 11.8:1 d.r.

(—)

(92), 17:1 d.r.

(69), 1.7:1 d.r.

CH_2Cl_2, −78°, 2-3 h

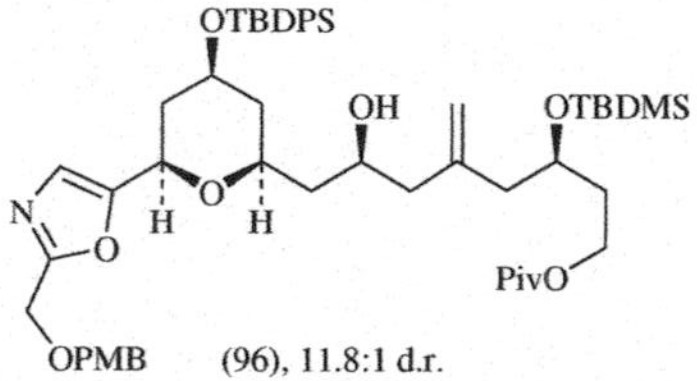

Toluene, −78° to rt, 1.5 h

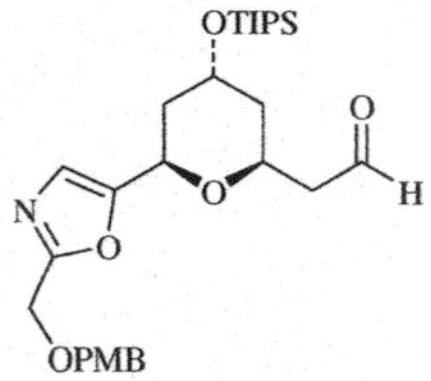

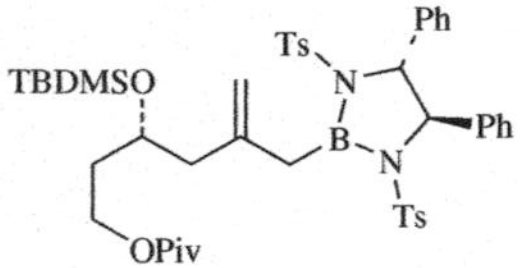

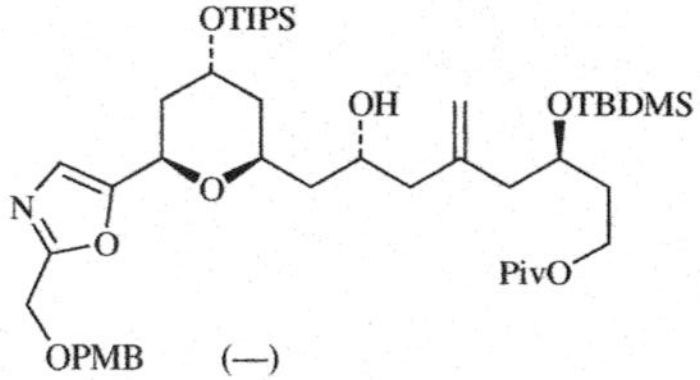

THF, −78°, 1 h

OTBDMS · PivO · OH · OTBDPS · OPMB

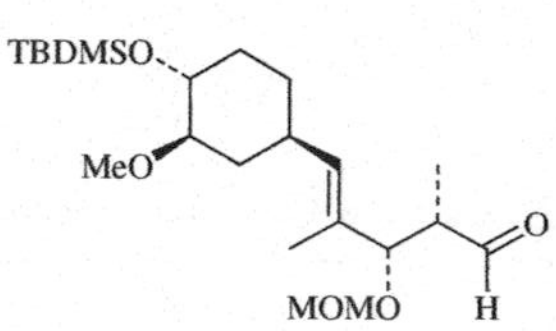

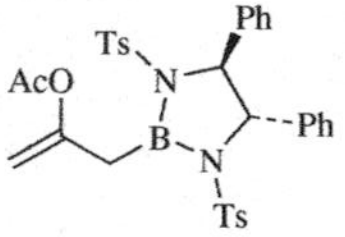

OTBDMS · PivO · OH · OTIPS · OPMB

OAc · OH · MOMO · TBDMSO · MeO

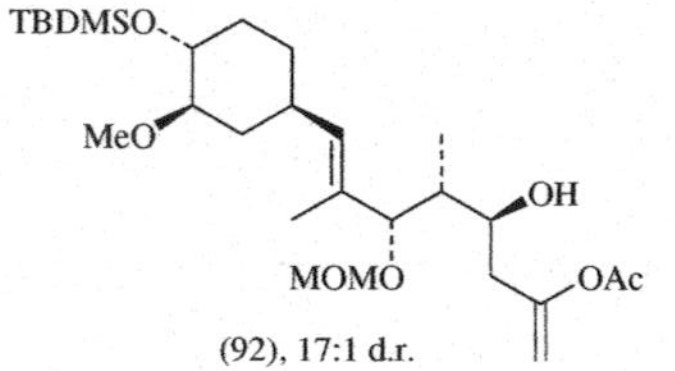

HO · OTMS · OAc · AcO

Ph · SO₂Ph · PhO₂S · N · B · OPiv · TBDMSO

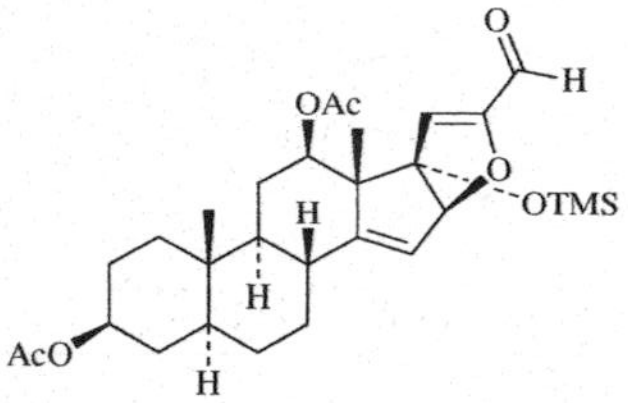

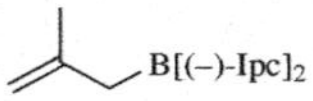

Ph · Ts · N · B · OPiv · TBDMSO

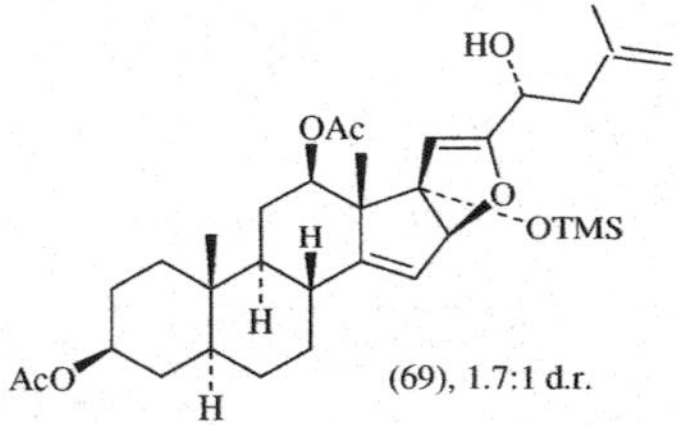

Ph · Ts · N · B · AcO

$B[(-)-Ipc]_2$

C_{13}

C_{22}

TABLE 5. ADDITION OF β-SUBSTITUTED REAGENTS (*Continued*)

Carbonyl Substrate	Allylborane Reagent	Conditions	Product(s) and Yield(s) (%), % ee	Refs.
C$_{26}$		THF, –70° to 0°, 30 min	(64)	728

a The reaction mixture was free of Mg^{2+} ions.

TABLE 6. ADDITION OF γ-SUBSTITUTED REAGENTS

Carbonyl Substrate	Allylborane Reagent	Conditions	Product(s) and Yield(s) (%), % ee	Refs.
C$_1$				
(1,3-dioxane)	(B-allyl-OPr-i borinate, isopropenyl)	Toluene, 80°, 12 h	(68), >98, >20:1 d.r.	729
C$_2$				
(acetaldehyde)	B[(−)-Ipc]$_2$, OMe	THF, −78°, 3 h	(57), 90, >99:1 d.r.	48
	B[(+)-Ipc]$_2$, OMe	THF, −78°, 3 h	(59), 92, >99:1 d.r.	48, 503
	B[(+)-Ipc]$_2$, OPMP	THF, −100°, 10 h	(—), 95, >99:1 d.r.	407
	B[(+)-Ipc]$_2$, OMEM	THF, −78°, 10 h	(—), 93, >99:1 d.r. **I** OMEM	388
	"	THF, −78°, 3 h	**I** (75), >95	505
	Ph$_2$N, B[(−)-Ipc]$_2$	THF, −78° to 0°, 3 h	Ph$_2$N (40), >95, >95:5 d.r. + (27) (OH, H NPh$_2$ cyclopropane)	730, 731

TABLE 6. ADDITION OF γ-SUBSTITUTED REAGENTS (*Continued*)

Carbonyl Substrate	Allylborane Reagent	Conditions	Product(s) and Yield(s) (%), % ee	Refs.
C₂	Ph₂N⌇⌇B[(+)-Ipc]₂	THF, −78° to 0°, 3 h	(41), >95, >95:5 d.r. + (28)	730, 731
	BEt₂ ⌇ BEt₂	−70° to 0°	(61)	732
	Ph⌇⌇BR₂	THF, −78° to rt, 4 h	R d.r. Cy (82), — >99:1 (−)-Ipc (75), 84 >99:1	733
	⌇B[(−)-Ipc]₂	1. Ether, −78° to 0°, 4 h 2. H₂O₂, NaOH	(75), 92, >97:3 d.r.	157
	PhMe₂Si⌇⌇B[(+)-Ipc]₂	THF, −78°, 4 h	(—)	734
	⌇B[(+)-Ipc]₂	THF, −78°, 12 h	(73), 91	138, 136a

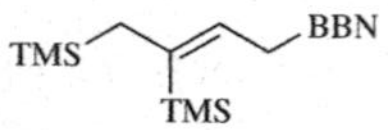

TMS

TMS

H

II

Work-Up	I + II	I:II
KOH	(61)	98:2
H₂SO₄	(50)	2:98

TMS

H

I

THF, −78°, 3 h

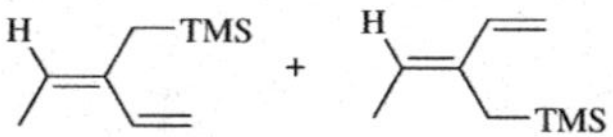

BBN

TMS

TMS

(70-80), 78:22 d.r.

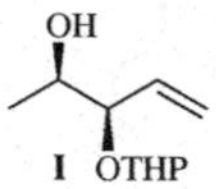

OH OTHP

I

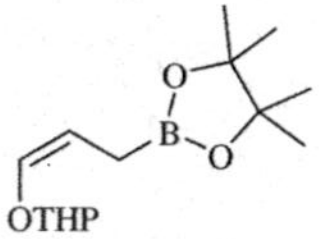

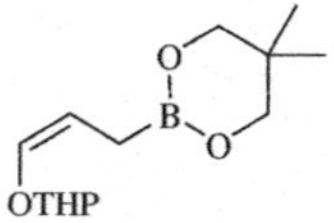

Neat, rt, 48 h

I (70-80); 78:22 d.r.

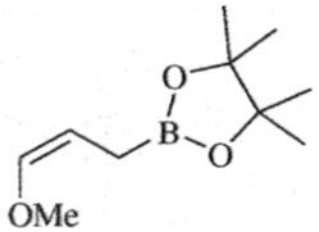

—

(85), 93:7 d.r.

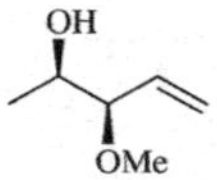

OH OMe

Neat, rt, 7 d

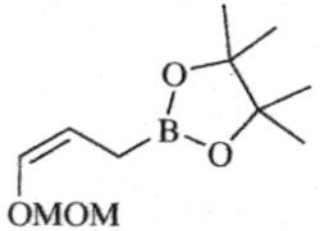

(89), 92:8 d.r.

OH OMOM

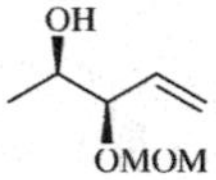

Neat, rt, 2 d

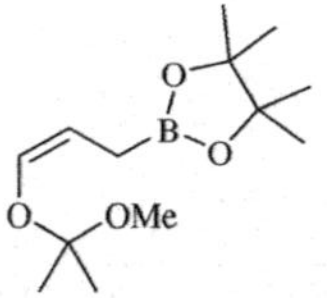

(86), 88:12 d.r.

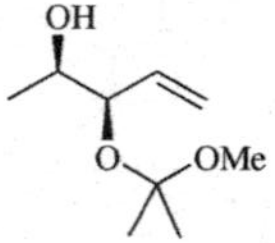

OMe

OH

Neat, rt, 2 d

TABLE 6. ADDITION OF γ-SUBSTITUTED REAGENTS (*Continued*)

Carbonyl Substrate	Allylborane Reagent	Conditions	Product(s) and Yield(s) (%), % ee	Refs.
C₂				
		Neat, rt, 2 d	(91), 93:7 d.r.	47
		Petroleum ether, rt, 36 h	(76), 95:5 d.r.	47, 36
		THF, rt, 16 h	(57), >99:1 d.r. **I**	79
	"	Cl₂Pd(PPh₃)₂, THF, 50°, 3 h; rt, 16 h	**I** (57), >99:1 d.r.	79
	SMe 8:92 E:Z	—	(86), 92:8 d.r.	737
	SEt 0.65 equiv 30:70 E:Z	—	(80), 9:1 d.r.	737

380

(TMS allyl boronate structure)	Petroleum ether	(product: OH, TMS) (92), 94:6 d.r.	47
(Me–Si–Me, CyO, OPr-i structure)	4 Å MS, toluene, −78°, 3-4 h	(product: OH, SiMe$_2$(OCy)) (94), 69, >99:1 d.r.	38, 40
(PhMe$_2$Si, OPr-i structure)	4 Å MS, toluene, −78°, 3-4 h	(product: OH, SiMe$_2$Ph) (95), 84, >99:1 d.r.	39, 40
(Ph dioxaborinane structure)	THF, rt, 16 h	(product: OH, Ph) **I** (61), >99:1 d.r.	79
"	1. IZn— B(O)... Ph, I, Cl$_2$Pd(PPh$_3$)$_2$, THF, 50°, 3 h 2. Aldehyde, rt, 16 h	**I** (61), >99:1 d.r.	79
(Cl chain, OCH(Pr-i)$_2$ structure)	Toluene/pentane (10:3), −100°	(product: OH, Cl chain) (70), 75	738

TABLE 6. ADDITION OF γ-SUBSTITUTED REAGENTS (Continued)

Carbonyl Substrate	Allylborane Reagent	Conditions	Product(s) and Yield(s) (%), % ee	Refs.
C₂		Petroleum ether, 20°, 3 d	(50), 70:30 d.r.	739
		Petroleum ether, 20°, 6 d	(61), 95:5 d.r.	739
		Neat, rt, 5-8 d	(83), 95:5 d.r.	55
		Neat, rt, 5-8 d	(83), 88:12 d.r.	55
		4 Å MS, toluene, −78°, 5 h	(88), 72, 96:4 d.r.	740
		4 Å MS, toluene, −78°, 5 h	(86), 67, 96:4 d.r.	740

	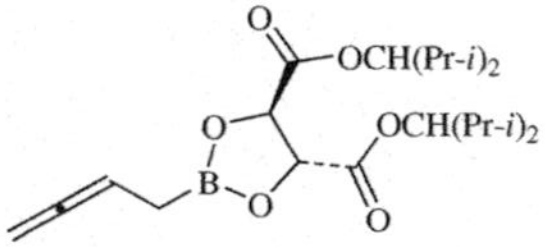O OCH(Pr-*i*)$_2$ O OCH(Pr-*i*)$_2$ B O O	4 Å MS, toluene, –78°, 72 h	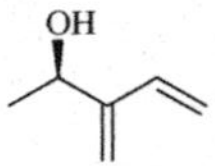OH (75), 61	741
	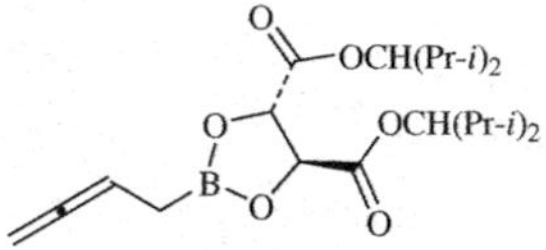O OCH(Pr-*i*)$_2$ O OCH(Pr-*i*)$_2$ B O O	4 Å MS, toluene, –78°, 72 h	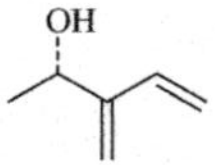OH (82), 60	741
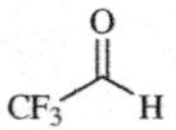O CF$_3$ H	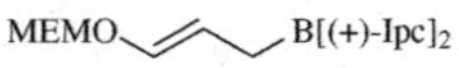MEMO B[(+)-Ipc]$_2$	Pentane/ether (1:1), –100°	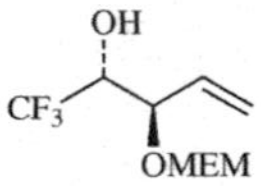OH CF$_3$ OMEM (75), 96, >19:1 d.r.	249
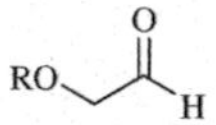O RO H R= 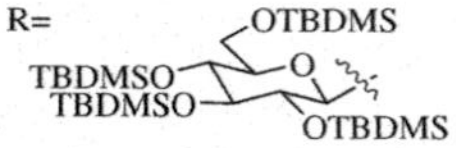OTBDMS TBDMSO TBDMSO O OTBDMS	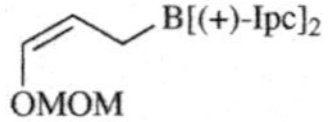B[(+)-Ipc]$_2$ OMOM	THF, –85°, 18 h	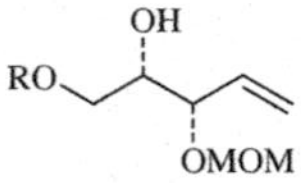OH RO OMOM (65)	742
R = Bn	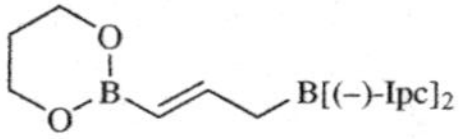O B O B[(–)-Ipc]$_2$	Ether, –78°, 2 h; rt, 24 h	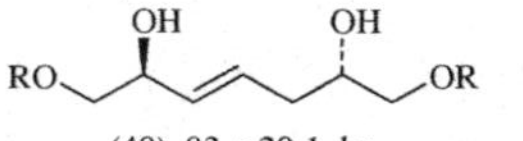OH OH RO OR (40), 93, >20:1 d.r.	175
	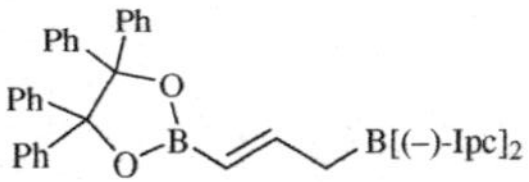Ph Ph Ph Ph O B O B[(–)-Ipc]$_2$	–78°, 2 h	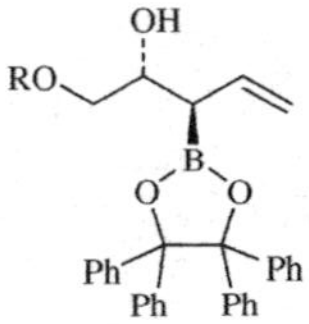OH RO B O O Ph Ph Ph Ph (—)	670

TABLE 6. ADDITION OF γ-SUBSTITUTED REAGENTS (Continued)

Carbonyl Substrate	Allylborane Reagent	Conditions	Product(s) and Yield(s) (%), % ee	Refs.

C₂

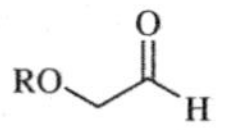

R = Bn, TBDPS

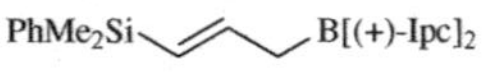

		THF, −78°, 4 h	R	734
			Bn (85), 88	
			TBDPS (70), 90	

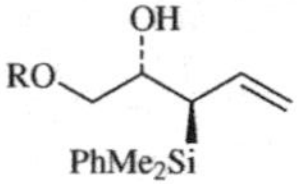

R = Bn

| | 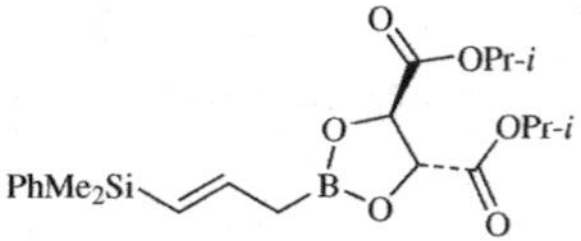 | OPr-i 4 Å MS, toluene, −78° | 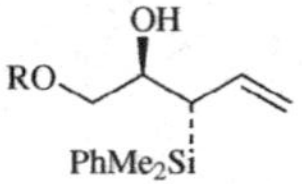 (99) | 743 |

| | | CH₂Cl₂, rt, 3 d | 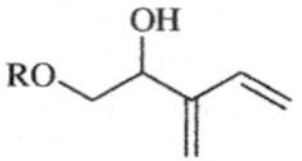 (59) | 741, 744 |

| | 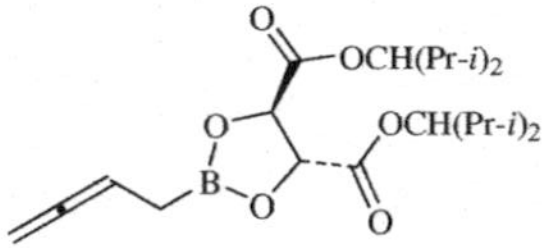 | 4 Å MS, toluene, −78°, 72 h | 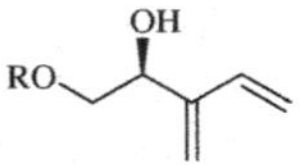 (36), 50 | 741 |

| | 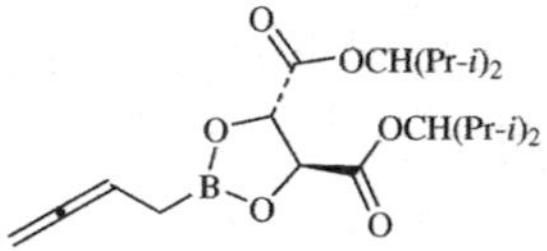 | 4 Å MS, toluene, −78°, 72 h | 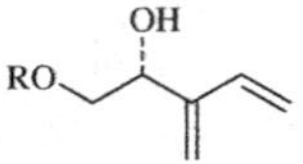 (29), 48 | 741 |

| | 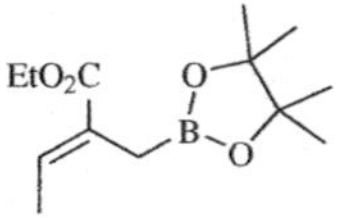 | Sc(OTf)₃, toluene, rt, 24 h | 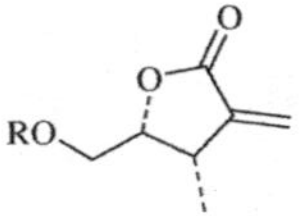 (33), >20:1 d.r. | 26 |

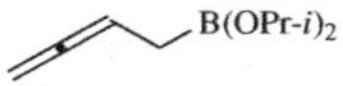

384

	Sc(OTf)$_3$, toluene, rt, 24 h	(53), 19:1 d.r.	26
R= PMB	1. , Pt(dba)$_2$, PCy$_3$, benzene, rt 2. Aldehyde, −78°, 3 h 3. NaOH, H$_2$O$_2$	(58), 66, >19:1 d.r.	185
	THF, −78°, 4 h	(85), 95	734
	THF, −78°, 3 h	(65), 88, >99:1 d.r.	48
	THF, −78°, 3 h	(68), 90, >99:1 d.r.	48

C$_3$

TABLE 6. ADDITION OF γ-SUBSTITUTED REAGENTS (*Continued*)

Carbonyl Substrate	Allylborane Reagent	Conditions	Product(s) and Yield(s) (%), % ee	Refs.
C₃ (acrolein)	B[(+)-Ipc]₂, OMOM	THF, −78°, 3 h	(62), 90, >99:1 d.r. I, OH/OMOM	48, 156
	"	—	I (66), >95	745
	B[(−)-Ipc]₂ (dioxaborinane)	1. Ether, −78° to 0°, 4 h; 2. H₂O₂, NaOH	(63), >97:3 d.r. OH/OH	157
	Ph₂N, B[(−)-Ipc]₂	THF, −78° to 0°, 3 h	(45), >95, >95:5 d.r. OH/Ph₂N	730, 731
	Ph₂N, B[(+)-Ipc]₂	THF, −78° to 0°, 3 h	(47), >95, >95:5 d.r. OH/Ph₂N	730, 731
	B[(+)-Ipc]₂ (prenyl)	THF, −78°, 12 h	(70), 95	138, 136a
(propanal)	B[(−)-Ipc]₂, OMe	THF, −78°, 3 h	(65), 88, >99:1 d.r. OH/OMe	48
	B[(+)-Ipc]₂, OMe	THF, −78°, 3 h	(68), 90, >99:1 d.r. OH/OMe	48

Reagent	Conditions	Product	Refs.
B[(−)-Ipc]₂ (crotyl, OSEM)	THF, −100° to rt	OH / OSEM (70), >95	189
B[(−)-Ipc]₂ (dioxaborinane allyl)	1. Ether, −78° to 0°, 4 h 2. H_2O_2, NaOH	OH (67), 90, >97:3 d.r.	157
Ph–N=C(Ph)– B[(−)-Ipc]₂	1. Ph–N=C(Ph), LDA, MeOB[(−)-Ipc]₂ 2. Aldehyde, THF, −78°, 3 h	N=C(Ph)Ph (48), >95:1 d.r.	746
B[(+)-Ipc]₂ (prenyl)	Ether, pentane, −100°	OH (81), 97	505
pinacol boronate, OMe	Neat, 20°, 2 d	OH / OMe (94), 92:8 d.r.	47, 33, 36
pinacol boronate, OMOM	Neat, 60°, 6 h	OH / OMOM (88), 89:11 d.r.	47, 36

TABLE 6. ADDITION OF γ-SUBSTITUTED REAGENTS (Continued)

Carbonyl Substrate	Allylborane Reagent	Conditions	Product(s) and Yield(s) (%), % ee	Refs.
C₃ propanal (see image)	(pinacol allylboronate, OMe)	Neat, rt	(88), 80:20 d.r.	47, 36
	(pinacol allylboronate, TMS)	Neat, 20°, 4 d	(94), 93:7 d.r.	47
	(pinacol allylboronate, OTHP)	Neat, rt, 48 h	**I** (70–80), >80:20 d.r.	34a
	(neopentyl glycol boronate, OTHP)	—	**I** (70–80), 80:20 d.r.	104
	(MeO allylboronate)	Petroleum ether, rt, 36 h	(68), 95:5 d.r.	47, 33
	(SR allylboronate)	—	see table below	737

E:Z / Equiv for SR reagent:
E:Z	Equiv
8:92	1.0
44:56	1.0
73:27	1.0
44:56	0.4

R		d.r.
Me	(91)	91:9
Et	(90)	56:44
Et	(94)	27:73
Et	(95)	91:9

Reagent	Conditions	Product	Refs.
B(OMe)₂ (cis-diene)	Ether, −75°	(31)	43
B(OMe)₂, Et	Ether, −75°	(44), 90:10 d.r.	43
B(OMe)₂ (diene)	Ether, −75°	(36)	43
pinacol boronate, TMS	Petroleum ether, rt, 3 d	(70), 94:6 d.r.	47
pinacol boronate, MeO	Petroleum ether, 20°, 9 d	(74), 98:2 d.r.	739
pinacol boronate, n-Bu	Neat, rt, 5-8 d	(70), 95:5 d.r.	55
pinacol boronate, Bu-n	Neat, rt, 5-8 d	(82), 90:10 d.r.	55

389

TABLE 6. ADDITION OF γ-SUBSTITUTED REAGENTS (Continued)

Carbonyl Substrate	Allylborane Reagent	Conditions	Product(s) and Yield(s) (%), % ee	Refs.
C₃				
	B(OPr-i)₂	rt		744
			Solvent: toluene, Time: 3 h, (84) Solvent: neat, Time: 1 h, (82)	
	B(OMe)₂	Ether, −75°	(34), 98:2 d.r.	43
	BE₂ / BE₂	−70° to 0°	(80)	732
	BBN	Pentane, rt, 2 h	(93) **I**	103
	"	1. HBBN, pentane, rt, 24 h 2. Ketone, rt, 2 h	**I** (80)	103
	B[(−)-Ipc]₂, OMe	THF, −78° to rt, 15 h	(CO)₆Co₂ OH, (76), >96, 94:6 d.r.	747
	MeO B[(−)-Ipc]₂	THF, −78° to rt, 15 h	(CO)₆Co₂ OH OMe, (20), >95, >97.5:2.5 d.r.	747

390

Substrate	Reagent	Conditions	Product (%), % ee	Refs.
R = Ac (RO—CH₂CH₂—CHO)	$\overset{I}{\underset{OMe}{\diagup}}$ FB(OBu-n)₂ ; B(OBu-n)₂, OMe	—	(59)	46
R = PMB	B[(+)-Ipc]₂, OMOM	−78° to 0°	(80), 90	748
R = TBDMS	B[(−)-Ipc]₂, Cl	Ether, −78°	(85), >95, >95:5 d.r.	631
R = Bn, PMB	PhMe₂Si⟶B[(+)-Ipc]₂	THF, −78°, 4 h	R: Bn (74), 90; PMB (72), 95	734
R = TBDMS	B[(−)-Ipc]₂	—	(—), 99	749
	B[(+)-Ipc]₂	Ether, −78° to rt, 12 h	(49), 95	749
R = PMB	B[(+)-Ipc]₂	Ether, pentane, −100°	I (82), 95	505
R = Bn	dioxaborolane, OMe	THF, cyclohexane, benzene, −100° to rt	(46), 88	545

TABLE 6. ADDITION OF γ-SUBSTITUTED REAGENTS (*Continued*)

Carbonyl Substrate	Allylborane Reagent	Conditions	Product(s) and Yield(s) (%), % ee	Refs.
C₃				
R = Ac	THPO	—	(71), 13:1 d.r.	52
R = Bn	*n*-Bu	4 Å MS, toluene, −78°, 5 h	(56), 85, 97:3 d.r.	740
	n-Bu	4 Å MS, toluene, −78°, 5 h	(57), 82, 97:3 d.r.	740
	Bu-*n*	4 Å MS, toluene, −78°, 5 h	(56), 73, 97:3 d.r.	740
	Bu-*n*	4 Å MS, toluene, −78°, 5 h	(46), 80, 97:3 d.r.	740
TBDPSO	EtO₂C	Neat, rt, >12 d	(75) **I**	59

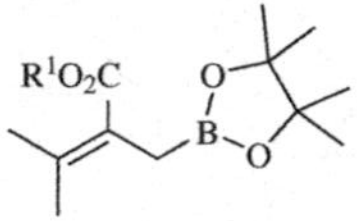

R¹O₂C ...

R¹ = (−)-8-phenylmenthyl

1. Neat, rt, 14 d
2. pTSA

I (80), 82   59

rt

"

R¹ = see table   59, 26

Solvent	Time	**I** R¹	R²	
neat	>12 d	Et	—	(92), 0
neat	8 d	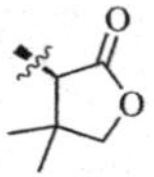	—	(89), 22
neat	14 d	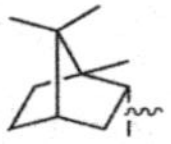 Ph	—	(64), −6
toluene	26 d	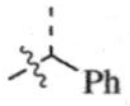	—	(70), 10
neat	8 d	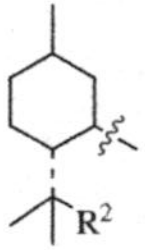 R²	H	(78), 7
toluene	36 d	"	Ph	(51), 80
toluene	14 d	"	2-Nph	(95), 82
CH₂Cl₂	14 d	"	2-Nph	(86), 75
toluene	14 d	"	4-MeOC₆H₄	(74), 56
toluene	14 d	"	4-PhC₆H₄	(24), 66
toluene	14 d	"	3,5-Me₂C₆H₃	(6), 62

TABLE 6. ADDITION OF γ-SUBSTITUTED REAGENTS (Continued)

Carbonyl Substrate	Allylborane Reagent	Conditions	Product(s) and Yield(s) (%), % ee	Refs.
C₃				
		THPO-⟨⟩-Li, ⟨pinacol boronate with Cl⟩	(50), 13:1 d.r. ⟨structure with OH, OTHP, PhS⟩	52
	⟨pinacol allylboronate, THPO⟩			
⟨PhS-CH₂CH₂-CHO⟩				
⟨aldehyde with OBn⟩	⟨pinacol allylboronate, OMe⟩	—	(—), 88:12 d.r. ⟨structure with OH, OMe, OBn⟩	275, 750
	⟨pinacol allylboronate, TMS⟩	Petroleum ether, 22°, 7 d	(87), 92, 82:18 d.r. ⟨structure with OH, OBn, TMS⟩	275
	⟨pinacol allylboronate, MeO⟩	Petroleum ether, 22°, 7 d	(97), 92, 57:43 d.r. ⟨structure with OH, OMe, OBn⟩	275
	⟨pinacol allylboronate, TMS-O⟩	Petroleum ether, 22°, 4 d	(94), 92, 60:40 d.r. ⟨structure with OH, OBn, TMS⟩	275
	⟨pinacol allylboronate, TMS⟩	Neat, rt, 4 d	(52), 58:42 d.r. ⟨structure with OH, OBn, TMS⟩	521

394

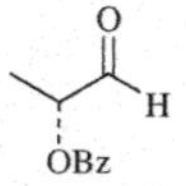

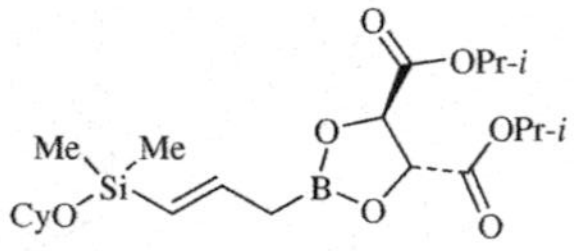

38, 40

(80), 84:16 d.r.

38, 40

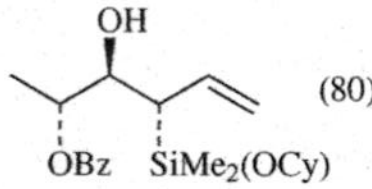

(90), 23:1 d.r.

739

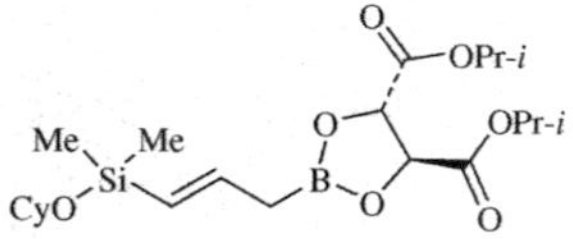

R	I+II	III
TBDMS	(24)	(57)
Bn	(15)	(82)
THP	(8)	(46)

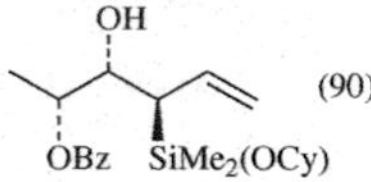

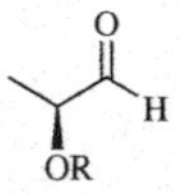

751, 680

(83), 50:50:50 d.r.

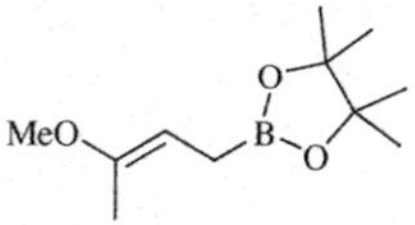

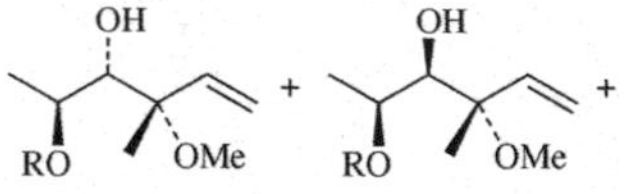

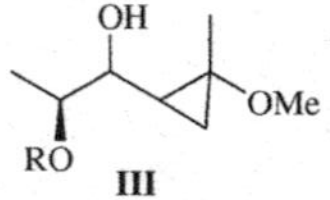

521

(71), >95:5 d.r.

4 Å MS, toluene, −78°, 3-4 h

4 Å MS, toluene, −78°, 3-4 h

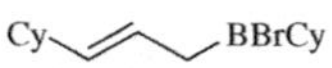

6-8 kbar

1. [propargyl bromide], HBCy₂, n-Bu₄NBr, −78° to 0°, 1 h
2. Aldehyde, 0° to rt, 1 h

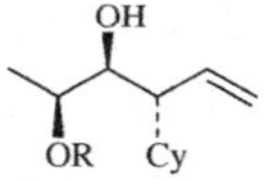

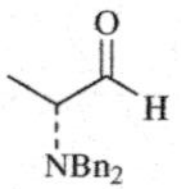

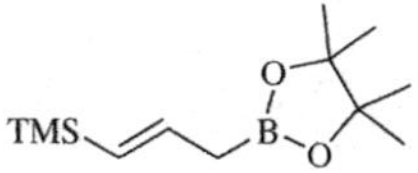

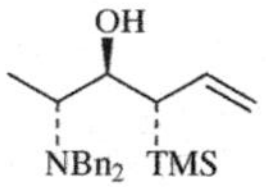

Neat, rt, 4 d

R = TBDMS, Bn, THP

R = TBDMS

TABLE 6. ADDITION OF γ-SUBSTITUTED REAGENTS (Continued)

Carbonyl Substrate	Allylborane Reagent	Conditions	Product(s) and Yield(s) (%), % ee	Refs.
C₃				
	$(i\text{-Pr})_2NMe_2Si$—...—B[(−)-Ipc]₂	1. Ether, −78°, 3.5 h 2. KF, H₂O₂	(57), >95:5 d.r.	752
	$(i\text{-Pr})_2NMe_2Si$—...—B[(+)-Ipc]₂	1. Ether, −78°, 3.5 h 2. KF, H₂O₂	I (45), 2:1 d.r.	752, 45, 511
	BBN ... Cl	Ether, THF, −95°, 6 h	(68), 95:5 d.r.	753, 754
	B[(+)-Ipc]₂ ... Cl	Ether, THF, −95°, 6 h	I (37), 62:38 d.r.	753
	B[(−)-Ipc]₂ ... Cl	Ether, THF, −95°, 6 h	I (72), >97:3 d.r.	753, 754
	TMS—...—B[(−)-Ipc]₂	1. KHCO₃ 2. KF, H₂O₂	(57), >98:2 d.r.	278
	TMS—...—B[(+)-Ipc]₂	1. KHCO₃ 2. KF, H₂O₂	I (45), 67:33 d.r.	278
	Ph—N=...—B[(−)-Ipc]₂, Ph	THF, −78°	(58), 1.2:1 d.r.	160

Aldehyde	Reagent	Conditions	Product	Refs.
TBDMSO–CHO (OBn)	(allyl imine) Ph–N=C(Ph)–CH₂CH=CH₂	1. Ph–N=C(Ph)–CH₂CH=CH₂, LDA, MeOB[(−)-Ipc]₂; 2. Aldehyde, THF, −78°, 3 h	**I** (48), 1.2:1 d.r.	746
		1. Ph–N=C(Ph)–CH₂CH=CH₂, LDA, MeOB[(+)-Ipc]₂; 2. Aldehyde, THF, −78°, 3 h	**I** (40), >95:5 d.r.	746
	"	THF, −78°	**I** (40), >95:5 d.r.	160
	n-Pr, B[(−)-Ipc]₂, SiMe₂Ph	THF, −78° to rt	TBDMSO···OH···SiMe₂Ph, BnO, n-Pr^i (56), 1.8:1 d.r.	206
	n-Pr, B[(−)-Ipc]₂, SiMe₂Bn	THF, −78° to rt	TBDMSO···OH···SiMe₂Bn, BnO, n-Pr^i (52), 1.8:1 d.r.	206
(dioxaspiro aldehyde)	B(OMe)₂, OMOM	THF, hexane, rt, 2 d	OH···OMOM (75-85), >20:1 d.r.	755

397

TABLE 6. ADDITION OF γ-SUBSTITUTED REAGENTS (Continued)

Carbonyl Substrate	Allylborane Reagent	Conditions	Product(s) and Yield(s) (%), % ee	Refs.
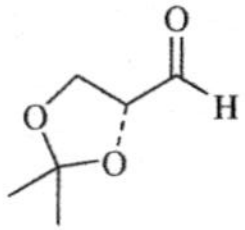	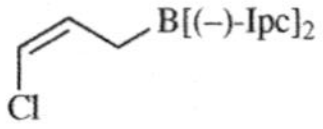 B[(–)-Ipc]₂ reagent	Ether, –95°, 4 h	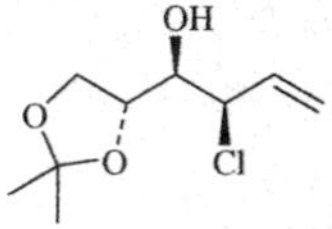 (65), 99.5:0.5 d.r.	756
	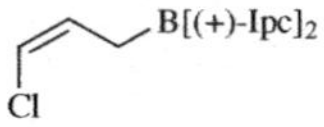 B[(+)-Ipc]₂ reagent	Ether, –95°, 4 h	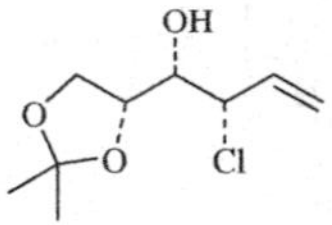 (68), 61:39 d.r.	756
	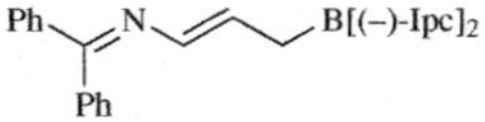 Ph-N=/B[(–)-Ipc]₂ reagent	1. 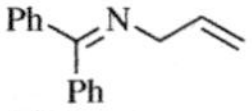 Ph\N=, Ph, LDA, MeOB[(–)-Ipc]₂, 2. Aldehyde, THF, –78°, 3 h	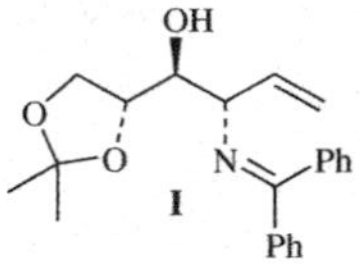 (43), >95:5 d.r. **I**	746
	"	THF, –78°	**I** (43), >95:5 d.r.	160
	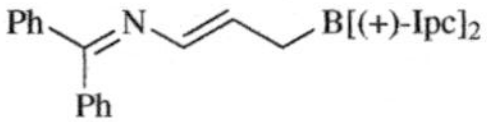 Ph-N=/B[(+)-Ipc]₂ reagent	1. 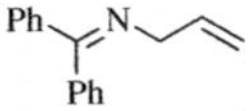 Ph\N=, Ph, LDA, MeOB[(+)-Ipc]₂, 2. Aldehyde, THF, –78°, 3 h	**I** (41), 2.7:1 d.r.	746
	"	THF, –78°	**I** (41), 2.7:1 d.r.	160

C₃

398

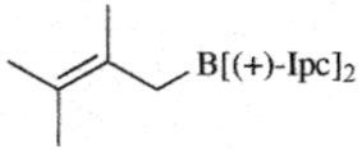

(73)

(55), 10:1 d.r.

(90), 23:1 d.r.   SiMe₂Ph

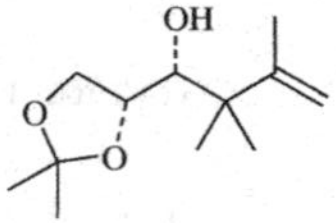

(96), 20:1 d.r.   SiMe₂Ph

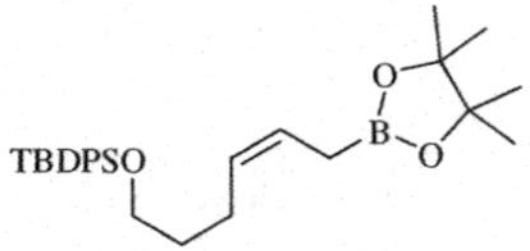

SiMe₂R

OTBDPS

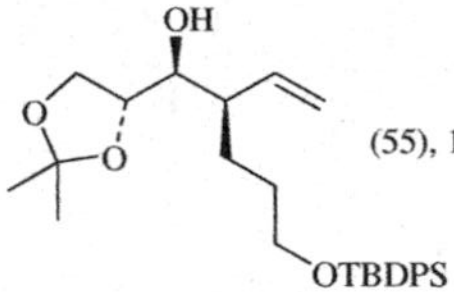

533

61

39, 40

39, 40

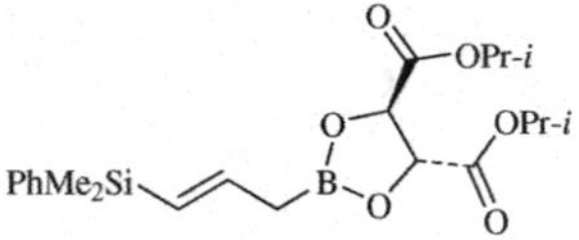

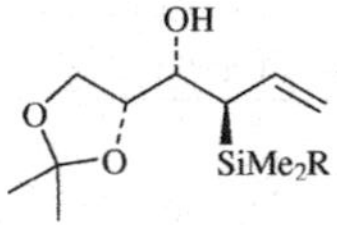

158, 38, 40

R		d.r.
2-(5-methyl)furyl	(93)	95:5
menthofuryl	(79)	>96:4
CyO	(82)	>95:5

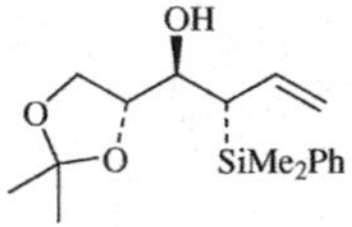

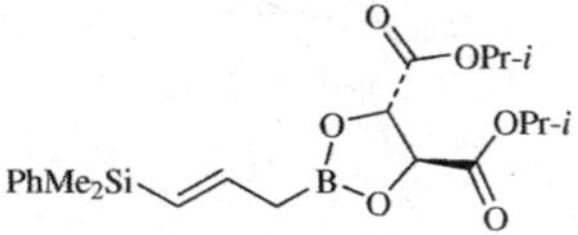

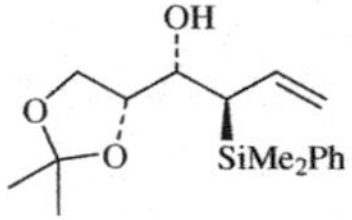

B[(+)-Ipc]₂

BF₃•Et₂O, THF, −78°, 1 h

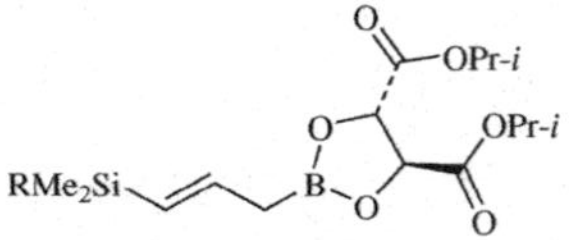

TBDPSO

CH₂Cl₂, rt, 12-24 h

PhMe₂Si   OPr-i   4 Å MS, toluene, −78°

PhMe₂Si   OPr-i   4 Å MS, toluene, −78°

RMe₂Si   OPr-i   4 Å MS, toluene, −78°

TABLE 6. ADDITION OF γ-SUBSTITUTED REAGENTS (*Continued*)

Carbonyl Substrate	Allylborane Reagent	Conditions	Product(s) and Yield(s) (%), % ee	Refs.

C₃

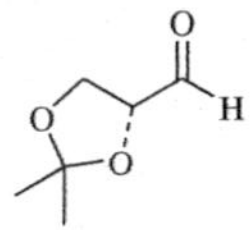

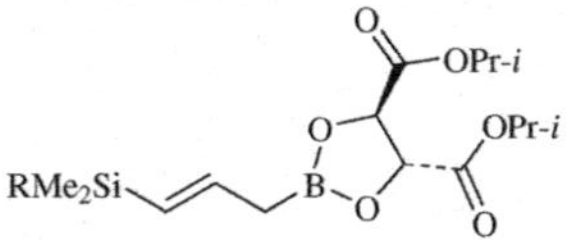

| | | 4 Å MS, toluene, −78° | | 158, 38, 40 |

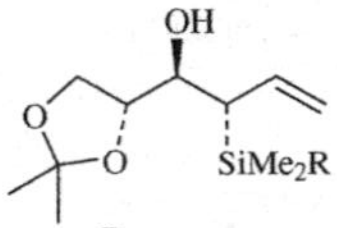

R	d.r.
2-(5-methyl)furyl	(90) 94:6
menthofuryl	(85) 92:8
CyO	(88) 85:15

R = menthofuryl — 4 Å MS, toluene, −78° — (79), >96:4 d.r. — 757

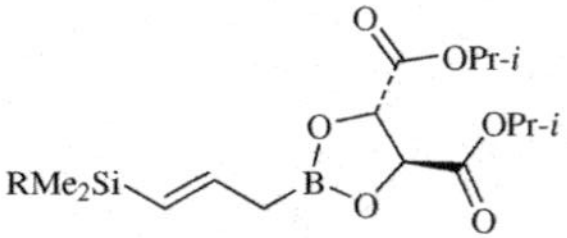

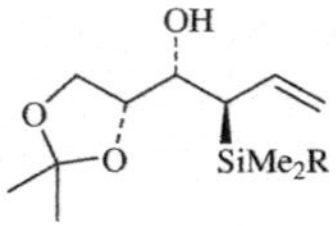

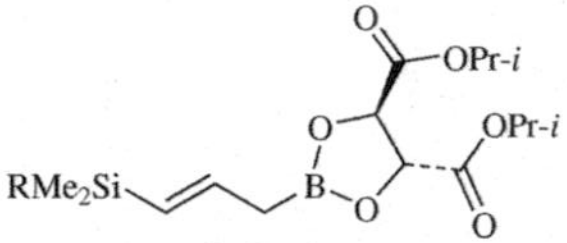

R = menthofuryl — 4 Å MS, toluene, −78° — (85), 92:8 d.r. — 757

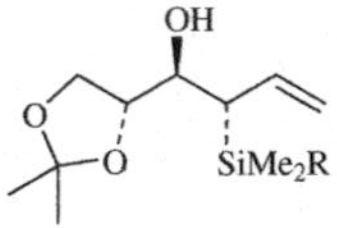

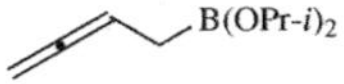

B(OPr-i)₂ — CH₂Cl₂, 12 h

Temp		d.r.
rt	(93)	87:13
0°	(—)	92:8

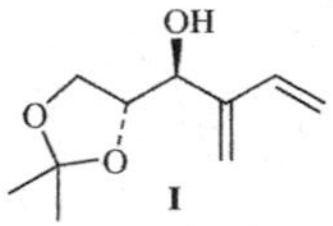

741, 744

" — Toluene, rt, 3 h — **I** (90), 89:11 d.r. — 741

400

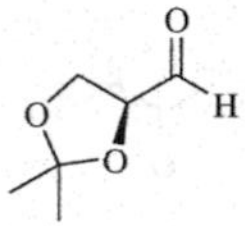

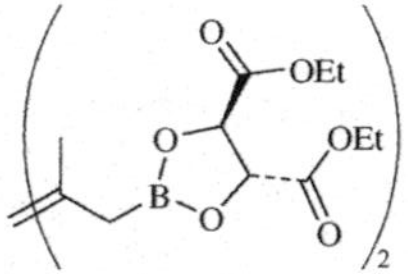

185

(71), >19:1 d.r.

1. −78°, 3 h
2. NaOH, H$_2$O$_2$

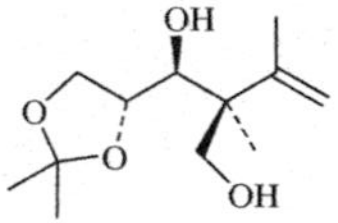

751, 680

(40), 85:15 d.r.

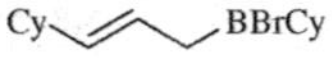

1. Br — propargyl bromide

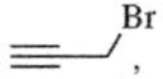

HBCy$_2$, n-Bu$_4$NBr, −78° to 0°, 1 h
2. Aldehyde, 0° to rt, 1 h

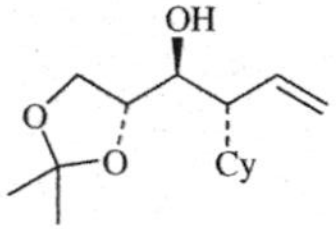

730, 731

(23), >95, 23:6 d.r.

(19)

+

THF, −78° to 0°, 3 h

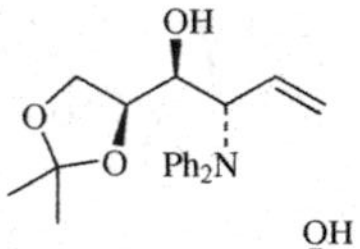

BBrCy

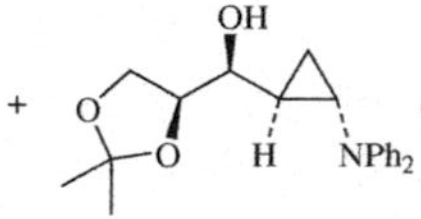

730, 731

(28), >95, >95:5 d.r.

(26)

+

THF, −78° to 0°, 3 h

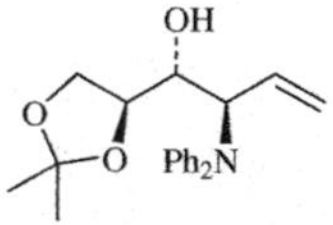

Ph$_2$N — B[(−)-Ipc]$_2$

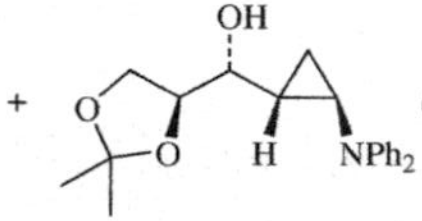

Ph$_2$N — B[(+)-Ipc]$_2$

TABLE 6. ADDITION OF γ-SUBSTITUTED REAGENTS (Continued)

Carbonyl Substrate	Allylborane Reagent	Conditions	Product(s) and Yield(s) (%), % ee	Refs.
C₃				
		1. –78°, 3 h 2. NaOH, H₂O₂	(66), 1:1 d.r.	185
		4 Å MS, toluene, –78°	(79), >99:1 d.r.	194, 758
		Petroleum ether, 6 kbar, 45°, 80 h	(85), 14:1 d.r.	104
C₄				
		THF, –78°	(65-68), 82	759
		THF, –78°	(85), 96	138, 140
		THF, –78°	(83), 96	138, 140

402

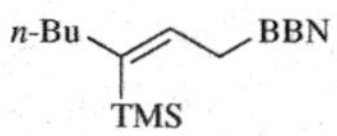

736

II

I + II (53), I:II = 93:7

1. THF, −78°, 3 h
2. NaOH

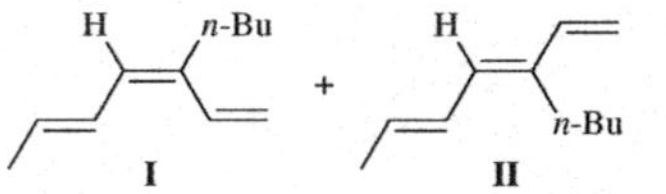

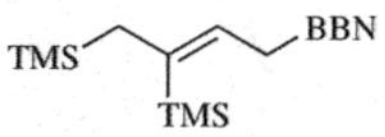

735

II

I + II (59), I:II = 98:2

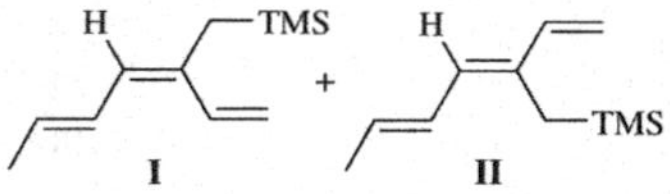

1. THF, −78°, 3 h
2. KOH

26

(61), 1:1 d.r.

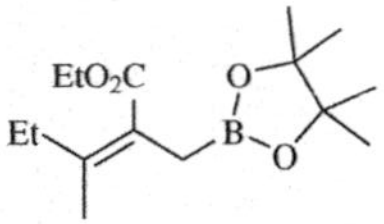

Sc(OTf)$_3$, toluene,
rt, 16-24 h

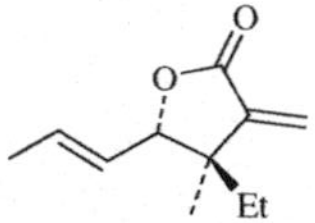

751, 680

(84), >98:2 d.r.

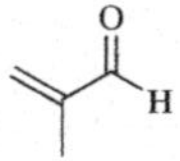

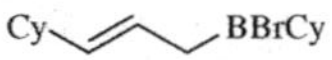

1.

HBCy$_2$, n-Bu$_4$NBr,
−78° to 0°, 1 h
2. Aldehyde, 0° to rt,
1 h

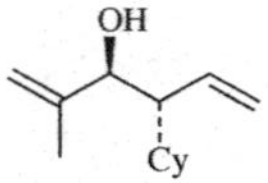

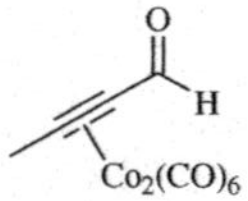

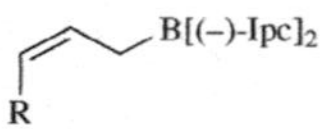

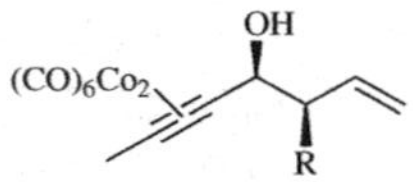

747

R		d.r.
OMe	(73), >96	>97.5:2.5
OMOM	(65), >96	>97.5:2.5

THF, −78° to rt,
15 h

TABLE 6. ADDITION OF γ-SUBSTITUTED REAGENTS (Continued)

Carbonyl Substrate	Allylborane Reagent	Conditions	Product(s) and Yield(s) (%), % ee	Refs.
C₄				
(structure: O=CH–C≡C–Co₂(CO)₆)	MeO⌒⌒B[(−)-Ipc]₂	THF, −78° to rt, 15 h	(44), >95, >97.5:2.5 d.r.	747
(butenal, O=CH–CH₂–CH=CH₂)	PhMe₂Si⌒⌒⌒B[(−)-Ipc]₂	THF, −78°	(77), 98	760
n-Pr–CHO	⌒⌒B(Pr-*n*)₂	Neat, 60°	(62)	238
	⌒⌒B[(+)-Ipc]₂	Ether, −78°, 12 h	(79), 92	138, 136a
	(pinacol boronate, OTHP)	Neat, rt, 48 h	(70–80), >80:20 d.r.	34a
	(tartrate boronate OCH(Pr-*i*)₂, Cl)	Toluene/pentane (10:3), −100°	(85), 96	738
	(tartrate boronate OCH(Pr-*i*)₂, Cl)	Toluene/pentane (10:3), −100°	(82), 93	738

PhMe₂Si–B (pinacol boronate), n-Bu	Neat, rt, 5-8 d	(76), 95:5 d.r., OH, n-Pr, Bu-n	55
Bu-n pinacol boronate	Neat, rt, 5-8 d	(79), 90:10 d.r., OH, n-Pr, Bu-n	55
PhMe₂Si pinacol boronate, Ph	PhMe₂Si–B, methyl iodocyclohexenone, Pd₂(dba)₃, EtOAc, 80°, 5 h	(70), >99:1 d.r., SiMe₂Ph, OH, n-Pr, Ph	761
Et, B·Et₂SePh	1. PhSe, LDA, THF, −78°, 30 min; 2. Et₃B, −78° to rt, 12 h	(85), 75:25 d.r., OH, n-Pr, Et	692
B(OBu-n)₂, OPh	OPh, FB(OBu-n)₂	OH, i-Pr, OPh I + OPh, OH, i-Pr II; I + II (72), I:II = 3:1	46
B[(+)-Ipc]₂, OMe	THF, −78°, 3 h	(57), 88, >99:1 d.r., OH, i-Pr, OMe	48

O, i-Pr, H

TABLE 6. ADDITION OF γ-SUBSTITUTED REAGENTS (Continued)

Carbonyl Substrate	Allylborane Reagent	Conditions	Product(s) and Yield(s) (%), % ee	Refs.
C$_4$ i-Pr–CHO	CH$_2$=CH–CH(OMe)–CH$_2$–B[(−)-Ipc]$_2$	THF, −78°, 3 h	OH, i-Pr, OMe (62), 88, >99:1 d.r.	48
	CH$_2$=CH–CH(Cl)–CH$_2$–B[(−)-Ipc]$_2$	1. allyl Cl, MeOB[(−)-Ipc]$_2$, LDA, ether, −78° 2. BF$_3$•OEt$_2$ 3. Aldehyde	OH, i-Pr, Cl (—), 77	159, 762
	(1,3,2-dioxaborinan-2-yl)CH$_2$CH=CH–B[(−)-Ipc]$_2$	Ether, −78°, 2 h; rt, 24 h	OH, i-Pr, Pr-i (34), 84, >20:1 d.r.	175
	"	Ether, −78°, 2 h	OH, i-Pr, B(OCH$_2$)$_2$ (—)	175
	(4,4,5,5-tetraphenyl-1,3,2-dioxaborolan-2-yl)CH=CH–CH$_2$–B[(−)-Ipc]$_2$	Ether, −78°, 2 h	OH, i-Pr, B(Ph)(Ph) (—)	175
	(CH$_3$)$_2$C=CH–CH$_2$–B[(+)-Ipc]$_2$	Ether, pentane, −100°	OH, i-Pr I (85), 95	505
	"	−78°, 12 h	I (73), 89	138, 136a

Reagent	Conditions	Product	Refs.
(boronate, OMe)	Neat, 20°, 2 d	OH, i-Pr, OMe (94), 89:11 d.r.	47, 33, 36
(boronate, OMOM)	Neat, rt, 2 d	OH, i-Pr, OMOM (90), 80:20 d.r.	47, 36
(boronate, OMe)	Neat, rt, 2 d	OH, i-Pr, OMe (57), 68:32 d.r.	47, 36
(boronate, OTHP)	—	OH, i-Pr, OTHP (70–80), 80:20 d.r	104
(boronate, TMS)	Neat, 20°, 14 d	OH, i-Pr, TMS (76), 89:11 d.r.	47
(boronate, TMS)	Petroleum ether, rt, 2 d	OH, i-Pr, TMS (86), >95:5 d.r.	47
(boronate, MeO)	Petroleum ether, rt, 36 h	OH, i-Pr, OMe (77), >98:2 d.r.	47, 33

407

TABLE 6. ADDITION OF γ-SUBSTITUTED REAGENTS (Continued)

Carbonyl Substrate	Allylborane Reagent	Conditions	Product(s) and Yield(s) (%), % ee	Refs.
C₄ ![i-Pr CHO]	(pinacolboronate, SR) Z:E / Equiv >95:5 / 1.0 56:44 / 0.4	—	(OH, i-Pr, SR) R / Z:E Me (90) / >95:5 Et (95) / 5:95	737
	(pinacolboronate, n-Bu)	Neat, 4 kbar, rt	(81), 99:1 d.r.	55
	(pinacolboronate, Bu-n)	Neat, 4 kbar, rt	(87), 92:8 d.r.	55
	(pinacolboronate, MeO)	Petroleum ether, 20°, >20 d	(31), 98:2 d.r. **I**	739
	"	Petroleum ether, 8 kbar, 46°, 5 h	**I** (70), 97:3 d.r.	739
	B(OPr-i)₂	Toluene, rt, 8 h	(80)	741, 744
	(boronate, OCH(Pr-i)₂, OCH(Pr-i)₂)	4 Å MS, toluene, −78°, 72 h	(79), 88	741

408

Reagent	Conditions	Product	Yield (d.r.)	Refs.
(diisopropyl tartrate allylboronate, OCH(Pr-i)$_2$)	4 Å MS, toluene, −78°, 72 h	(OH, i-Pr, 2-methylene-3-butenyl)	(82), 88	741
(diisopropyl tartrate allylboronate, OPr-i)	4 Å MS, toluene, −78°, 72 h	(OH, i-Pr, 2-methylene-3-butenyl)	(80), 81	741
(pinacol boronate, EtO$_2$C, Bu-n)	Sc(OTf)$_3$, toluene, rt, 24 h	(α-methylene-γ-butyrolactone, i-Pr, n-Bu)	(32), >20:1 d.r.	26
BBrCy (Cy crotyl)	1. propargyl Br, HBCy$_2$, n-Bu$_4$NBr, −78° to 0°, 1 h 2. Aldehyde. 0° to rt, 1 h	(OH, i-Pr, Cy)	(72), >98:2 d.r.	751, 680
BBr(C$_5$H$_9$-c), c-C$_5$H$_9$	1. propargyl Br, HB(C$_5$H$_9$-c)$_2$, n-Bu$_4$NBr, −78° to 0°, 1 h 2. Aldehyde. 0° to rt, 1 h	(OH, i-Pr, C$_5$H$_9$-c)	(60), >98:2 d.r.	751, 680
(pinacol boronate, TMS)	Neat, rt, 4 d	(OH, Cl, TMS)	(90), 65:35 d.r.	521

(aldehyde: O=CH, Cl, ethyl)

TABLE 6. ADDITION OF γ-SUBSTITUTED REAGENTS (Continued)

Carbonyl Substrate	Allylborane Reagent	Conditions	Product(s) and Yield(s) (%), % ee	Refs.
C₄	TMS, BBN, TMS	THF, −78°, 3 h	**I** + **II** (68), **I:II** = 55:45	735, 736
	B[(+)-Ipc]₂ / OMEM	—	OMEM (69), 94, >99:1 d.r.	388
	B[(−)-Ipc]₂ / Cl	Ether, −95°, 4 h	Cl (55), 95:5 d.r.	651
	B[(+)-Ipc]₂ / Cl	Ether, −95°, 4 h	Cl (58), 86:14 d.r.	651
	B(OMe)₂ / OMOM	THF, rt	OMOM (40), 70:30 d.r.	602
	RMe₂Si / OPr-i / OPr-i	4 Å MS, toluene, −78°	SiMe₂R	158, 38, 40

R		d.r.
2-(5-methyl)furyl	(67)	>96:4
menthofuryl	(82)	94:6
CyO	(76)	>95:5

410

R		d.r.
2-(5-methyl)furyl	(58)	67:33
menthofuryl	(80)	55:45
CyO	(70)	64:36

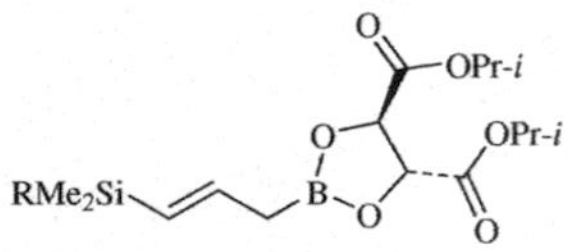

(82), 96:4 d.r.

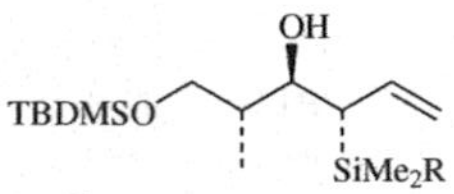

(80), 55:45 d.r.

(83), 8:1 d.r.

(60), >95:5 d.r.

(55), 85:15 d.r.

R = menthofuryl

R = menthofuryl

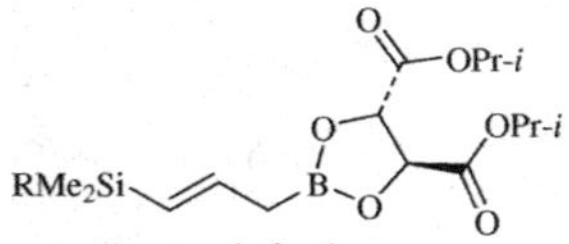

B[(+)-Ipc]₂

B[(+)-Ipc]₂

B[(−)-Ipc]₂

R = PMB

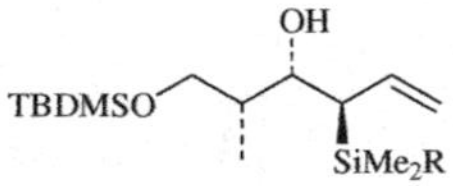

R = TES

R = TBDMS

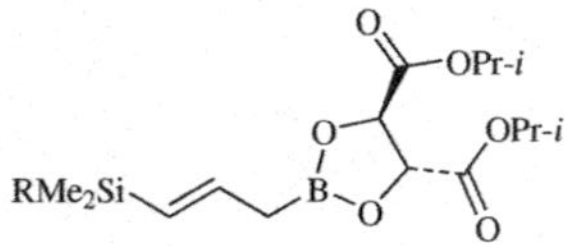

4 Å MS, toluene, −78°

4 Å MS, toluene, −78°

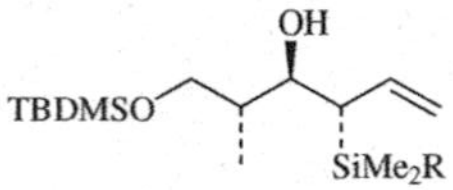

4 Å MS, toluene, −78°

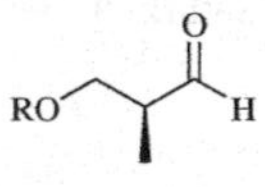

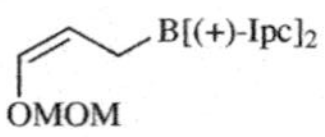

THF, −78° to rt

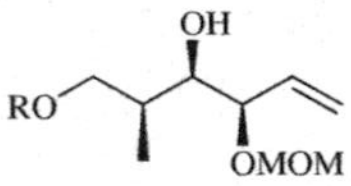

THF, −89°, 18 h

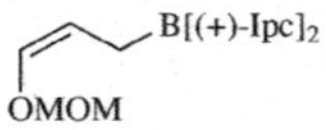

Ether, −95°, 4 h

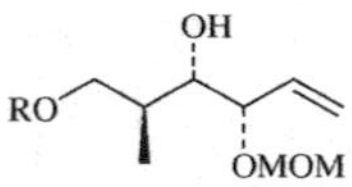

158, 38, 40

757

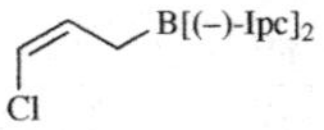

757

763

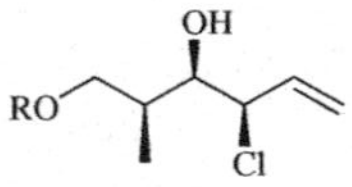

764

651

411

TABLE 6. ADDITION OF γ-SUBSTITUTED REAGENTS (Continued)

Carbonyl Substrate	Allylborane Reagent	Conditions	Product(s) and Yield(s) (%), % ee	Refs.
C₄				
RO—CH₂—CH(CH₃)—CHO; R = TBDMS	(Z)-CH₂=... B[(+)-Ipc]₂ (Cl-substituted)	Ether, –95°, 4 h	(56), 95:5 d.r.	651
R = PMB	B[(+)-Ipc]₂	Ether, pentane, –100°	(90), 96:4 d.r.	505
R = TBDPS	PhMe₂Si—, OPr-i, O-B-O, CO₂ tartrate ester "	4 Å MS, toluene, –78°	**I** (75)	743
R = TBDMS	PhMe₂Si—, OPr-i, O-B-O, CO₂ tartrate ester	4 Å MS, toluene, –78°	**I** (91), >20:1 d.r.	39, 40
		4 Å MS, toluene, –78°	(92), 1.5:1 d.r.	39, 40
TBDPSO—CH(CH₃)—CH₂—CHO	B[(–)-Ipc]₂ (Cl-substituted)	Ether, –95°, 4 h	(68), 93:7 d.r.	765

1. B(NMe₂)₂, OMe, OMe, O-B-O, –78° to rt, 1.5 h
2. L.A., –78° to rt, 3 h

	Solvent		d.r.
L.A.			
FeCl₃	THF	(61)	87:13
Ti(OPr-i)₄	ether	(84)	72:28
ZrCl₄	THF	(72)	77:23

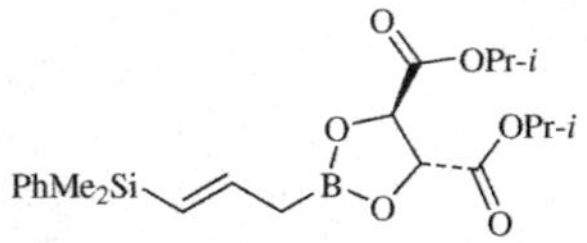

SiMe₂Ph

(82), 15:1 d.r. — 40, 39

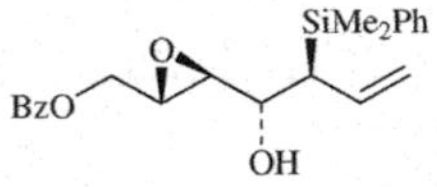

SiMe₂Ph

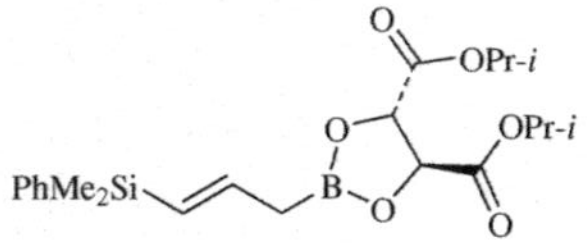

(89), 8:1 d.r. — 40, 39

SiMe₂(OPh)

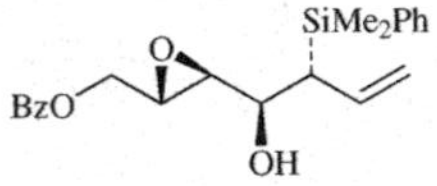

(53), 2:1 d.r. — 38, 40

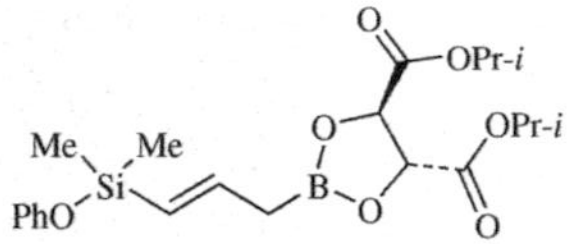

SiMe₂(OPh)

(80), 13:1 d.r. — 38, 40

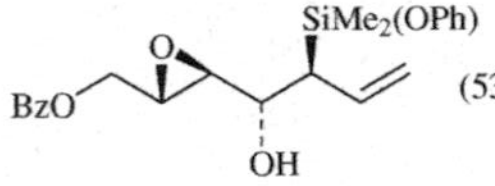

OMe OH

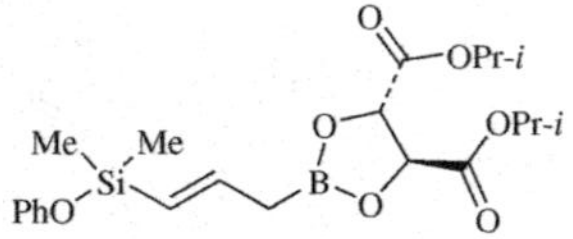

I (93), 97.5:2.5 d.r. — 505

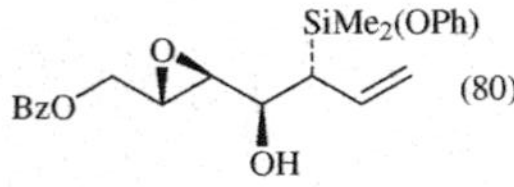

I (82), 2:1 d.r. — 53

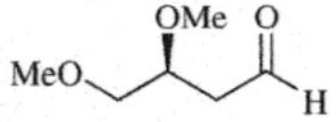

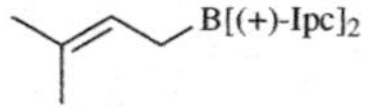

Conditions: 4 Å MS, toluene, −78°

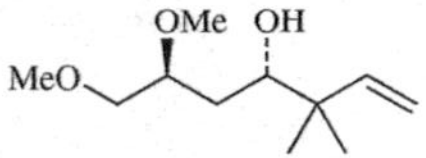

Ether, pentane, −100°

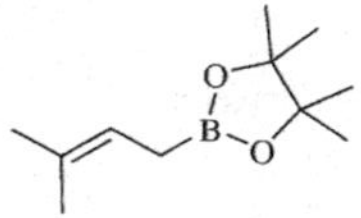

Petroleum ether, 4 kbar, 3 d

TABLE 6. ADDITION OF γ-SUBSTITUTED REAGENTS (Continued)

Carbonyl Substrate	Allylborane Reagent	Conditions	Product(s) and Yield(s) (%), % ee	Refs.
C₄				
	B(OMe)₂ reagent with OMe	THF, −78°	(83) product **I**	766, 767
	pinacol boronate with OMe	CH₂Cl₂, −78° to rt, 24-48 h	**I** (70-75), 95:5 d.r.	118, 767, 768
	B[(−)-Ipc]₂ with OMEM	THF, −95°, 3 h; rt	product with OMEM (87), 95, >99:1 d.r.	769, 770
	B[(−)-Ipc]₂ with OMOM	THF, −78°	product with OMOM (60), >96.5:3.5 d.r.	771
	B[(+)-Ipc]₂ reagent (diphenyl boronate, cis-allyl)	CH₂Cl₂, −78°	product (diphenyl pinacol-type boronate) R: PMB (—), t-Bu (—)	201

C_5 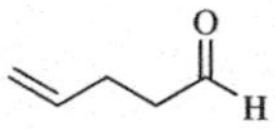	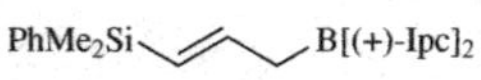	THF, –78°	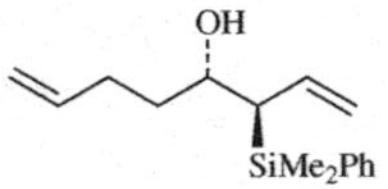 (86), 94	760
	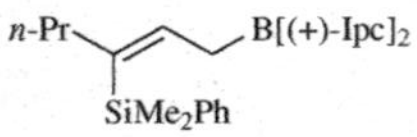	THF	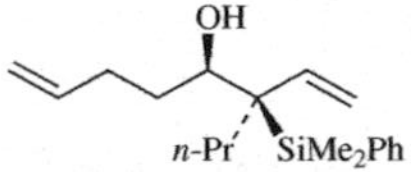	760

Temp	Time		d.r.
66°	6 h	(76)	3:1
rt	6 h	(77)	3:1
–23°	6 h	(79)	8:1
–50°	14 h	(67)	10:1

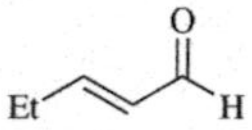	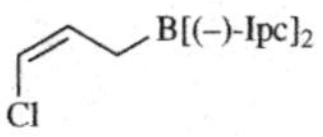	1. , MeOB[(–)-Ipc]$_2$, LiNCy$_2$, ether, –78° 2. BF$_3$•OEt$_2$ 3. Aldehyde	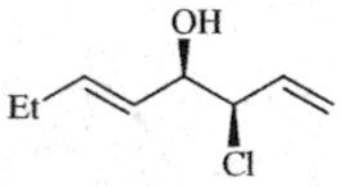 (78), 99, 99:1 d.r.	159, 762
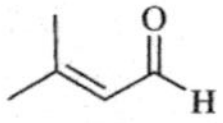	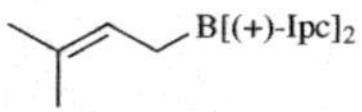	–78°, 12 h	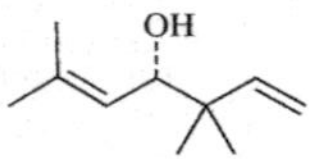 (85), 96	138
	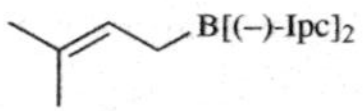	–78°, 12 h	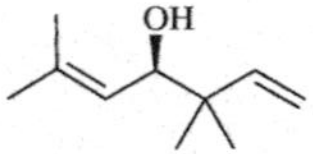 (83), 96	138
	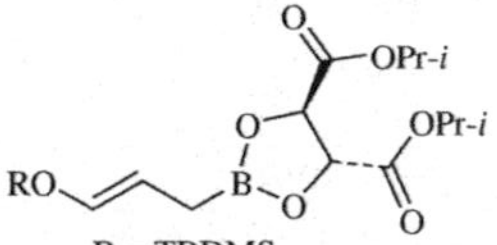 R = TBDMS	4 Å MS, toluene, –78°, 5 h	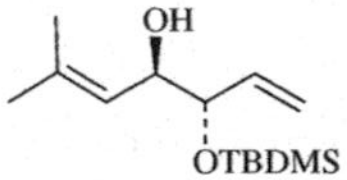 (84), 69	88

TABLE 6. ADDITION OF γ-SUBSTITUTED REAGENTS (Continued)

Carbonyl Substrate	Allylborane Reagent	Conditions	Product(s) and Yield(s) (%), % ee	Refs.
C₅				
	$R = CH_2CO_2Me$	–78°, 5 h	(78), 82	90
		THF, –78° to rt, 18 h	(81), 89:11 d.r.	772
		THF, –78° to rt, 18 h	(89), 95:5 d.r.	772
		4 Å MS, toluene, –78°, 5 h	I (83), 73, 97:3 d.r.	740
		4 Å MS, toluene, –78°, 5 h	(77), 71, 96:4 d.r.	740
		Sc(OTf)₃, toluene, rt, 6-24 h	(62), 19:1 d.r.	24, 26

416

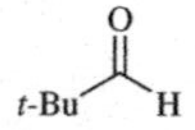	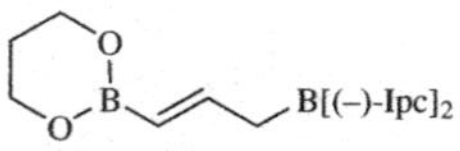	1. Ether, –78° to 0°, 4 h 2. H_2O_2, NaOH	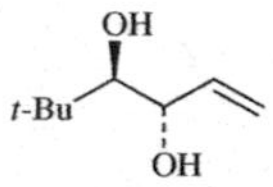 (59), >95, >97:3 d.r.	157
	"	Ether, –78°, 2 h; rt, 24 h	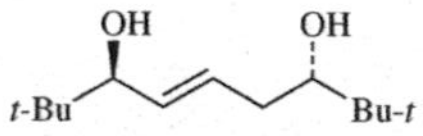 (34), 92, >20:1 d.r.	175
	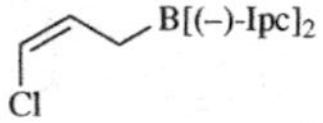	1. , MeOB[(–)-Ipc]$_2$, LiNCy$_2$, THF, –78° 2. BF_3•OEt_2 3. Aldehyde	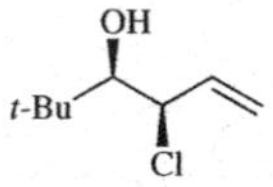 (65), 78	159
	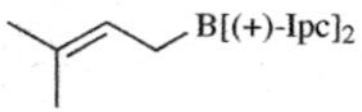	Ether, pentane, –100°	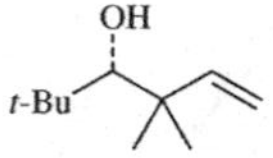 (88), 95	505
	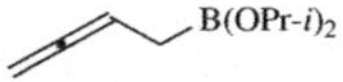	Toluene, rt, 3 d	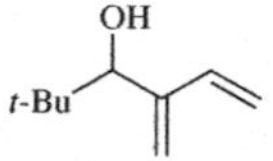 (60)	741, 744
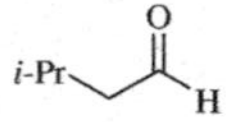	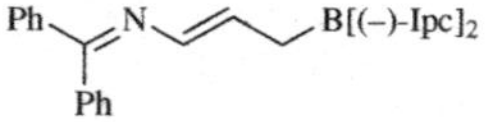	1. 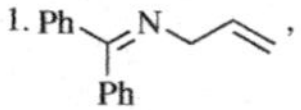, LDA, MeOB[(–)-Ipc]$_2$, 2. Aldehyde, THF, –78°, 3 h	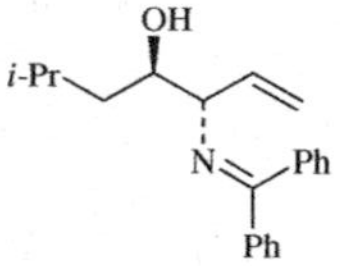 (41), >95:5 d.r.	746
	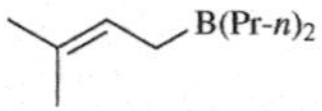	Neat, 60°	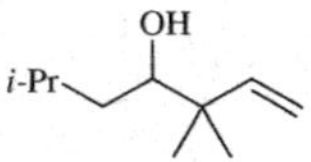 (65)	238

TABLE 6. ADDITION OF γ-SUBSTITUTED REAGENTS (Continued)

Carbonyl Substrate	Allylborane Reagent	Conditions	Product(s) and Yield(s) (%), % ee	Refs.
C₅		4 Å MS, toluene, −78°, 5 h	(93), 84, 96:4 d.r.	740
		4 Å MS, toluene, −78°, 5 h	(85), 72, 96:4 d.r.	740
		4 Å MS, toluene, −78°, 5 h	(72), 63, 96:4 d.r.	740
		4 Å MS, toluene, −78°, 5 h	(76), 66, 96:4 d.r.	740
		Sc(OTf)₃, toluene, rt, 24 h	(46), >20:1 d.r.	26
		Sc(OTf)₃, toluene, rt, 6-24 h	(53), 19:1 d.r.	24, 26

418

				Ref.
	B[(−)-Ipc]₂ with (Z)-ClCH₂CH=CHCH₂–	Ether, −95°, 4 h	(59), 85:15 d.r.	651
	B[(+)-Ipc]₂	Ether, −95°, 4 h	(57), 95:5 d.r.	651
	pinacol boronate, TMS	Neat, rt, 4 d	(82), 72:28 d.r.	539
	pinacol boronate, R₃Si	Neat, rt, 4 d		
	pinacol boronate, R	Pentane, 4 kbar, rt, 3 d		539

R (SiR₃):

R		d.r.	Ref.
Me	(79)	73:27	539, 521
Et	(70)	74:26	539

R:

R		d.r.	Ref.
TIPS	(50)	72:28	539
t-Bu	(66)	75:25	

Conditions	Product	Ref.
−70° to 0°	(70)	732
Ether, −95°, 4 h	(61), 85:15 d.r.	651

TABLE 6. ADDITION OF γ-SUBSTITUTED REAGENTS (Continued)

Carbonyl Substrate	Allylborane Reagent	Conditions	Product(s) and Yield(s) (%), % ee	Refs.
C₅				
		Ether, −95°, 4 h	(60), 99:1 d.r.	651
		—	(—)	202
		THF, ether, −78°	(60), >95, >97.5:2.5 d.r.	511
		THF, ether, −78°	(49), >95, >19:1 d.r.	513
		THF, −78° to rt, 3 h	(73)	773
		THF, −78°	(47)	515

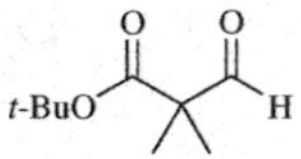

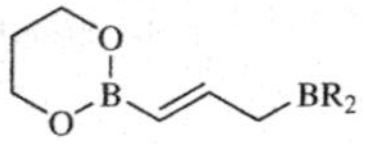

R	Solvent	Yield	ee
(+)-Ipc	ether	77	85
(+)-Ipc	CH$_2$Cl$_2$	74	85
(+)-Ipc	toluene	69	83
2-d-Icr	ether	36	>95

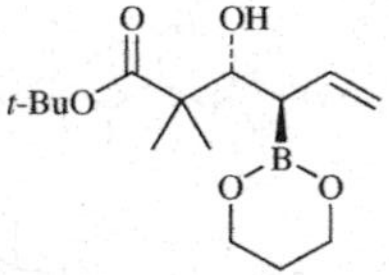

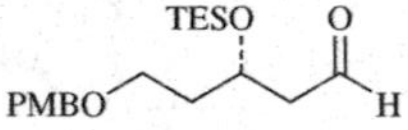

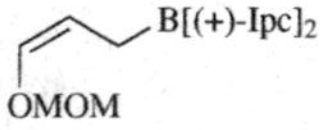

203

−78°

(91), >10:1 d.r.

458

THF, −95° to rt, 3 h

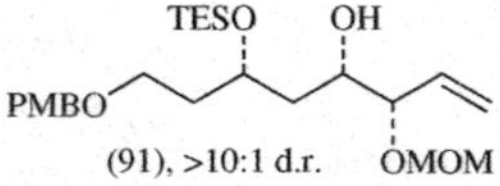

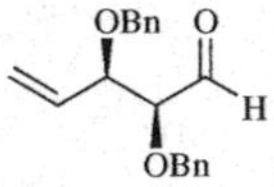

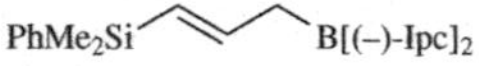

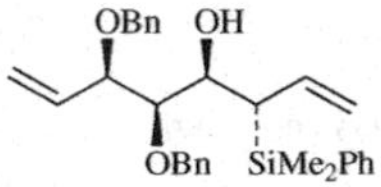

(83), 9:1 d.r.

774

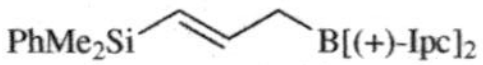

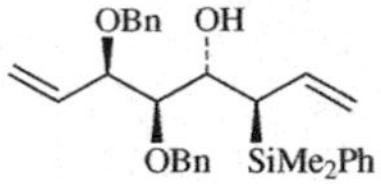

—

(79), 1:1 d.r.

774

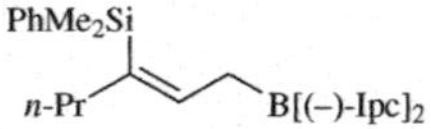

—

760

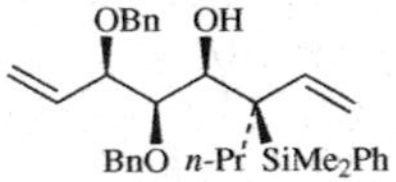

THF

Temp	d.r.	
−30°	(90)	69:19
−78°	(79)	77:16

TABLE 6. ADDITION OF γ-SUBSTITUTED REAGENTS (Continued)

Carbonyl Substrate	Allylborane Reagent	Conditions	Product(s) and Yield(s) (%), % ee	Refs.
C₅		—		545
			R d.r. Syn:Anti	
			(CH₂)₂OTBDMS (85) 19:4 2:21	
			(86) 8:20 21:6	
	B[(−)-Ipc]₂ OMOM	THF, −78° to −23°, 12.5 h	(73)	775
	B[(+)-Ipc]₂ OMe	−78°	(79), >99:1 d.r.	776
	B[(+)-Ipc]₂ OMe	−78°	(60), >99:1 d.r.	776
	(i-Pr)₂NMe₂Si⌒⌒B[(−)-Ipc]₂	1. Ether, −78°, 3 h 2. KF, H₂O₂	(44), >95:5 d.r.	45

C$_6$

45	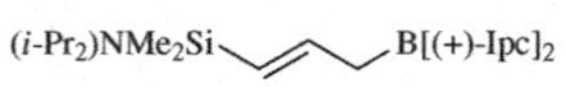 (47), >95:5 d.r.	1. Ether, −78°, 3 h 2. KF, H$_2$O$_2$	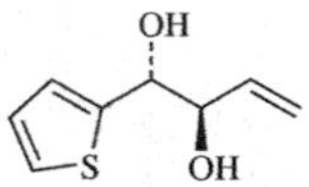 $(i\text{-Pr}_2)\text{NMe}_2\text{Si}$ —— B[(+)-Ipc]$_2$
761	(74), >99:1 d.r.	PhMe$_2$Si—B, **I** Pd$_2$(dba)$_3$, EtOAc. 80°, 5 h	
26	(45), 2.5:1 d.r.	Sc(OTf)$_3$, toluene, rt, 16-24 h	EtO$_2$C, Bu-n
683	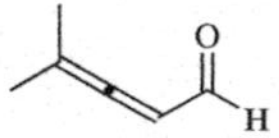 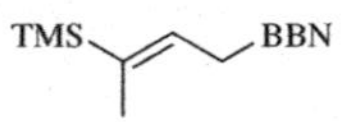 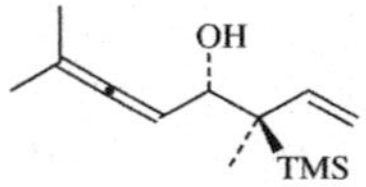 (94), >95:5 d.r.	—	TMS
46	(65)	OMe FB(OBu-n)$_2$	B(OBu-n)$_2$, OMe
46	**I** + **II** (61), **I:II** = 3.5:1	OPh FB(OBu-n)$_2$	B(OBu-n)$_2$, OPh

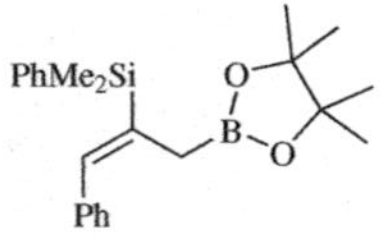

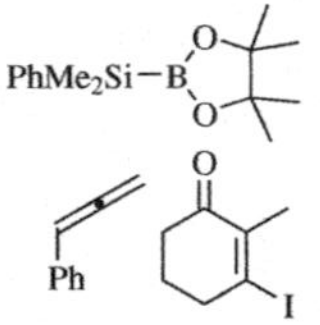

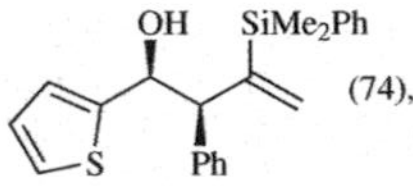

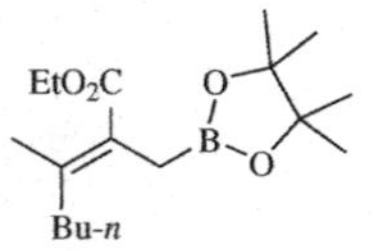

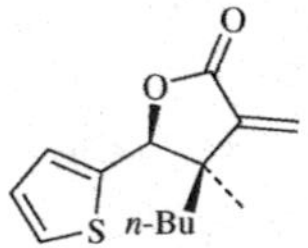

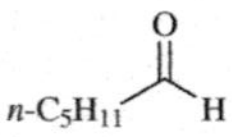

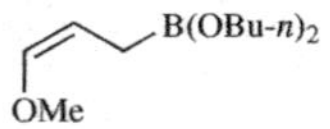

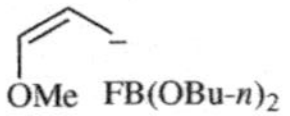

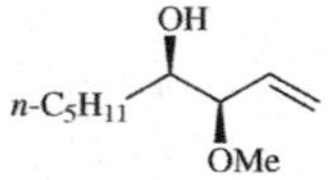

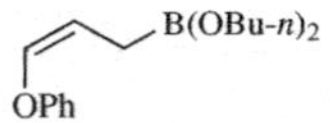

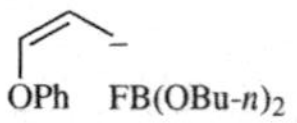

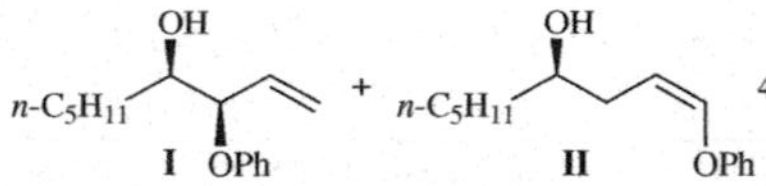

n-C$_5$H$_{11}$

n-C$_5$H$_{11}$

TABLE 6. ADDITION OF γ-SUBSTITUTED REAGENTS (Continued)

Carbonyl Substrate	Allylborane Reagent	Conditions	Product(s) and Yield(s) (%), % ee	Refs.
C$_6$				
		—	 (59), 97, >99:1 d.r.	250
		THF, −78° to 0°, 3 h	 (43), >95, >95:5 d.r. (25)	730, 731
		THF, −78° to 0°, 3 h	 (40), >95, >95:5 d.r. (25)	730, 731
		THF		777

R	Temp	Time		d.r.
Me	rt	3 h	(93)	3:97
n-Bu	−15°	6 h	(90)	96:4
n-Bu	rt	3 h	(86)	70:30
n-Bu	50°	7 h	(91)	9:91
CH$_2$TMS	0°	4 h	(81)	94:6
CH$_2$TMS	50°	20 h	(78)	17:83
CH$_2$TMS	reflux	52 h	(73)	4:96
i-Pr	rt	4 h	(92)	97:3

736

n-Bu—BBN / TMS — THF, rt, 1 h →

$$n\text{-}C_5H_{11} \quad \text{H} \quad \text{Bu-}n \quad \mathbf{I} \quad + \quad n\text{-}C_5H_{11} \quad \text{H} \quad \text{Bu-}n \quad \mathbf{II}$$

Work-Up	I + II	I:II
NaOH	(81)	96:4
H$_2$SO$_4$	(70)	3:97

736

t-Bu—BBN / TMS — THF, rt, 1 h →

$$n\text{-}C_5H_{11} \quad \text{H} \quad \text{Bu-}t \quad \mathbf{I} \quad + \quad n\text{-}C_5H_{11} \quad \text{H} \quad \text{Bu-}t \quad \mathbf{II}$$

Work-Up	I + II	I:II
NaOH	(55)	41:59
NaOEt	(79)	29:71
H$_2$SO$_4$	(69)	72:28

777

TMS / R^1 — R^2 — THF →

$$n\text{-}C_5H_{11} \quad \text{OH} \quad R^1 \quad \text{TMS}$$

R^1	R^2	Temp	Time		d.r.
Me	BBN	rt	3 h	(89)	98:2
Me	BCy$_2$	rt	3 h	(93)	97:3
n-Bu	BBN	rt	4 h	(89)	97:3
n-Bu	BCy$_2$	−15°	6 h	(90)	4:96
n-Bu	BCy$_2$	rt	3 h	(86)	30:70
n-Bu	BBN	50°	7 h	(91)	91:9
CH$_2$TMS	BBN	rt	4 h	(93)	>99:1
CH$_2$TMS	BCy$_2$	0°	4 h	(81)	6:94
CH$_2$TMS	BCy$_2$	50°	20 h	(78)	83:17
CH$_2$TMS	BCy$_2$	reflux	52 h	(73)	94:6
i-Pr	BBN	rt	24 h	(75)	88:12
i-Pr	BCy$_2$	rt	4 h	(92)	3:97

TABLE 6. ADDITION OF γ-SUBSTITUTED REAGENTS (Continued)

Carbonyl Substrate	Allylborane Reagent	Conditions	Product(s) and Yield(s) (%), % ee		Refs.

C₆

Carbonyl substrate: $n\text{-}C_5H_{11}$CHO (aldehyde)

Reagent 1: Ph / TMS substituted allyl-BBN — Conditions: THF, rt, 1 h

Products I and II

Work-Up	I + II	I:II
NaOH	(67)	96:4
NaOEt	(80)	95:5
H₂SO₄	(84)	5:95

Refs. 736

Reagent 2: i-Pr / TMS substituted allyl-BBN — Conditions: THF, rt, 1 h

Work-Up	I + II	I:II
NaOH	(77)	88:12
NaOEt	(82)	88:12
H₂SO₄	(65)	11:89

Refs. 736

Reagent 3: TMS / TMS substituted allyl-BBN — Conditions: THF, rt, 1 h

Work-Up	I + II	I:II
KOH	(82)	97:3
NaOEt	(85)	97:3
H₂SO₄	(87)	0.5:99.5

Refs. 735, 736

Reagent 4: methyl / TMS substituted allyl-BBN — Conditions: THF, rt, 1 h

Work-Up	I + II	I:II
NaOH	(73)	95:5
NaOEt	(72)	98:2
H₂SO₄	(65)	3:97

Refs. 736

426

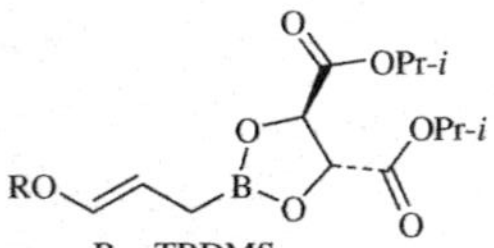

88

740

740

761

732

(86), 82

(75), 79, 96:4 d.r.

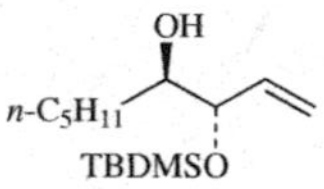

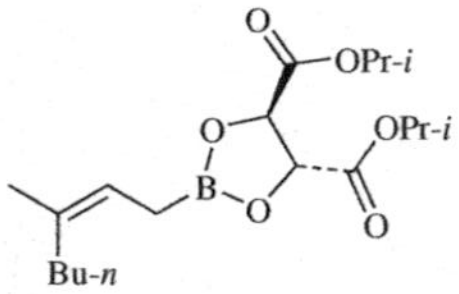

(73), 73, 96:4 d.r.

(67), >99:1 d.r.

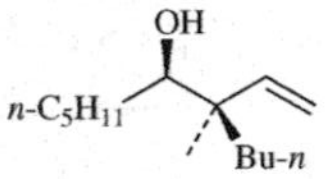

(83)

4 Å MS, toluene, −78°, 5 h

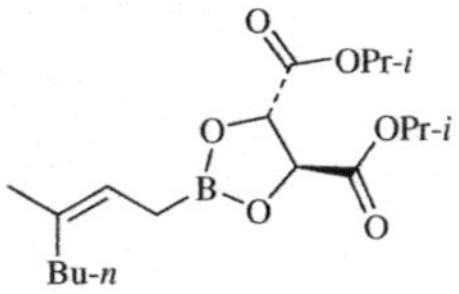

4 Å MS, toluene, −78°, 5 h

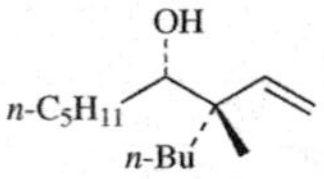

4 Å MS, toluene, −78°, 5 h

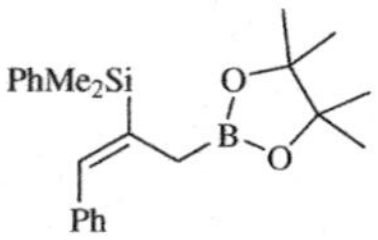

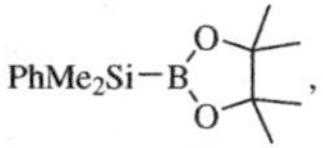

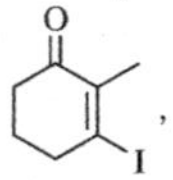

Pd₂(dba)₃, EtOAc, 80°, 5 h

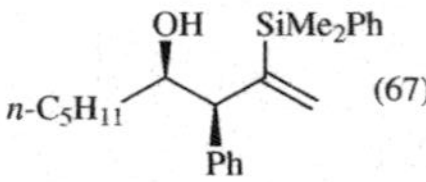

−70° to 0°

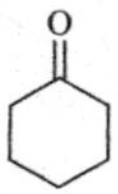

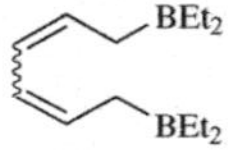

R = TBDMS

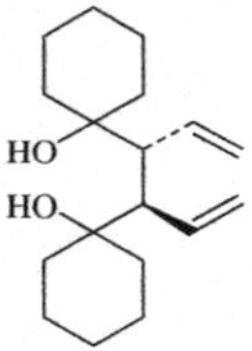

TABLE 6. ADDITION OF γ-SUBSTITUTED REAGENTS *(Continued)*

Carbonyl Substrate	Allylborane Reagent	Conditions	Product(s) and Yield(s) (%), % ee	Refs.
C$_6$				
	BBN / TMS	1. THF, rt, 1 h 2. NaOH	(67)	736
	BBN / TMS	THF, rt, 1 h	Work-Up KOH (58) H$_2$SO$_4$ (61)	735, 736
	B(OMe) pinacol	Neat, 10 d	(77), 94:6 d.r.	47, 36
	B(OBu-n)$_2$ / OPh	FB(OBu-n)$_2$ / OPh	**I** + **II** (50), **I**:**II** = 1.85:1	46
	n-Pr / PhMe$_2$Si B[(−)-Ipc]$_2$	THF, −50°	(73), 83:10 d.r.	760

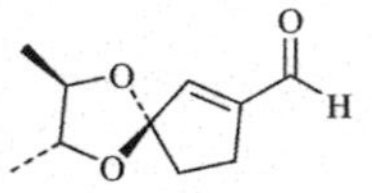

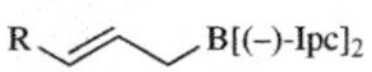

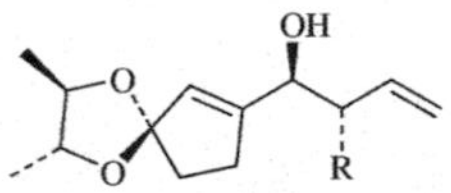

343

R		d.r.
Ph	(—), 89	98:2
(CH$_2$)$_3$	(—), 95	97.5:2.5

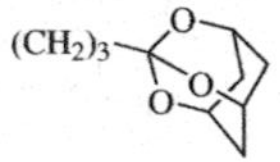

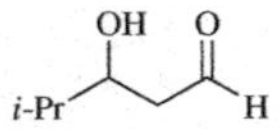

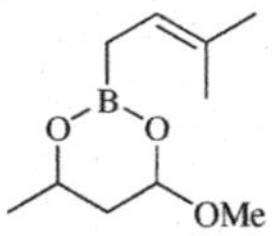

R—B[(—)-Ipc]$_2$

299

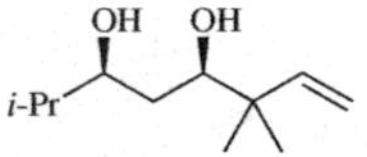

L.A.		d.r.
ZrCl$_4$	(98)	70:30
B(OMe)$_3$	(96)	78:22

1. B(NMe$_2$)$_2$, ether, −78° to rt, 1.5 h
2. L.A., −78° to rt, 3 h

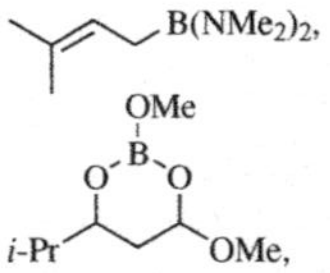

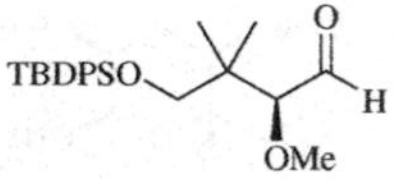

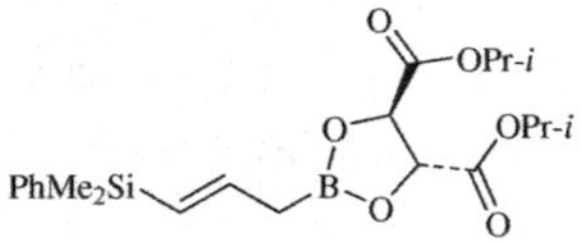

194, 758

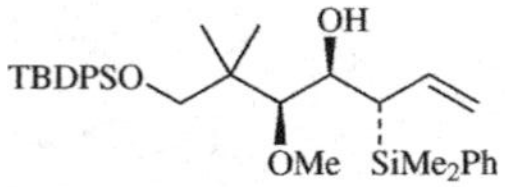

(86), >99:1 d.r.

4 Å MS, toluene, −78°

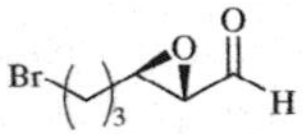

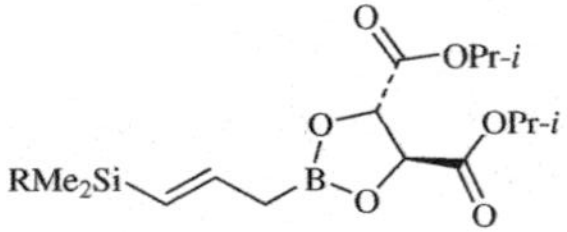

158

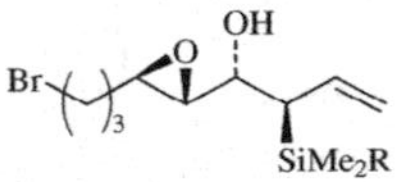

R		d.r.
2-(5-methyl)furyl	(69)	90:10
menthofuryl	(81)	90:10
CyO	(78)	67:33

4 Å MS, toluene, −78°

TABLE 6. ADDITION OF γ-SUBSTITUTED REAGENTS *(Continued)*

Carbonyl Substrate	Allylborane Reagent	Conditions	Product(s) and Yield(s) (%), % ee	Refs.
C₆				
	RMe₂Si— reagent with OPr-i, OPr-i boronate	4 Å MS, toluene, −78°	R: 2-(5-methyl)furyl (68) 90:10 d.r.; menthofuryl (82) 96:4 d.r.	158
	B[(−)-Ipc]₂, OTHP	THF, −78° to rt, 3 h	(52), >97.5:2.5 d.r.	778
	Ph₂C=N / B[(−)-Ipc]₂	1. Ph—C(Ph)=N—allyl, LDA, MeOB[(−)-Ipc]₂, −78°, 3 h 2. Aldehyde, THF, −78°, 3 h	I (49), 91, >95:5 d.r.	746
	"	THF, −78°	I (49), 91, >95:5 d.r.	160
	Ph—C(Ph)=N / B[(+)-Ipc]₂	1. Ph—C(Ph)=N—allyl, LDA, MeOB[(+)-Ipc]₂, −78°, 3 h 2. Aldehyde, THF, −78°, 3 h	I (51), 90, >95:5 d.r.	746
	"	THF, −78°	I (51), 90, >95:5 d.r.	160

430

C$_7$

Ph–CHO

Reagent	Conditions	Product	Refs.
B[(+)-Ipc]$_2$, OMe	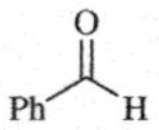 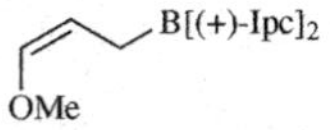 THF, −78°, 3 h	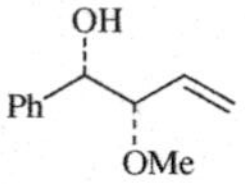 (72), 90, >99:1 d.r.	48
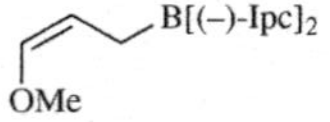 B[(−)-Ipc]$_2$, OMe	THF, −78°, 3 h	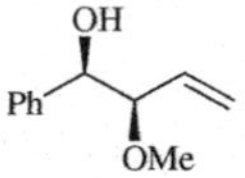 (75), 90, >99:1 d.r.	48
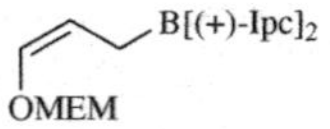 B[(+)-Ipc]$_2$, OMEM	THF, −100°	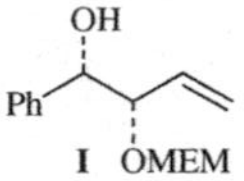 **I** (71), 98	407, 250
"	THF, −78°, 10 h	**I** (—), 95, >99:1 d.r.	388
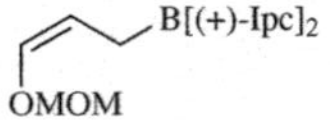 B[(+)-Ipc]$_2$, OMOM	THF, −78°	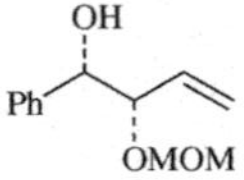 (62)	779
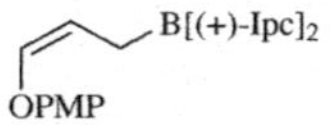 B[(+)-Ipc]$_2$, OPMP	THF, −100°, 10 h	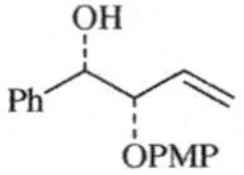 (76), 96, >99:1 d.r.	407
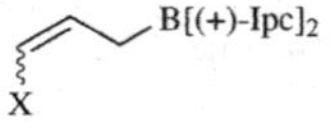 B[(+)-Ipc]$_2$, X	1. 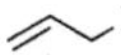 X, LiNCy$_2$, MeOBI[(+)-Ipc]$_2$ ether, −78° 2. BF$_3$•OEt$_2$ 3. Aldehyde	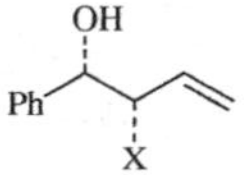 see table below	159

X		d.r.
Cl	(75), 97	98:2
Br	(68), 93	94:6

431

TABLE 6. ADDITION OF γ-SUBSTITUTED REAGENTS (Continued)

Carbonyl Substrate	Allylborane Reagent	Conditions	Product(s) and Yield(s) (%), % ee	Refs.
C7				
Ph—CHO	[structure] Cl / B[(+)-Ipc]₂	1. [allyl-Cl], LDA, MeOB[(+)-Ipc]₂, ether, −78° 2. BF₃•OEt₂ (x equiv) 3. Aldehyde	[product OH, Ph, Cl] x : (—), 84 at 2.5 ; (—), 95 at 1.33	159
	[structure] X / B[(−)-Ipc]₂	1. [allyl-X], MeOB[(−)-Ipc]₂, LiNCy₂, −78° 2. BF₃•OEt₂ 3. Aldehyde	[product OH, Ph, X] X / Solvent / d.r.: Cl, ether (78), 98, 98:2; Cl, THF (77), 76, 99:1; Br, ether (77), 95, 96:4	159
	Ph₂N—[structure]—B[(−)-Ipc]₂	THF, −78° to 0°, 3 h	[product OH, Ph, Ph₂N] (48), >95, >95:5 d.r. + [cyclopropane OH, Ph, H, NPh₂] (15)	730, 731
	Ph₂N—[structure]—B[(+)-Ipc]₂	THF, −78° to 0°, 3 h	[product OH, Ph, Ph₂N] (47), >95, >95:5 + [cyclopropane OH, Ph, H, NPh₂] (12)	730

Substrate	Conditions	Product		Ref.
Ph⟶BEt₂	Ph⟶OCOPh, BEt₃, Pd(PPh₃)₄, THF, rt, 23 h	OH, Ph, **I**, Ph (74), 2.3:1 d.r.		483
"	OCOPh, Ph, BEt₃, Pd(PPh₃)₄, THF, rt, 14 h	**I** (83), 2.2:1 d.r.		483
Ph⟶BR₂	THF, −78° to rt, 4 h	OH, Ph, Ph	R: Cy (90),— d.r. >99:1 ; (−)-Ipc (84),84 >99:1	733
Ph⟶BEt₂	OCOPh, Ph, BEt₃, Pd(PPh₃)₄, THF, rt, 14 h	OH, Ph, Ph (77), 2.3:1 d.r.		483
⟶BEt₂ (Ph chain)	OCOPh, Ph, BEt₃, Pd(PPh₃)₄, THF, rt, 4 h; 45°, 21 h	OH, Ph, Ph (76), 1.6:1 d.r.		483
Cy⟶BR₂	THF, −78° to rt, 4 h	OH, Ph, Cy	R: Cy (77),— d.r. 96:4 ; (−)-Ipc (80),80 4:1	733

TABLE 6. ADDITION OF γ-SUBSTITUTED REAGENTS (Continued)

Carbonyl Substrate	Allylborane Reagent	Conditions	Product(s) and Yield(s) (%), % ee	Refs.
C₇				
Ph—CHO (O, H)	Cy~~~BCy₂	CCl₄, 0°	(—), 96:4 d.r.	733
	n-Bu~~~BR₂	THF, −78° to rt, 4 h	R / d.r. / Cy (81), — / 87:13 / (−)-Ipc (78), 78 / 88:12	733
	n-C₆H₁₃~~~BR₂	THF, −78° to rt, 4 h	R / d.r. / Cy (82), — / 88:12 / (−)-Ipc (81), 74 / 9:1	733
	Ph~N~~~B[(−)-Ipc]₂ (Ph)	THF, −78°	(53), 93, >95:5 d.r.	160
	Ph~N~~~B[(+)-Ipc]₂ (Ph)	1. Ph~N~~~, Ph LDA, MeOB[(+)-Ipc]₂; 2. Aldehyde, THF, −78°, 3 h	(52), 90, >95:5 d.r. I	746
	"	THF, −78°	I (52), 90, >95:5 d.r.	160

434

Reagent	Conditions	Product	Yield (%)	Refs.
[structure: CH₂=CH–CH=CH–CH₂–B[(−)-Ipc]₂ type allyl with Ph–N=C(Ph)] B[(−)-Ipc]₂	1. Ph–N=C(Ph)–N , LDA, MeOB[(−)-Ipc]₂, 2. Aldehyde, THF, −78°, 3 h	[structure with OH, Ph, N=C(Ph)Ph]	(53), 93, >95:5 d.r.	746
(i-Pr)₂NMe₂Si⌇⌇⌇⌇B[(−)-Ipc]₂	1. Ether, −78°, 3 h 2. KF, H₂O₂	[structure OH, OH, Ph]	(50), >95:5 d.r.	45, 752
(i-Pr)₂NMe₂Si⌇⌇⌇⌇B[(+)-Ipc]₂	1. Ether, −78°, 3 h 2. KF, H₂O₂	[structure OH, OH, Ph]	(47), >95:5 d.r.	45, 752
PhMe₂Si⌇⌇⌇⌇B[(+)-Ipc]₂	THF, −78°, 4 h	[structure OH, Ph, SiMe₂Ph]	(81), 93	734
[dioxaborolane with Ph groups, B] B[(−)-Ipc]₂	Ether, −78°, 2 h	[structure OH, Ph, B dioxaborolane with Ph groups]	(—)	175
[dioxaborinane O–B–O] B[(−)-Ipc]₂	1. Ether, −78° to 0°, 4 h 2. H₂O₂, NaOH	[structure OH, OH, Ph]	(76), >95, >97:3 d.r.	157

435

TABLE 6. ADDITION OF γ-SUBSTITUTED REAGENTS (Continued)

Carbonyl Substrate	Allylborane Reagent	Conditions	Product(s) and Yield(s) (%), % ee	Refs.
C₇	B[(−)-Ipc]₂	Ether, −78°, 2 h	(—)	175
	"	Ether, −78°, 2 h; rt, 24 h	(40), 93, >20:1 d.r.	175
	B[(+)-Ipc]₂	1. Ether, −78° to 0°, 4 h 2. H₂O₂, NaOH	(80), >95, >97:3 d.r.	157
	"	Ether, −78°, 2 h; rt, 24 h	(42), 95, >20:1 d.r.	175
	B[(−)-Ipc]₂	1. Cl, MeOB[(−)-Ipc]₂, LiNCy₂, ether 2. BF₃•OEt₂ (x equiv) 3. Aldehyde	I + II, III	159, 762

First scheme

1. [allyl]–X,
   MeOBBN,
   LDA, THF
2. BF$_3$·OEt$_2$
3. Aldehyde

X = Cl, Br

x	Temp	X	(I+II):III	II:I	II % ee
2.63	-78°	Cl	98:2	91:9	78
2.63	-78°	Br	96:4	9:1	74
2.63	-95°	Cl	99:1	99:1	88
2.65	-95°	Br	98:2	94:6	86
2.65	-95°	Cl	>99:1	99:1	87
1.33	-95°	Cl	>99:1	99:1	98
1.66	-95°	Cl	>99:1	99:1	97
2.63	-95° to -78°	Cl	>99:1	99:1	86
1.33	-95° to -78°	Cl	>99:1	99:1	85

159

I + II + III

X	Temp	I + II + III	II:(I+III)
Cl	-78°	(79)	85:15
Br	-78°	(76)	88:12
Cl	-95°	(77)	91:9

Second scheme

1. [allyl]–Cl,
   MeOBBN,
   LiNCy$_2$, THF
2. BF$_3$·OEt$_2$ (x equiv)
3. Aldehyde

X = Cl

I + II + III, II:I = 99:1

x	Temp	(I+II):III
2.63	-95°	95:5
1.33	-95° to -78°	99:1
1.33	-95° to -62°	99:1
1.33	-95° to -41°	>99:1
1.33	-95°	>99:1
1.5	-95°	96:4

159

TABLE 6. ADDITION OF γ-SUBSTITUTED REAGENTS (Continued)

Carbonyl Substrate	Allylborane Reagent	Conditions	Product(s) and Yield(s) (%), % ee	Refs.
C₇ PhCHO	structure with B[(+)-Ipc]₂	Ether, pentane, −100°	OH...Ph (92), 95	505
	t-Bu / BBN / TMS	THF, rt, 1 h	I + II (t-Bu, Ph / H, Ph, Bu-t)	736
	n-Bu / BBN / TMS	THF, rt, 1 h	I + II	736
	TMS / BBN / TMS	THF, rt, 1 h	I + II	735, 736

Row 2 work-up data:

Work-Up	I + II	I:II
NaOH	(70)	41:59
NaOEt	(83)	29:71
H₂SO₄	(68)	62:38

Row 3 work-up data:

Work-Up	I + II	I:II
NaOH	(81)	98:2
H₂SO₄	(75)	2:98

Row 4 work-up data:

Work-Up	I + II	I:II
KOH	(88)	98:2
H₂SO₄	(83)	2:98

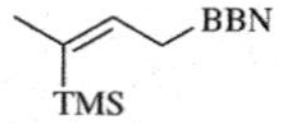

THF, rt, 1 h

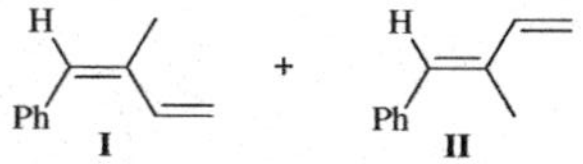

Work-Up	I + II	I:II
NaOH	(71)	98:2
H$_2$SO$_4$	(69)	2:98

736

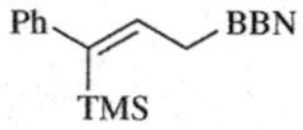

THF, rt, 1 h

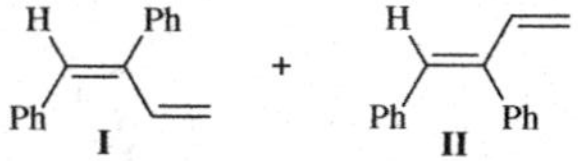

Work-Up	I + II	I:II
NaOH	(58)	>99.5:0.5
H$_2$SO$_4$	(56)	<0.5:99.5

736

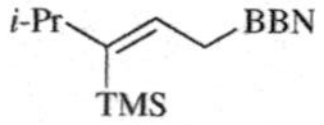

THF, rt, 1 h

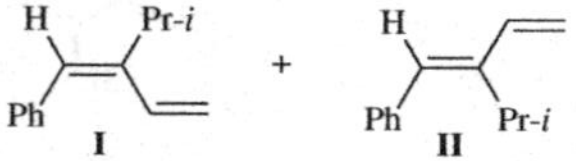

Work-Up	I + II	I:II
NaOH	(85)	91:9
H$_2$SO$_4$	(84)	11:89

736

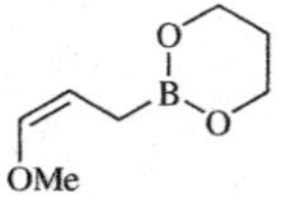

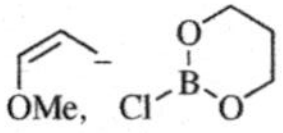

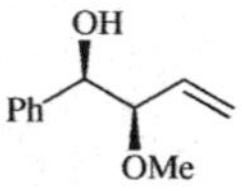

(95)

46

439

TABLE 6. ADDITION OF γ-SUBSTITUTED REAGENTS (*Continued*)

Carbonyl Substrate	Allylborane Reagent	Conditions	Product(s) and Yield(s) (%), % ee	Refs.
C7 Ph–CHO	(pinacol boronate, OMe)	Neat, 20°, 2 d	(86), >95:5 d.r. OH / Ph / OMe	47, 33, 36
	(dimethyl dioxaborinane, OMEM)	(cyclic boronate, OMOM, Cl–B–O)	(59) OH / Ph / OMEM	46
	(pinacol boronate, OMOM)	Neat, 60°, 4 h	(82), >95:5 d.r. OH / Ph / OMOM	47, 36
	(pinacol boronate, OTHP)	Neat, rt, 48 h	(70-80), 93:7 d.r. OH / Ph / OTHP **I**	34
	(dimethyl dioxaborinane, OTHP)	—	**I** (70-80); 93:7 d.r.	104
	(pinacol boronate, OMe / O–C(CH3)2)	Neat, rt	(93), 94:6 d.r. OH / Ph / OMe	47, 36

440

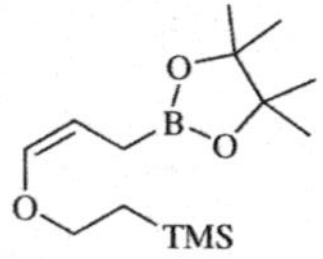

(98), >95:5 d.r.   TMS   47

OH / Ph

Neat, 20°, 2 d

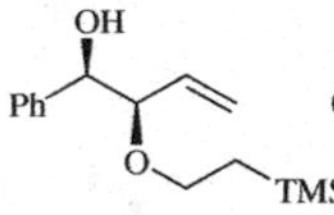

OH / **II** / OPh

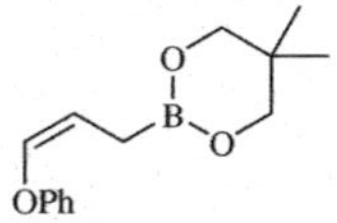

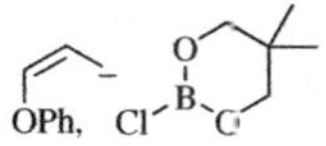

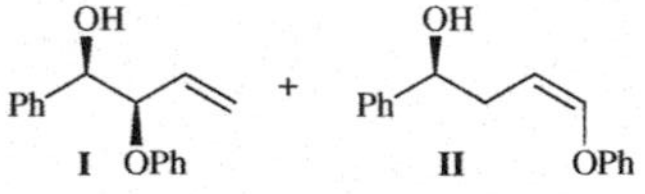

+

I / OPh

I + II (80), **I:II** = 4.7:1   46

Ph

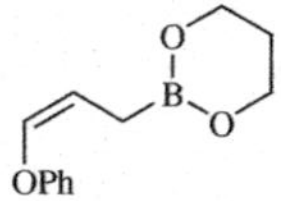

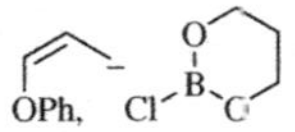

OPh, Cl–B

I + II (69), **I:II** = 3.3:1   46

OPh, Cl–B

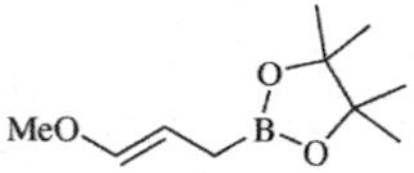

---

(87), 95:5 d.r.   OMe   47, 33

OH / Ph

Petroleum ether, 20°, 36 h

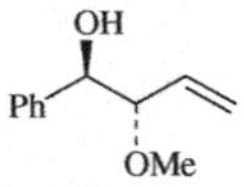

MeO

---

(92), 94:6 d.r.   TMS   47

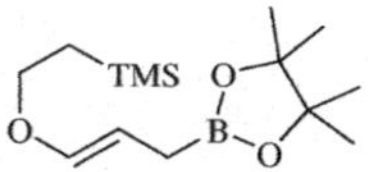

OH / Ph

Petroleum ether, rt, 2 d

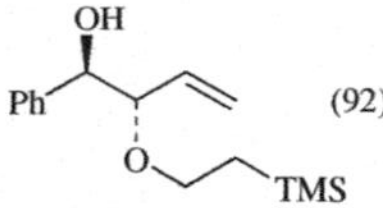

TMS

---

(67), 96:4 d.r.   OTBDMS   780

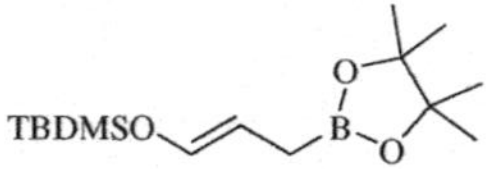

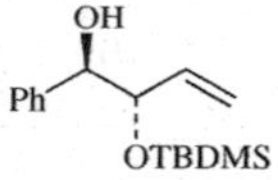

OH / Ph

—

TBDMSO

TABLE 6. ADDITION OF γ-SUBSTITUTED REAGENTS (Continued)

Carbonyl Substrate	Allylborane Reagent	Conditions	Product(s) and Yield(s) (%), % ee	Refs.
C₇ Ph–CHO	(structure, R = TBDMS)	4 Å MS, toluene, −78°, 5 h	(82), 67	88
	(structure)	Pd₂(dba)₃, DMSO, toluene, 20°, 21 h	(59), 45	349
	Z:E Equiv >95:5 1.0 30:70 0.65	—	R d.r. Me (95) 98:2 Et (95) 5:95	737
	(structure, SiMe₂Ph)	PhMe₂Si–B(pinacol), Pt(C₂H₄)(PPh₃)₂, octane, 50°	(63), 95:5 d.r. **I**	781
	"	CH₂Cl₂, rt	**I** (85)	72

442

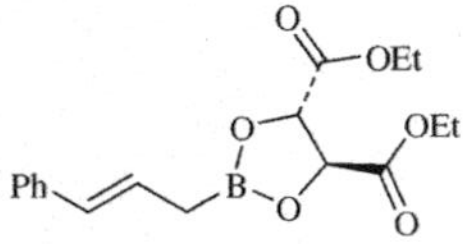

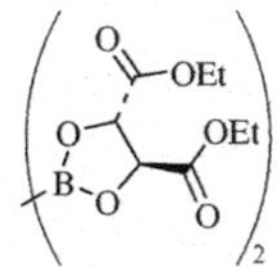

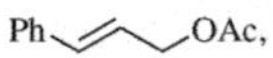

349

(83), 43

x		d.r.
5	(73)	3.8:1
2	(72)	3.6:1

184

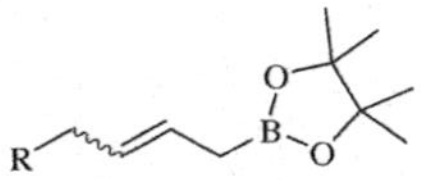

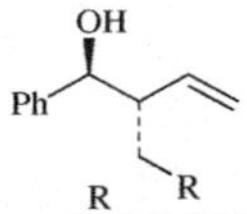

R	x		d.r.
TBDMSO	5	(44–67)	4:1
BnO	5	(60)	4.7:1
BnO	2	(63)	3.2:1
Cl	5	(78)	4.9:1
Cl	2	(79)	4.6:1

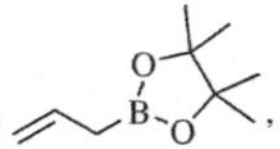

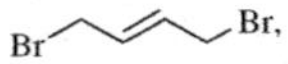

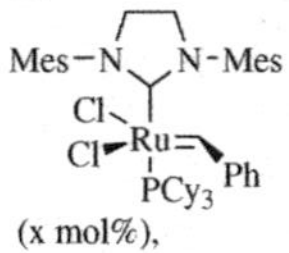

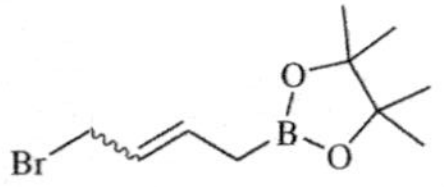

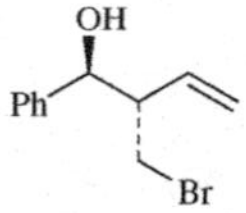

184

Ph, Pd$_2$(dba)$_3$, DMSO, toluene, 20°, 21 h

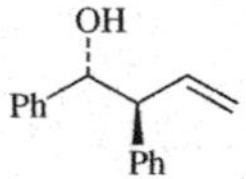

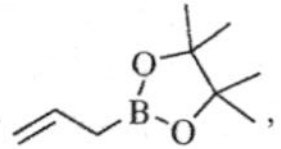

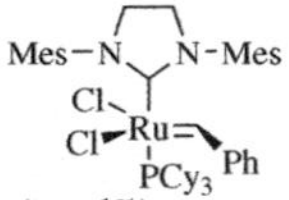

1. [reagent], Br, Mes–N N–Mes, Cl–Ru=Ph, Cl PCy$_3$ (x mol%), CH$_2$Cl$_2$, 40°
2. Aldehyde, rt

1. [reagent], R, Mes–N N–Mes, Cl–Ru=Ph, Cl PCy$_3$ (x mol%), CH$_2$Cl$_2$, 40°
2. Aldehyde, rt

TABLE 6. ADDITION OF γ-SUBSTITUTED REAGENTS (Continued)

Carbonyl Substrate	Allylborane Reagent	Conditions	Product(s) and Yield(s) (%), % ee	Refs.
C₇	 R = CH₂CO₂Me	−78°, 5 h	 (30), 49	90
		Toluene/pentane (10:3), −100°	 (80), 83	738
		Neat, 60°, 5 h	 (52)	178
		1. (y equiv), **A**, **B** CH₂Cl₂, 40° 2. Aldehyde, rt		184

	x	y	Cat.		d.r.
	1.5	3	**A**	(32)	1.8:1
	1.5	3	**B**	(75)	4.5:1
	1.5	3	**B**	(75)	4.5:1
	1.5	0.5	**B**	(57)	4.7:1
	0.75	3	**B**	(75)	4.5:1

x equiv

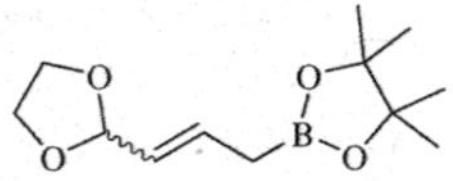

		d.r.
x		
5	(68)	>20:1
2	(69)	>20:1

1. Mes−N N−Mes, Cl Cl Ru=Ph PCy₃
(x mol%), CH₂Cl₂, 40°
2. Aldehyde, rt

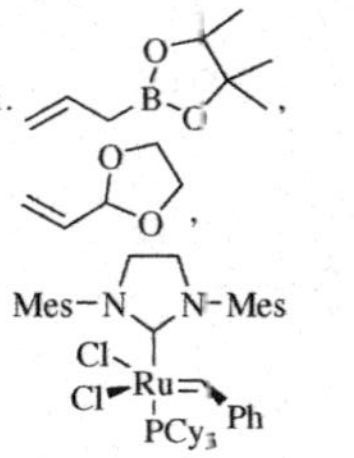

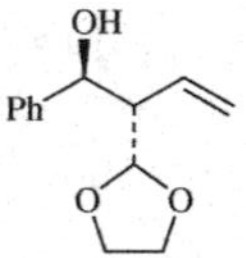

OH, Ph (product)

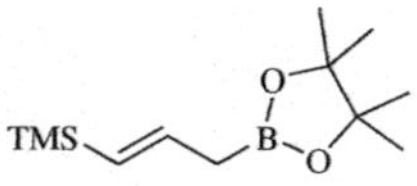

184

Ether, rt, 1-3 d

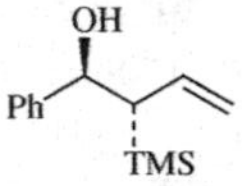

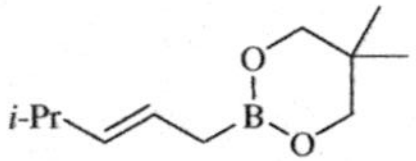

OH, Ph, TMS (89)

32

1. Cl i-PrMgCl, toluene, −15°, 30 min
2. Aldehyde, rt, 24 h

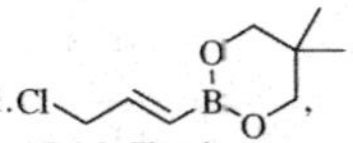

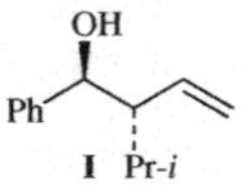

OH, Ph, **I** Pr-i (44), >99:1 d.r.

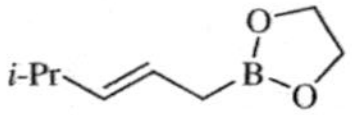

87

1. Cl i-PrMgCl, THF, −15° to rt
2. Aldehyde, 24 h

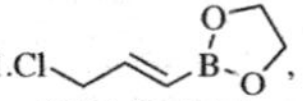

I (30), 99:1 d.r.

87

TABLE 6. ADDITION OF γ-SUBSTITUTED REAGENTS (Continued)

Carbonyl Substrate	Allylborane Reagent	Conditions	Product(s) and Yield(s) (%), % ee	Refs.
C₇ Ph–CHO	(pinacol allylboronate)	1. R¹, R² (x equiv), Mes–N N–Mes Grubbs II catalyst (y mol%), CH₂Cl₂, 40° 2. Aldehyde, rt	R¹ / R² / x / y (yields below)	184
			740
			740
			772

Product R¹/R² table:

R¹	R²	x	y	
t-Bu	H	5	5	(35)
Me₂COH	H	20	5	(51)
Me₂COH	H	20	2	(17)
—(CH₂)₄—		3	5	(49)
—(CH₂)₄—		10	5	(58)
—(CH₂)₅—		10	5	(34)
Ph	H	20	5	(66)
2-BrC₆H₄	H	20	5	(45)
2-BrC₆H₄	H	5	5	(25)
2-BrC₆H₄	H	10	5	(60)

Additional entries:

Allylborane Reagent	Conditions	Product and Yield	Refs.
(tartrate-derived boronate, n-Bu)	4 Å MS, toluene, –78°, 5 h	(99), 37, 98:2 d.r.	740
(tartrate-derived boronate, n-Bu)	4 Å MS, toluene, –78°, 5 h	(96), 40, 96:4 d.r.	740
B(OPr-i)₂, Bu-n	THF, –78° to rt, 18 h	(96), 88:12 d.r.	772

Reagent	Conditions	Product	Yield (%), d.r.	Refs.
$n\text{-Bu}$ / B(OPr-i)$_2$	THF, −78° to rt, 18 h	OH, Ph, Bu-n, Me (**I**)	(95), 97:3 d.r.	772
n-Bu pinacol boronate	Neat, rt, 5-8 d	**I**	(71), 99:1 d.r.	55
pinacol boronate, Bu-n	Neat, rt, 5-8 d	OH, Ph, n-Bu	(69), 93:7 d.r.	55
pinacol boronate, OMe	Petroleum ether, 20°, 2 d	OH, Ph, MeO	(55), 73:27 d.r.	739
MeO pinacol boronate	Petroleum ether, 20°, 9 d	OH, Ph, MeO	(70), 90:10 d.r.	739
B(OPr-i)$_2$	Toluene, rt, 0.5 h	OH, Ph	(85)	741, 744
OCH(Pr-i)$_2$ tartrate boronate	4 Å MS, toluene, −78°, 72 h	OH, Ph	(89), 38	741

TABLE 6. ADDITION OF γ-SUBSTITUTED REAGENTS (Continued)

Carbonyl Substrate	Allylborane Reagent	Conditions	Product(s) and Yield(s) (%), % ee	Refs.
C₇				
		4 Å MS, toluene, –78°, 72 h	(86), 37	741
		4 Å MS, toluene, –78°, 72 h	I (90), 49	741
		4 Å MS, toluene, –78°, 72 h	(95), 43	741
		—	(80), 9:1 d.r.	52
		Pd(PPh₃)₄	I (89)	550
		(CH₂)₂CH=C(Me)₂, Pd(PPh₃)₄	(77)	550

782

782

782

MeO$_2$C (pinacol boronate, alkene)

RO$_2$C (pinacol boronate, alkene)
R = (+)-menthyl

MeO$_2$C (bis-ethyl ester boronate)
OEt OEt

I (lactone with Ph)

See table

Solvent	Temp	Time	d.r.
THF	rt	7 d (64)	84:16
THF	70°	4 d (83)	85:15
toluene	100°	1 d (87)	87:13

THF, rt, 5 d   I (50), 25, 87:13 d.r.

See table

Solvent	Temp	Time		d.r.
DMF	rt	2 d	(80), 26	94:6
DMF	0°	3 d	(52), 27	95:5
DMF/THF	−25°	6 d	(76), 25	96:4
DMF/THF	−78°	6 d	(40), 20	94:6
DMF/THF	−25°	7 d	(56), 21	98:2
THF	0°	2 d	(75), 22	98:2
THF[b]	0°	2 d	(69), 19	98:2
CH$_2$Cl$_2$	0°	2 d	(64), 16	98:2
4 Å MS, CH$_2$Cl$_2$	0°	2 d	(67), 22	98:2
toluene	0°	2 d	(74), 25	98:2
4 Å MS, toluene	0°	2 d	(76), 12	98:2

449

TABLE 6. ADDITION OF γ-SUBSTITUTED REAGENTS (Continued)

Carbonyl Substrate	Allylborane Reagent	Conditions	Product(s) and Yield(s) (%), % ee	Refs.
C_7		See table	**I** Solvent / Temp / Time / (yield) / d.r. THF, rt, 7 d, (75), 10, 79:21 toluene, 100°, 3 d, (80), 8, 83:17	782
		Toluene, 80°, 16–120 h	**I** (60), >20:1 d.r.	59, 26
		  Pd(PPh$_3$)$_4$	(81)	550
	  R = (–)-8-phenylmenthyl	1. Neat, rt, 14 d 2. pTSA	(55), >95	59
		L.A., rt, 6–24 h	**I** L.A. / Solvent Sc(OTf)$_3$ / CH$_2$Cl$_2$ (72) Sc(OTf)$_3$ / toluene (93) Cu(OTf)$_2$ / toluene (67) Yb(OTf)$_3$ / toluene (91)	24, 26

Toluene, rt, >12 d   **I** (89), >20:1 d.r.   59, 26

x equiv

RO₂C pinacol boronate

R
Me
Me
Me
Me
(–)-menthyl
(–)-menthyl
(–)-menthyl
(–)-menthyl

Catalyst, toluene   30

x	Catalyst	Temp	Time	
0.9	none	rt	24 h	(<5)
0.9	CF$_3$CO$_2$H	rt	24 h	(77)
0.9	Tf$_2$NH	rt	24 h	(99)
0.9	TfOH	rt	24 h	(99)
0.9	TfOH	0°	16 h	(78)
0.9	Sc(OTf)$_3$	0°	16 h	(<5)
1.5	TfOH	0°	16 h	(96)
2.0	TfOH	0°	16 h	(99)

PhMe₂Si–B(pinacol),

Pt(C₂H₄)(PPh₃)₂, octane   781

I

Temp	Time		d.r.
120°	2 h	(74)	81:19
80°	4 h	(78)	95:5
50°	8 h	(61)	98:2
80°	4 h	(63)	93:7
80°	4 h	(85)	96:4

CH₂Cl₂, rt   **I** (87)   72

TABLE 6. ADDITION OF γ-SUBSTITUTED REAGENTS (Continued)

Carbonyl Substrate	Allylborane Reagent	Conditions	Product(s) and Yield(s) (%), % ee	Refs.
C₇ 		1. , Pt(dba)₂, PCy₃, benzene, rt 2. Aldehyde, −78°, 3 h 3. NaOH, H₂O₂	(65), 33, >19:1 d.r. 	185
		PhMe₂Si–B , Pt(C₂H₄)(PPh₃)₂, octane, 80°	(60), 99:1 d.r. 	781
		1. Br , HBCy₂, n-Bu₄NBr, −78° to 0°, 1 h 2. Aldehyde, 0° to rt, 1 h	(68), >98:2 d.r. 	680
		1. Br , HB(Thex)₂, n-Bu₄NBr, −78° to 0°, 1 h 2. Aldehyde, 0° to rt, 1 h	(40), >98:2 d.r. 	680

1. ,

   HB(C$_5$H$_9$-c)$_2$,
   n-Bu$_4$NBr,
   −78° to 0°, 1 h

2. Aldehyde, 0° to rt,
   1 h

(45), >98:2 d.r.

751, 680

---

1. ,

   HBR$_2$;
   n-Bu$_4$NBr (x equiv),
   −78°, time 1

2. Aldehyde, 0° to rt,
   time 2

751

x	Time 1	Time 2	R	
none	0.5 h	2 h	Cy	(22)
none	2 h	2 h	Cy	(63)
none	12 h	2 h	Cy	(73)
0.1	1 h	2 h	Cy	(85)
none	1 h	3 h	siamyl	(1)
none	12 h	3 h	siamyl	(26)
0.1	12 h	3 h	siamyl	(46)

---

1. ,

   HB(siamyl)$_2$,
   n-Bu$_4$NBr,
   −78° to 0°, time1

2. Aldehyde, 0° to rt,
   1 h

	Time 1		d.r.
	1 h	(13)	>98:2
	15 h	(38)	>98:2

680

TABLE 6. ADDITION OF γ-SUBSTITUTED REAGENTS (Continued)

Carbonyl Substrate	Allylborane Reagent	Conditions	Product(s) and Yield(s) (%), % ee	Refs.
C₇		crotyl M⁺, BR₃, −70°, 1 h		783

R	M	Solvent		d.r.
Et	Li	ether	(90)	82:18
Et	MgBr	ether	(95)	50:50
Et	Li	THF	(91)	83:17
Et	Li	ether, HMPA	(83)	83:17
n-Bu	Li	ether	(89)	83:17
s-Bu	Li	ether	(85)	61:39

Carbonyl Substrate	Allylborane Reagent	Conditions	Product(s) and Yield(s) (%), % ee	Refs.
	Ph₂N–⟋⟍–B[(−)-Ipc]₂	THF, −78° to 0°, 3 h	(—) (11), >95, >95:5 d.r.	730
	Pd₂(dba)₃, DMSO, toluene, 20°, 91 h		(74) I	349

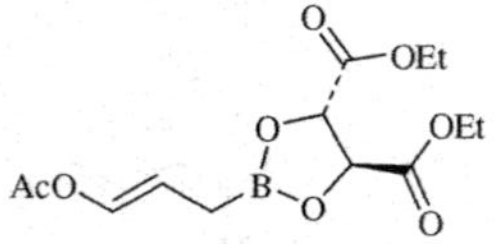

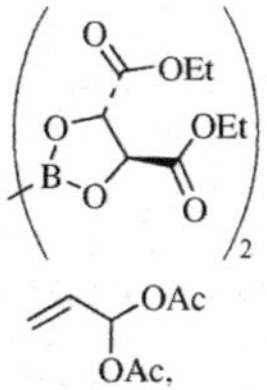

349

I (67), 50

OEt, OEt
OAc,
Pd$_2$(dba)$_3$, DMSO,
toluene, 20°, 21 h

AcO, OEt, OEt

Ph, OAc,
Pd$_2$(dba)$_3$, DMSO,
40°, 21 h

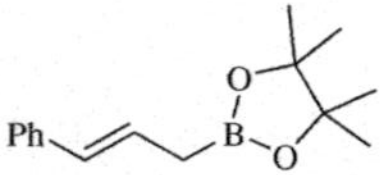

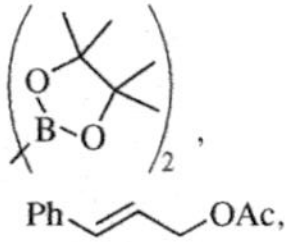

Ph

(86), 33:1 d.r.

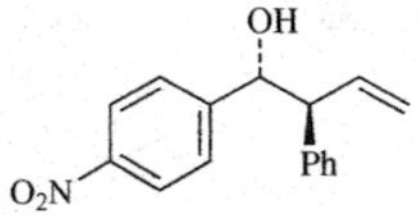

349, 636

OH, Ph, O$_2$N

CO$_2$Et, OAc,
Pd$_2$(dba)$_3$, DMSO,
toluene, 20°, 24 h

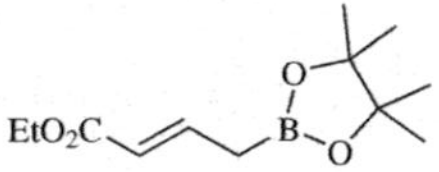

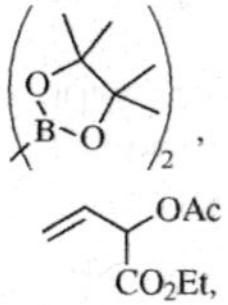

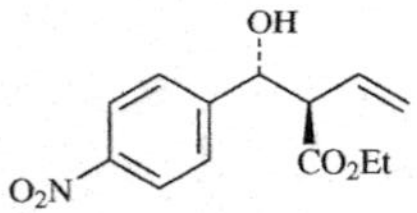

EtO$_2$C

(75), 8:1 d.r.

349

OH, CO$_2$Et, O$_2$N

TABLE 6. ADDITION OF γ-SUBSTITUTED REAGENTS (Continued)

Carbonyl Substrate	Allylborane Reagent	Conditions	Product(s) and Yield(s) (%), % ee	Refs.

C₇ : C_7

4-nitrobenzaldehyde (O₂N–C₆H₄–CHO)

First reagent (pinacol boronate, H₂N(O)C-substituted):
Conditions: C(O)NH₂, Pd₂(dba)₃, DMSO, toluene, 20°, 21 h
Product: (61), 8:1 d.r. — 349

Second reagent (pinacol boronate, AcO-substituted):
Conditions: OAc, Pd₂(dba)₃

Solvent	Temp	Time			d.r.	
DMSO	40°	21 h	(59)		>99:1	636
DMSO, toluene	20°	93 h	(69)		—	349

Third reagent (pinacol boronate, R-substituted):
Conditions: OAc, R, Pd₂(dba)₃, DMSO

R	Time	Temp			d.r.	
C₆H₄OMe-p	21 h	40°	(83)		10:1	636
CO₂Et	21 h	40°	(75)		8:1	
C(O)NH₂	21 h	20°	(61)		8:1	
OAc	91 h	20°	(74)		>99:1	

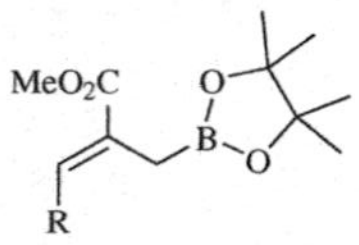

784

MeO$_2$C—(boronate), R

1. [pinacol boronate], OAc, CO$_2$Me, R, Pd$_2$(dba)$_3$, toluene, rt
2. BF$_3$•OEt$_2$ (30 mol%), rt, 2 d

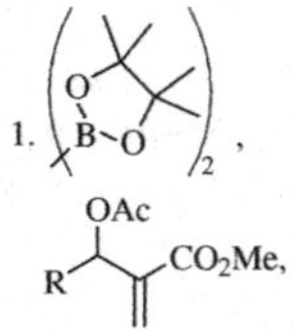

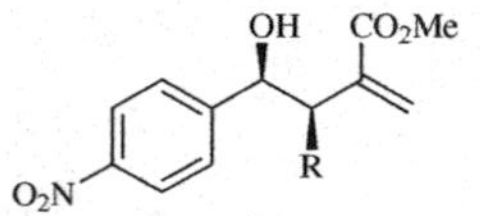

Product: OH, CO$_2$Me, R, O$_2$N

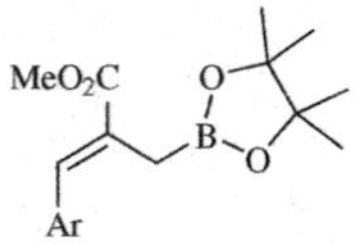

R	
Ph	(81)
n-C$_8$H$_{17}$	(59)

785

MeO$_2$C—(boronate), Ar

BF$_3$•SiO$_2$, toluene, rt, 24 h

Product: OH, CO$_2$Me, Ar, O$_2$N

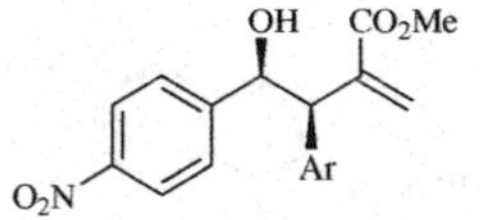

Ar	
4-MeC$_6$H$_4$	(80)
4-ClC$_6$H$_4$	(73)
4-MeOC$_6$H$_4$	(84)
1-Nph	(68)
2-ClC$_6$H$_4$	(69)
2-furyl	(56)
n-C$_8$H$_{17}$	(64)

26

EtO$_2$C—(boronate), Ph, Me

Toluene, 60-80°, 16-120 h

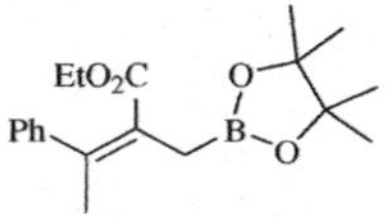

Product: lactone, Ph, Me, O$_2$N (26)

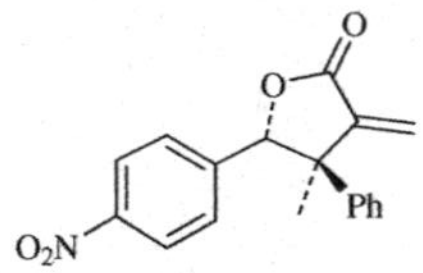

TABLE 6. ADDITION OF γ-SUBSTITUTED REAGENTS (*Continued*)

Carbonyl Substrate	Allylborane Reagent	Conditions	Product(s) and Yield(s) (%), % ee	Refs.
C₇				
		Toluene, 80°, 16-120 h	(75), >20:1 d.r.	59, 26
		Toluene, 80°, 16-120 h	(76), >20:1 d.r.	59, 26
		CH₂Cl₂, 40°, 48 h	(67), >20:1 d.r.	59, 26
		Toluene, rt, 12 d	(26), 15:1 d.r.	59
		n-Bu₄NI (0.1 equiv), CH₂Cl₂/H₂O (1:1), rt, 5 h	(89)	784

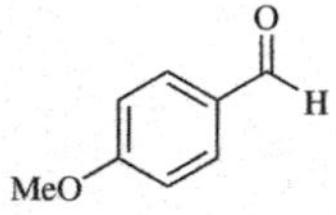

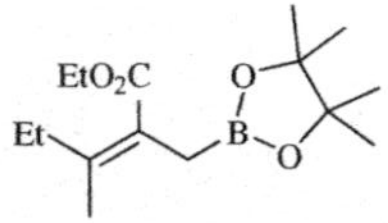

(55), 20:1 d.r.   59, 26

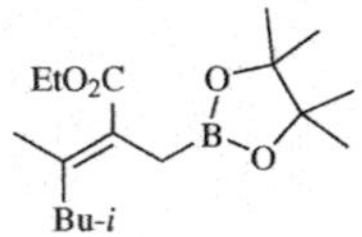

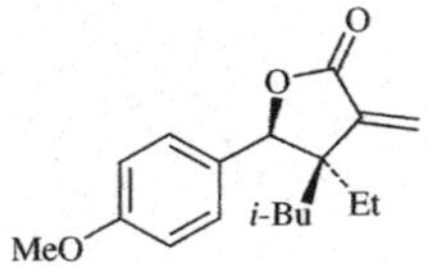

(65), >20:1 d.r.   59, 26

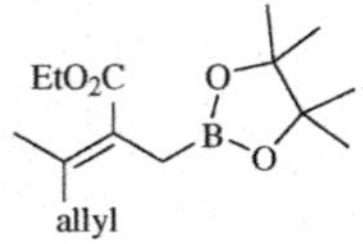

(48)   59, 26

785

Toluene, 110°, 16-24 h

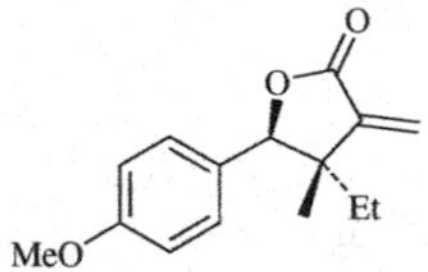

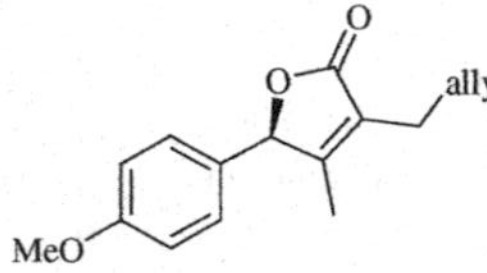

Toluene, 110°, 16-24 h

Toluene, 110°, 16-24 h

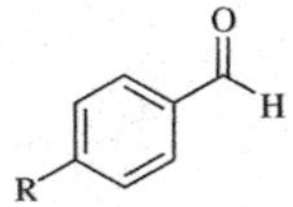

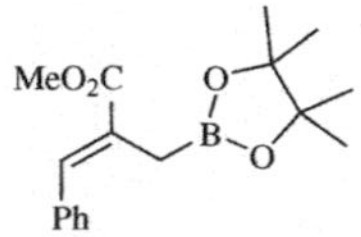

Catalyst, toluene

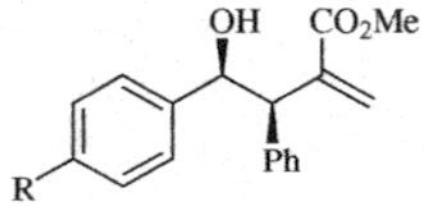

R	Catalyst	Temp	Time	
O$_2$N	none	rt	192 h	(74)
O$_2$N	Sc(OTf)$_3$	rt	62 h	(48)
O$_2$N	BF$_3$•OEt$_2$	rt	48 h	(81)
O$_2$N	BF$_3$	rt	24 h	(83)
H	BF$_3$	rt	48 h	(67)
MeO	none	90°	48 h	(62)
Cl	BF$_3$	rt	24 h	(78)

459

TABLE 6. ADDITION OF γ-SUBSTITUTED REAGENTS (Continued)

Carbonyl Substrate	Allylborane Reagent	Conditions	Product(s) and Yield(s) (%), % ee	Refs.
C$_7$				

Row 1 product table:

R^1	R^2		d.r.
H	Ph	(96)	>99:1
MeO	Ph	(55)	>99:1
Cl	Ph	(92)	>99:1
H	4-MeC(O)C$_6$H$_4$	(86)	>99:1
H	4-MeOC$_6$H$_4$	(87)	>99:1
H	4-EtO$_2$CC$_6$H$_4$	(70)	>99:1
H	n-Bu	(92)	95:5
H	Cy	(75)	97:3
H	c-Pr	(85)	97:3

Conditions (row 1): PhMe$_2$Si–B(pin), cyclohexenone iodide, R^2 allyl, Pd$_2$(dba)$_3$, EtOAc, 80°, 5 h

Refs (row 1): 761

Row 2 product table:

R^1		d.r.
MeO	(70)	19:1
O$_2$N	(81)	>20:1

Conditions (row 2): Toluene, 80°, 16-120 h

Refs (row 2): 59, 26

26

Scheme 1

Reactant: EtO$_2$C–CH=C(R^2)(R^3) with pinacol boronate

Conditions: Sc(OTf)$_3$, toluene, rt, 6–24 h

R^1	R^2	R^3		d.r.
Br	Me	Me	(62)	>20:1
AcO	Et	Me	(57)	>20:1
O$_2$N	Me	n-Bu	(58)	>20:1
MeO	Et	Me	(66)	2:1
MeO	Me	n-Bu	(39)	1.8:1
MeS	Et	n-Bu	(69)	2.5:1
Me	Me	n-Bu	(75)	>20:1

26

Scheme 2

Conditions: 1. Toluene, rt, 14 d; 2. pTSA, 4 h

R	
H	(55), >95
Br	(40), >95

692

Scheme 3

Reactant: B$^-$R$^2_2$SePh

Conditions: 1. PhSe–, LDA, THF, –78°, 30 min; 2. BR$^2_3$, –78° to rt, 12 h

I + II

R^1	R^2	I+II	I:II	d.r.
H	Et	(89)	91:9	76:24
H	n-Bu	(92)	100:0	88:12
Me	Et	(92)	100:0	64:36
MeO	Et	(81)	92:8	86:14
O$_2$N	Et	(93)	100:0	82:18

TABLE 6. ADDITION OF γ-SUBSTITUTED REAGENTS (Continued)

Carbonyl Substrate	Allylborane Reagent	Conditions	Product(s) and Yield(s) (%), % ee	Refs.
C₇				

(R-substituted benzaldehyde)	(pinacol allylborane, Ph / SiMe₂Ph)	PhMe₂Si–B(pinacol), Ph (butadienyl), Pt(C₂H₄)(PPh₃)₂, octane, 120°		781

R		d.r.
H	(79)	99:1
4-MeO	(77)	99:1
2-MeO	(83)	99:1

(3-iodobenzaldehyde)	(EtO₂C, Bu-n pinacol allylborane)	Sc(OTf)₃, toluene, rt, 6–24 h	(55), 19:1 d.r.	24, 26

(bromo-methylenedioxy-methoxy benzaldehyde)	B[(+)-Ipc]₂ (prenyl)	Ether, –78° [a]	(97), 90, 4:3 d.r.	786, 787

(Br, MeO, MeO, OMe substituted benzaldehyde)	(MeO₂C pinacol allylborane)	1. Toluene, 110°, 2 d 2. pTSA, rt, 6 h	(74)	30

30

(57), >19:1 d.r.

TfOH (20 mol%),
toluene, 0°, 16 h

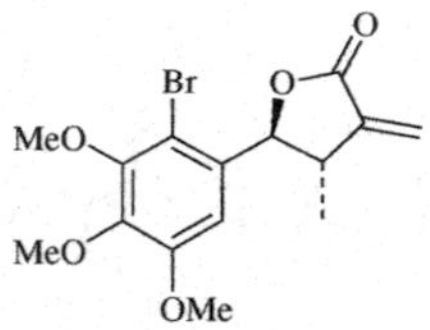

"

250

(59), 97, >99:1 d.r.

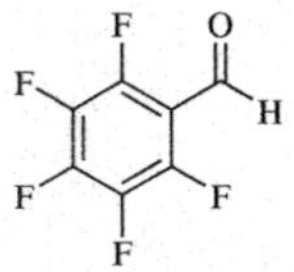

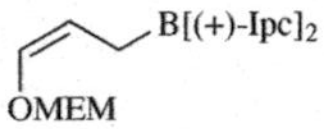

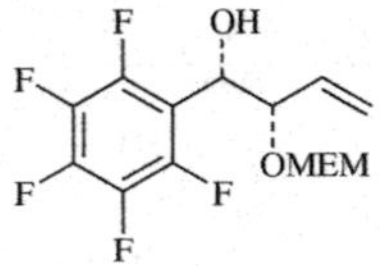

—

B[(+)-Ipc]₂

OMEM

491

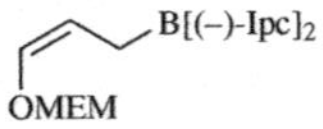

(65), 95, >95:5 d.r.

THF, −100° to −78°,
24 h

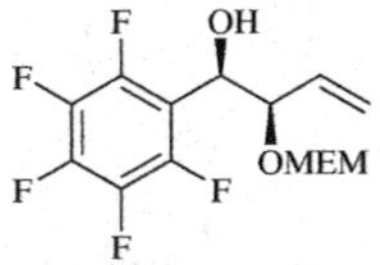

B[(−)-Ipc]₂

OMEM

249

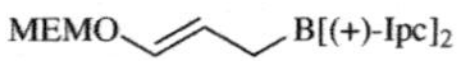

(65), 95, >19:1 d.r.

Pentane/ether (1:1),
−100°

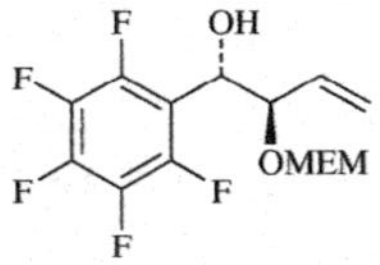

MEMO

B[(+)-Ipc]₂

TABLE 6. ADDITION OF γ-SUBSTITUTED REAGENTS (Continued)

Carbonyl Substrate	Allylborane Reagent	Conditions	Product(s) and Yield(s) (%), % ee	Refs.
C₇	MeO₂C — Ph (pinacol boronate)	BF₃, toluene, rt, 24 h	(87)	785
	EtO₂C — Bu-n (pinacol boronate)	Sc(OTf)₃, toluene, rt, 6–24 h	(53), 19:1 d.r.	24, 26
	BBN reagent (X)	1. ⟋⟍X, MeOBBN 2. LDA, BF₃•OEt₂, THF 3. Aldehyde	I + II + III	159
	B[(−)-Ipc]₂ (X)	1. ⟋⟍X, MeOB[(−)-Ipc]₂, LiNCy₂, ether, −78° 2. BF₃•OEt₂ 3. Aldehyde		159, 762

For reagent III series:

X	Temp	II:(I+III)
Br	−78°	(78) 89:11
Cl	−95°	(76) 95:5

For B[(−)-Ipc]₂ series:

X		d.r.
Cl	(72), 95	98:2
Br	(71), 94	94:6

Reagent	Conditions	Product	Ref.
X—[CH₂CH=CHCH₂]—B[(+)-Ipc]₂	1. CH₂=CHCH₂X, MeOB[(+)-Ipc]₂, LiNCy₂, ether, –78° 2. BF₃•OEt₂ 3. Aldehyde	OH, Cy, X structure X (yield), % ee — d.r. Cl (78), 96 — 98:2 Br (70), 94 — 95:5	159
Cl—[CH₂CH=CHCH₂]—B[(+)-Ipc]₂	1. CH₂=CHCH₂Cl, MeOB[(+)-Ipc]₂, base, ether, –78° 2. BF₃•OEt₂ 3. Aldehyde	(—) Base — % ee — d.r. LDA — 96 — 97:3 LiTMP — 94 — >99:1 LiCy(Pr-i) — 93 — 97:3	159
Ph—N=C(Ph)—[CH=CHCH₂]—B[(–)-Ipc]₂	1. Ph—N=C(Ph)—CH₂CH=CH₂, LDA, MeOB[(–)-Ipc]₂, 2. Aldehyde, THF, –78°, 3 h	OH, Cy, N=C(Ph)Ph structure **I** (56), 91, >95:5 d.r.	746
"	THF, –78°	**I** (56), 91, >95:5 d.r.	160
Ph—N=C(Ph)—[CH=CHCH₂]—B[(+)-Ipc]₂	1. Ph—N=C(Ph)—CH₂CH=CH₂, LDA, MeOB[(+)-Ipc]₂, 2. Aldehyde, THF, –78°, 3 h	OH, Cy, N=C(Ph)Ph structure **I** (61), 93, >95:5 d.r.	746
"	THF, –78°	**I** (61), 93, >95:5 d.r.	160

TABLE 6. ADDITION OF γ-SUBSTITUTED REAGENTS (Continued)

Carbonyl Substrate	Allylborane Reagent	Conditions	Product(s) and Yield(s) (%), % ee	Refs.
C$_7$ Cy–CHO	$(i\text{-Pr})_2\text{NMe}_2\text{Si}$ ⟶ B[(–)-Ipc]$_2$	1. Ether, –78°, 3.5 h 2. KF, H$_2$O$_2$	Cy, OH ⋯ OH (**I**) (45), >95.5 d.r.	752
	"	1. Ether, –78° 2. KF, H$_2$O$_2$	**I** (63), >95.5 d.r.	45
	PhMe_2Si ⟶ B[(+)-Ipc]$_2$	THF, –78°, 4 h	Cy, OH, SiMe$_2$Ph (85), 91	734
	(dioxaborolane, Ph$_4$) B[(–)-Ipc]$_2$	Ether, –78°, 2 h	Cy, OH, B(O$_2$C$_2$Ph$_4$) (—)	175
	(1,3,2-dioxaborinane) B[(–)-Ipc]$_2$	Ether, –78°, 2 h	Cy, OH, B(O$_2$C$_3$H$_6$) (—)	175
	"	1. Ether, –78° to 0°, 4 h 2. H$_2$O$_2$, NaOH	Cy, OH ⋯ OH (70), 94, >97:3 d.r.	157

	1. Cl⟍⟍ PhMgCl, toluene, −15°, 30 min 2. Aldehyde: rt, 24 h	(49), >99:1 d.r. 	87
R = CH$_2$CO$_2$Me	−78°, 5 h	(60), 65 	90
	Toluene/pentane (10:3), −100°	$\dfrac{n}{\begin{matrix}1 & (82), 98\\ 2 & (80), 97\end{matrix}}$ 	738
R = TBDMS	4 Å MS, toluene, −78°, 5 h	(85), 91 	88
	4 Å MS, toluene, −78°	$\dfrac{R}{\begin{matrix}\text{2-(5-methyl)furyl} & (83), 82\\ \text{menthofuryl} & (87), 81\end{matrix}}$ 	158
	4 Å MS, toluene, −78°	$\dfrac{R}{\begin{matrix}\text{CyO} & (95), 72\\ \text{Ph} & (88), 87\end{matrix}}$ 	158

TABLE 6. ADDITION OF γ-SUBSTITUTED REAGENTS (*Continued*)

Carbonyl Substrate	Allylborane Reagent	Conditions	Product(s) and Yield(s) (%), % ee	Refs.
C₇				
(Cy-CHO)		OAc, Pd₂(dba)₃, DMSO, toluene, 20°, 69 h	(77), 53	349
		1. Cl---, PhMgCl, THF, –15°, 24 h 2. Aldehyde, rt, 24 h	(36), 96:4 d.r.	87
	B(OPr-i)₂, Bu-n	THF, –78° to rt, 18 h	(82), 89:11 d.r.	772
	n-Bu---B(OPr-i)₂	THF, –78° to rt, 18 h	(84), 95:5 d.r.	772
	B(OPr-i)₂	rt	Solvent / Time: CH₂Cl₂ 30 h (91); toluene 4 h (91); neat 1 h (90)	744, 741
	OCH(Pr-i)₂	4 Å MS, toluene, –78°, 72 h	(94), 89	741

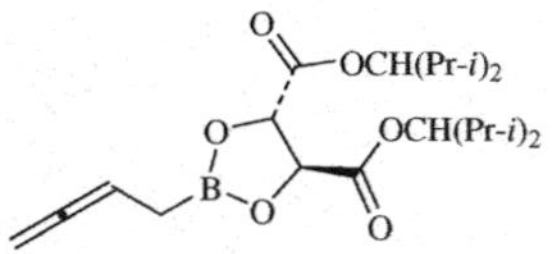

(92), 90   741

I (94), 84   741

(60), 96:4 d.r.   781

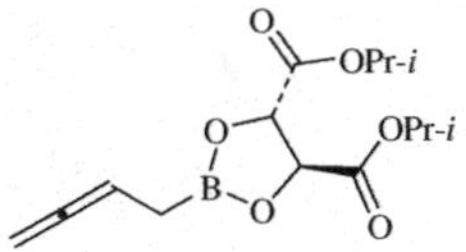

(65), 74, >19:1 d.r.   185

4 Å MS, toluene, −78°, 72 h

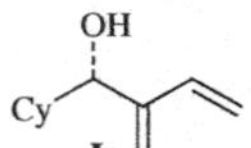

4 Å MS, toluene, −78°, 72 h

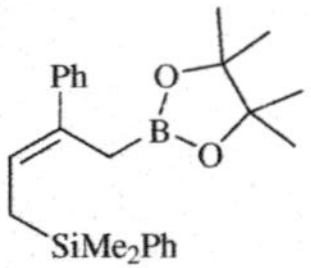

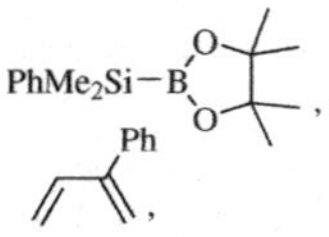

Pt(C₂H₄)(PPh₃)₂, octane, 120°

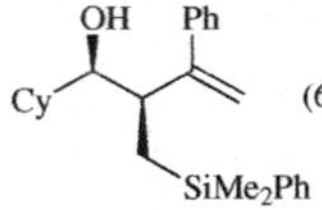

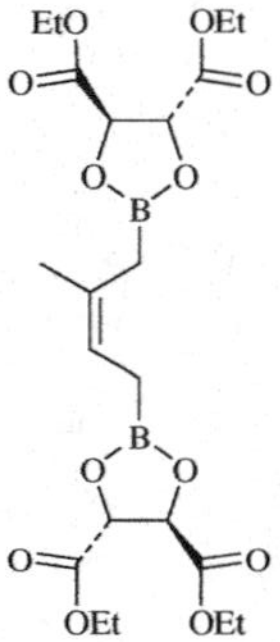

1.

2. Aldehyde, −78°, 3 h

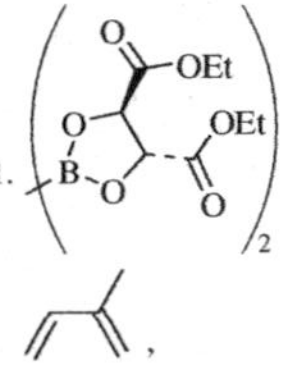

Pt(C₂H₄)(PPh₃)₂, benzene, 80°

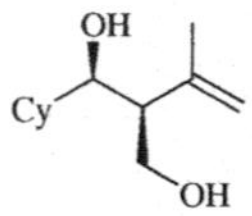

3. NaOH, H₂O₂

TABLE 6. ADDITION OF γ-SUBSTITUTED REAGENTS (*Continued*)

Carbonyl Substrate	Allylborane Reagent	Conditions	Product(s) and Yield(s) (%), % ee	Refs.
C₇				
		Sc(OTf)₃, toluene, rt, 24 h	(32), >20:1 d.r.	26
		Cu(OTf)₂, toluene, 60°, 16 h	(54), 19:1 d.r.	24, 26
		1. Pt(dba)₂, PCy₃, benzene, rt 2. Aldehyde, −78°, 3 h 3. NaOH, H₂O₂	(72), 74, >19:1 d.r.	185
		1. Pt(dba)₂, PCy₃, benzene, rt 2. Aldehyde, −78°, 3 h 3. NaOH, H₂O₂	**I**	185

R	
Me	(21), 72
Et	(68), 74
Pr-*i*	(38), 58
CH(Pr-*i*)₂	(17), 60

Substrate:

$n\text{-}C_6H_{13}\text{CHO}$

Reagent / Conditions	Product	Yield (d.r.)	Refs.
$(i\text{-Pr})_2\text{NMe}_2\text{Si}$—/—$\text{B}[(-)\text{-Ipc}]_2$; 1. Ether, $-78°$, 3 h; 2. KF, H$_2$O$_2$	OH, OH diol, $n\text{-}C_6H_{13}$	(52), >95:5 d.r.	45
dioxaborolane (CyO–SiMe$_2$, Me, Me, OPr-i, OPr-i); 4 Å MS, toluene, $-78°$, 3–4 h	SiMe$_2$(OCy), OH, $n\text{-}C_6H_{13}$	(86), 64, >99:1 d.r.	38, 40
dioxaborolane (PhMe$_2$Si, OPr-i, OPr-i); 4 Å MS, toluene, $-78°$, 3–4 h	SiMe$_2$Ph, OH, $n\text{-}C_6H_{13}$	(95), 81, >99:1 d.r.	40
PhMe$_2$Si–B pinacol boronate, Ph; Pt((C$_2$H$_4$)(PPh$_3$)$_2$), octane, 120°	Ph, SiMe$_2$Ph, OH, $n\text{-}C_6H_{13}$	(71), 93:7 d.r.	781
1. propargyl bromide, HBCy$_2$, n-Bu$_4$NBr, $-78°$ to 0°, 1 h; 2. Aldehyde, 0° to rt, 1 h; Cy–BBrCy	Cy, OH, $n\text{-}C_6H_{13}$	d.r. (82) —; (82) >98:2	751, 680

TABLE 6. ADDITION OF γ-SUBSTITUTED REAGENTS (*Continued*)

Carbonyl Substrate	Allylborane Reagent	Conditions	Product(s) and Yield(s) (%), % ee	Refs.
C_7				
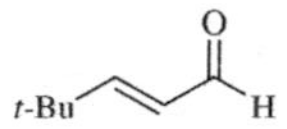	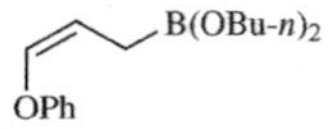	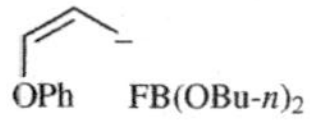	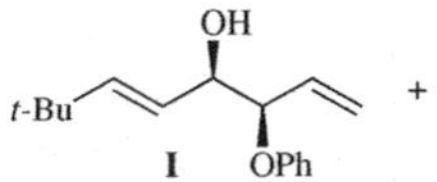 + 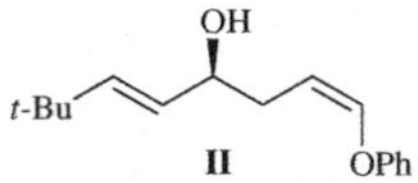 **I** + **II** (73), **I**:**II** = 37.3:1	46
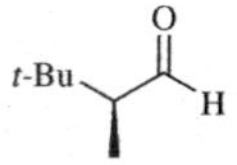	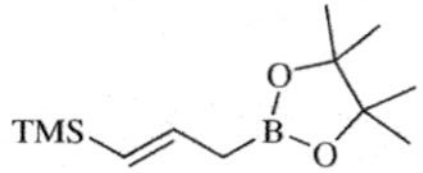	Neat, rt, 4 d	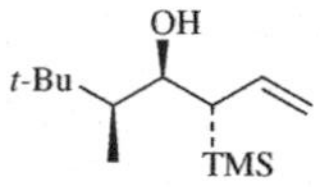 (48), 89:11 d.r.	521
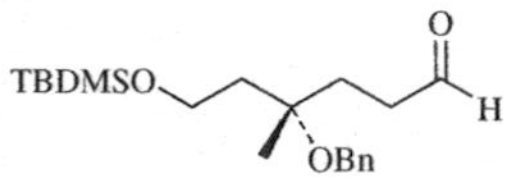	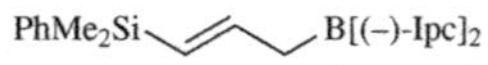	THF, –78°, 4 h	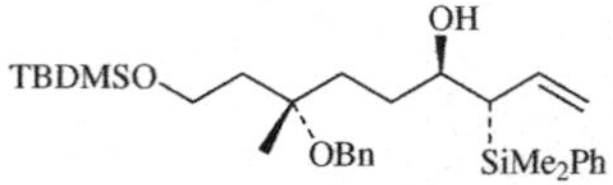 (77), >90, 9:1 d.r.	734, 788
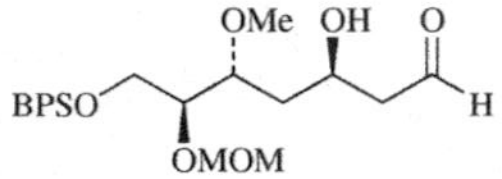	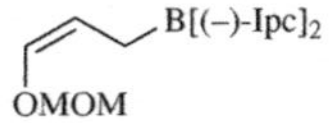	THF, –78°	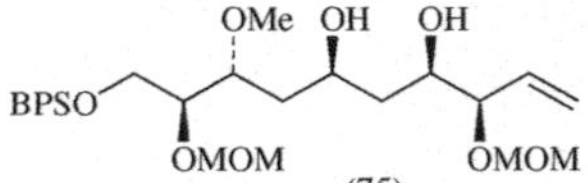 (75)	335
C_8				
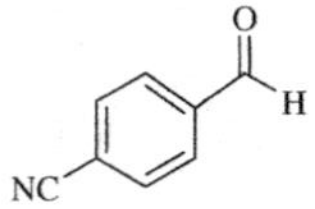	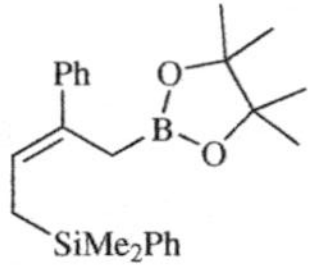	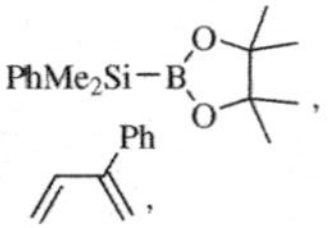 $Pt(C_2H_4)(PPh_3)_2$, octane, 120°	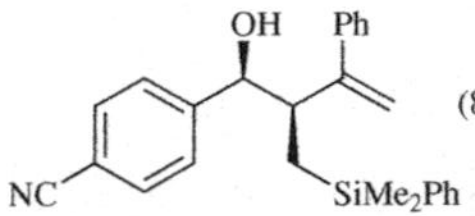 (80), 99:1 d.r.	781

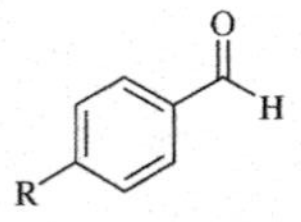

785

761

730

730

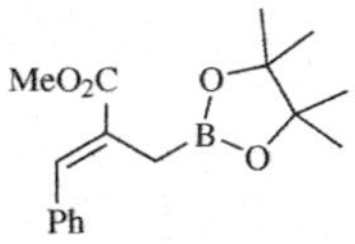

BF$_3$, toluene, rt

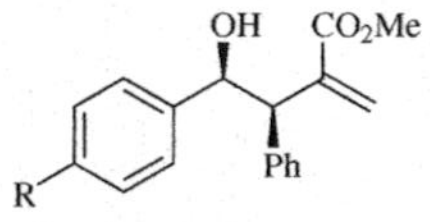

R	Time	
CF$_3$	24 h	(84)
NC	20 h	(81)

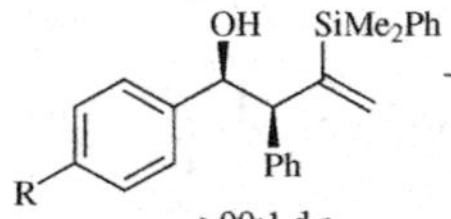

>99:1 d.r.

R		
MeO$_2$C	(93)	
Me	(70)	
NC	(85)	

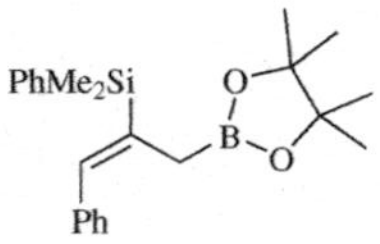

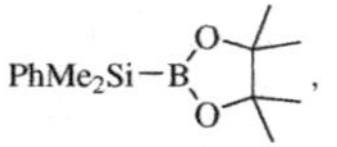

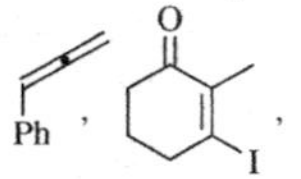

Pd$_2$(dba)$_3$, EtOAc, 80°, 5 h

THF, −78° to 0°, 3 h

(24)

(46), >95, >95:5 d.r.

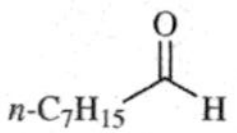

THF, −78° to 0°, 3 h

(25)

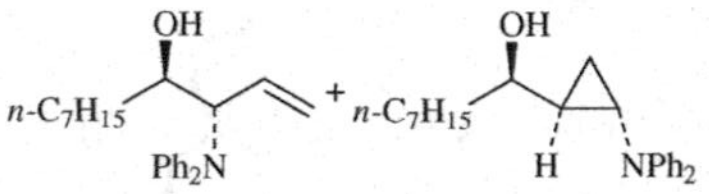

(47), >95, >95:5 d.r.

R = CF$_3$, NC

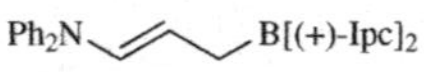

R = MeO$_2$C, Me, NC

B[(−)-Ipc]$_2$

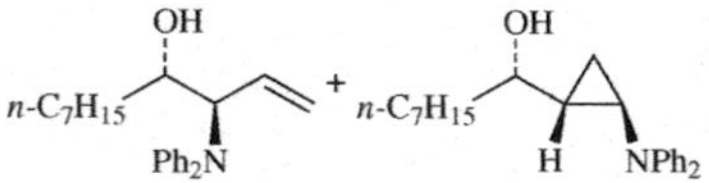

B[(+)-Ipc]$_2$

n-C$_7$H$_{15}$

TABLE 6. ADDITION OF γ-SUBSTITUTED REAGENTS (Continued)

Carbonyl Substrate	Allylborane Reagent	Conditions	Product(s) and Yield(s) (%), % ee	Refs.
C_8				
$n\text{-}C_7H_{15}$CHO	TMS (pinacol allylboronate)	Ether, rt, 1-3 d	OH, $n\text{-}C_7H_{15}$, TMS (79)	32
	(di-OEt tartrate boronate)$_2$	1. (di-OEt tartrate boronate)$_2$, isopropenyl; Pt(dba)$_2$, PCy$_3$, benzene, rt 2. Aldehyde, −78°, 3 h 3. NaOH, H$_2$O$_2$	OH, $n\text{-}C_7H_{15}$, OH (68), 70, >19:1 d.r.	185
cycloheptadiene-CHO	BBN, Cl	Ether, −95° to 0°, 6 h	HO, Cl (57), 97:3 d.r.	789
	B[(−)-Ipc]$_2$, Cl	1. Ether, −95° to rt, 6 h 2. 8-HQ 3. DBU	(epoxide vinyl) (−), 90 + (epoxide vinyl) (−), 77	790
	B[(+)-Ipc]$_2$, Cl	1. Ether, −95° to rt, 6 h 2. 8-HQ 3. DBU	(epoxide vinyl) (−), 95 + (epoxide vinyl) (−), 88	790

474

	B[(−)-Ipc]₂	1. Ether, −95° to rt, 6 h 2. 8-HQ 3. DBU	(—), 96	790
	B[(+)-Ipc]₂	1. Ether, −95° to rt, 6 h 2. 8-HQ 3. DBU	(—), 97	790
	B[(−)-Ipc]₂	1. Ether, −95° to rt, 6 h 2. 8-HQ 3. DBU	(—), 90	790
	B[(+)-Ipc]₂	1. Ether, −95° to rt, 6 h 2. 8-HQ 3. DBU	(—), >97	790
	BBN	1. Cl, MeOBBN, LDA, THF, −78° 2. BF₃•OEt₂ 3. Aldehyde	I + II + III (86), II:(I+III) = 87:13	159

TABLE 6. ADDITION OF γ-SUBSTITUTED REAGENTS (Continued)

Carbonyl Substrate	Allylborane Reagent	Conditions	Product(s) and Yield(s) (%), % ee	Refs.
C₈	B[(–)-Ipc]₂ (with Cl)	1. Cl, MeOB[(–)-Ipc]₂, LiNCy₂, ether, –78° 2. BF₃•OEt₂ 3. Aldehyde	(85), 90, 99:1 d.r.	159, 762
C₉	PhMe₂Si, Ph pinacol boronate	PhMe₂Si–B(pin), Ph • I (with cyclohexenone), Pd₂(dba)₃, EtOAc, 80°, 5 h	(60), >99:1 d.r. SiMe₂Ph	761
	B[(–)-Ipc]₂ (with Cl)	1. Cl, MeOB[(–)-Ipc]₂, LiNCy₂, ether, –78° 2. BF₃•OEt₂ 3. Aldehyde	(70), 98, 99:1 d.r.	159, 762
n-C₈H₁₇CHO	B[(–)-Ipc]₂ (with Cl)	1. Cl, MeOB[(–)-Ipc]₂, LDA, ether, –78° 2. BF₃•OEt₂ (x equiv) 3. Aldehyde	x: 2.5 → (–), 93; 1.33 → (–), 92	159

476

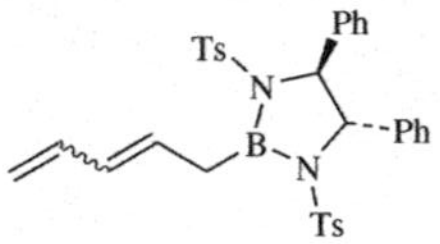

791

185

747

747

747

(78), 98

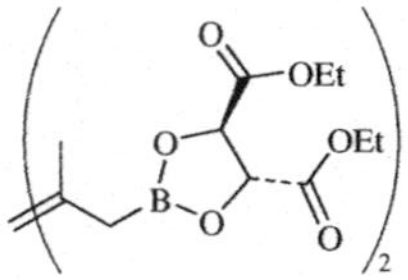

(60)

(5-10), 65

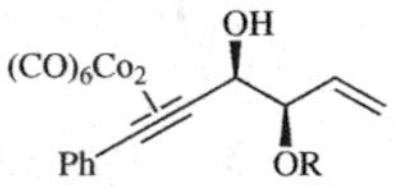

R		d.r.
Me	(75), >98	>97.5:2.5
MOM	(62), >96	>97.5:2.5

(75), >98, >97.5:2.5 d.r.

CH_2Cl_2, −78°

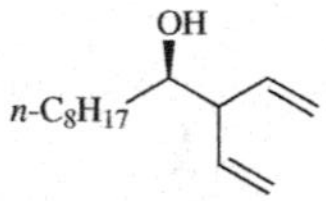

1. [structure] Pt(dba)$_2$, PCy$_3$; benzene, rt
2. Aldehyde, −78°, 3 h
3. Me$_2$Zn, H$_2$O

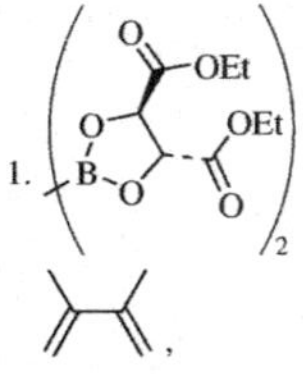

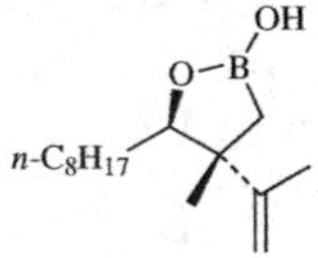

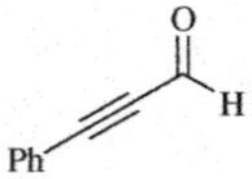

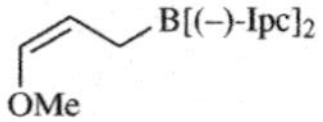

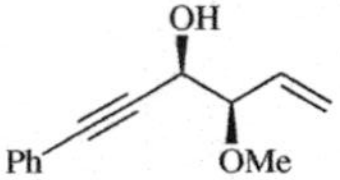

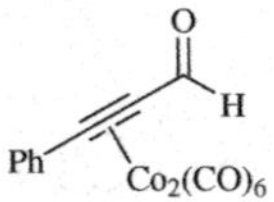

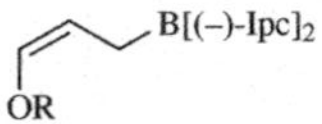

THF, −78°

THF, −78° to rt, 15 h

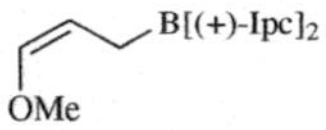

THF, −78° to rt, 15 h

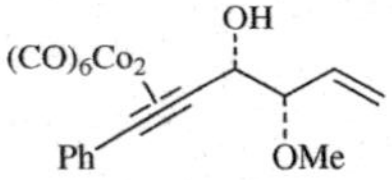

B[(−)-Ipc]$_2$

B[(−)-Ipc]$_2$

B[(+)-Ipc]$_2$

TABLE 6. ADDITION OF γ-SUBSTITUTED REAGENTS (Continued)

Carbonyl Substrate	Allylborane Reagent	Conditions	Product(s) and Yield(s) (%), % ee	Refs.
C₉				
(Ph–C≡C–CO₂(CO)₆, CHO)	MeO⟶⟶B[(−)-Ipc]₂	THF, −78° to rt, 15 h	(50), >95, >97.5:2.5 d.r.	747
(Ph–CH=CH–CHO)	B[(−)-Ipc]₂ (OMEM)	Ether, −78°, 5 h	(76), 94, >99:1 d.r.	292
	B[(−)-Ipc]₂ (Cl)	1. ⟶⟶Cl, MeOB[(−)-Ipc]₂, LiNCy₂, ether, −78° 2. BF₃•OEt₂ 3. Aldehyde	(85), 98, 98:2 d.r.	159
	B[(+)-Ipc]₂	Ether, pentane, −100°	(90), 87	505
	(O–B–O, OTHP)	(OTHP, Cl⁻, O–B–O)	I + II (76), I:II = 5.9:1	46
	(THPO, Bpin)	(THPO⟶Li; pin–B–CH₂Cl)	(62), 13:1 d.r.	52

Reagent	Conditions	Product	Reference
R = TBDMS	4 Å MS, toluene, −78°, 5 h	(98), 73	88
B(OPr-i)$_2$, Bu-n	THF, −78° to rt, 18 h	(95), 90:10 d.r.	772
B(OPr-i)$_2$, n-Bu	THF, −78° to rt, 18 h	(96), 95:5 d.r.	772
	4 Å MS, toluene, −78°, 5 h	**I** (98), 71, 98:2 d.r.	740
	4 Å MS, toluene, −78°, 5 h	(93), 70, 96:4 d.r.	740
	4 Å MS, toluene, −78°, 5 h	(100), 62, 96:4 d.r.	740

TABLE 6. ADDITION OF γ-SUBSTITUTED REAGENTS (Continued)

Carbonyl Substrate	Allylborane Reagent	Conditions	Product(s) and Yield(s) (%), % ee	Refs.
C₉				
		4 Å MS, toluene, −78°, 5 h	(100), 68, 96:4 d.r.	740
		CH₂Cl₂, rt, 5 h	(95)	744
		1. Pt(dba)₂, PCy₃, benzene, rt; 2. Aldehyde, −78°, 3 h; 3. NaOH, H₂O₂	(59), 40, >19:1 d.r.	185
		Sc(OTf)₃, toluene, rt, 16-24 h	(—), 2.8:1 d.r.	26

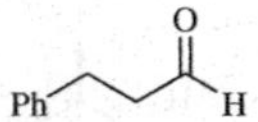

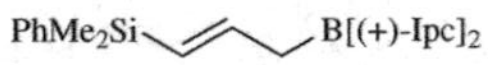

THF, −78°, 4 h (76), 95 734

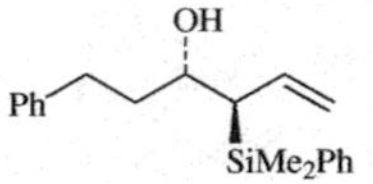

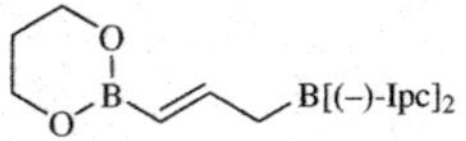

Ether, −78°, 2 h; rt, 24 h (40), 93, >20:1 d.r. 175

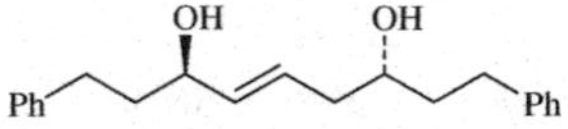

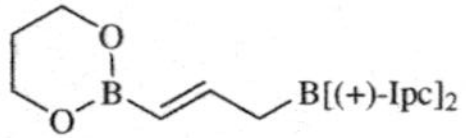

Ether, −78°, 2 h; rt, 24 h (42), 95, >20:1 d.r. 175

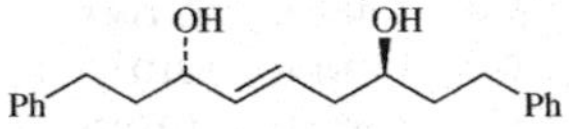

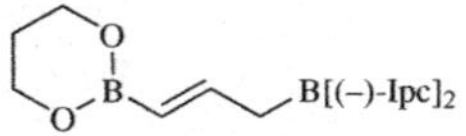

Ether, −78°, 2 h (—) 175

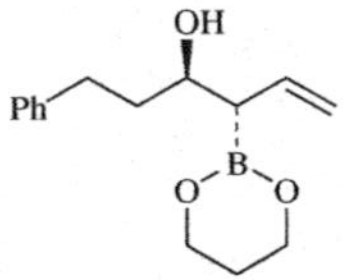

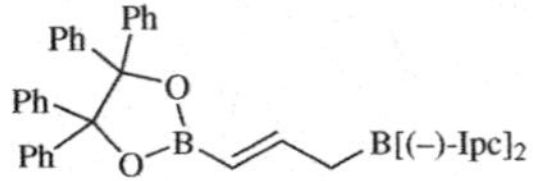

Ether, −78°, 2 h (—) 175

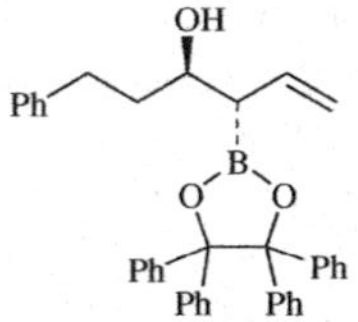

Ether, −78°, 2 h; rt, 24 h (72), 91, >14:1 d.r. 175

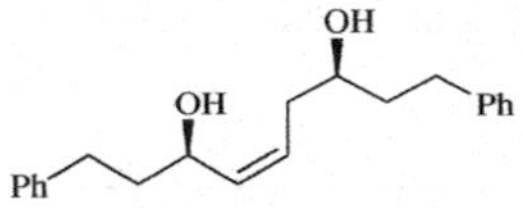

TABLE 6. ADDITION OF γ-SUBSTITUTED REAGENTS (*Continued*)

Carbonyl Substrate	Allylborane Reagent	Conditions	Product(s) and Yield(s) (%), % ee	Refs.
C₉				
(PhCH₂CH₂CHO)	(boronate ester, OPr-i, OMOM)	4 Å MS, toluene, −78° to 0°, 1 h	(70), 90	792
	(boronate ester, OPr-i, PhMe₂Si)	4 Å MS, toluene, −78°	(75), 75	743
	(pinacol boronate, EtO₂C, Et)	L.A., rt, 24 h		26, 24

L.A.	Solvent	
Sc(OTf)₃	toluene	(64)
Cu(OTf)₂	toluene	(63)
Yb(OTf)₃	toluene	(66)
Sc(OTf)₃	CH₂Cl₂	(62)
Cu(OTf)₂	CH₂Cl₂	(52)
Yb(OTf)₃	CH₂Cl₂	(38)

(PhCH(CH₃)CHO)	(pinacol boronate, TMS)	Neat, rt, 4 d	(—), 90:10 d.r.	521

482

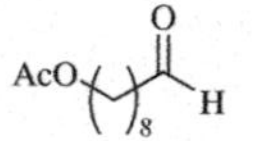

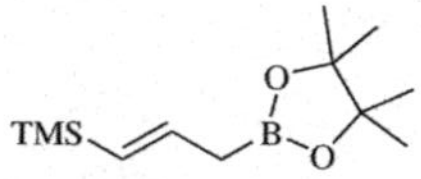

Ether, rt, 1-3 d — (92) — 32

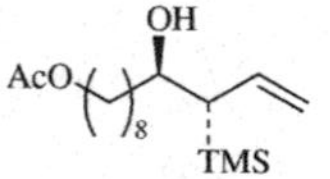

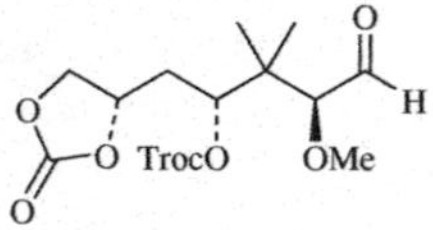

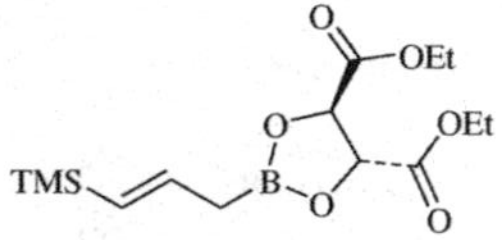

4 Å MS, toluene, −78° — (89), >98:2 d.r. — 193

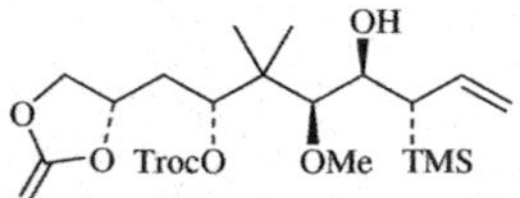

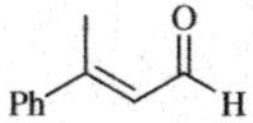

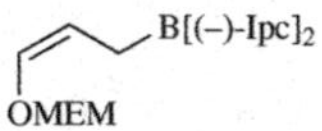

Ether, −78°, 5 h — (78), 93, >99:1 d.r. — 292

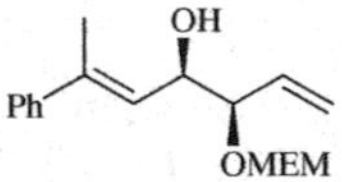

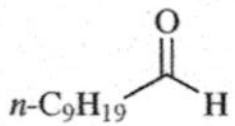

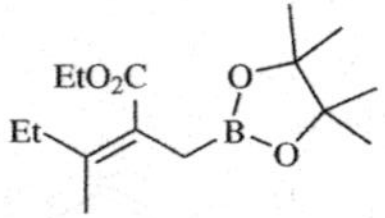

Toluene, rt, >12 d — (68), 18:1 d.r. — 59

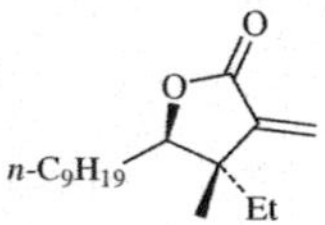

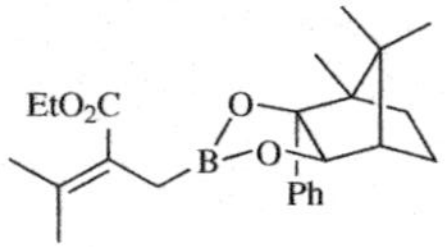

1. Neat, rt, 14 d
2. pTSA — (—), 60 — 59, 26

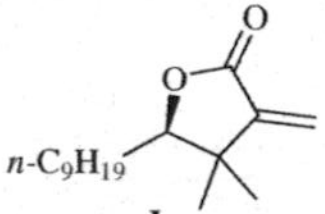

L.A., toluene, rt, 24 h — 26

L.A.	
Sc(OTf)$_3$	(53), 3
Cu(OTf)$_2$	(31), 13
Yb(OTf)$_3$	(37), 21

C$_{10}$

TABLE 6. ADDITION OF γ-SUBSTITUTED REAGENTS (Continued)

Carbonyl Substrate	Allylborane Reagent	Conditions	Product(s) and Yield(s) (%), % ee	Refs.
C_{10} $n\text{-}C_9H_{19}\text{CHO}$	EtO$_2$C / R (pinacol allylboronate)	Toluene	R table: Me, 60–80°, 16–120 h, (40); Et, 110°, 16–24 h, (68)	26
	R = (−)-8-phenylmenthyl	L.A., toluene, rt, 24 h	L.A.: Sc(OTf)$_3$ (29), −10; Cu(OTf)$_2$ (93), 71; Yb(OTf)$_3$ (30), 31	26
	R = (−)-8-phenylmenthyl (Ph)	1. Neat, rt, 14 d 2. pTSA	(—), 10	59, 26
	R = (−)-8-phenylmenthyl (Ph)	1. Neat, rt, 14 d 2. pTSA	(75), 98	59, 26
	R = (−)-8-(2-naphthyl)menthyl	1. Toluene, rt, 14 d 2. pTSA, 3 h	(—), 82	26

Products: 5-(n-C$_9$H$_{19}$)-3-methylene-4,4-disubstituted-γ-butyrolactone

Et—⟋⟍—B⁻Et₂SePh

1. PhSe⟋⟍ ,
LDA, THF, −78°,
30 min
2. Et₃B, −78° to rt,
12 h

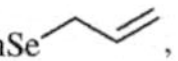

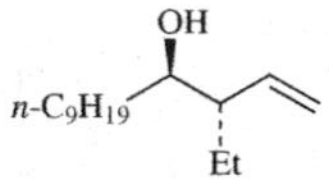

OH
n-C₉H₁₉ ⟋⟍ Et

(83), 61:39 d.r.

692

RO⟋⟍OR⟋⟍OR (R = TBDMS)

Ph Ph / Ph Ph dioxaborolane–B—⟋⟍B[(+)-Ipc]₂

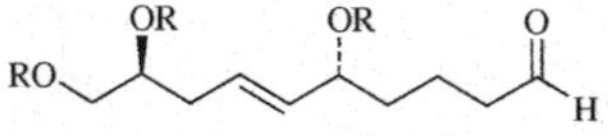

CH₂Cl₂, MeOH,
−78° to rt

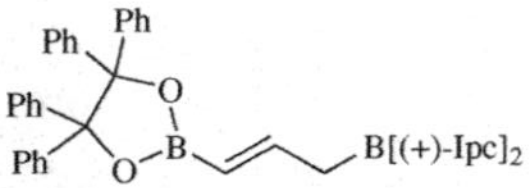

OR⟋⟍OR⟋⟍OH—B(pinacol)

(63), 10:1 d.r.

202

OTBS OAc / O / PMBO — CHO

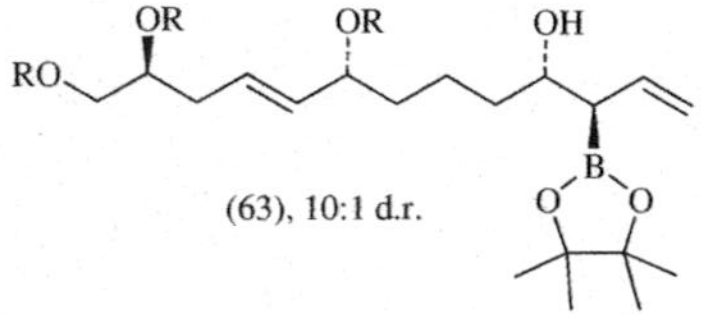

TMSCl

B(OPr-i)₂

OTBS OAc / OH / O / PMBO

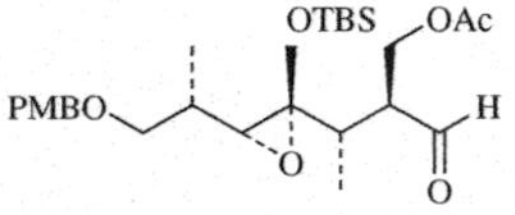

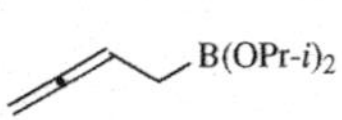

(40)

793

1-naphthaldehyde (CHO)

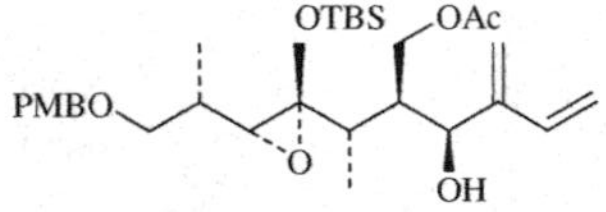

EtO₂C / Me / Bu-n / B(pinacol)

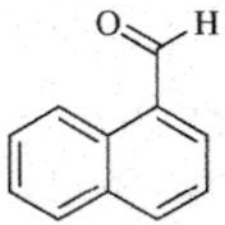

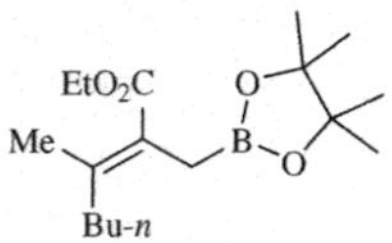

Sc(OTf)₃, toluene,
rt, 24 h

O=⟋ lactone Bu-n Me naphthyl

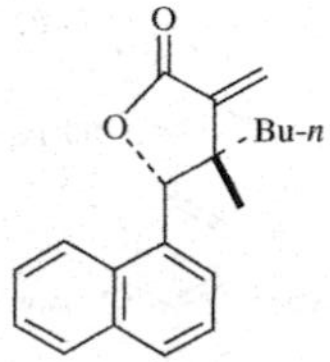

(44)

26

O=CH / OMe OMe / OPMB (C₁₂)

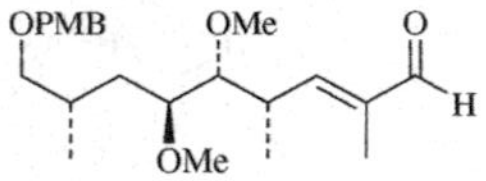

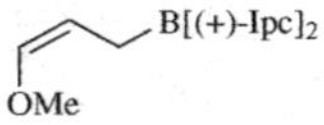

⟋⟍B[(+)-Ipc]₂
OMe

THF, −78° to −20°,
2 h

OMe / OH OMe / OMe / OPMB

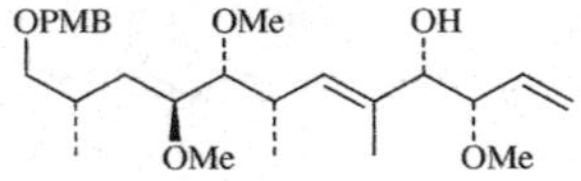

(89.5), 76:13.5 d.r.

794, 795

C₁₁

C₁₂

485

TABLE 6. ADDITION OF γ-SUBSTITUTED REAGENTS (Continued)

Carbonyl Substrate	Allylborane Reagent	Conditions	Product(s) and Yield(s) (%), % ee	Refs.
C_{13}				
		Ether, –90°	(81), 95, >98:2 d.r.	796
		Sc(OTf)$_3$ (10 mol%), rt, 3 d	(12)	797
	"	90°, 3 d	I (19)	797
	"	BF$_3$•SiO$_2$, rt, 3 d	I (15)	797
		Ether, THF, –78° to –20°, 23 h	(96), >97.5:2.5 d.r.	468
		OPr-i 4 Å MS, toluene, –78°	(80)	788

C_{17} 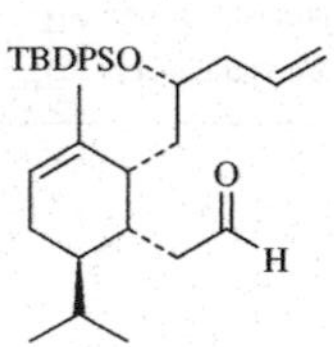	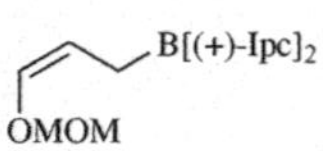	THF, –78°	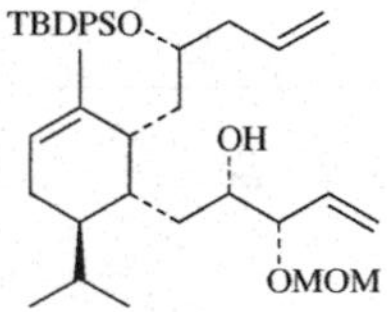 (70), 10:1 d.r.	452, 453
	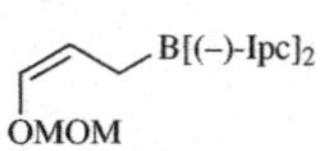	Ether, THF, –78° to rt, 3 h	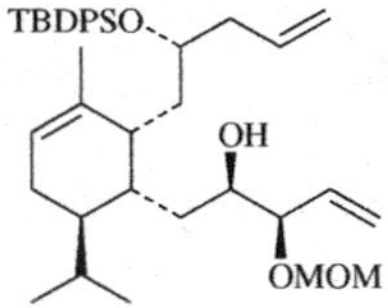 (78), 99:1 d.r.	452, 453
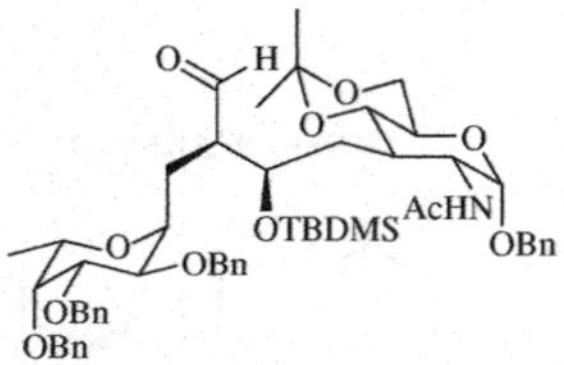	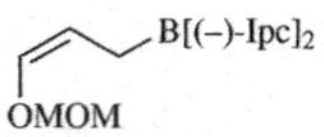	THF, –78° to 10°, 16 h	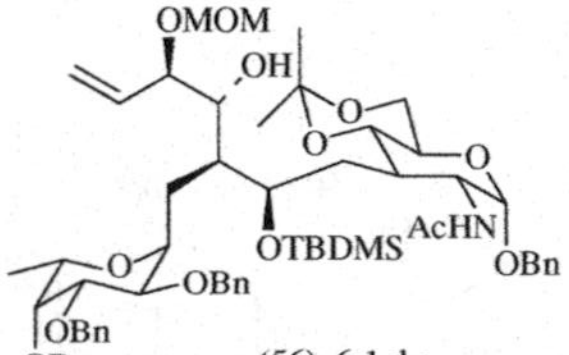 (56), 6:1 d.r.	469
C_{18} 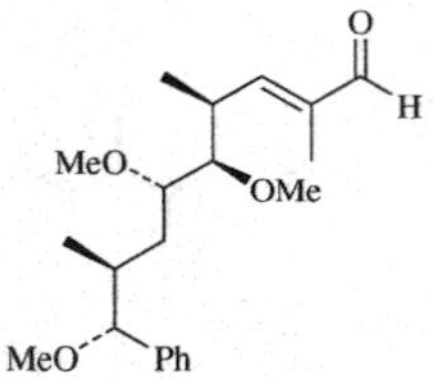	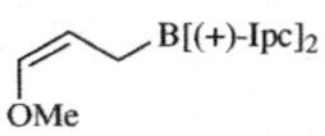	—	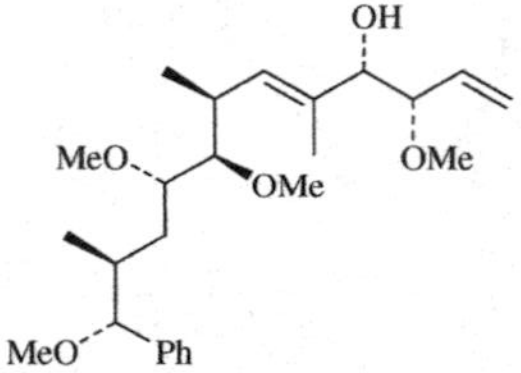 (77)	798

TABLE 6. ADDITION OF γ-SUBSTITUTED REAGENTS (Continued)

Carbonyl Substrate	Allylborane Reagent	Conditions	Product(s) and Yield(s) (%), % ee	Refs.
C₂₀				
R¹ = TIPS, R² = MOM	1. Toluene, 0°, 1 h 2. KH, THF, 0°, 10 min		(80), >20:1 Z:E	799
R¹ = TIPS, R² = MOM	1. Toluene, 0°, 1 h 2. KH, THF, 0°, 10 min		(70), >20:1 Z:E	800
C₂₅		1. CH₂Cl₂, rt, 1 h 2. KOBu-t, THF, rt, 1 h	(83), 15:1 Z:E	801

a The reaction mixture was free of Mg²⁺ ions.

b The starting material was prepared using different conditions.

TABLE 7. ADDITION OF ALLENYL AND PROPARGYL REAGENTS

Carbonyl Substrate	Allylborane Reagent	Conditions	Product(s) and Yield(s) (%), % ee	Refs.
C_1	R^1, R^2, $B(R^1)_2$	1. R^2 —O—OMe, n-BuLi, THF 2. BR^1_3, $-78°$ 3. BCl_3, $-78°$ to rt	R^2 —, R^1, HO—	R^1 R^2 n-Bu Me (45); Ph Ph (68); n-C$_6$H$_{13}$ Ph (84); n-Bu Ph (89)	802
	C_6H_{13}-n, $B(C_6H_{13}$-$n)_2$	1. —O—OMe, n-BuLi, THF 2. $B(C_6H_{13}$-$n)_3$, $-78°$ 3. BCl_3, $-78°$ to rt	HO— C_6H_{13}-n	(71)	802
	C_6H_{13}-n, BOMe(Thex), Ph	1. Ph —O—OMe, n-BuLi, THF 2. (n-C$_6$H$_{13}$)BOMe(Thex), $-78°$ 3. BCl_3, $-78°$ to rt	Ph, HO— C_6H_{13}-n	(65)	802
	n-Pr —, $B(Bu$-$n)_2$	Ether, $0°$	OH, Pr-n	(65)	803
C_2	Bu-n, $B(Bu$-$n)_2$	1. —O—OMe, n-BuLi, THF 2. $B(Bu$-$n)_3$, $-78°$ 3. BCl_3, $-78°$ to rt	HO— Bu-n	(60), 56:44 d.r.	802

TABLE 7. ADDITION OF ALLENYL AND PROPARGYL REAGENTS (*Continued*)

Carbonyl Substrate	Allylborane Reagent	Conditions	Product(s) and Yield(s) (%), % ee	Refs.
C₂				
(acetaldehyde)	R—≡—B[(−)-Ipc]₂	Ether, −100°, 2 h	R: Me (68), 89; n-Pr (74), 87	804
	TMS—≡—B[(−)-Ipc]₂	Ether, −100°, 3 h	(72), 87 (TMS product)	805
	TMS / B (9-BBN) with ≡—TMS	THF, −78°, 3 h	(71), 94 (TMS product)	806
CCl₃CHO	n-Pr—CH=C=CH—B(Bu-n)₂	Ether, 0°	(72), CCl₃ with Pr-n	803
	TMS / B (9-BBN)	Ether, −78°, 3 h	(74), 94	144
C₃ (acrolein)	C₅H₉-c—CH=C=CH—B(C₅H₉-c)₂	1. Li—≡—CH₂Cl, B(C₅H₉-c)₃, THF, −90° 2. Aldehyde, −78° to rt, 1.5 h	(75) C₅H₉-c	807
	c-C₅H₉—≡—CH₂—B(C₅H₉-c)₂	1. Li—≡—CH₂Cl, B(C₅H₉-c)₃, THF, −90° to rt 2. Aldehyde, −78° to rt, 1.5 h	(84) with C₅H₉-c	807

490

Aldehyde: EtCHO (O=CH-Et)

Reagent	Conditions	Product	Yield (%)	Refs.
allyl–BBN	Ether, rt, 15 min	(OH, Et alkyne)	(82)	808, 809
R–BR₂ (allenyl)	1. Li–C≡C–CH₂Cl, BR₃, THF, –90°; 2. Aldehyde, –78° to rt, 1.5 h	(OH, Et, R)	R = c-C₅H₉ (89); n-C₆H₁₃ (85)	807
n-Pr–C≡C–CH=CH–B(Bu-n)₂	Ether, 0°	(OH, Et, Pr-n)	(70)	803
R–C≡C–BR₂	1. Li–C≡C–CH₂Cl, BR₃, THF, –90° to rt; 2. Aldehyde, –78° to rt, 1.5 h	(OH, Et, allene R)	R = Cy (77); c-C₅H₉ (80); n-C₆H₁₃ (73)	807
R–B(OBu-n)₂	Ether	(OH, Et, R alkyne)	R = Et (60); n-Pr (78)	810
(Pr-n)–CH=CH–B(OBu-n)₂	Ether	(OH, Et, Pr-n)	(70)	810

TABLE 7. ADDITION OF ALLENYL AND PROPARGYL REAGENTS (*Continued*)

Carbonyl Substrate	Allylborane Reagent	Conditions	Product(s) and Yield(s) (%), % ee	Refs.
C$_3$ Et—CHO	Ph—C≡C—CH$_2$—B(OBu-n)$_2$	Ether	(52) (with OH, Et, Ph)	810
	Ph—C≡C—CH(CH$_3$)—B(OBu-n)$_2$	Ether	(50) (with OH, Et, Ph)	810
	C$_6$H$_{13}$-n ... B(Thex)—Cl	1. Li—C≡C—C$_6$H$_{13}$-n, (Thex)B—Cl, THF, −90° 2. Aldehyde, −78° to rt, 1.5 h	C$_6$H$_{13}$-n (76) (with OH, Et)	807
C$_3$ (CH$_3$)$_2$C=O	C$_6$H$_{13}$-n ... B(C$_6$H$_{13}$-n)$_2$ (with Ph)	1. Ph—C≡C—C(CH$_3$)$_2$—O—OMe, n-BuLi, THF 2. B(C$_6$H$_{13}$-n)$_3$, −78° 3. BCl$_3$, −78° to rt	C$_6$H$_{13}$-n (62) (with HO, Ph)	802
	TMS—C≡C—CH$_2$—BBN	THF, rt, 1.5 h	(86) (with HO, TMS)	811

C$_4$

Reagent conditions	Product	Ref.
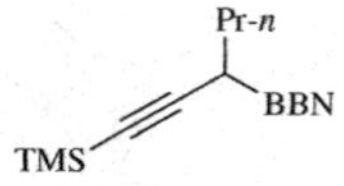 THF, rt, 1.5 h	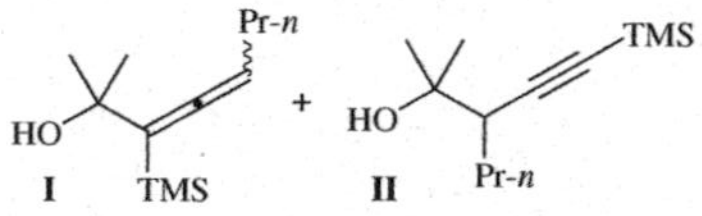 **I + II** (71), **I:II** = 91:9	811
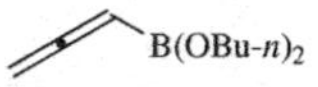 Ether	 (55)	810
Ether, 3 h	(95)	598
	Temp d.r. −78° 85:15 0° 83:17	598
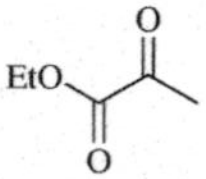 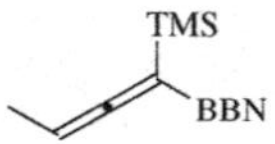 Ether, −78°, 3 h	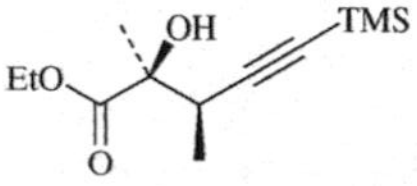 (90), 75:25 d.r.	598
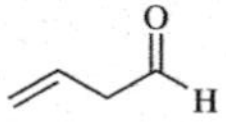 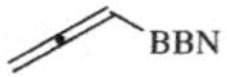  Sc(OTf)$_3$ (7.5 mol%), THF, 0°, 4 h	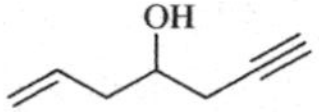 (62)	812
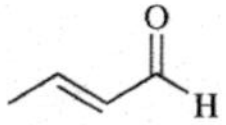 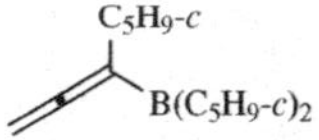 1. 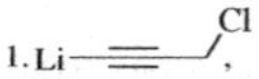 Li—CH$_2$C(Cl)=CH$_2$, B(C$_5$H$_9$-c)$_3$, THF, −90° 2. Aldehyde, −78° to rt, 1.5 h	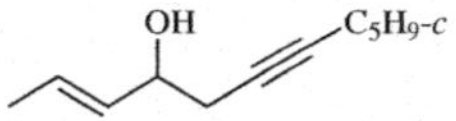 (84)	807
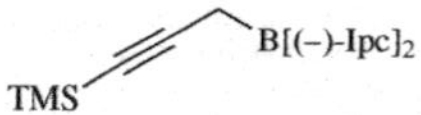 Ether, −100°, 3 h	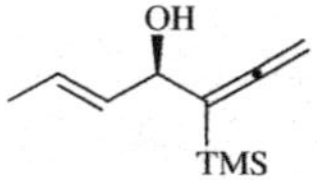 (68, 87)	805

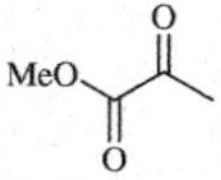

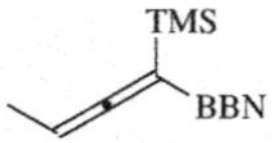

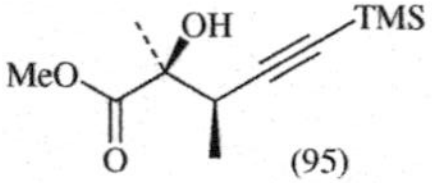

493

TABLE 7. ADDITION OF ALLENYL AND PROPARGYL REAGENTS (*Continued*)

Carbonyl Substrate	Allylborane Reagent	Conditions	Product(s) and Yield(s) (%), % ee	Refs.
C$_4$				
	c-C$_5$H$_9$ ··· B(C$_5$H$_{9}$-c)$_2$	1. Li ··· Cl , B(C$_5$H$_9$-c)$_3$, THF, −90° to rt 2. Aldehyde, −78° to rt, 1.5 h	(84)	807
	TMS ··· BBN	THF, rt, 1.5 h	(79)	811
	TMS—B, TMS	THF, −78°, 3 h	(87), 97	806
	B(OBu-n)$_2$	Ether	(30)	810
	Et, Ts-N-B-N-Ts, Ph, Ph	CH$_2$Cl$_2$, −78°, 2-5 h	(75-80), 97-98	153
	TMS—B	Ether, −78°, 3 h	(82), 94	144

494

Reagent	Conditions	Product (yield), ratio	Ref.
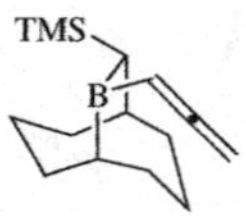 (B, TMS, TMS)	Ether, −78°, 3 h	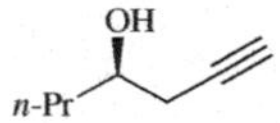 OH, n-Pr (alkynyl) (80), 93	144
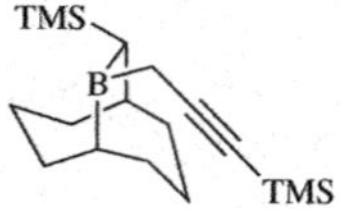 (B, TMS, TMS, TMS)	THF, −78°, 3 h	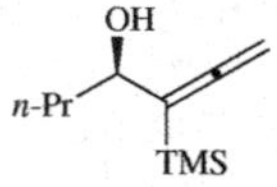 OH, n-Pr, TMS (87), 98	806
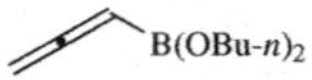 B(OBu-n)$_2$	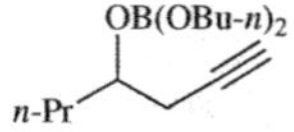 —	OB(OBu-n)$_2$, n-Pr (50–60)	2
"	Ether 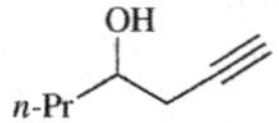	OH, n-Pr (60)	810
BBN	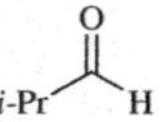 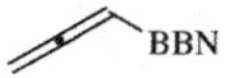 Ether, rt, 15 min	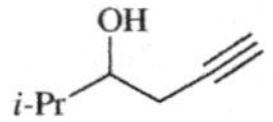 OH, i-Pr (88)	808, 809
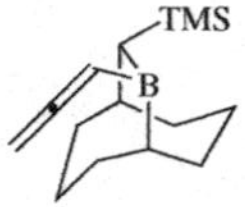 (B, TMS)	Ether, −78°, 3 h	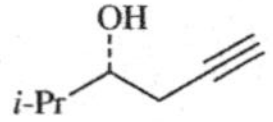 OH, i-Pr (81), 93	144
 n-Pr—B(Bu-n)$_2$	Ether, 0°	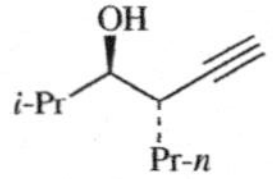 OH, i-Pr, Pr-n (78)	803
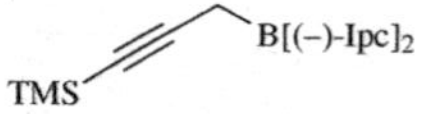 TMS—B[(−)-Ipc]$_2$	Ether, −100°, 3 h	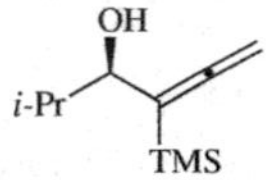 OH, i-Pr, TMS (76), 99	805

i-Pr—CHO

495

TABLE 7. ADDITION OF ALLENYL AND PROPARGYL REAGENTS (*Continued*)

Carbonyl Substrate	Allylborane Reagent	Conditions	Product(s) and Yield(s) (%), % ee	Refs.
C$_4$				
i-Pr–CHO	R–C≡C–CH$_2$–B[(−)-Ipc]$_2$	Ether, −100°, 2 h	R: Me (72), 96; n-Pr (80), 96; Ph (70), 96	804
	c-C$_5$H$_9$ substituted propargyl B(C$_5$H$_9$-c)$_2$	1. Li–C≡C–CH$_2$Cl, B(C$_5$H$_9$-c)$_3$, THF, −90° to rt; 2. Aldehyde, −78° to rt, 1.5 h	(78)	807
	TMS–C≡C–CH$_2$–BBN	THF, rt, 1.5 h	(82)	811
	TMS–BBN (TMS–C≡C substituted)	THF, −78°, 3 h	(77), 97	806
	Pr-n substituted TMS–C≡C–BBN	THF, rt, 1.5 h	I + II (74), I:II = 88:12	811
	Ts–N,N-Ts diphenyl pyrazolidine boron allenyl reagent	CH$_2$Cl$_2$, −78°, 2.5 h	(76), 94	153

Aldehyde/Ketone	Reagent	Conditions	Product (yield)	Refs.
(methyl vinyl ketone)	BBN (allenyl-9-BBN)	Ether, rt, <5 min	OH (71)	809
(methyl ethyl ketone)	BBN	Ether, rt, <5 min	OH (89)	809
i-PrCH₂CHO	OCH(Pr-i)₂ / OCH(Pr-i)₂ tartrate boronate, x equiv	Toluene, −78°, 20 h	OH, i-Pr; x 1.5 (78), 99; 3.0 (74), 99	813
t-BuCHO	BBN	Ether, rt	OH, t-Bu; Time 15 min (89); 90 min (88)	808, 809
t-BuCHO	TMS–B (9-BBN deriv)	Ether, −78°, 3 h	OH, t-Bu (75), 94	144
	$C_5H_9\text{-}c$, $B(C_5H_9\text{-}c)_2$	1. Li— ≡—CH₂Cl, B(C₅H₉-c)₃, THF, −90°; 2. Aldehyde, −78° to rt, 1.5 h	OH, t-Bu, $C_5H_9\text{-}c$ (86)	807
	TMS— ≡ —B[(−)-Ipc]₂	Ether, −100°, 3 h	OH, t-Bu, TMS (75), 92	805

C_5

TABLE 7. ADDITION OF ALLENYL AND PROPARGYL REAGENTS (*Continued*)

Carbonyl Substrate	Allylborane Reagent	Conditions	Product(s) and Yield(s) (%), % ee	Refs.
C₅				
(t-Bu, CHO)	(c-C₅H₉ allenylborane, B(C₅H₉-c)₂)	1. Li ≡≡ Cl, B(C₅H₉-c)₃, THF, −90° to rt, 2. Aldehyde, −78° to rt, 1.5 h	(79)	807
	(BBN reagent, TMS)	THF, rt, 1.5 h	(85)	811
	(TMS 9-BBN, TMS)	THF, −78°, 3 h	(80), 98	806
	(B(OBu-n)₂)	Ether, 40°, 48 h	(50)	814
	(tartrate boronate, OPr-i)	Toluene, −78°	(47), 93	123
	(Ts,N–B–N,Ts, Ph, Ph)	CH₂Cl₂, −78°, 2.5 h	(74), 98	153

Reagent	Conditions	Product(s)	Reference

| | Toluene, −78°, 46 h | (53), 88 | 123 |

| | CH₂Cl₂, −78°, 2.5 h | (78), >99 | 153 |

| | Ether, rt | Time 1 h (88); 15 min (87) | 812 / 808, 809 |

| | Ether, rt, <5 min | (87) | 809 |

| | THF, rt, 1.5 h | (91) | 811 |

| | THF, rt, 1.5 h | I + II (75), I:II = 83:17 | 811 |

| | DMSO, 150°, 2 h | (60) | 814 |

TABLE 7. ADDITION OF ALLENYL AND PROPARGYL REAGENTS (Continued)

Carbonyl Substrate	Allylborane Reagent	Conditions	Product(s) and Yield(s) (%), % ee	Refs.
C₅				
(ethyl 4-oxopentanoate)	TMS—≡—BBN	THF, rt, 2 h	(90)	815
(HO—aldehyde)	Bu-n / B(Bu-n)₂	1. R—OTHP, n-BuLi, THF; 2. B(Bu-n)₃, −78°; 3. BCl₃, −78° to rt	R: Me (65) d.r. 60:40; Ph (55) 60:40	802
(2-furaldehyde)	—=•= BBN	Ether, rt, 15 min	(79)	808, 809
	tartrate borolane (OEt)	Toluene, −78°, 46 h	(50), 70	123
C₆				
(n-Pr enal)	tartrate borolane (OR)	Toluene, −78° to rt	R: Et (39-66), 62-66; Pr-i (39), —	123
(n-C₅H₁₁ CHO)	TMS—≡—BBN	THF, rt, 1.5 h	(82)	811

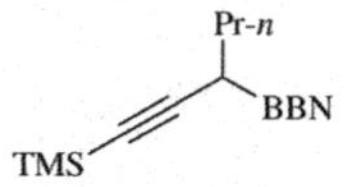

THF, rt, 1.5 h

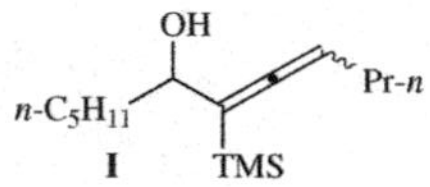

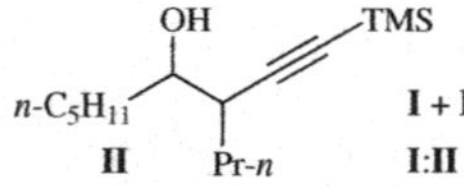

I + **II** (78), **I:II** = 88:12

811

813

x		
1.5	(81), 94	(75), 88
5.0	(72), 97	(63), >95

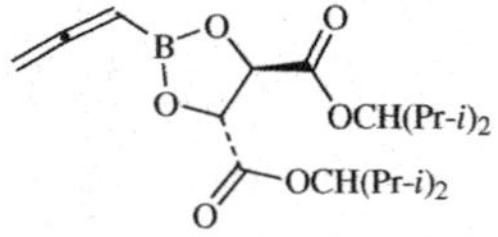

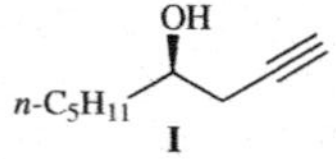

Toluene, −78°, 20 h

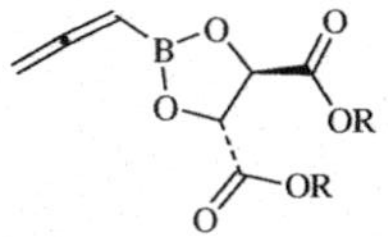

R	Temp	
	−78° to −20°	
Et		
Pr-i	−78°	

Toluene

123

(81), 91

CH₂Cl₂, −78°, 2.5 h

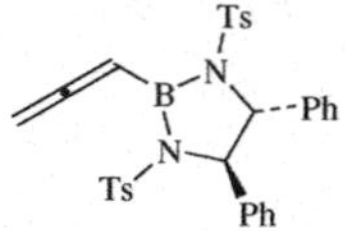

153

(75–80), 97–98

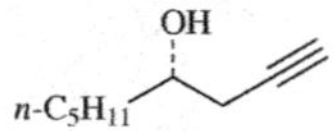

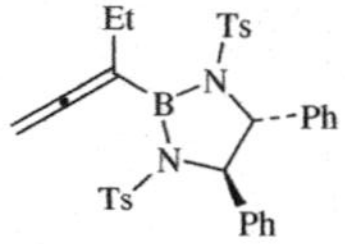

CH₂Cl₂, −78°, 2.5 h

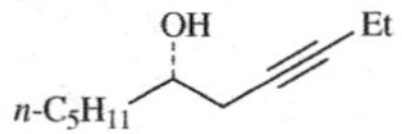

153

(82), >99

CH₂Cl₂, −78°, 2.5 h

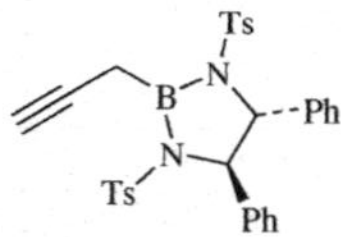

153, 816

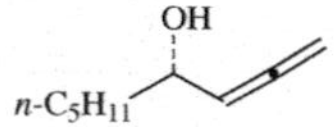

x equiv

TABLE 7. ADDITION OF ALLENYL AND PROPARGYL REAGENTS (*Continued*)

Carbonyl Substrate	Allylborane Reagent	Conditions	Product(s) and Yield(s) (%), % ee	Refs.
C₆ (cyclohexanone)	(allenyl)BBN	Ether, rt, 1 h	(94)	812
	"	Ether, rt, 15 min	**I** (88)	808, 809
	TMS-propargyl-BBN	THF, rt, 1.5 h	(91)	811
	Pr-*n* / TMS-BBN	THF, rt, 1.5 h	**I** + **II** (76), **I**:**II** = 91:9	811
	B(OBu-*n*)₂	Ether, 40°, 12 h	(68)	814
C₇ (*t*-Bu ketone)	BBN	Ether, rt, 1 h	(88)	812
Ph–CHO	BBN	Ether, rt, 15 min	(82)	808, 809

502

Reagent	Conditions	Product	Yield	Refs.
(9-BBN type, TMS-allenyl borane)	Ether, −78°, 3 h	(propargyl alcohol, Ph)	(78), 93	144
$C_5H_9\text{-}c$, $B(C_5H_9\text{-}c)_2$ (allenyl borane)	1. Li≡—Cl , $B(C_5H_9\text{-}c)_3$, THF, −90° 2. Aldehyde, −78° to rt, 1.5 h	(alcohol, Ph, $C_5H_9\text{-}c$)	(86)	807
$n\text{-}Pr$ —CH=CH— $B(Bu\text{-}n)_2$	Ether, 0°	(alcohol, Ph, Pr-n)	(79)	803
TMS—CH= —BBN	THF, rt, 1.5 h	(alcohol, Ph, TMS)	(88)	811
$c\text{-}C_5H_9$—CH= —$B(C_5H_9\text{-}c)_2$	1. Li≡—Cl , $B(C_5H_9\text{-}c)_3$, THF, −90° to rt 2. Aldehyde, −78° to rt, 1.5 h	(alcohol, Ph, $C_5H_9\text{-}c$)	(85)	807
R—CH= —$BI[(−)\text{-}Ipc]_2$	Ether, −100°	(alcohol, Ph, R)	R Time: n-Pr 2 h (77), 87; TMS 3 h (74), 89	804 805
TMS, B (9-BBN type with TMS-alkyne)	THF, −78°, 3 h	(alcohol, Ph, TMS)	(60), 98	806

TABLE 7. ADDITION OF ALLENYL AND PROPARGYL REAGENTS (*Continued*)

Carbonyl Substrate	Allylborane Reagent	Conditions	Product(s) and Yield(s) (%), % ee	Refs.
C_7		THF, rt, 1.5 h	$I + II$ (72), $I:II = 87:13$	811
	$B(OBu-n)_2$	Ether, 40°, 3 h	(72)	814
	R = Et, Pr-*i*	Toluene	I R — Temp — Time Et — −78° to −20° — 21 h — (52), 60 Pr-*i* — −78° — 44 h — (43), 79	123
	" R = CH(Pr-*i*)$_2$, C$_{10}$H$_{21}$-*n*	Toluene, −78°, 20 h	I R CH(Pr-*i*)$_2$ — (<50), 73 C$_{10}$H$_{21}$-*n* — (<50), 79	813
		Toluene, −78°, 44 h	(70), 63	123
		CH$_2$Cl$_2$, −78°, 2.5 h	I (76), 96	153

504

Reagent/Borane	Conditions	Product	Yield (%)	Refs.
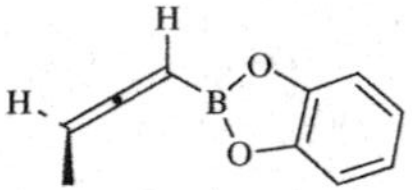 (catecholboryl dienyl, H)	CDCl$_3$, −78°	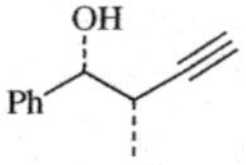 OH, Ph (propargyl)	(43), 34, 3.4:1.4 d.r.	817
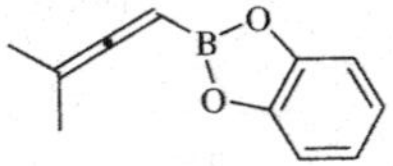 (catecholboryl, 3-methyl)	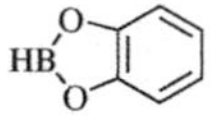 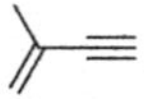 , Pd(PPh$_3$)$_4$	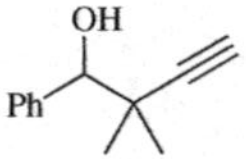 OH, Ph, gem-dimethyl	(57)	550
(Ts–N–B–N–Ts, Ph, Ph, propargyl)	CH$_2$Cl$_2$, −78°, 2.5 h	OH, Ph (allyl)	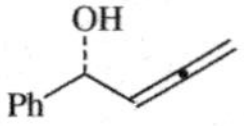 (72), >99	153
allyl-BBN	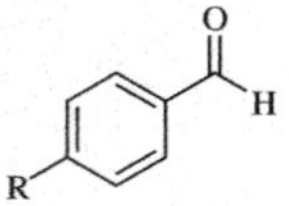 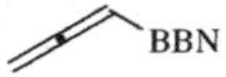 Ether, rt, 1 h	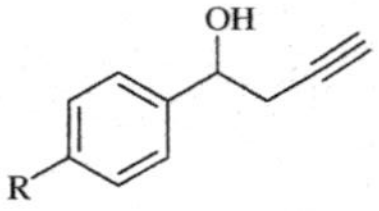 OH, aryl-R	R: H (96); MeO (99); O$_2$N (94)	812
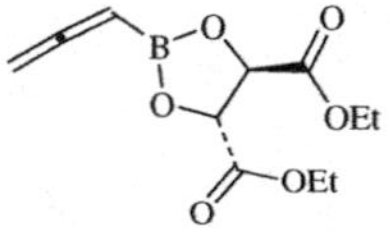 (diethyl tartrate boronate, allenyl)	Toluene, −78° to −20°, 20 h	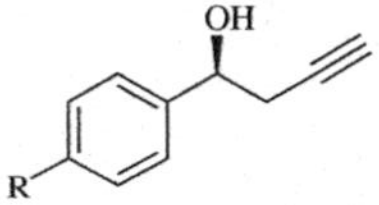 OH, aryl-R (propargyl)	R: O$_2$N (58), 72; MeO (66), 72	123

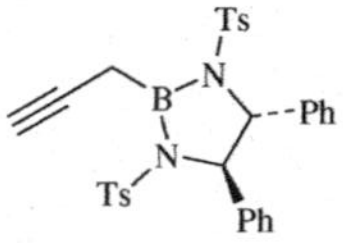

R = H, MeO, O$_2$N

R = O$_2$N, MeO

TABLE 7. ADDITION OF ALLENYL AND PROPARGYL REAGENTS (*Continued*)

Carbonyl Substrate	Allylborane Reagent	Conditions	Product(s) and Yield(s) (%), % ee	Refs.
C₇		Ether, rt, 1 h	(91)	812
		1. Li——⟨Cl⟩, B(C₅H₉-*c*)₃, THF, –90° 2. Aldehyde, –78° to rt, 1.5 h	(88)	807
		1. Li——⟨Cl⟩, B(C₅H₉-*c*)₃, THF, –90° to rt 2. Aldehyde, –78° to rt, 1.5 h	(86)	807
		Ether, –100°, 3 h	(78), 96	805
		Toluene, –78° to rt, 20 h	x 1.0 (85), 86 1.5 (56), 92	123
		Toluene, –78°, 44 h	(74), 87	123

x	
1.0	(70), 91
1.5	(81), 95

R	
Et	(81), 91
i-Pr	(88), 92
c-C$_5$H$_9$	(42), 91
menthyl	(37), 93
c-C$_{10}$H$_{19}$	(85), 98

x	
1.0	(88), 98
3.0	(89), 99

Reagent	Conditions	Product	Yield	Ref.
(OPr-i)$_2$ diester, allyl boronate	Toluene, −78°, 24 h	OH, Cy propargyl alcohol	(42), 90	123
(OPr-i)$_2$ diester, x equiv	Toluene, −78°, 22 h	OH, Cy	(see x table)	123
(OR)$_2$ diester	Toluene, −78°, 20 h	OH, Cy	(see R table)	813
O(menthyl) diester	Toluene, −78°, 20 h	OH, Cy	(69), 92	813
OCH(Pr-i)$_2$ diester, x equiv	Toluene, −78°, 20 h	OH, Cy	(see x table)	813
OCH(Pr-i)$_2$ diester	Toluene, −78°, 20 h	OH, Cy	(82), 99	813

TABLE 7. ADDITION OF ALLENYL AND PROPARGYL REAGENTS (Continued)

Carbonyl Substrate	Allylborane Reagent	Conditions	Product(s) and Yield(s) (%), % ee	Refs.
C7		CH$_2$Cl$_2$, −78°, 2.5 h	(82), 92	153
		CH$_2$Cl$_2$, −78°, 2.5 h	(78), >99	153
		CH$_2$Cl$_2$, −78°, 2.5 h	(78), 95	153
C8		Ether, rt	Time: 15 min (86); 1 h (97)	808, 809 / 812
		THF, rt, 1.5 h	(93)	811
		THF, rt, 1.5 h	I + II (88), I:II = 54:46	811

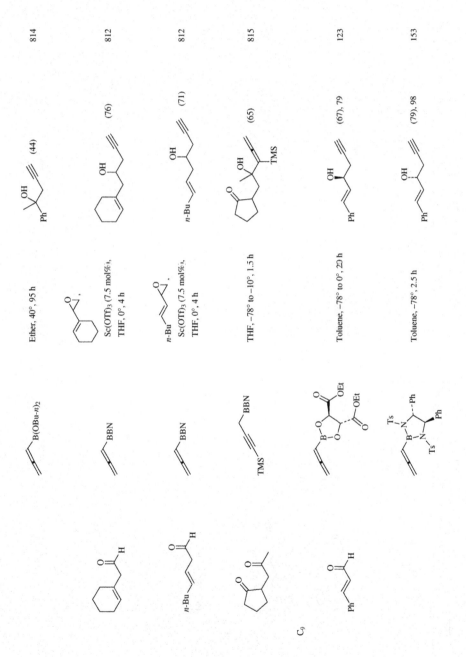

				814
			(44)	
Ether, 40°, 95 h		B(OBu-*n*)$_2$		
Sc(OTf)$_3$ (7.5 mol%), THF, 0°, 4 h		BBN	(76)	812
Sc(OTf)$_3$ (7.5 mol%), THF, 0°, 4 h		BBN	(71)	812
THF, −78° to −10°, 1.5 h		BBN, TMS	(65)	815
Toluene, −78° to 0°, 20 h			(67), 79	123
Toluene, −78°, 2.5 h			(79), 98	153

C$_9$

509

TABLE 7. ADDITION OF ALLENYL AND PROPARGYL REAGENTS (Continued)

Carbonyl Substrate	Allylborane Reagent	Conditions	Product(s) and Yield(s) (%), % ee	Refs.
C9				
		Toluene, −78°, 2.5 h	(74), >99	153
		THF, −78° to −15°, 1.5 h	(86)	815
		THF, rt, 2 h	(82)	815
C10				
		Ether, rt, 1 h	(99)	812
R = H, OMe, NO2		Sc(OTf)$_3$ (7.5 mol%), THF, 0°, 4 h	$\begin{array}{l|l} R & \\ \hline H & (73) \\ OMe & (77) \\ NO_2 & (98) \end{array}$	812
R = H, NO2		BF$_3$•OEt$_2$ (15 mol%), THF, 20°, 4 h	I $\begin{array}{l|l} R & \\ \hline H & (73) \\ NO_2 & (22) \end{array}$	812

510

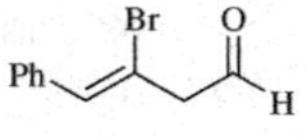

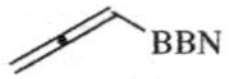

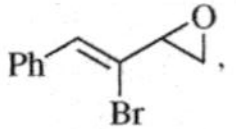

	(83)	812
I		
I (70)		812
	x	813
	1.5 (74), 92	
	5.0 (90), 99	
	(67), 98	813
	(52), >95	123
	(82)	815

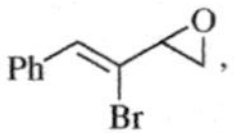

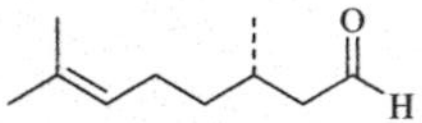

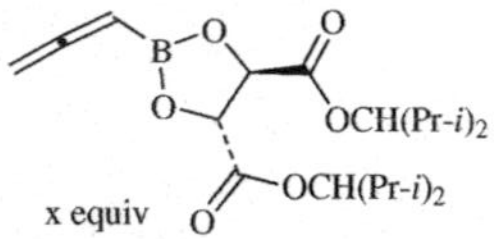

Sc(OTf)₃ (7.5 mol%), THF, 0°, 4 h

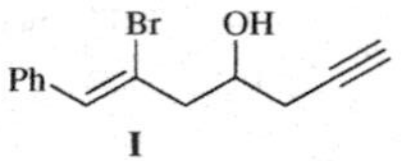

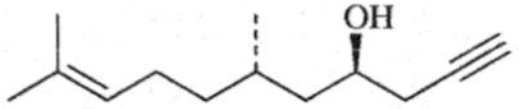

BF₃•OEt₂ (15 mol%), THF, 20°, 4 h

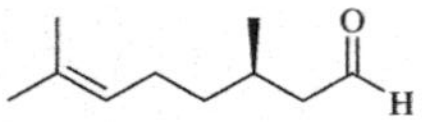

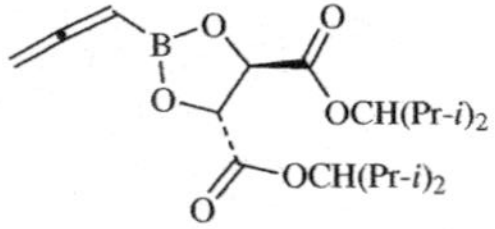

Toluene, −78°, 20 h

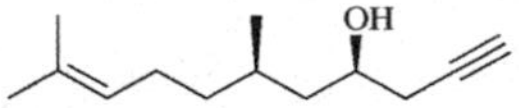

Toluene, −78°, 20 h

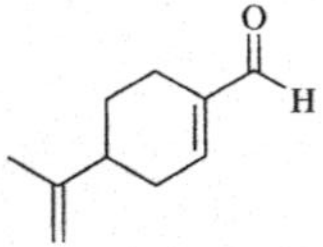

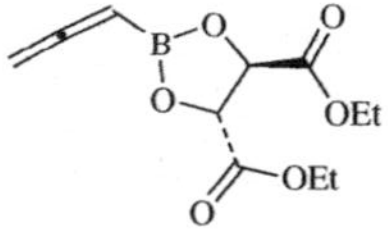

Toluene, −78°, 22 h

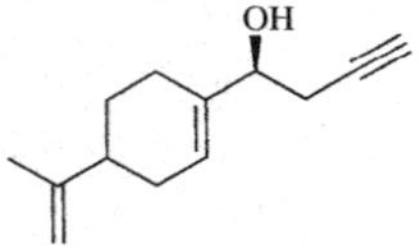

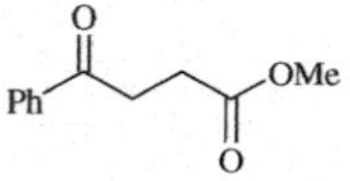

1. THF, rt, 2 h
2. BF₃•OEt₂

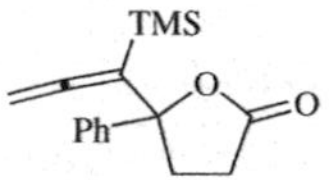

TABLE 7. ADDITION OF ALLENYL AND PROPARGYL REAGENTS (Continued)

Carbonyl Substrate	Allylborane Reagent	Conditions	Product(s) and Yield(s) (%), % ee	Refs.
C_{11}	BBN	$Sc(OTf)_3$ (7.5 mol%), THF, 0°, 4 h	(77) **I**	812
	"	$BF_3 \cdot OEt_2$ (15 mol%), THF, 20°, 4 h	**I** (78)	812
	BBN	$Sc(OTf)_3$ (7.5 mol%), THF, 0°, 4 h	d.r. R CH_2OBn (89) 1:1 Me (70) 1.2:1	812
C_{13}	BBN	Ether, rt, <5 min	(90)	809
C_{16}	BBN	$Sc(OTf)_3$ (7.5 mol%), THF, 0°, 4 h	(93)	812

TABLE 8. ADDITION OF CYCLIC REAGENTS

Carbonyl Substrate	Allylborane Reagent	Conditions	Product(s) and Yield(s) (%), % ee	Refs.
C$_1$		–60° to 0°	(50)	545
		–60° to 0°	(91)	545
		1. 100° 2. DDQ	(29), >95:5 d.r.	818
C$_2$		Ether, –78° to rt	(94), >97.5:2.5 d.r.	819
	B[(–)-Ipc]$_2$	THF, –78° to rt	(66), 94, >99:1 d.r.	69, 820
	B[(–)-Ipc]$_2$	THF, –78°	(36), 93, >99:1 d.r.	820
	B[(–)-Ipc]$_2$	THF, –78°	(24), 95, >99:1 d.r.	820

513

TABLE 8. ADDITION OF CYCLIC REAGENTS (Continued)

Carbonyl Substrate	Allylborane Reagent	Conditions	Product(s) and Yield(s) (%), % ee	Refs.
C$_2$ CH$_3$CHO	$\overset{}{\text{B(Pr-}n)_2}$ (methylenecyclobutyl)	$-70°$ to $0°$	**I** + **II** (65-80), **I:II** = 53:47	545
	$\overset{}{\text{B(Pr-}n)_2}$ (methylenecyclopentyl)	$-60°$ to $0°$	(70)	545
	$\overset{}{\text{B(Pr-}n)_2}$ (methylenecyclohexyl)	$-60°$ to $0°$	(74)	545
	B—Ph (dimethyl borolene)	Ether, $-78°$ to rt	(94), >97.5:2.5 d.r.	819
CCl$_3$CHO	B—Ph (vinyl borolane)	Ether, $-78°$ to rt	(95), >97.5:2.5 d.r.	819
OHC—CHO	B[(−)-Ipc]$_2$ (cyclohexenyl)	THF, $-78°$, 4 h	(81), >98, 98:2 d.r.	255

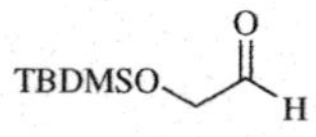

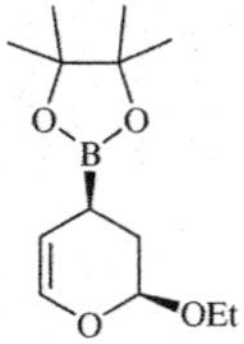

150	
819	
819	
820	
819	

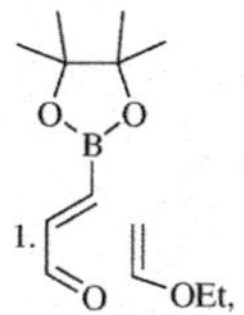

1.

1 (1 mol%), 4 Å MS, neat
2. Aldehyde, 45°, 48 h

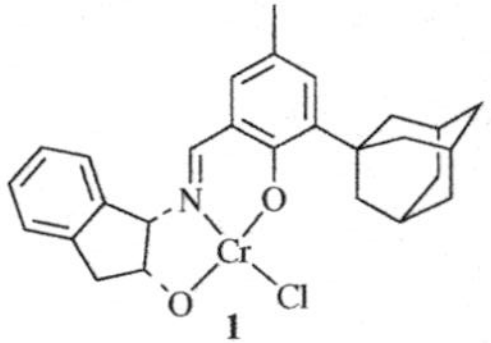

1

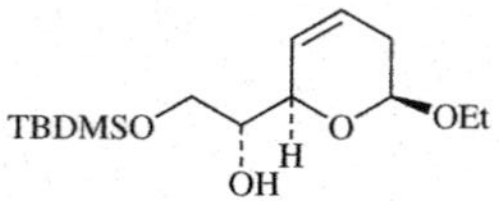

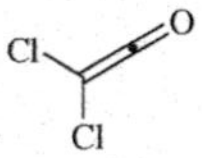

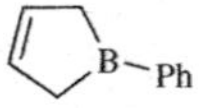

(82)

(41)

(94), >97.5:2.5 d.r.

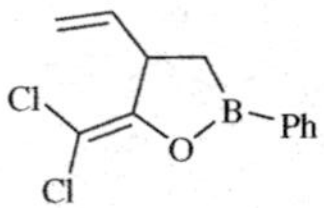

(58), 92, >99:1 d.r.

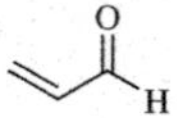

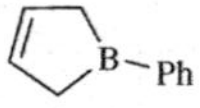

(92), >97.5:2.5 d.r.

Ether, −78° to rt

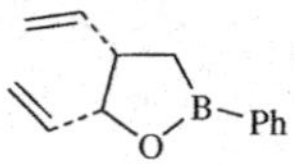

Ether, −78° to rt

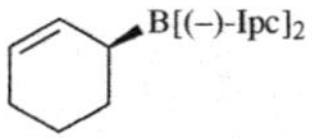

THF, −78° to rt

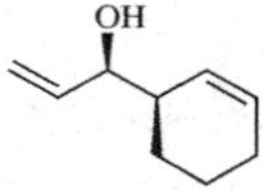

Ether, −78° to rt

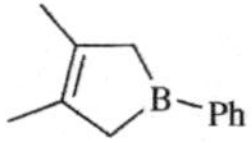

B—Ph

B—Ph

B[(−)-Ipc]₂

B—Ph

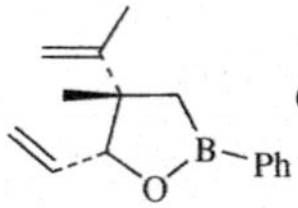

TBDMSO

Cl

C₃

515

TABLE 8. ADDITION OF CYCLIC REAGENTS (*Continued*)

Carbonyl Substrate	Allylborane Reagent	Conditions	Product(s) and Yield(s) (%), % ee	Refs.
C₃				
		1. 40° 2. DDQ	 (71), >95:5 d.r.	818
		THF, −78° to rt	 (65), 93, >99:1 d.r.	820
		—		601

R¹	R²	I + II	I:II
Thex	Cy	(70)	98:2
Thex	Cy	(85)	97:3
Thex	n-C₆H₁₃	(72)	54:46
Cy	Cy	(89)	54:46

| | | toluene, 80°, 70 h |
(62) | 199 |

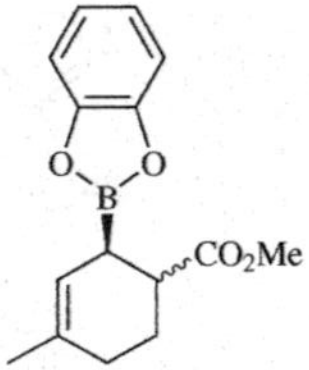

821

183

821

89

89

CO₂Me (40)

I

I (22)

CO₂Me (8)

N–Ph (66)

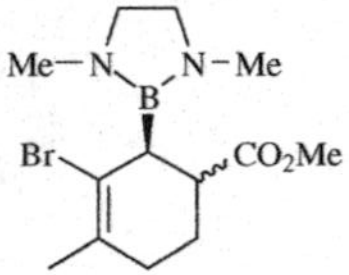

(70)

Et₃N, CH₂Cl₂

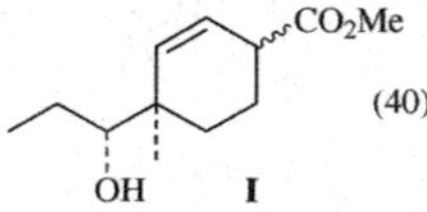

60°, 12 h

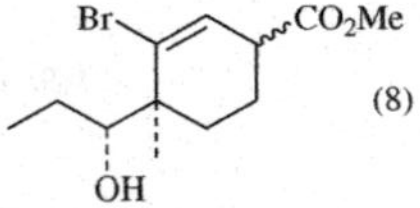

CH₂Cl₂, −78° to 20°

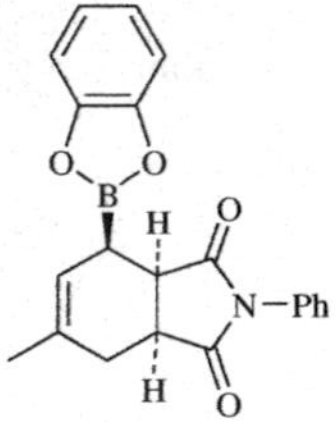

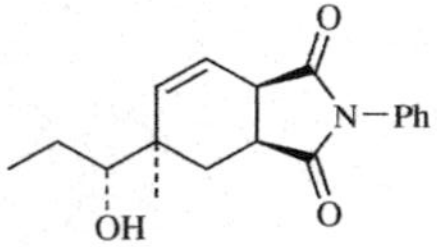

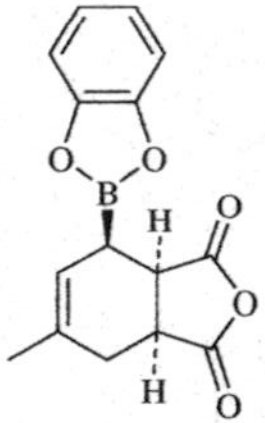

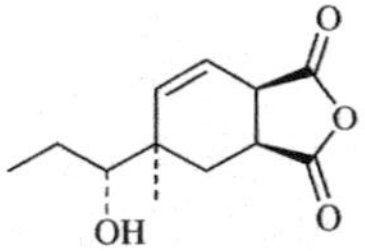

rt

80°

TABLE 8. ADDITION OF CYCLIC REAGENTS (*Continued*)

Carbonyl Substrate	Allylborane Reagent	Conditions	Product(s) and Yield(s) (%), % ee	Refs.
C₃				
		1. 40° 2. DDQ	 (61), >95:5 d.r.	818
	B–Ph (cyclopentene)	Ether, –78° to rt	 (88)	819
	BBN (cyclohexene)	Pentane, rt, 2 h	 **I** (86)	103
	"	1. [cyclohexadiene], HBBN, pentane, rt, 24 h 2. Ketone, rt, 2 h	**I** (68)	103
	B(Pr-*n*)₂ (methylenecyclobutane)	–70° to 0°	 **I** + **II** (65-80), **I:II** = >100:1	545
	B(Pr-*n*)₂ (methylenecyclopentane)	–60° to 0°	 (87)	545

C4

Substrate	Reagent	Conditions	Product	Ref.
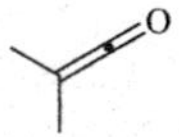 acetone (methyl ketone)	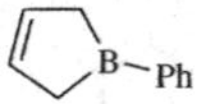 B–Ph	Ether, −78° to rt	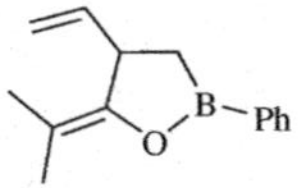 B–Ph (90)	819
acetone (methyl ketone)	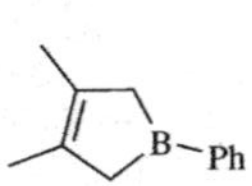 B–Ph	Ether, −78° to rt	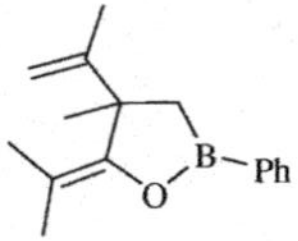 B–Ph (44)	819
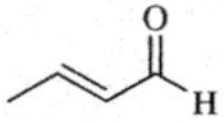 crotonaldehyde	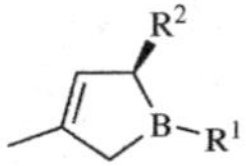 R²—B–R¹	—	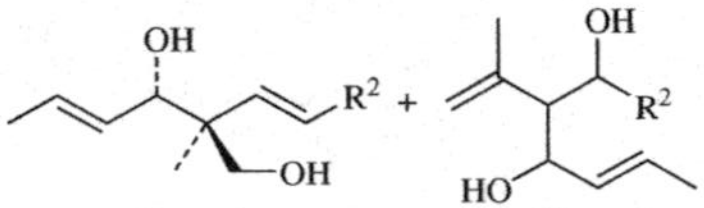 I + II — see table below	601
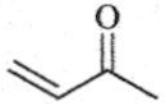 methyl vinyl ketone	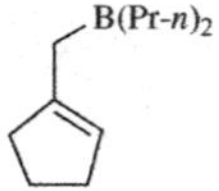 B(Pr-n)₂	−60° to 0°	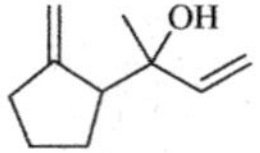 (75)	545
methyl vinyl ketone	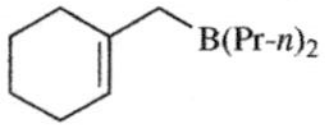 B(Pr-n)₂	−60° to 0°	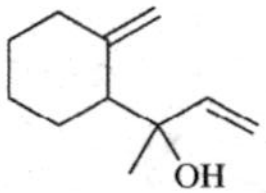 (69)	545
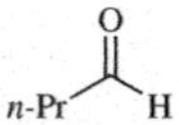 butyraldehyde (n-Pr CHO)	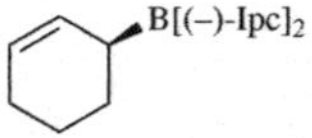 B[(−)-Ipc]₂	THF, −78° to rt	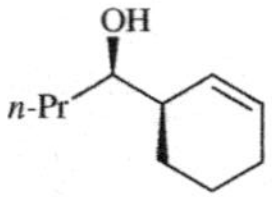 (69), 90, >99:1 d.r.	820

I + II

R^1	R^2	I + II	I:II
Thex	Cy	(82)	99:1
Thex	Cy	(85)	94:6
Thex	$n\text{-}C_6H_{13}$	(84)	53:47

TABLE 8. ADDITION OF CYCLIC REAGENTS (Continued)

Carbonyl Substrate	Allylborane Reagent	Conditions	Product(s) and Yield(s) (%), % ee	Refs.
C₄ *i*-Pr CHO		—		601
			I **II**	
			R¹ R² **I + II** **I:II**	
			Thex Cy (85) 97:3	
			Thex C₅H₉-*c* (81) 53:47	
		rt	(67)	89
		80°	(75)	89
		Benzene, 20°, 4 h	(76), 9:1 d.r.	822

520

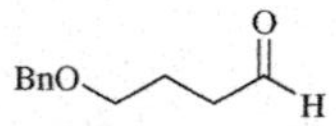

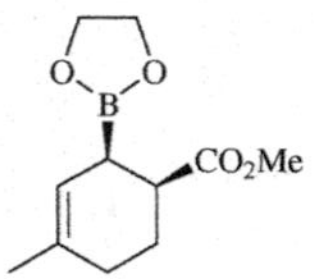

x	Pressure	Temp	Time	
1.5	—	80°	15 h	(29)
1.5	10 kbar	rt	24 h	(22)
2	—	80°	16 h	(38)
3	—	80°	16 h	(50)

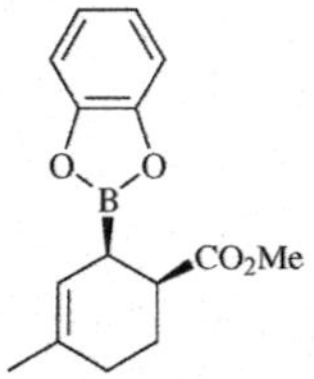

1. [structure], 80°

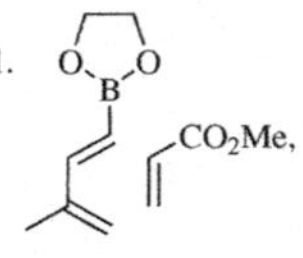

2. Aldehyde, benzene, rt

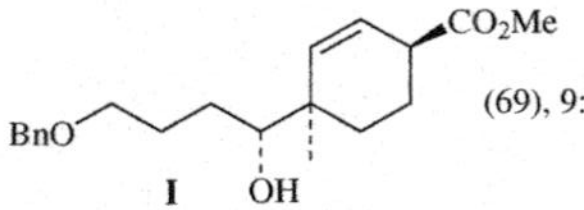

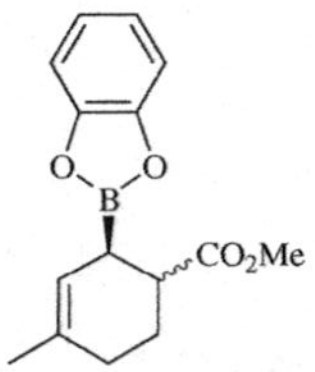

Benzene, 20°, 4 h

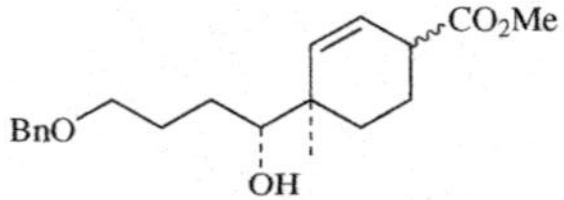

Benzene, 20°, 4 h

See table

I (69), 9:1 d.r. 823

I (67), 9:1 d.r. 824

I (63), 9:1 d.r. 821, 822

183, 821

x equiv

TABLE 8. ADDITION OF CYCLIC REAGENTS (*Continued*)

Carbonyl Substrate	Allylborane Reagent	Conditions	Product(s) and Yield(s) (%), % ee	Refs.
C₄				
		80°, 15.5 h	(33)	183, 821
		1. 80°	(49), 70	823
		2. Aldehyde, benzene, rt		
		Benzene, 20°, 4 h	(67), 9:1 d.r.	822

C$_5$

	1. OEt-B(pinacol) / OEt 1^a (1 mol%), 4 Å MS, neat 2. Aldehyde, 45°, 48 h	(pinacol boronate, OEt pyran product) **(61)**	150
(cyclopentanone)	B(Pr-n)$_2$ (cyclobutenyl)	**I** (methylenecyclobutyl-cyclopentanol) + **II** (methylenecyclopentyl...)	545
	−70° to 0°	**I + II** (65-80), **I:II** = >100:1	
	B(Pr-n)$_2$ (cyclohexenyl)	(cyclohexanol) **(91)**	545
	−60° to 0°		
(n-Bu CHO)	B[(−)-Ipc]$_2$ (cyclohexenyl)	(n-Bu, OH cyclohexenyl) (67), 90, >99:1 d.r.	820
	THF, −78° to rt		
(i-Pr CH$_2$CHO)	1. OEt-B(pinacol) / OEt 1^a (1 mol%), 4 Å MS, neat 2. Aldehyde, 45°, 24 h	(OEt pyran product, i-Pr, OH) **(81)**	150

TABLE 8. ADDITION OF CYCLIC REAGENTS (*Continued*)

Carbonyl Substrate	Allylborane Reagent	Conditions	Product(s) and Yield(s) (%), % ee	Refs.
C$_5$				
		toluene, 80°, 72 h	(50)	181
		toluene, 80°, 24 h	(88)	825
		Ether, −78° to rt	(95), >97.5:2.5 d.r.	819
		Ether, −78° to rt	(94), >97.5:2.5 d.r.	819

524

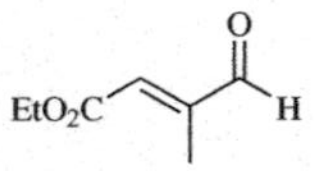

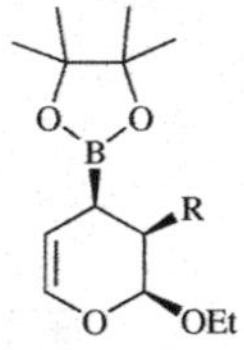

198

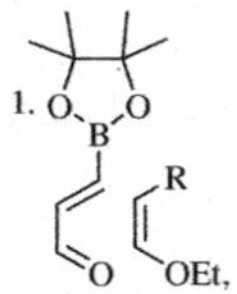

R		d.r.
C$_7$H$_{15}$-n	(74), —	—
CH$_2$CH=C(Me)$_2$	(76), 95	99:1

1^a (3 mol%), BaO, neat, 20°, 5 h
2. Aldehyde, 110°, 48 h

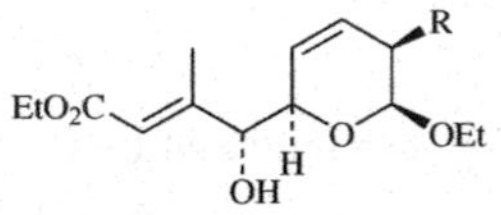

819

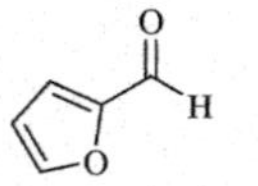

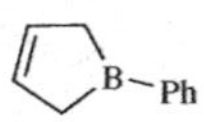

Ether, −78° to rt

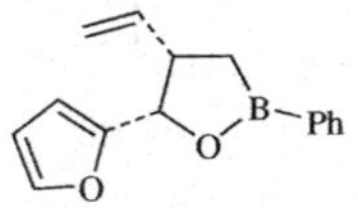

(96, >97.5:2.5 d.r.)

545

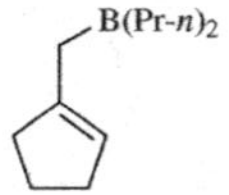

−60° to 0°

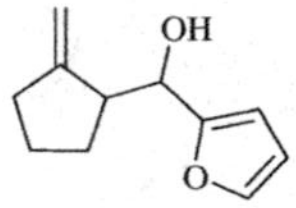

(40)

819

Ether, −78° to rt

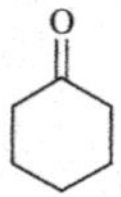

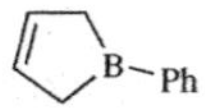

(92)

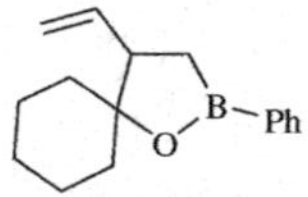

545

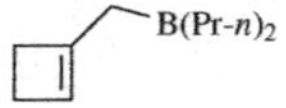

−70° to 0°

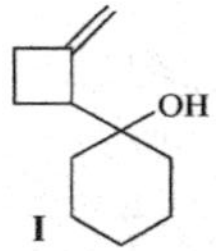

I + II (65-80), I:II = 9:1

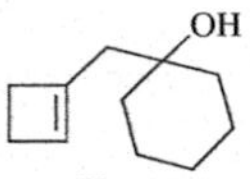

C$_6$

525

TABLE 8. ADDITION OF CYCLIC REAGENTS (*Continued*)

Carbonyl Substrate	Allylborane Reagent	Conditions	Product(s) and Yield(s) (%), % ee	Refs.
C₇ Ph—CHO	B—Ph (dihydroborole)	Ether, −78° to rt	(97), >97.5:2.5 d.r.	819
	B[(−)-Ipc]₂ (cyclohexenyl)	THF, −78° to rt	(71), 94, >99:1 d.r.	820
	B[(−)-Ipc]₂ (cycloheptenyl)	THF, −78°	(38), 95, >99:1 d.r.	820
	B[(−)-Ipc]₂ (cyclooctenyl)	THF, −78°	(28), 93, >99:1 d.r.	820
	B[(−)-Ipc]₂	THF, −78°	(40), 79, 24:1 d.r.	826
	B(Pr-*n*)₂	−70° to 0°	I + II (65–80), I:II = 3:2	545
	B—Ph	Ether, −78° to rt	(94), >97.5:2.5 d.r.	819

526

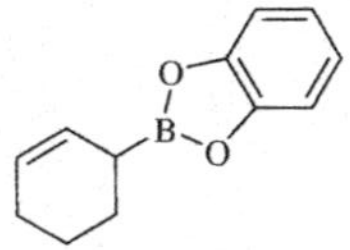

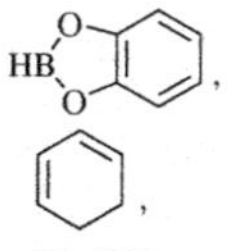

550

(92)

181

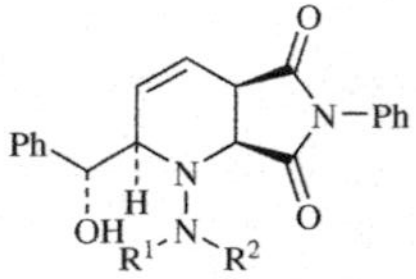

R^1	R^2	
H	Ph	(46)
Me	Ph	(46)
H	Ac	(42)

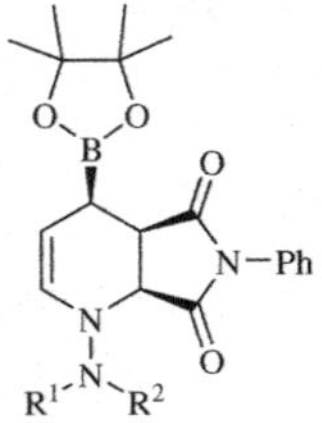

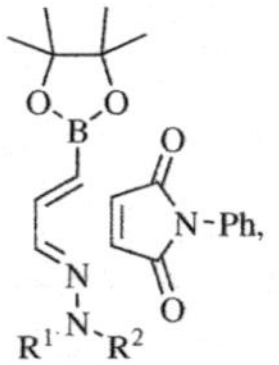

181

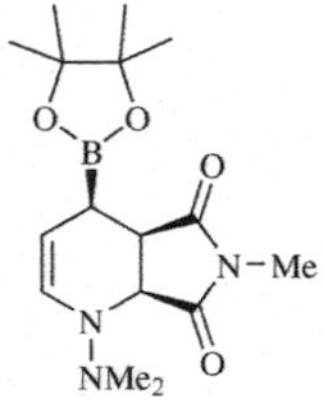

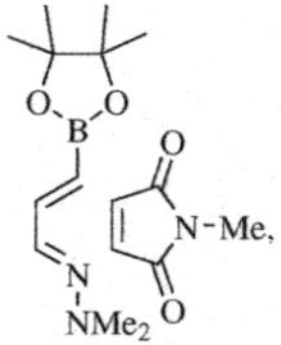

N–Me (50)

199

R		
Pr-n	(52)	
Bu-t	(69)	

Rh$_4$(CO)$_{12}$

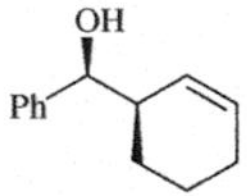

toluene, 80°, 72 h

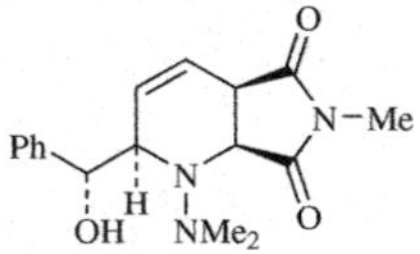

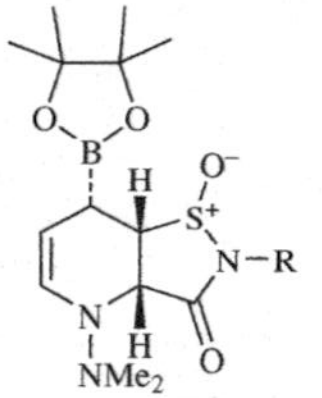

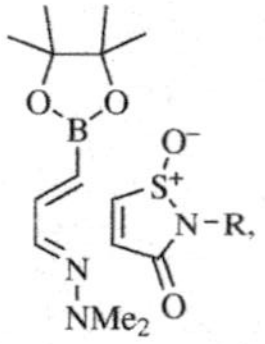

toluene, 80°, 72 h

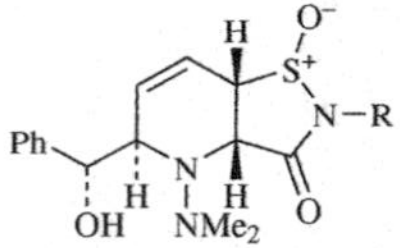

toluene, 80°, 70 h

TABLE 8. ADDITION OF CYCLIC REAGENTS (*Continued*)

Carbonyl Substrate	Allylborane Reagent	Conditions	Product(s) and Yield(s) (%), % ee	Refs.
C₇				
		toluene, 80°, 70 h	(65)	199
		toluene, 80°, 72 h		182 181, 182
		—	(90), >100:1	52

For the second product entry:

R		d.r.
H	(55)	9:1
Me	(55)	>97.5:2.5

Y	Temp	
O	80°	(80)
NPh	rt	(69)

89

(59), >95:5 d.r.

818

(90), 3:1 d.r.

827

R		
H	(85), 95	151
MeO	(84), 96	
Cl	(77), 93	
F	(80), 96	

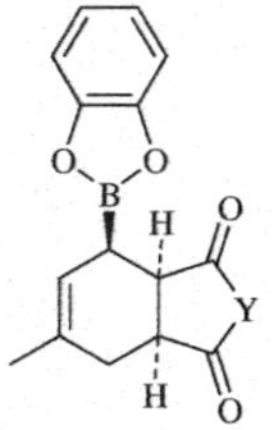

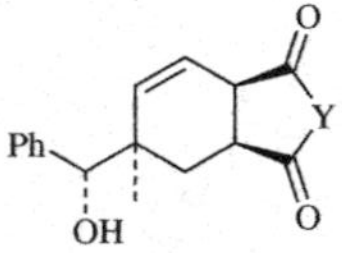

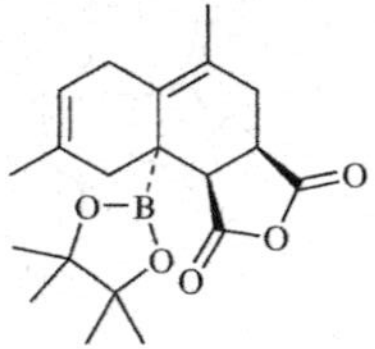

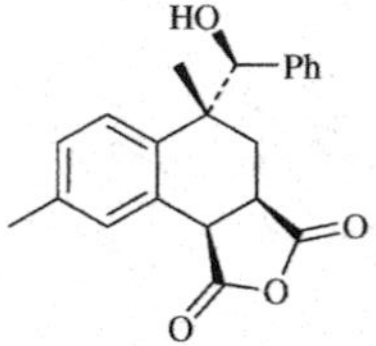

R		
H	(82)	150
O₂N	(92)	
MeO	(81)	

See table

1. 40°
2. DDQ

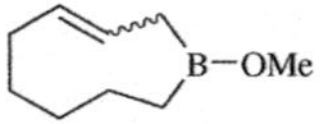

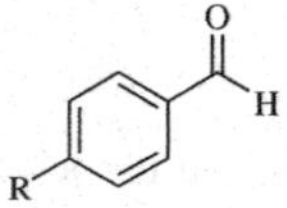

1. Ether, 0°, 20 h
2. NaOH, H₂O₂, THF

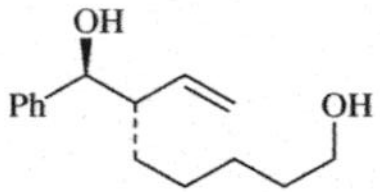

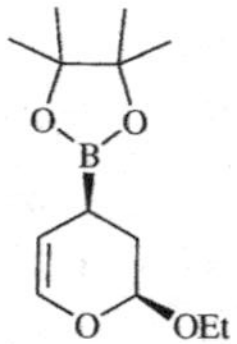

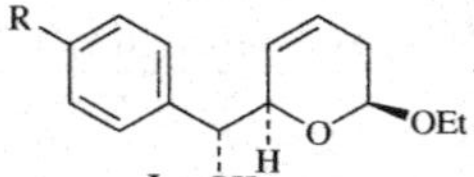

Neat, 70°

1ᵃ (1 mol%), 4 Å MS, neat
2. Aldehyde, 25–45°, 24 h

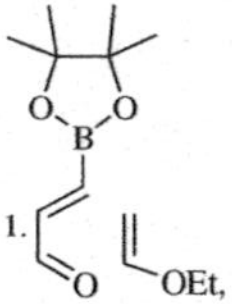

R = H, MeO, Cl, F

R = H, O₂N, MeO

529

TABLE 8. ADDITION OF CYCLIC REAGENTS (*Continued*)

Carbonyl Substrate	Allylborane Reagent	Conditions	Product(s) and Yield(s) (%), % ee	Refs.

C₇

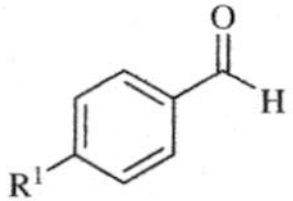

R^1 = H, Br

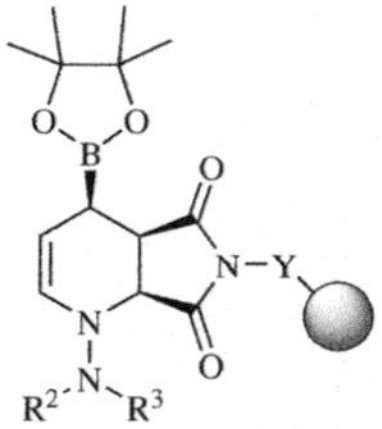

$Y = C_6H_4CO_2$-p

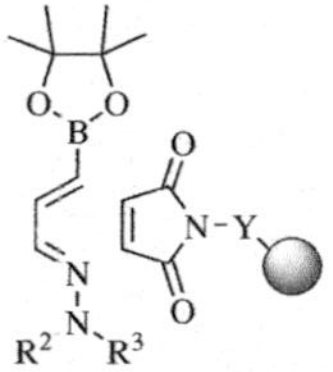

toluene, 80°, 72 h

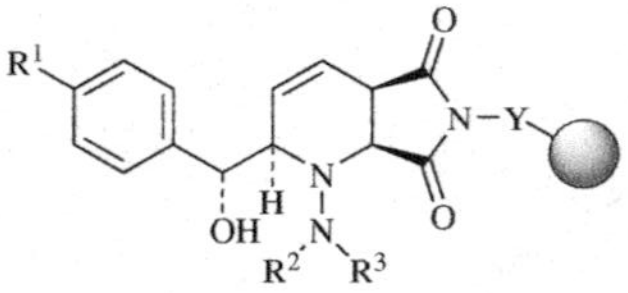

R^1	R^2	R^3	
H	Me	Me	(62)
Br	Me	Me	(54)
H	Me	H	(60)

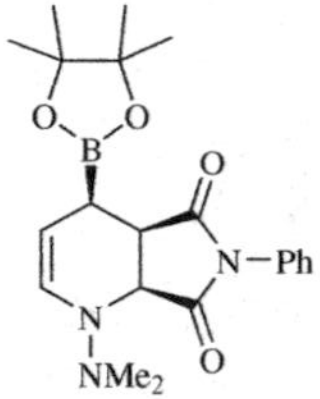

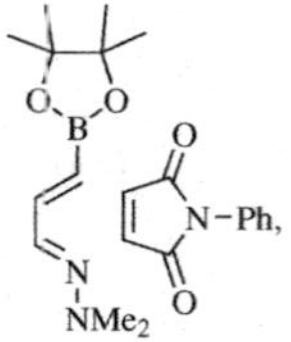

182

R^1 = H, O₂N, MeO

toluene, 80°, 72 h

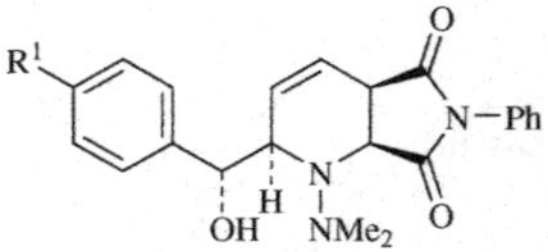

R^1	
H	(47)
O₂N	(48)
MeO	(52)

181

530

825

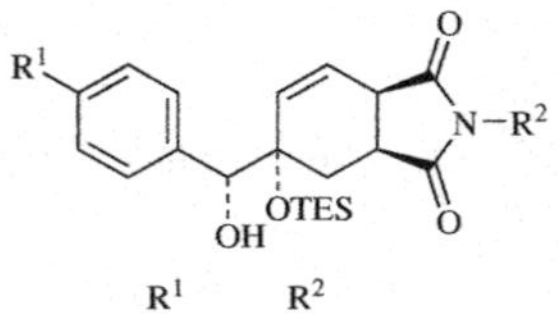

R[1]	R[2]	
H	Ph	(76)
O₂N	Ph	(92)
MeO	Ph	(82)
Br	Ph	(93)
O₂N	Me	(89)
MeO	Me	(67)
Br	Me	(78)

R¹ = see table

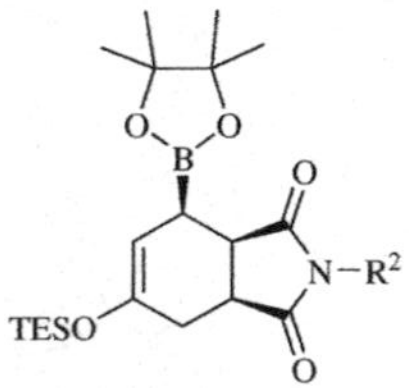

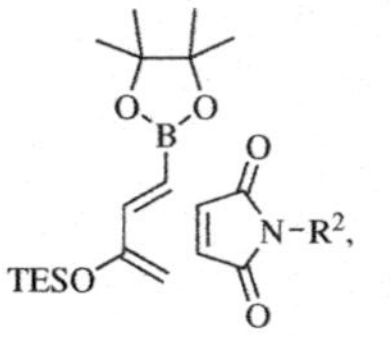

toluene, 80-100°, 16-24 h

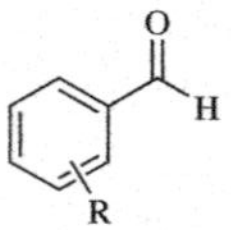

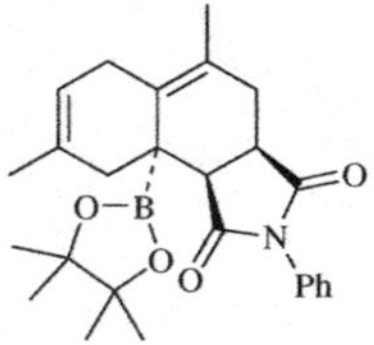

818

R	
o-NO₂	(67)
m-NO₂	(77)
p-NO₂	(48)
p-Cl	(56)

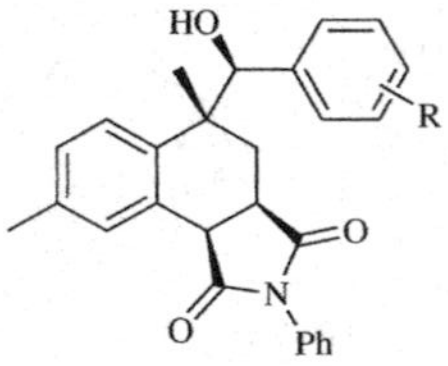

>95:5 d.r.

1. 40°
2. DDQ

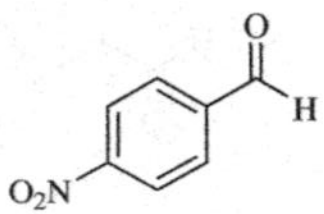

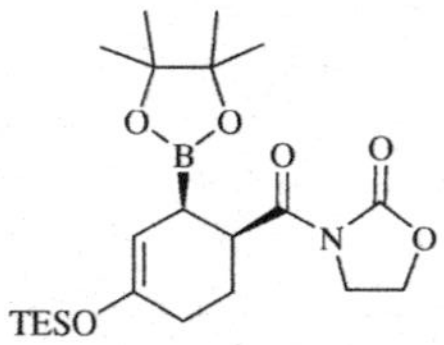

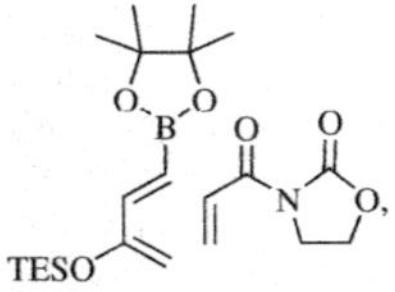

825

(71), 4:1 d.r.

toluene, 110°, 18 h

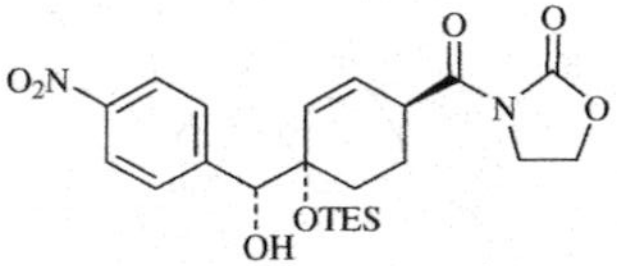

TABLE 8. ADDITION OF CYCLIC REAGENTS (*Continued*)

Carbonyl Substrate	Allylborane Reagent	Conditions	Product(s) and Yield(s) (%), % ee	Refs.
C₇				
		1. 40° 2. DDQ	 (83), >95:5 d.r.	818
		1. 40° 2. DDQ	 (52), >95:5 d.r.	818
		1. 40° 2. DDQ	 (30), >95:5 d.r.	818

828

R¹	R²	
H	H	(77)
MeO	H	(70)
Cl	H	(66)
F	H	(81)
O₂N	H	(95)
H	O₂N	(80)

1. $\overset{O}{\diagdown}\text{B}\diagup\overset{\diagup}{\diagdown}$ OEt

CH₂Cl₂, rt, 24 h
2. Reflux, 12 h

182

R¹	R²	R³	R⁴	R⁵	
H	H	Me	Me	Ph	(75)
O₂N	H	Me	Me	Me	(50)
MeO	H	Me	Me	Ph	(52)
H	H	Me	Me	Ph	(48)
H	H	H	Ph	Ph	(76)
H	H	H	p-CF₃C₆H₄	Me	(77)
H	H	H	p-MeOC₆H₄	Me	(55)
H	MeO	Me	Ph	Ph	(65)
H	H	H	Ph	Ph	(68)
H	H	H	Ac	Ph	(42)
H	H	H	Boc	Me	(61)

toluene, 80°, 72 h

TABLE 8. ADDITION OF CYCLIC REAGENTS (*Continued*)

Carbonyl Substrate	Allylborane Reagent	Conditions	Product(s) and Yield(s) (%), % ee	Refs.
C₇				
		1. CH₂Cl₂, rt, 24 h 2. Reflux, 12 h	(45)	828
		toluene, 80°, 72 h	(39)	182
		–70° to 0°	I + II (65-80), I:II = 98:2	545
		Ether, –78° to rt	(53)	819

534

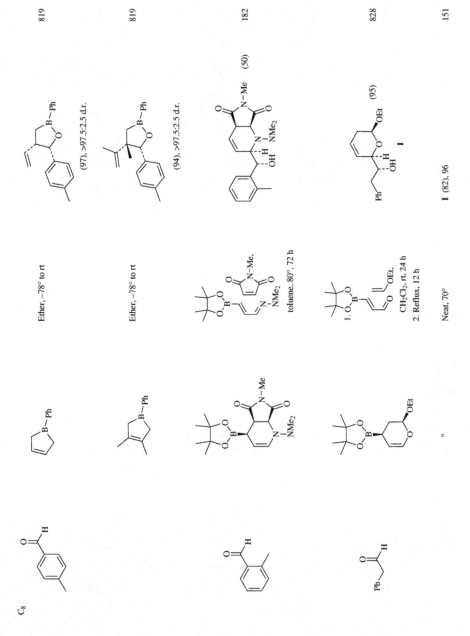

C₈

819

819

182

828

151

TABLE 8. ADDITION OF CYCLIC REAGENTS (*Continued*)

Carbonyl Substrate	Allylborane Reagent	Conditions	Product(s) and Yield(s) (%), % ee	Refs.
C8				
		Neat, 70°	(82), 96	151
	"	70^c	I Solvent / Time: toluene 48 h (78); neat 10 h (65)	829
C9				
		1. 1^d (1 mol%), 4 Å MS, neat 2. Aldehyde, 40°, 24 h	(81)	150
		1. CH$_2$Cl$_2$, rt, 24 h 2. Reflux, 12 h	(92)	828
"	"	Neat, 70°	I (87), 96	151

536

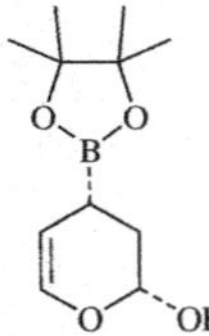

151

(87), 96

OEt

II

I + II (65-80), I:II = >100:1

I

Neat, 70°

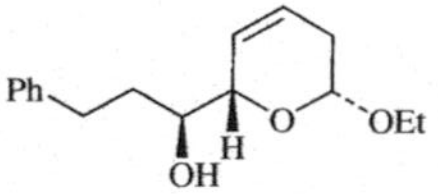

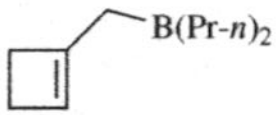

545

OH

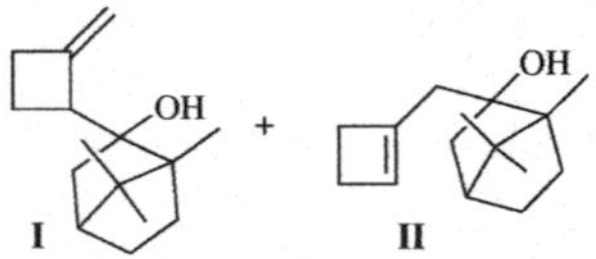

−70° to 0°

830

SPh

OTBDPS

(73)

O

1. THF, 0° to rt, 5.5 h
2. NaOH, 4 h

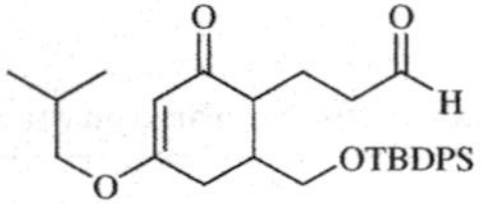

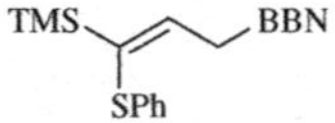

150

OEt

H

OH

(89)

$n\text{-}C_{10}H_{21}$

1^a (1 mol%), 4 Å MS, neat

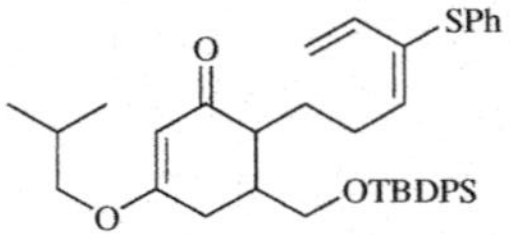

OEt

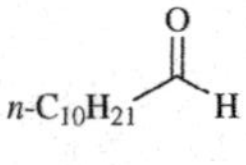

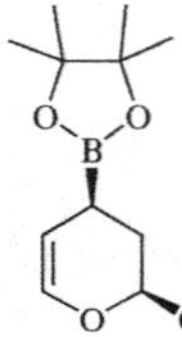

2. Aldehyde, 45°, 24 h

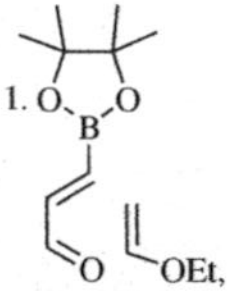

C_{10}

OEt

B(Pr-n)$_2$

TMS

BBN

SPh

H

O

OTBDPS

O

OEt

C_{11}

$n\text{-}C_{10}H_{21}$

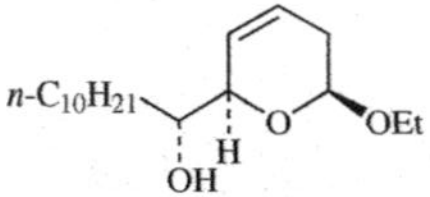

H

O

TABLE 8. ADDITION OF CYCLIC REAGENTS (*Continued*)

Carbonyl Substrate	Allylborane Reagent	Conditions	Product(s) and Yield(s) (%), % ee	Refs.
C$_{14}$				
		Ether, −78° to rt	(90)	819
		Ether, −78° to rt	(94)	819

^{a} The structure for catalyst **1** appears on the third page of Table 8.

TABLE 9. INTRAMOLECULAR ADDITIONS

Carbonyl Substrate	Conditions	Product(s) and Yield(s) (%), % ee	Refs.
C$_7$			
	HCl, H$_2$O, THF, 1 d	(48)	82
	TFA, H$_2$O, THF, 1 d	(55)	82
	HCl, H$_2$O, THF, 1 d	(66), 53:47 d.r.	82
	Yb(OTf)$_3$, H$_2$O, 90°, 2 h	Solvent CH$_2$Cl$_2$ (58) THF (28) DMF (18)	173

TABLE 9. INTRAMOLECULAR ADDITIONS (*Continued*)

Carbonyl Substrate	Conditions	Product(s) and Yield(s) (%), % ee	Refs.
C_7 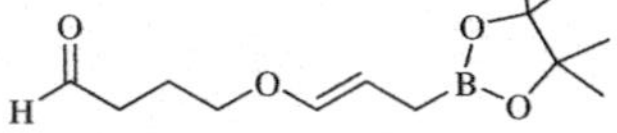	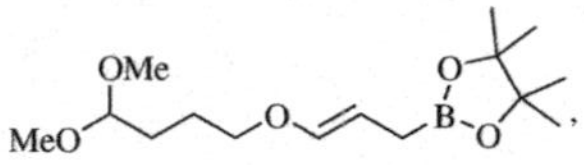 L.A., H_2O, CH_3CN, 90°, 2 h	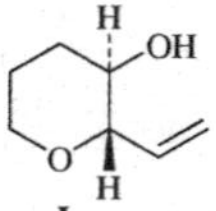 **I** L.A. / d.r.: $LiBF_4$ (56) — CuOTf (44) — AgOTf (42) — $Sm(OTf)_3$ (55) — $Er(OTf)_3$ (58) — $Yb(OTf)_3$ (77) >96:4	173
	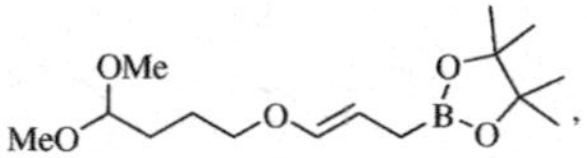 $Yb(OTf)_3$, H_2O, CH_3CN, rt, 12 h	**I** (66)	177
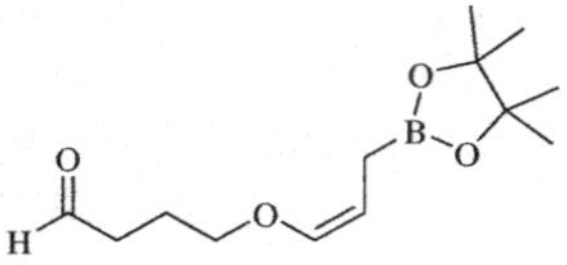	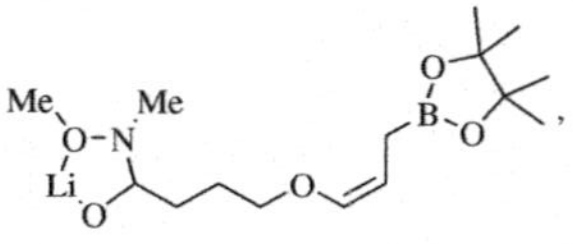 rt, 30 min	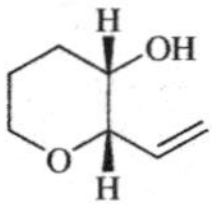 **I** Conditions: pH 7 buffer, THF (60) H_2O (pH 7) (70-74)	831 832
	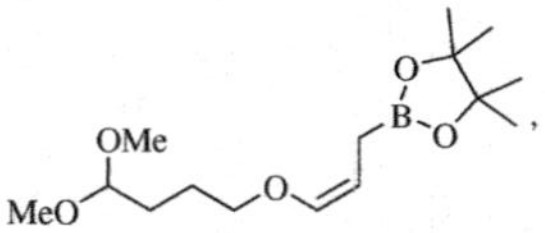 $Yb(OTf)_3$, H_2O, CH_3CN, 90°, 2 h	**I** (61), >93.5:6.5 d.r.	173

Acetal Cleavage

dppe, Pd(OTf)$_2$, acetone, rt		(21)
SnCl$_2$ · 2H$_2$O, CH$_2$Cl$_2$, rt		(30)
DMSO, dioxane, H$_2$O, 100°		(56)
LiBF$_4$, 2% H$_2$O/CH$_3$CN, rt		(60)

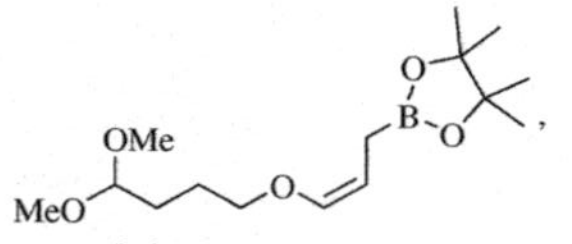

832

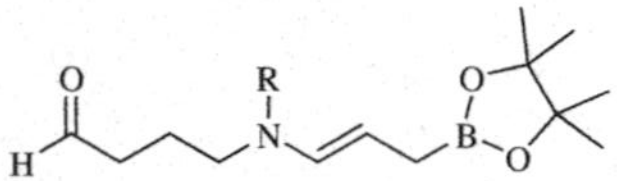

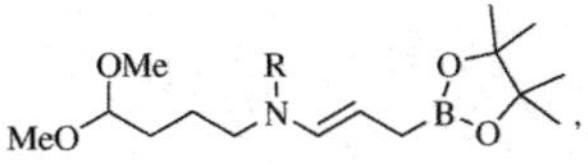

R	
Ts	(80)
Cbz	(66)

179

(66)

178

L.A.	Solvent	
LiBF$_4$	CH$_3$CN	(43)
Yb(OTf)$_3$	CH$_2$Cl$_2$	(45)

178

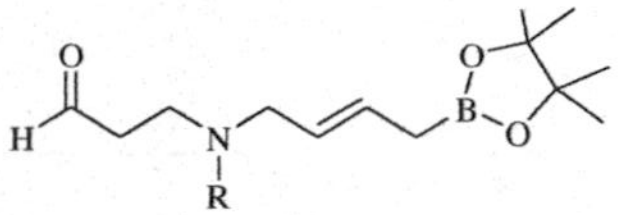

R	
Boc	(51)
CO$_2$Me	(49)

172

acetal cleavage

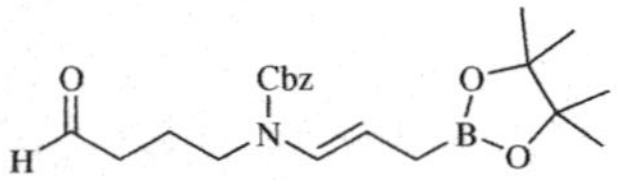

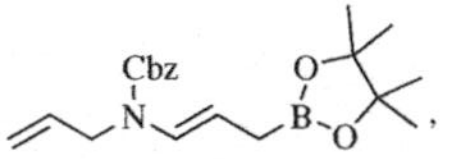

Rh(CO)$_2$acac, BIPHEPHOS,
H$_2$/CO (5 kbar), 60°, 4 d

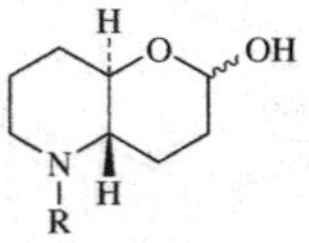

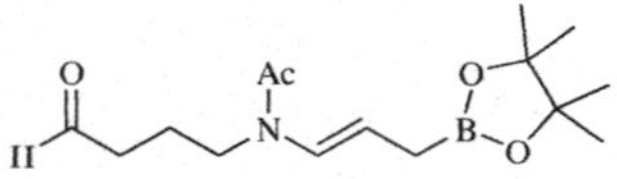

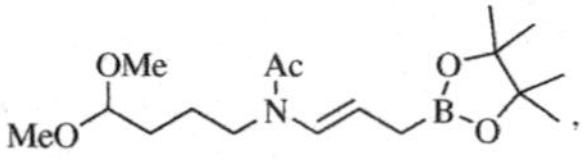

Rh(CO)$_2$acac, BIPHEPHOS,
H$_2$/CO (5 kbar), THF, 65°, 5 d

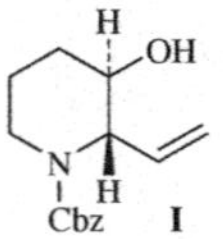

L.A., H$_2$O

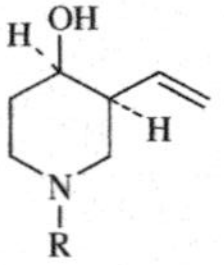

CH$_3$CN, rt, 1 d

541

TABLE 9. INTRAMOLECULAR ADDITIONS (*Continued*)

Carbonyl Substrate	Conditions	Product(s) and Yield(s) (%), % ee	Refs.
C_7			
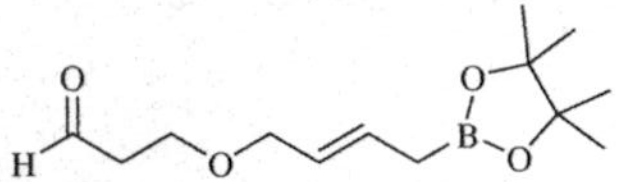	CH_3CN, rt, 1 d	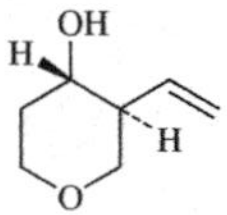 (70), 94:6 d.r.	172
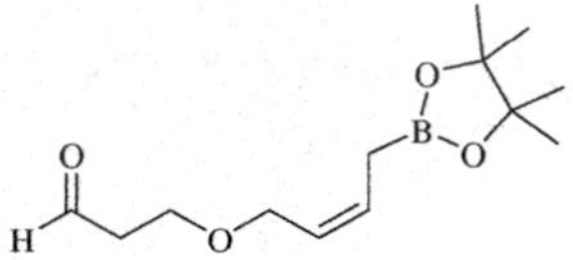	CH_3CN, rt, 1 d	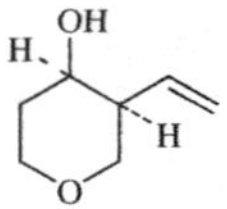 (72)	172
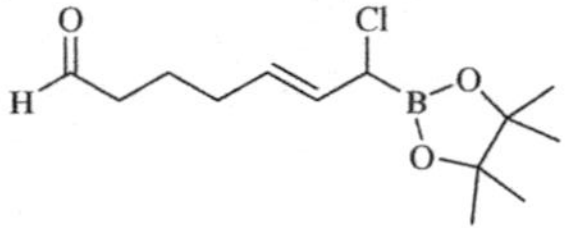	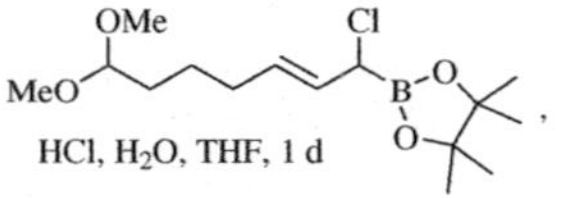, HCl, H_2O, THF, 1 d	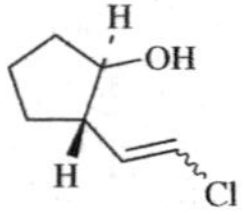 (17), 20:80 E:Z	82
C_8			
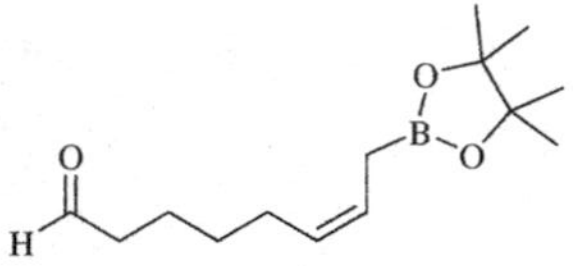	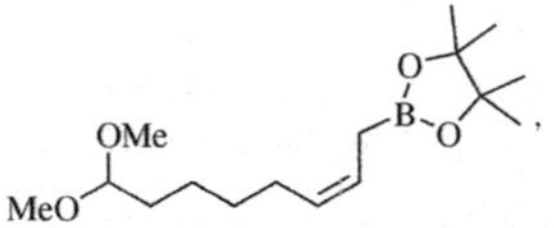, acetal cleavage	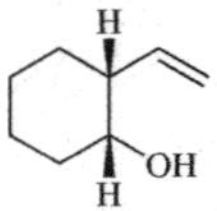 Acetal Cleavage dppe, $Pt(OTf)_2$, acetone (17) TFA, H_2O, $CHCl_3$, 0° (55)	80
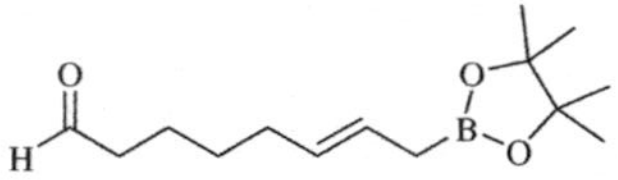	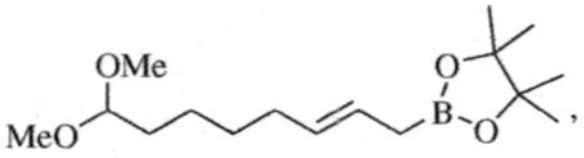, acetal cleavage	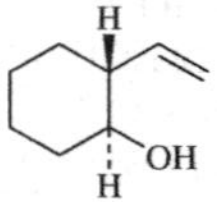 Acetal Cleavage TFA, H_2O, $CHCl_3$, 0° (51) $SnCl_2 \cdot 2H_2O$, CH_2Cl_2 (51) TMSOTf, acetone (9) $LiBF_4$, H_2O, CH_3CN (74)	80

(35), >99:1 d.r.

(46), 70:30 d.r.

(57), 53:47 d.r.

(75), 51:49 d.r.

(54), >99:1 d.r.

TFA, CHCl$_3$/H$_2$O (4:1), 0°, 3 h

TFA, CHCl$_3$/H$_2$O (4:1), 0°, 3 h

TFA, CHCl$_3$/H$_2$O (4:1), 0°, 3 h

TFA, CHCl$_3$/H$_2$O (4:1), 0°, 3 h

TFA, CHCl$_3$/H$_2$O (4:1), 0°, 3 h

TABLE 9. INTRAMOLECULAR ADDITIONS (*Continued*)

Carbonyl Substrate	Conditions	Product(s) and Yield(s) (%), % ee	Refs.
C_8			
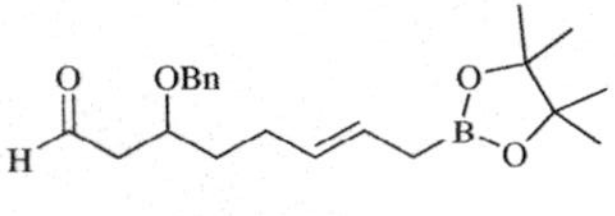	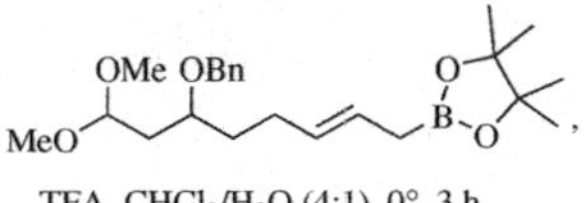 TFA, $CHCl_3/H_2O$ (4:1), 0°, 3 h	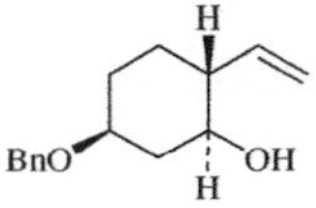 (58), 51:49 d.r.	171
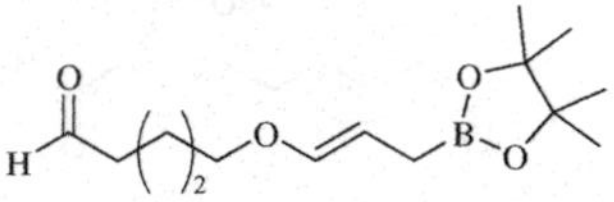	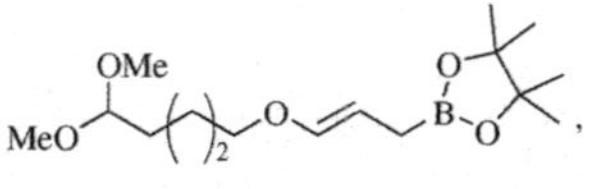 $Yb(OTf)_3$, H_2O, CH_3CN, 90°, 2 h	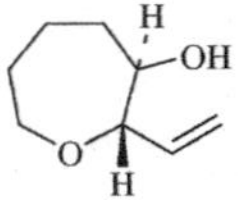 (56), >98.5:1.5 d.r.	173
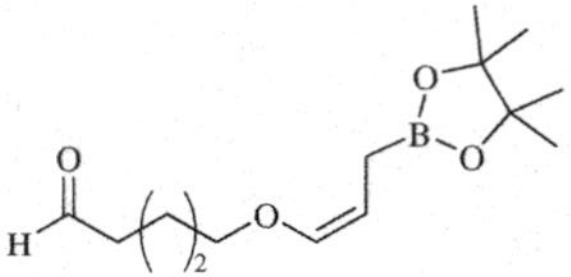	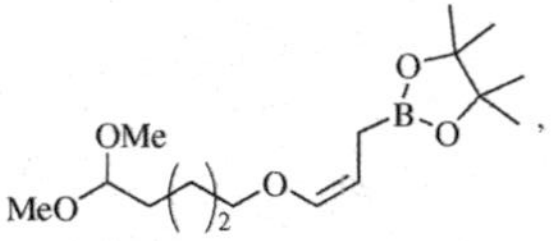 $Yb(OTf)_3$, H_2O, CH_3CN, 90°, 2 h	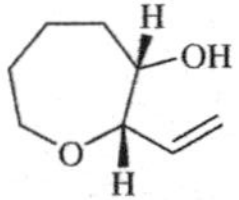 (66), >90.5:9.5 d.r.	173
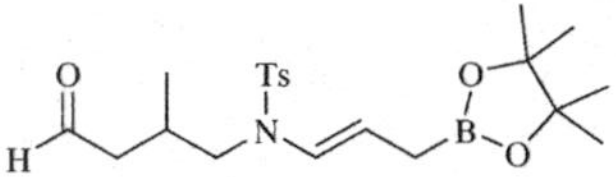	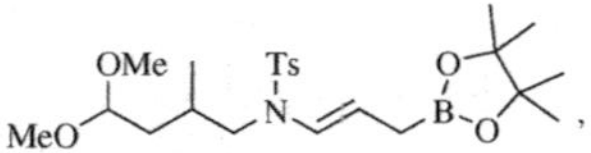 $Rh(CO)_2acac$, BIPHEPHOS, H_2, CO, 5 kbar, 60°, 4 d	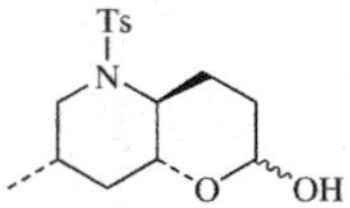 (83), 97:3 d.r.	179
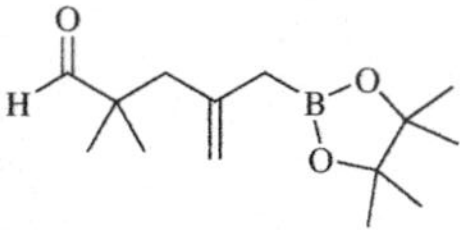	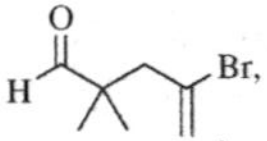 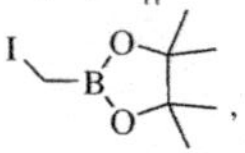 $Pd(OAc)_2$, PPh_3 (3 equiv)[a]	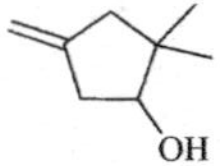 (47)	174

C9

1. H

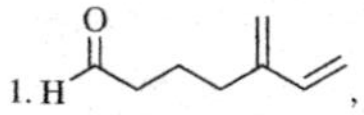

B₂(catechol)₂, (Ph₃P)₂Pt(C₂H₄),
THF, 60°, 24 h

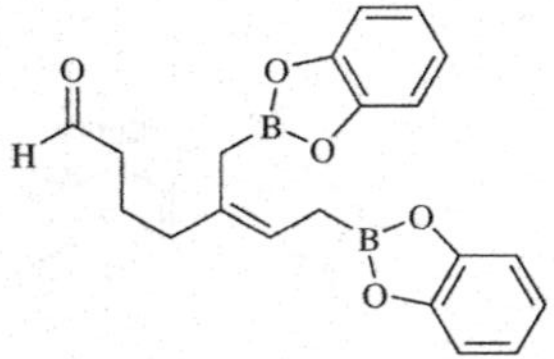

2. NaOH, H₂O₂

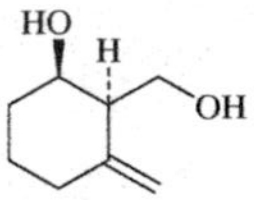

(10), >19:1 d.r.   186

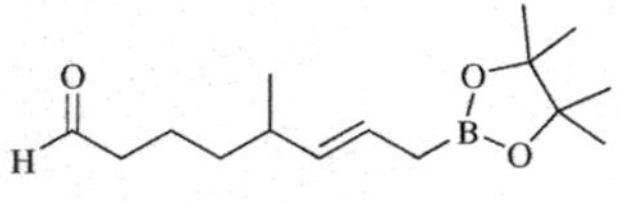

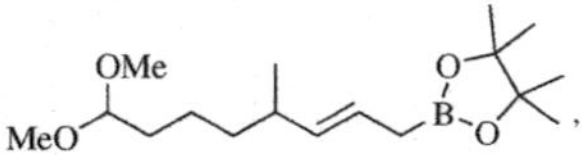

TFA, CHCl₃/H₂O (4:1), 0°, 3 h   (66), 85:15 d.r.   833

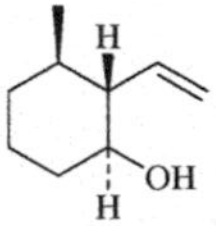

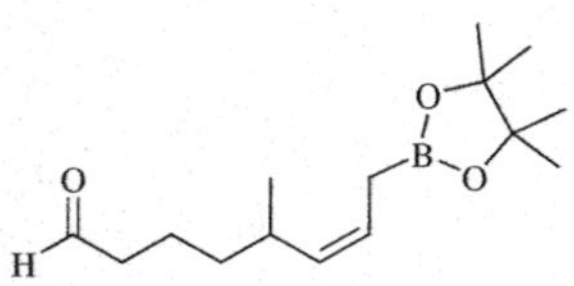

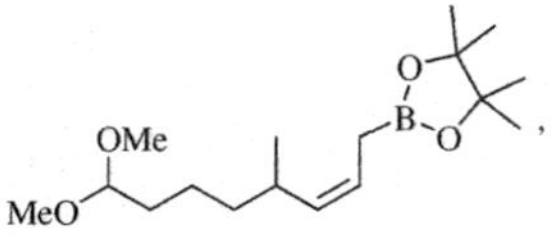

TFA, CHCl₃/H₂O (4:1), 0°, 3 h   (46), >99:1 d.r.   833

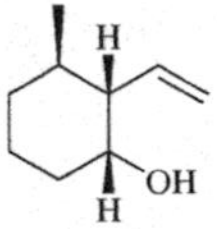

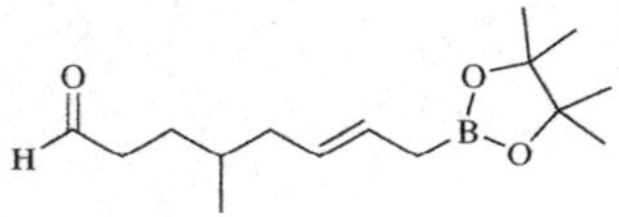

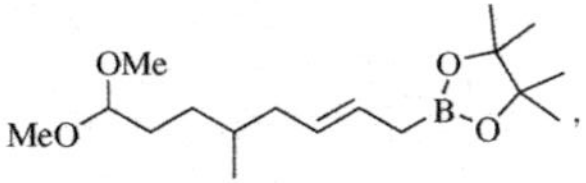

TFA, CHCl₃/H₂O (4:1), 0°, 3 h   (65), 93:7 d.r.   833

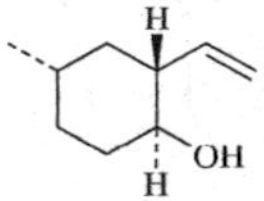

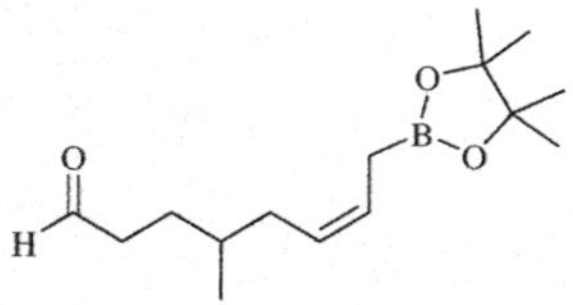

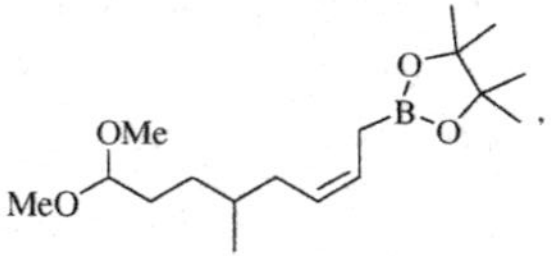

TFA, CHCl₃/H₂O (4:1), 0°, 3 h   (42), >99:1 d.r.   833

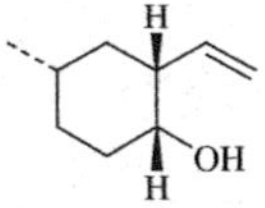

TABLE 9. INTRAMOLECULAR ADDITIONS (Continued)

Carbonyl Substrate	Conditions	Product(s) and Yield(s) (%), % ee	Refs.
	TFA, CHCl$_3$/H$_2$O (4:1), 0°, 3 h	(44), 95:5 d.r.	833
	TFA, CHCl$_3$/H$_2$O (4:1), 0°, 3 h	(47), >99:1 d.r.	833
	pH 7 buffer, THF, rt, 30 min	(35)	834
	Pd(dba)$_2$, AsPh$_3$ (2 equiv)^a, toluene, 100°	(62)	180

C$_9$

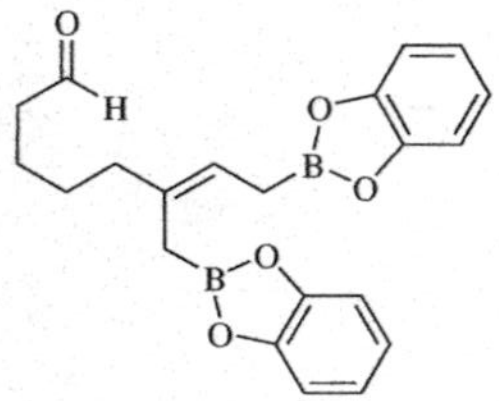

186

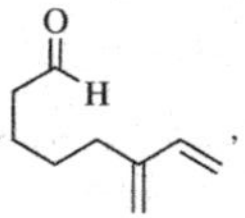

(56), >19:1 d.r.

$B_2(catechol)_2$, $(Ph_3P)_2Pt(C_2H_4)$, THF, 60°, 24 h

2. NaOH, H_2O_2

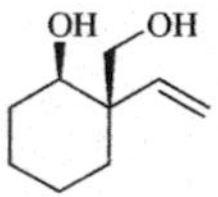

834

(38)

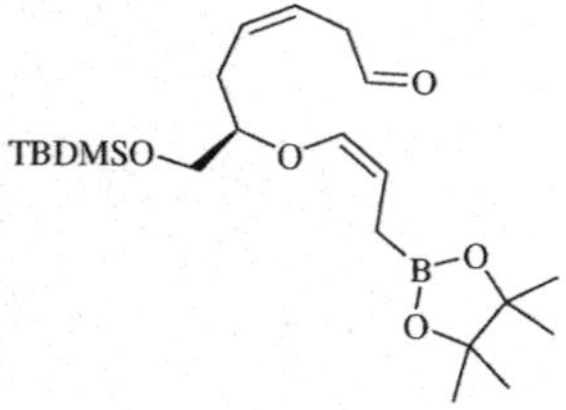

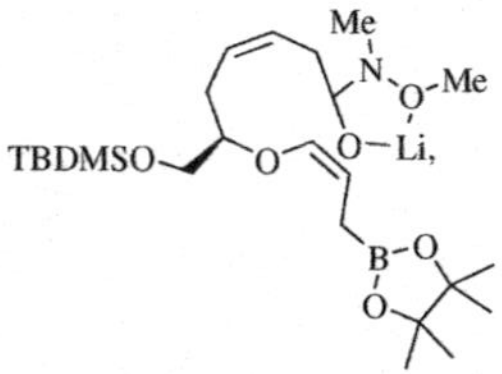

pH 7 buffer, THF, rt, 16 h

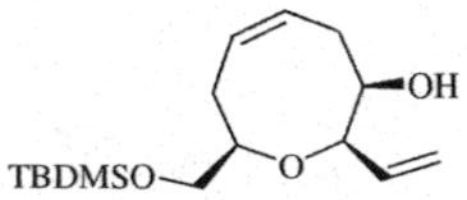

831

(51)

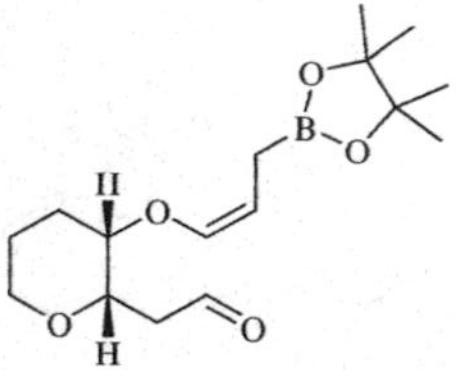

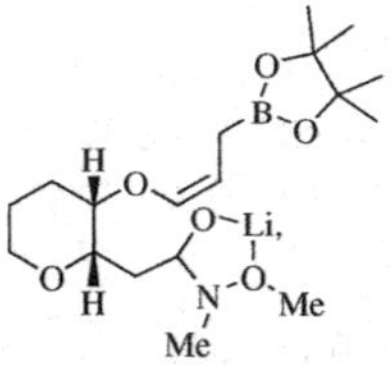

pH 7 buffer, THF, rt, 30 min

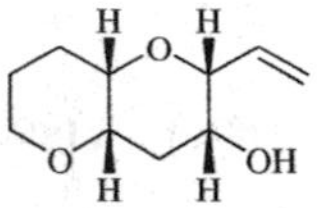

C_{10}

TABLE 9. INTRAMOLECULAR ADDITIONS (*Continued*)

Carbonyl Substrate	Conditions	Product(s) and Yield(s) (%), % ee	Refs.
C₁₀			

C_{10}

LiBF₄, H₂O, CH₃CN. rt, 1 d — (76) — 177

Pd(dba)₂, AsPh₃ (2 equiv)[a], toluene, 100° — **I** (82) — 180

Pd(OAc)₂, PPh₃ (3 equiv)[a] — **I** (71) — 174

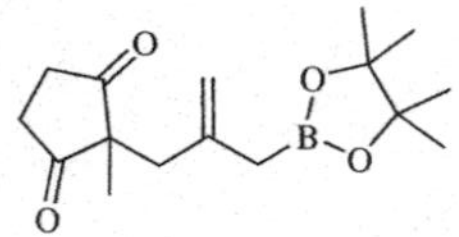

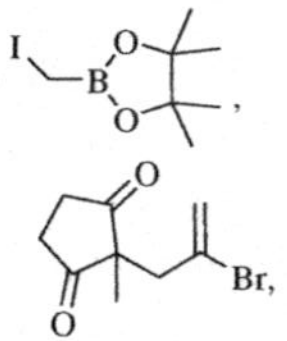

174

(50)

Pd(OAc)$_2$, dppf, PPh$_3$ (3 equiv)^a

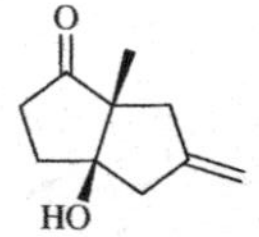

180

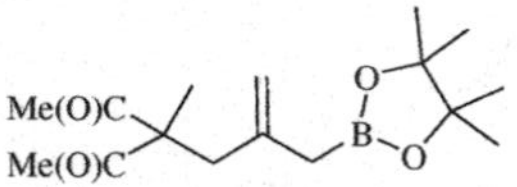

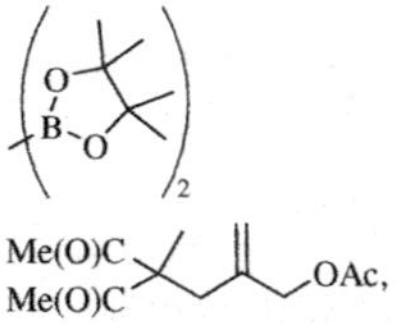

(71)

Pd(dba)$_2$, AsPh$_3$ (2 equiv)^a,
toluene, 100°

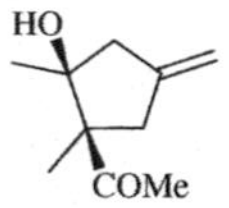

180

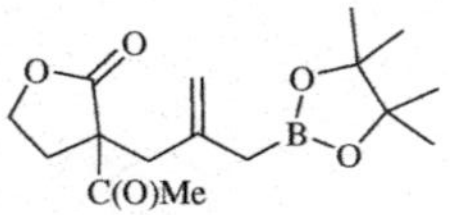

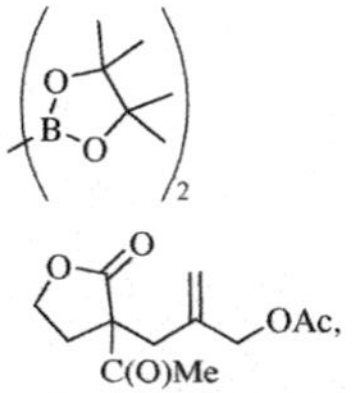

(52)

Pd(dba)$_2$, AsPh$_3$ (2 equiv)^a,
toluene, 100°

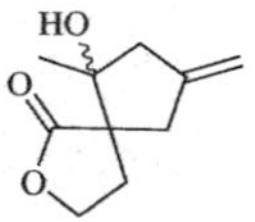

549

TABLE 9. INTRAMOLECULAR ADDITIONS (*Continued*)

Carbonyl Substrate	Conditions	Product(s) and Yield(s) (%), % ee	Refs.
C$_{10}$			
	, B_2(catechol)$_2$, (Ph$_3$P)$_2$Pt(C$_2$H$_4$), THF, 60°, 24 h 2. NaOH, H$_2$O$_2$	(44), >19:1 d.r.	186
	, B_2(catechol)$_2$, (Ph$_3$P)$_2$Pt(C$_2$H$_4$), THF, 60°, 24 h 2. NaOH, H$_2$O$_2$	(57), >19:1 d.r.	186
	, B_2(catechol)$_2$, (Ph$_3$P)$_2$Pt(C$_2$H$_4$), THF, 60°, 24 h 2. NaOH, H$_2$O$_2$	(64), >19:1 dr	186

C$_{11}$

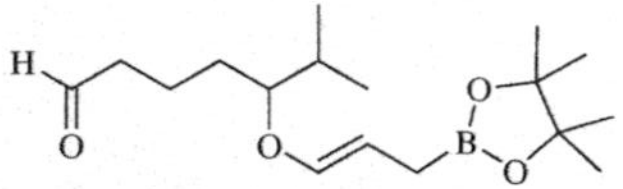

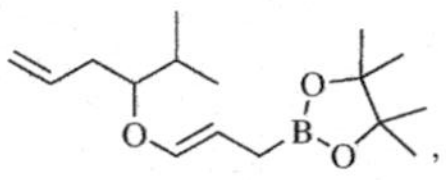

Rh(CO)$_2$acac, BIPHEPHOS,
H$_2$/CO (5 kbar), 65°, 3 d

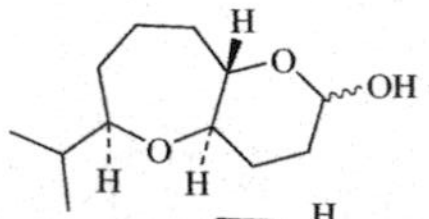

I + II (48), I:II = 1:1

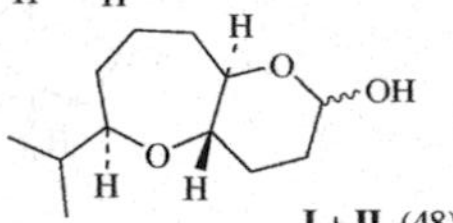

177

177

Rh(CO)$_2$acac, BIPHEPHOS,
H$_2$/CO, 65°, 3 d

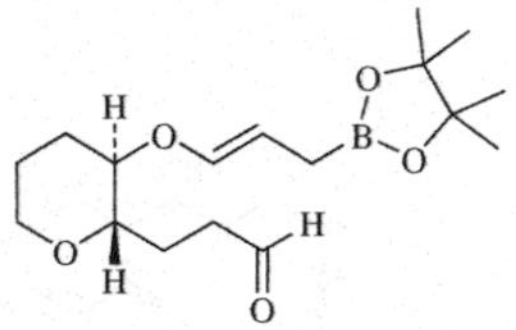

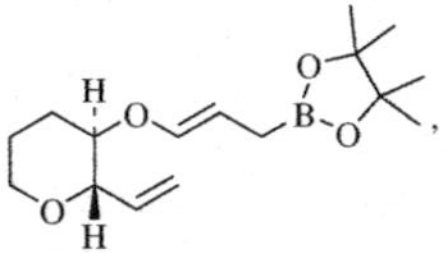

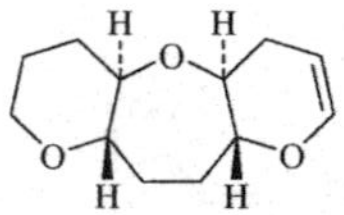

I + II (52), I:II = 1:1

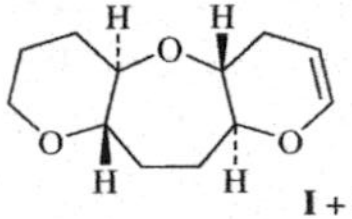

177

IZn ,
Br, Pd(PPh$_3$)$_4$, 70°, 3 h
dioxane/THF (3:1)

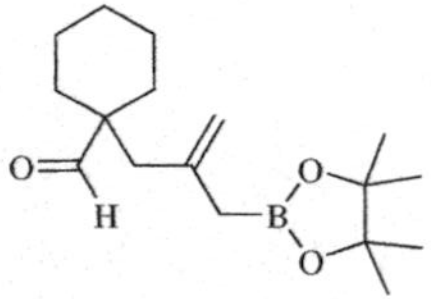

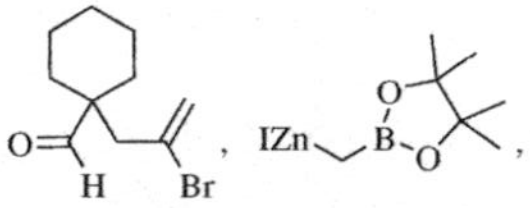

(67)

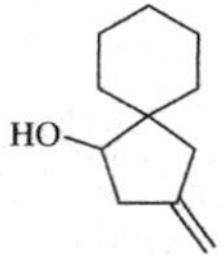

79

Br, Pd(OAc)$_2$, PPh$_3$ (3 equiv)[a]

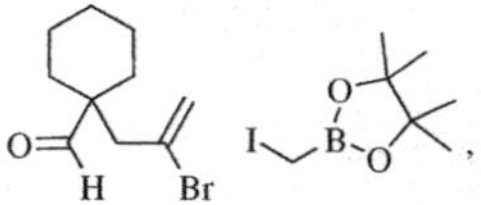

I (78)

174

TABLE 9. INTRAMOLECULAR ADDITIONS (*Continued*)

Carbonyl Substrate	Conditions	Product(s) and Yield(s) (%), % ee	Refs.
	 Pd(dba)$_2$, AsPh$_3$ (2 equiv)^a, toluene, 100°	(77)	180
	 Pd(OAc)$_2$, PPh$_3$ (3 equiv)^a	(71)	174
	 Pd(dba)$_2$, AsPh$_3$ (2 equiv)^a, toluene, 100°	(72)	180

C$_{11}$

C$_{12}$

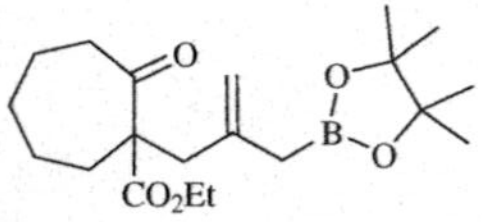

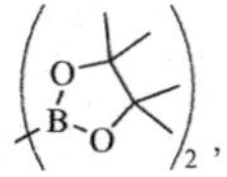

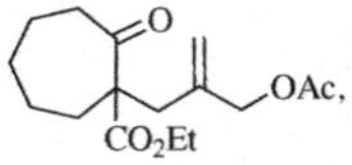

$\left(\begin{array}{c}\text{pinacol boronate}\end{array}\right)_2$,

Pd(dba)$_2$, AsPh$_3$ (2 equiv)^a, toluene, 100°

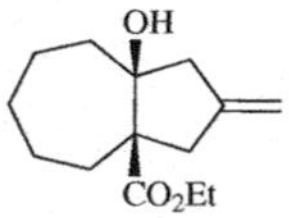

(71)

180

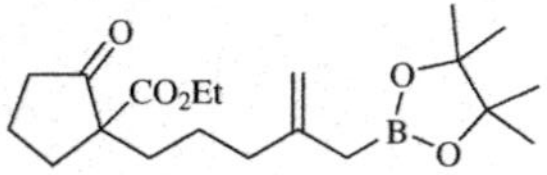

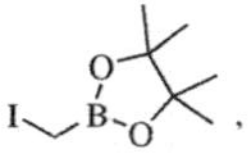

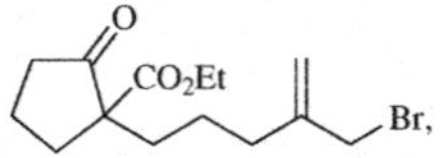

Pd(OAc)$_2$, PPh$_3$ (3 equiv)^a

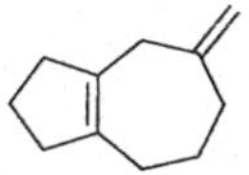

(64)

174

C$_{13}$

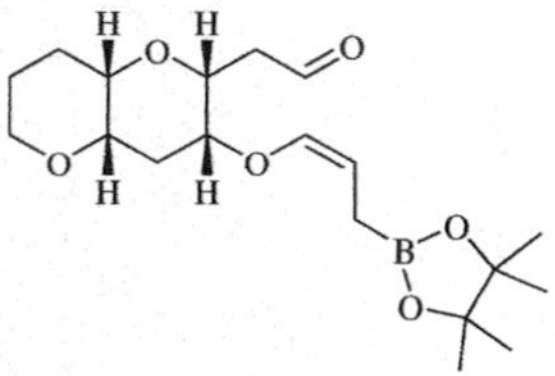

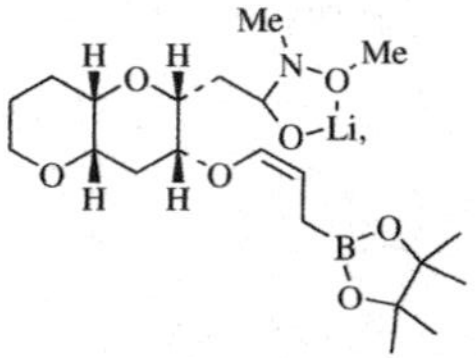

pH 7 buffer, THF, rt, 30 min

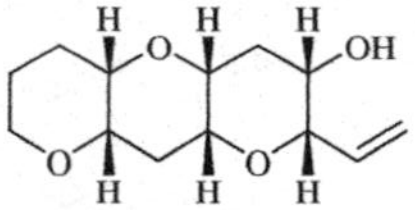

(46)

831

TABLE 9. INTRAMOLECULAR ADDITIONS (*Continued*)

Carbonyl Substrate	Conditions	Product(s) and Yield(s) (%), % ee	Refs.
C₁₃			

^{a} Equivalents are relative to catalyst.

554

REFERENCES

[1] Mikhailov, B. M.; Bubnov, Y. N. *Izv. Akad. Nauk SSSR, Ser. Khim.* **1964**, 1874.

[2] Favre, E.; Gaudemar, M. *C. R. Hebd. Seances Acad. Sci. Ser. C* **1966**, *263*, 1543.

[3] Hoffmann, R. W.; Zeiss, H.-J. *Angew. Chem., Int. Ed. Engl.* **1979**, *18*, 306.

[4] Hoffmann, R. W. *Pure Appl. Chem.* **1988**, *60*, 123.

[5] Hoffmann, R. W.; Niel, G.; Schlapbach, A. *Pure Appl. Chem.* **1990**, *62*, 1993.

[6] Yamamoto, Y.; Asao, N. *Chem. Rev.* **1993**, *93*, 2207.

[7] Hoffmann, R. W. In *Stereoselective Organic Synthesis*; Trost, B. M., Ed.; Blackwell Scientific Publications: Cambridge, U.K., 1994, pp 259–274.

[8] Roush, W. R. In *Methoden der Organischen Chemie (Houben-Weyl)*; Georg Thieme: Stuttgart, 1995; Vol. E21b, pp 1410–1486.

[9] Brown, H. C.; Ramachandran, P. V. *J. Organomet. Chem.* **1995**, *500*, 1.

[10] Denmark, S. E.; Almstead, N. G. In *Modern Carbonyl Chemistry*; Otera, J., Ed.; Wiley-VCH: Weinheim, 2000, pp 299–402.

[11] Chemler, S. R.; Roush, W. R. In *Modern Carbonyl Chemistry*; Otera, J., Ed.; Wiley-VCH: Weinheim, 2000, pp 403–490.

[12] Ramachandran, P. V. *Aldrichim. Acta* **2002**, *35*, 23.

[13] Kennedy, J. W. J.; Hall, D. G. *Angew. Chem., Int. Ed.* **2003**, *42*, 4732.

[14] Kennedy, J. W. J.; Hall, D. G. In *Boronic Acids*; Hall, D. G., Ed.; Wiley-VCH: Weinheim, 2005, pp 241–277.

[15] Denmark, S. E.; Weber, E. J. *Helv. Chim. Acta* **1983**, *66*, 1655.

[16] Li, Y.; Houk, K. N. *J. Am. Chem. Soc.* **1989**, *111*, 1236.

[17] Gennari, C.; Fioravanzo, E.; Bernardi, A.; Vulpetti, A. *Tetrahedron* **1994**, *50*, 8815.

[18] Vulpetti, A.; Gardner, M.; Gennari, C.; Bernardi, A.; Goodman, J. M.; Paterson, I. *J. Org. Chem.* **1993**, *58*, 1711.

[19] Omoto, K.; Fujimoto, H. *J. Org. Chem.* **1998**, *63*, 8331.

[20] Gajewski, J. J.; Bician, W.; Brichford, N. L.; Henderson, J. L. *J. Org. Chem.* **2002**, *67*, 4236.

[21] Brown, H. C.; Racherla, U. S.; Pellechia, P. J. *J. Org. Chem.* **1990**, *55*, 1868.

[22] Hoffmann, R. W.; Zeiss, H.-J. *J. Org. Chem.* **1981**, *46*, 1309.

[23] Corey, E. J.; Rohde, J. J. *Tetrahedron Lett.* **1997**, *38*, 37.

[24] Kennedy, J. W. J.; Hall, D. G. *J. Am. Chem. Soc.* **2002**, *124*, 11586.

[25] Ishiyama, T.; Ahiko, T.-a.; Miyaura, N. *J. Am. Chem. Soc.* **2002**, *124*, 12414.

[26] Kennedy, J. W. J.; Hall, D. G. *J. Org. Chem.* **2004**, *69*, 4412.

[27] Lachance, H.; Lu, X. S.; Gravel, M.; Hall, D. G. *J. Am. Chem. Soc.* **2003**, *125*, 10160.

[28] Gravel, M.; Lachance, H.; Lu, X. S.; Hall, D. G. *Synthesis* **2004**, 1290.

[29] Rauniyar, V.; Hall, D. G. *J. Am. Chem. Soc.* **2004**, *126*, 4518.

[30] Yu, S.-H.; Ferguson, M. J.; McDonald, R.; Hall, D. G. *J. Am. Chem. Soc.* **2005**, *127*, 12808.

[31] Blais, J.; L'Honoré, A.; Soulié, J.; Cadiot, P. *J. Organomet. Chem.* **1974**, *78*, 323.

[32] Tsai, D. J. S.; Matteson, D. S. *Tetrahedron Lett.* **1981**, *22*, 2751.

[33] Hoffmann, R. W.; Kemper, B. *Tetrahedron Lett.* **1981**, *22*, 5263.

[34] Hoffmann, R. W.; Kemper, B. *Tetrahedron* **1984**, *40*, 2219.

[34a] Hoffmann, R. W.; Kemper, B.; Metternich, R.; Lehmeier, T. *Liebigs Ann. Chem.* **1985**, 2246.

[35] Stürmer, R. *Angew. Chem., Int. Ed. Engl.* **1990**, *29*, 59.

[36] Hoffmann, R. W.; Kemper, B. *Tetrahedron Lett.* **1982**, *23*, 845.

[37] Roush, W. R.; Ando, K.; Powers, D. B.; Palkowitz, A. D.; Halterman, R. L. *J. Am. Chem. Soc.* **1990**, *112*, 6339.

[38] Roush, W. R.; Grover, P. T.; Lin, X. F. *Tetrahedron Lett.* **1990**, *31*, 7563.

[39] Roush, W. R.; Grover, P. T. *Tetrahedron Lett.* **1990**, *31*, 7567.

[40] Roush, W. R.; Grover, P. T. *Tetrahedron* **1992**, *48*, 1981.

[41] Hoffmann, R. W. *Angew. Chem., Int. Ed. Engl.* **1982**, *21*, 555.

[42] Schlosser, M.; Rauchschwalbe, G. *J. Am. Chem. Soc.* **1978**, *100*, 3258.

[43] Fujita, K.; Schlosser, M. *Helv. Chim. Acta* **1982**, *65*, 1258.

[44] Brown, H. C.; Bhat, K. S. *J. Am. Chem. Soc.* **1986**, *108*, 293.

[45] Barrett, A. G. M.; Malecha, J. W. *J. Org. Chem.* **1991**, *56*, 5243.

[46] Wuts, P. G. M.; Bigelow, S. S. *J. Org. Chem.* **1982**, *47*, 2498.
[47] Hoffmann, R. W.; Kemper, B.; Metternich, R.; Lehmeier, T. *Liebigs Ann. Chem.* **1985**, 2246.
[48] Brown, H. C.; Jadhav, P. K.; Bhat, K. S. *J. Am. Chem. Soc.* **1988**, *110*, 1535.
[49] Hoffmann, R. W.; Feussner, G.; Zeiss, H.-J.; Schulz, S. *J. Organomet. Chem.* **1980**, *187*, 321.
[50] Corey, E. J.; Yu, C.-M.; Kim, S. S. *J. Am. Chem. Soc.* **1989**, *111*, 5495.
[51] Beckmann, E.; Hoppe, D. *Synthesis* **2005**, 217.
[52] Wuts, P. G. M.; Thompson, P. A.; Callen, G. R. *J. Org. Chem.* **1983**, *48*, 5398.
[53] Hoffmann, R. W.; Schlapbach, A. *Tetrahedron* **1992**, *48*, 1959.
[54] Brown, H. C.; De Lue, N. R.; Yamamoto, Y.; Maruyama, K.; Kasahara, T.; Murahashi, S.-i.; Sonoda, A. *J. Org. Chem.* **1977**, *42*, 4088.
[55] Hoffmann, R. W.; Schlapbach, A. *Liebigs Ann. Chem.* **1990**, 1243.
[56] Matteson, D. S.; Majumdar, D. *J. Am. Chem. Soc.* **1980**, *102*, 7588.
[57] Nyzam, V.; Belaud, C.; Villiéras, J. *Tetrahedron Lett.* **1993**, *34*, 6899.
[58] Nyzam, V.; Belaud, C.; Zammattio, F.; Villiéras, J. *Bull. Soc. Chim. Fr.* **1997**, *134*, 583.
[59] Kennedy, J. W. J.; Hall, D. G. *J. Am. Chem. Soc.* **2002**, *124*, 898.
[60] Zhu, N.; Hall, D. G. *J. Org. Chem.* **2003**, *68*, 6066.
[61] Roush, W. R.; Peseckis, S. M.; Walts, A. E. *J. Org. Chem.* **1984**, *49*, 3429.
[62] Matteson, D. S. *Tetrahedron* **1998**, *54*, 10555.
[63] Ditrich, K.; Bube, T.; Stürmer, R.; Hoffmann, R. W. *Angew. Chem., Int. Ed. Engl.* **1986**, *25*, 1028.
[64] Hoffmann, R. W.; Ditrich, K.; Köster, G.; Stürmer, R. *Chem. Ber.* **1989**, *122*, 1783.
[65] Stürmer, R.; Hoffmann, R. W. *Synlett* **1990**, 759.
[66] Hoffmann, R. W.; Landmann, B. *Chem. Ber.* **1986**, *119*, 2013.
[67] Hoffmann, R. W.; Dresely, S.; Lanz, J. W. *Chem. Ber.* **1988**, *121*, 1501.
[68] Hoffmann, R. W.; Dresely, S. *Chem. Ber.* **1989**, *122*, 903.
[69] Brown, H. C.; Jadhav, P. K.; Bhat, K. S. *J. Am. Chem. Soc.* **1985**, *107*, 2564.
[70] Yamamoto, Y.; Fujikawa, R.; Yamada, A.; Miyaura, N. *Chem. Lett.* **1999**, 1069.
[71] Ishiyama, T.; Yamamoto, M.; Miyaura, N. *Chem. Commun.* **1996**, 2073.
[72] Suginome, M.; Matsuda, T.; Yoshimoto, T.; Ito, Y. *Org. Lett.* **1999**, *1*, 1567.
[73] Pelz, N. F.; Woodward, A. R.; Burks, H. E.; Sieber, J. D.; Morken, J. P. *J. Am. Chem. Soc.* **2004**, *126*, 16328.
[74] Ishiyama, T.; Ahiko, T.-A.; Miyaura, N. *Tetrahedron Lett.* **1996**, *37*, 6889.
[75] Olsson, V. J.; Sebelius, S.; Selander, N.; Szabó, K. J. *J. Am. Chem. Soc.* **2006**, *128*, 4588.
[76] Ito, H.; Kawakami, C.; Sawamura, M. *J. Am. Chem. Soc.* **2005**, *127*, 16034.
[77] Murata, M.; Watanabe, S.; Masuda, Y. *Tetrahedron Lett.* **2000**, *41*, 5877.
[78] Falck, J. R.; Bondlela, M.; Ye, J.; Cho, S.-D. *Tetrahedron Lett.* **1999**, *40*, 5647.
[79] Watanabe, T.; Miyaura, N.; Suzuki, A. *J. Organomet. Chem.* **1993**, *444*, C1.
[80] Hoffmann, R. W.; Sander, T.; Hense, A. *Liebigs Ann. Chem.* **1993**, 771.
[81] Sadhu, K. M.; Matteson, D. S. *Organometallics* **1985**, *4*, 1687.
[82] Hoffmann, R. W.; Niel, G. *Liebigs Ann. Chem.* **1991**, 1195.
[83] Hoffmann, R. W.; Dresely, S. *Angew. Chem., Int. Ed. Engl.* **1986**, *25*, 189.
[84] Brown, H. C.; Rangaishenvi, M. V.; Jayaraman, S. *Organometallics* **1992**, *11*, 1948.
[85] Pietruszka, J.; Schöne, N. *Angew. Chem., Int. Ed.* **2003**, *42*, 5638.
[86] Pietruszka, J.; Schöne, N. *Eur. J. Org. Chem.* **2004**, 5011.
[87] Lombardo, M.; Morganti, S.; Tozzi, M.; Trombini, C. *Eur. J. Org. Chem.* **2002**, 2823.
[88] Yamamoto, Y.; Miyairi, T.; Ohmura, T.; Miyaura, N. *J. Org. Chem.* **1999**, *64*, 296.
[89] Vaultier, M.; Truchet, F.; Carboni, B.; Hoffmann, R. W.; Denne, I. *Tetrahedron Lett.* **1987**, *28*, 4169.
[90] Yamamoto, Y.; Takahashi, M.; Miyaura, N. *Synlett* **2002**, 128.
[91] Bubnov, Y. N.; Gurskii, M. E.; Gridnev, I. D.; Ignatenko, A. V.; Ustynyuk, Y. A.; Mstislavsky, V. I. *J. Organomet. Chem.* **1992**, *424*, 127.
[92] Kramer, G. W.; Brown, H. C. *J. Organomet. Chem.* **1977**, *132*, 9.
[93] Zweifel, G.; Horng, A. *Synthesis* **1973**, *67*, 672.
[94] Jutzi, P.; Seufert, A. *Chem. Ber.* **1979**, *112*, 2481.

[95] Hancock, K. G.; Kramer, J. D. *J. Am. Chem. Soc.* **1973**, *95*, 6463.

[96] Hancock, K. G.; Kramer, J. D. *J. Organomet. Chem.* **1974**, *64*, C29.

[97] Le Serre, S.; Guillemin, J. C. *Organometallics* **1997**, *16*, 5844.

[98] Batey, R. A.; Thadani, A. N.; Smil, D. V. *Tetrahedron Lett.* **1999**, *40*, 4289.

[99] Batey, R. A.; Thadani, A. N.; Smil, D. V.; Lough, A. J. *Synthesis* **2000**, 990.

[100] Thadani, A. N.; Batey, R. A. *Org. Lett.* **2002**, *4*, 3827.

[101] Lautens, M.; Maddess, M. L.; Sauer, E. L. O.; Ouellet, S. G. *Org. Lett.* **2002**, *4*, 83.

[102] Hoffmann, R. W.; Sander, T. *Chem. Ber.* **1990**, *123*, 145.

[103] Kramer, G. W.; Brown, H. C. *J. Org. Chem.* **1977**, *42*, 2292.

[104] Metternich, R.; Hoffmann, R. W. *Tetrahedron Lett.* **1984**, *25*, 4095.

[105] Wang, Z.; Meng, X.-J.; Kabalka, G. W. *Tetrahedron Lett.* **1991**, *32*, 4619.

[106] Wang, Z.; Meng, X.-J.; Kabalka, G. W. *Tetrahedron Lett.* **1991**, *32*, 5677.

[107] Pace, R. D.; Kabalka, G. W. *J. Org. Chem.* **1995**, *60*, 4838.

[108] Kabalka, G. W.; Yang, K.; Wang, Z. *Synth. Commun.* **2001**, *31*, 511.

[109] Wang, Z.; Meng, X. J.; Kabalka, G. W. *Tetrahedron Lett.* **1991**, *32*, 1945.

[110] Kabalka, G. W.; Narayana, C.; Reddy, N. K. *Tetrahedron Lett.* **1996**, *37*, 2181.

[111] Buynak, J. D.; Geng, B.; Uang, S.; Strickland, J. B. *Tetrahedron Lett.* **1994**, *35*, 985.

[112] Cherest, M.; Felkin, H.; Prudent, N. *Tetrahedron Lett.* **1968**, *9*, 2199.

[113] Roush, W. R. *J. Org. Chem.* **1991**, *56*, 4151.

[114] Hoffmann, R. W.; Zeiss, H.-J.; Ladner, W.; Tabche, S. *Chem. Ber.* **1982**, *115*, 2357.

[115] Hoffmann, R. W.; Weidmann, U. *Chem. Ber.* **1985**, *118*, 3966.

[116] Anh, N. T.; Eisenstein, O. *Nouv. J. Chim.* **1977**, *1*, 61.

[117] Cornforth, J. W.; Cornforth, R. H.; Mathew, K. K. *J. Chem. Soc.* **1959**, 112.

[118] Roush, W. R.; Adam, M. A.; Walts, A. E.; Harris, D. J. *J. Am. Chem. Soc.* **1986**, *108*, 3422.

[119] Gung, B. W.; Xue, X. W. *Tetrahedron: Asymmetry* **2001**, *12*, 2955.

[120] Hoffmann, R. W.; Herold, T. *Angew. Chem., Int. Ed. Engl.* **1978**, *18*, 768.

[121] Herold, T.; Schrott, U.; Hoffmann, R. W.; Schnelle, G.; Ladner, W.; Steinbach, K. *Chem. Ber.* **1981**, *114*, 359.

[122] Hoffmann, R. W.; Herold, T. *Chem. Ber.* **1981**, *114*, 375.

[123] Haruta, R.; Ishiguro, M.; Ikeda, N.; Yamamoto, H. *J. Am. Chem. Soc.* **1982**, *104*, 7667.

[124] Roush, W. R.; Walts, A. E.; Hoong, L. K. *J. Am. Chem. Soc.* **1985**, *107*, 8186.

[125] Roush, W. R.; Hoong, L. K.; Palmer, M. A. J.; Park, J. C. *J. Org. Chem.* **1990**, *55*, 4109.

[126] Roush, W. R.; Halterman, R. L. *J. Am. Chem. Soc.* **1986**, *108*, 294.

[127] Gung, B. W.; Xue, X. W.; Roush, W. R. *J. Am. Chem. Soc.* **2002**, *124*, 10692.

[128] Roush, W. R.; Park, J. C. *J. Org. Chem.* **1990**, *55*, 1143.

[129] Roush, W. R.; Park, J. C. *Tetrahedron Lett.* **1990**, *31*, 4707.

[130] Roush, W. R.; Park, J. C. *Tetrahedron Lett.* **1991**, *32*, 6285.

[131] Roush, W. R.; Banfi, L. *J. Am. Chem. Soc.* **1988**, *110*, 3979.

[132] Roush, W. R.; Grover, P. T. *J. Org. Chem.* **1995**, *60*, 3806.

[133] Reetz, M. T.; Zierke, T. *Chem. Ind. (London)* **1988**, 663.

[134] Chen, Y.; Eltepu, L.; Wenthworth, J. P. *Tetrahedron Lett.* **2004**, *45*, 8285.

[135] Wu, T. R.; Shen, L.; Chong, J. M. *Org. Lett.* **2004**, *6*, 2701.

[136] Brown, H. C.; Jadhav, P. K. *J. Am. Chem. Soc.* **1983**, *105*, 2092.

[136a] Jadhav, P. K.; Bhat, K. S.; Perumal, P. T.; Brown, H. C. *J. Org. Chem.* **1986**, *51*, 432.

[137] Brown, H. C.; Bhat, K. S. *J. Am. Chem. Soc.* **1986**, *108*, 5919.

[138] Brown, H. C.; Jadhav, P. K. *Tetrahedron Lett.* **1984**, *25*, 1215.

[139] Racherla, U. S.; Brown, H. C. *J. Org. Chem.* **1991**, *56*, 401.

[140] Jadhav, P. K.; Bhat, K. S.; Perumal, P. T.; Brown, H. C. *J. Org. Chem.* **1986**, *51*, 432.

[141] Short, R. P.; Masamune, S. *J. Am. Chem. Soc.* **1989**, *111*, 1892.

[142] Garcia, J.; Kim, B. M.; Masamune, S. *J. Org. Chem.* **1987**, *52*, 4831.

[143] Burgos, C. H.; Canales, E.; Matos, K.; Soderquist, J. A. *J. Am. Chem. Soc.* **2005**, *127*, 8044.

[144] Lai, C.; Soderquist, J. A. *Org. Lett.* **2005**, *7*, 799.

[145] Canales, E.; Prasad, K. G.; Soderquist, J. A. *J. Am. Chem. Soc.* **2005**, *127*, 11572.

[146] Hoffmann, R. W.; Landmann, B. *Chem. Ber.* **1986**, *119*, 1039.

[147] Hoffmann, R. W.; Dresely, S. *Tetrahedron Lett.* **1987**, *28*, 5303.
[148] Hoffmann, R. W.; Weidmann, U. *J. Organomet. Chem.* **1980**, *195*, 137.
[149] Hoffmann, R. W.; Dresely, S.; Hildebrandt, B. *Chem. Ber.* **1988**, *121*, 2225.
[150] Gao, X.; Hall, D. G. *J. Am. Chem. Soc.* **2003**, *125*, 9308.
[151] Deligny, M.; Carreaux, F.; Toupet, L.; Carboni, B. *Adv. Synth. Catal.* **2003**, *345*, 1215.
[152] Gao, X.; Hall, D. G.; Deligny, M.; Favre, A.; Carreaux, F.; Carboni, C. *Chem.—Eur. J.* **2006**, *13*, 3132.
[153] Corey, E. J.; Yu, C.-M.; Lee, D.-H. *J. Am. Chem. Soc.* **1990**, *112*, 878.
[154] Tamao, K.; Nakajo, E.; Ito, Y. *J. Org. Chem.* **1987**, *52*, 957.
[155] Wuts, P. G. M.; Bigelow, S. S. *J. Chem. Soc., Chem. Commun.* **1984**, 736.
[156] Coleman, R. S.; Kong, J.-S. *J. Am. Chem. Soc.* **1998**, *120*, 3538.
[157] Brown, H. C.; Narla, G. *J. Org. Chem.* **1995**, *60*, 4686.
[158] Hunt, J. A.; Roush, W. R. *J. Org. Chem.* **1997**, *62*, 1112.
[159] Hu, S.; Jayaraman, S.; Oehlschlager, A. C. *J. Org. Chem.* **1996**, *61*, 7513.
[160] Barrett, A. G. M.; Seefeld, M. A.; Williams, D. J. *J. Chem. Soc., Chem. Commun.* **1994**, 1053.
[161] Chataigner, I.; Zammattio, F.; Lebreton, J.; Villiéras, J. *Synlett* **1998**, 275.
[162] Ishihara, K.; Kaneeda, M.; Yamamoto, H. *J. Am. Chem. Soc.* **1994**, *116*, 11179.
[163] Rauniyar, V.; Hall, D. G. *Angew. Chem., Int. Ed.* **2006**, *45*, 2426.
[164] Wada, R.; Oisaki, K.; Kanai, M.; Shibasaki, M. *J. Am. Chem. Soc.* **2004**, *126*, 8910.
[165] Masamune, S.; Choy, W.; Petersen, J. S.; Sita, L. R. *Angew. Chem., Int. Ed. Engl.* **1985**, *24*, 1.
[166] Brown, H. C.; Bhat, K. S.; Randad, R. S. *J. Org. Chem.* **1987**, *52*, 319.
[167] Brown, H. C.; Bhat, K. S.; Randad, R. S. *J. Org. Chem.* **1989**, *54*, 1570.
[168] Ogawa, A. K.; Armstrong, R. W. *J. Am. Chem. Soc.* **1998**, *120*, 12435.
[169] Roush, W. R.; Palkowitz, A. D.; Palmer, M. A. J. *J. Org. Chem.* **1987**, *52*, 316.
[170] Roush, W. R.; Hoong, L. K.; Palmer, M. A. J.; Straub, J. A.; Palkowitz, A. D. *J. Org. Chem.* **1990**, *55*, 4117.
[171] Sander, T.; Hoffmann, R. W. *Liebigs Ann. Chem.* **1993**, 1193.
[172] Hoffmann, R. W.; Hense, A. *Liebigs Ann.* **1996**, 1283.
[173] Yamamoto, Y.; Kurihara, K.; Yamada, A.; Takahashi, M.; Takahashi, Y.; Miyaura, N. *Tetrahedron* **2003**, *59*, 537.
[174] Watanabe, T.; Sakai, M.; Miyaura, N.; Suzuki, A. *J. Chem. Soc., Chem. Commun.* **1994**, 467.
[175] Flamme, E. M.; Roush, W. R. *J. Am. Chem. Soc.* **2002**, *124*, 13644.
[176] Barrett, A. G. M.; Braddock, D. C.; de Koning, P. D.; White, A. J. P.; Williams, D. J. *J. Org. Chem.* **2000**, *65*, 375.
[177] Hoffmann, R. W.; Krüger, J.; Bruckner, D. *New J. Chem.* **2001**, *25*, 102.
[178] Hoffmann, R. W.; Brückner, D.; Gerusz, V. J. *Heterocycles* **2000**, *52*, 121.
[179] Hoffmann, R. W.; Brückner, D. *New J. Chem.* **2001**, *25*, 369.
[180] Ahiko, T.-a.; Ishiyama, T.; Miyaura, N. *Chem. Lett.* **1997**, 811.
[181] Tailor, J.; Hall, D. G. *Org. Lett.* **2000**, *2*, 3715.
[182] Touré, B. B.; Hoveyda, H. R.; Tailor, J.; Ulaczyk-Lesanko, A.; Hall, D. G. *Chem. Eur. J.* **2003**, *9*, 466.
[183] Six, Y.; Lallemand, J.-Y. *Tetrahedron Lett.* **1999**, *40*, 1295.
[184] Goldberg, S. D.; Grubbs, R. H. *Angew. Chem., Int. Ed.* **2002**, *41*, 807.
[185] Morgan, J. B.; Morken, J. P. *Org. Lett.* **2003**, *5*, 2573.
[186] Ballard, C. E.; Morken, J. P. *Synthesis* **2004**, 1321.
[187] Woodward, A. R.; Burks, H. E.; Chan, L. M.; Morken, J. P. *Org. Lett.* **2005**, *7*, 5505.
[188] Nicolaou, K. C.; Ninkovic, S.; Sarabia, F.; Vourloumis, D.; He, Y.; Vallberg, H.; Finlay, M. R. V.; Yang, Z. *J. Am. Chem. Soc.* **1997**, *119*, 7974.
[189] Bratz, M.; Bullock, W. H.; Overman, L. E.; Takemoto, T. *J. Am. Chem. Soc.* **1995**, *117*, 5958.
[190] Roush, W. R.; Wada, C. K. *J. Am. Chem. Soc.* **1994**, *116*, 2151.
[191] Williams, D. R.; Plummer, S. V.; Patnaik, S. *Angew. Chem., Int. Ed.* **2003**, *42*, 3934.
[192] Denmark, S. E.; Yang, S.-M. *J. Am. Chem. Soc.* **2002**, *124*, 15196.
[193] Marron, T. G.; Roush, W. R. *Tetrahedron Lett.* **1995**, *36*, 1581.
[194] Roush, W. R.; Marron, T. G. *Tetrahedron Lett.* **1993**, *34*, 5421.

[195] Scheidt, K. A.; Tasaka, A.; Bannister, T. D.; Wendt, M. D.; Roush, W. R. *Angew. Chem., Int. Ed.* **1999**, *38*, 1652.

[196] Smith, A. B., III; Adams, C. M.; Barbosa, S. A. L.; Degnan, A. P. *J. Am. Chem. Soc.* **2003**, *125*, 350.

[197] Andersen, M. W.; Hildebrandt, B.; Dahmann, G.; Hoffmann, R. W. *Chem. Ber.* **1991**, *124*, 2127.

[198] Gao, X.; Hall, D. G. *J. Am. Chem. Soc.* **2005**, *127*, 1628.

[199] Touré, B. B.; Hall, D. G. *Angew. Chem., Int. Ed.* **2004**, *43*, 2001.

[200] Touré, B. B.; Hall, D. G. *J. Org. Chem.* **2005**, *69*, 8429.

[201] Hicks, J. D.; Flamme, E. M.; Roush, W. R. *Org. Lett.* **2005**, *7*, 5509.

[202] Flamme, E. M.; Roush, W. R. *Org. Lett.* **2005**, *7*, 1411.

[203] Owen, R. M.; Roush, W. R. *Org. Lett.* **2005**, *7*, 3941.

[204] Ramachandran, P. V.; Reddy, M. V. R.; Brown, H. C. *Tetrahedron Lett.* **2000**, *41*, 583.

[205] Nelson, S. G.; Cheung, W. S.; Kassick, A. J.; Hilfiker, M. A. *J. Am. Chem. Soc.* **2002**, *124*, 13654.

[206] Lambert, W. T.; Roush, W. R. *Org. Lett.* **2005**, *7*, 5501.

[207] Hoffmann, R. W.; Schäfer, F.; Haeberlin, E.; Rohde, T.; Körber, K. *Synthesis* **2000**, 2060.

[208] Hoffmann, R. W.; Rohde, T.; Haeberlin, E.; Schäfer, F. *Org. Lett.* **1999**, *1*, 1713.

[209] Denmark, S. E.; Fu, J. *Chem. Rev.* **2003**, *103*, 2763.

[210] *Modern Aldol Reactions*; Mahrwald, R., Ed.; Wiley-VCH: Weinheim, 2004.

[211] Cowden, C. J.; Paterson, I. *Org. React.* **1997**, *51*, 1.

[212] Mukaiyama, T.; Kobayashi, S. *Org. React.* **1994**, *46*, 1.

[213] Konig, K.; Neuman, W. P. *Tetrahedron Lett.* **1967**, 493.

[214] Keck, G. E.; Tarbet, K. H.; Geraci, L. S. *J. Am. Chem. Soc.* **1993**, *115*, 8467.

[215] Costa, A. L.; Piazza, M. G.; Tagliavini, E.; Trombini, C.; Umani-Ronchi, A. *J. Am. Chem. Soc.* **1993**, *115*, 7001.

[216] Faller, J. W.; Sams, D. W. T.; Liu, X. *J. Am. Chem. Soc.* **1996**, *118*, 1217.

[217] Kim, J. G.; Waltz, K. M.; Garcia, I. F.; Kwiatkowski, D.; Walsh, P. J. *J. Am. Chem. Soc.* **2004**, *126*, 12580.

[218] Yanagisawa, A.; Nakashima, H.; Ishiba, A.; Yamamoto, H. *J. Am. Chem. Soc.* **1996**, *118*, 4723.

[219] Yanagisawa, A.; Nakashima, H.; Ishiba, A.; Yamamoto, H. *Synlett* **1997**, 88.

[220] Marshall, J. A. *Chem. Rev.* **1996**, *96*, 31.

[221] Furuta, K.; Mouri, M.; Yamamoto, H. *Synlett* **1991**, 561.

[222] Ishihara, K.; Mouri, M.; Gao, Q.; Maruyama, T.; Furuta, K.; Yamamoto, H. *J. Am. Chem. Soc.* **1993**, *115*, 11490.

[223] Gauthier, D. R.; Carreira, E. M. *Angew. Chem., Int. Ed. Engl.* **1996**, *35*, 2363.

[224] Bode, J. W.; Gauthier, D. R.; Carreira, E. M. *Chem. Commun.* **2001**, 2560.

[225] Wadamoto, M.; Ozasa, N.; Yanagisawa, A.; Yamamoto, H. *J. Org. Chem.* **2003**, *68*, 5593.

[226] Yamasaki, S.; Fujii, K.; Wada, R.; Kanai, M.; Shibasaki, M. *J. Am. Chem. Soc.* **2002**, *124*, 6536.

[227] Denmark, S. E.; Fu, J. *J. Am. Chem. Soc.* **2001**, *123*, 9488.

[228] Denmark, S. E.; Fu, J. *Chem. Commun.* **2003**, 167.

[229] Masse, C. E.; Panek, J. S. *Chem. Rev.* **1995**, *95*, 1293.

[230] Kubota, K.; Leighton, J. L. *Angew. Chem., Int. Ed.* **2003**, *42*, 946.

[231] Hackman, B. M.; Lombardi, P. J.; Leighton, J. L. *Org. Lett.* **2004**, *6*, 4375.

[232] Duthaler, R. O.; Hafner, A. *Chem. Rev.* **1992**, *92*, 807.

[233] Xia, G.; Yamamoto, H. *J. Am. Chem. Soc.* **2006**, *128*, 2554.

[234] Arnauld, T.; Barrett, A. G. M.; Seifried, R. *Tetrahedron Lett.* **2001**, *42*, 7899.

[235] Cavallaro, C. L.; Herpin, T.; McGuiness, B. F.; Shimshock, Y. C.; Dolle, R. E. *Tetrahedron Lett.* **1999**, *40*, 2711.

[236] Micalizio, G. C.; Schreiber, S. L. *Angew. Chem., Int. Ed.* **2002**, *41*, 3272.

[237] Hoffmann, R. W.; Dresely, S. *Synthesis* **1988**, 103.

[238] Mikhailov, B. M.; Bubnov, Y. N.; Tsyban, A. V.; Grigoryan, M. S. *J. Organomet. Chem.* **1978**, *154*, 131.

[239] Brown, H. C.; Jadhav, P. K. *J. Org. Chem.* **1984**, *49*, 4089.

[240] Brown, H. C.; Randad, R. S.; Bhat, K. S.; Zaidlewicz, M.; Racherla, U. S. *J. Am. Chem. Soc.* **1990**, *112*, 2389.

[241] Brown, H. C.; Kulkarni, S. V.; Racherla, U. S. *J. Org. Chem.* **1994**, *59*, 365.

[242] Brown, H. C.; Kulkarni, S. V.; Racherla, U. S.; Dhokte, U. P. *J. Org. Chem.* **1998**, *63*, 7030.

[243] Redlich, H.; Schneider, B.; Hoffmann, R. W.; Geueke, K. J. *Liebigs Ann. Chem.* **1983**, 393.

[244] Herold, T.; Hoffmann, R. W. *Angew. Chem., Int. Ed. Engl.* **1978**, *17*, 768.

[245] Itsuno, S.; Watanabe, K.; El-Shehawy, A. A. *Adv. Synth. Catal.* **2001**, *343*, 89.

[246] Itsuno, S.; El-Shehawy, A. A.; Sarhan, A. A. *React. Funct. Polym.* **1998**, *37*, 283.

[247] Reddy, M. V. R.; Yucel, A. J.; Ramachandran, P. V. *J. Org. Chem.* **2001**, *66*, 2512.

[248] Kumar, D. J. S.; Madhavan, S.; Ramachandran, P. V.; Brown, H. C. *Tetrahedron: Asymmetry* **2000**, *11*, 4629.

[249] Ramachandran, P. V.; Padiya, K. J.; Rauniyar, V.; Reddy, M. V. R.; Brown, H. C. *Tetrahedron Lett.* **2004**, *45*, 1015.

[250] Ramachandran, P. V.; Prabhudas, B.; Chandra, J. S.; Reddy, M. V. R. *J. Org. Chem.* **2004**, *69*, 6294.

[251] Reddy, M. V. R.; Brown, H. C.; Ramachandran, P. V. *J. Organomet. Chem.* **2001**, *624*, 239.

[252] Bonini, C.; Chiummiento, L.; Pullez, M.; Solladie, G.; Colobert, F. *J. Org. Chem.* **2004**, *69*, 5015.

[253] Paterson, I.; Coster, M. J.; Chen, D. Y. M.; Gibson, K. R.; Wallace, D. J. *Org. Biomol. Chem.* **2005**, *3*, 2410.

[254] Ramachandran, P. V.; Chen, G. M.; Brown, H. C. *Tetrahedron Lett.* **1997**, *38*, 2417.

[255] Chen, G. M.; Ramachandran, P. V. *Tetrahedron: Asymmetry* **1997**, *8*, 3935.

[256] Bubnov, Y. N.; Lavrinovich, L. I.; Zykov, A. Y.; Ignatenko, A. V. *Mendeleev Commun.* **1992**, 86.

[257] Ramachandran, P. V.; Krzeminski, M. P.; Reddy, M. V. R.; Brown, H. C. *Tetrahedron: Asymmetry* **1999**, *10*, 11.

[258] Trost, B. M.; Frederiksen, M. U.; Papillon, J. P. N.; Harrington, P. E.; Shin, S.; Shireman, B. T. *J. Am. Chem. Soc.* **2005**, *127*, 3666.

[259] Smith, A. B., III; Ott, G. R. *J. Am. Chem. Soc.* **1996**, *118*, 13095.

[260] Roush, W. R.; Palkowitz, A. D. *J. Org. Chem.* **1989**, *54*, 3009.

[261] Reddy, Y. K.; Falck, J. R. *Org. Lett.* **2002**, *4*, 969.

[262] Mikhailov, B. M.; Bubnov, Y. N. *Russ. Chem. Bull.* **1964**, *13*, 1774.

[263] Zhou, W.; Liang, S.; Yu, S.; Luo, W. *J. Organomet. Chem.* **1993**, *452*, 13.

[264] Paterson, I.; Wallace, D. J.; Gibson, K. R. *Tetrahedron Lett.* **1997**, *38*, 8911.

[265] White, J. B.; Blakemore, P. R.; Browder, C. C.; Hong, J.; Lincoln, C. M.; Nagornyy, P. A.; Robarge, L. A.; Wardrop, D. J. *J. Am. Chem. Soc.* **2001**, *123*, 8593.

[266] Dineen, T. A.; Roush, W. R. *Org. Lett.* **2004**, *6*, 2043.

[267] Smith, A. B., III; Safonov, I. G.; Corbett, R. M. *J. Am. Chem. Soc.* **2002**, *124*, 11102.

[268] Smith, A. B., III; Minbiole, K. P.; Verhoest, P. R.; Schelhaas, M. *J. Am. Chem. Soc.* **2001**, *123*, 10942.

[269] Diez-Martin, D.; Kotecha, N. R.; Ley, S. V.; Mantegani, S.; Menendez, J. C.; Organ, H. M.; White, A. D.; Bank, B. J. *Tetrahedron* **1992**, *48*, 7899.

[270] Nicolaou, K. C.; Ahn, K. H. *Tetrahedron Lett.* **1989**, *30*, 1217.

[271] Smith, A. B., III; Lin, Q. L.; Nakayama, K.; Boldi, A. M.; Brook, C. S.; McBriar, M. D.; Moser, W. H.; Sobukawa, M.; Zhuang, L. *Tetrahedron Lett.* **1997**, *38*, 8675.

[272] Crimmins, M. T.; Washburn, D. G. *Tetrahedron Lett.* **1998**, *39*, 7487.

[273] Wu, Y. S.; Esser, L.; De Brabander, J. K. *Angew. Chem., Int. Ed.* **2000**, *39*, 4308.

[274] Shi, Y.; Peng, L. F.; Kishi, Y. *J. Org. Chem.* **1997**, *62*, 5666.

[275] Hoffmann, R. W.; Metternich, R.; Lanz, J. W. *Liebigs Ann. Chem.* **1987**, 881.

[276] Thadani, A. N.; Batey, R. A. *Tetrahedron Lett.* **2003**, *44*, 8051.

[277] Roush, W. R.; Hunt, J. A. *J. Org. Chem.* **1995**, *60*, 798.

[278] Yakelis, N. A.; Roush, W. R. *J. Org. Chem.* **2003**, *68*, 3838.

[279] Pattenden, G.; Plowright, A. T. *Tetrahedron Lett.* **2000**, *41*, 983.

[280] Pattenden, G.; Gonzalez, M. A.; Little, P. B.; Millan, D. S.; Plowright, A. T.; Tornos, J. A.; Ye, T. *Org. Biomol. Chem.* **2003**, *1*, 4173.

[280a] Pattenden, G.; Plowright, A. T.; Tornos, J. A.; Ye, T. *Tetrahedron Lett.* **1998**, *39*, 6099.

[281] Deng, L.; Scharer, O. D.; Verdine, G. L. *J. Am. Chem. Soc.* **1997**, *119*, 7865.

[282] Jingensons, A.; Marinozzi, M.; Pellicciari, R. *Tetrahedron* **2005**, *61*, 373.

[283] Nicolaou, K. C.; Groneberg, R. D.; Stylianides, N. A.; Miyazaki, T. *J. Chem. Soc., Chem. Commun.* **1990**, 1275.

[284] Groneberg, R. D.; Miyazaki, T.; Stylianides, N. A.; Schulze, T. J.; Stahl, W.; Schreiner, E. P.; Suzuki, T.; Iwabuchi, Y.; Smith, A. L.; Nicolaou, K. C. *J. Am. Chem. Soc.* **1993**, *115*, 7593.

[285] Nicolaou, K. C.; Koide, K.; Bunnage, N. E. *Chem. Eur. J.* **1995**, *1*, 454.

[286] Nicolaou, K. C.; Bunnage, M. E.; Koide, K. *J. Am. Chem. Soc.* **1994**, *116*, 8402.

[287] Nicolaou, K. C.; Pihko, P. M.; Diedrichs, N.; Zou, N.; Bernal, F. *Angew. Chem., Int. Ed.* **2001**, *40*, 1262.

[288] Hoffmann, R. W.; Endesfelder, A.; Zeiss, H. J. *Carbohydr. Res.* **1983**, *123*, 320.

[289] Barrett, A. G. M.; Beall, J. C.; Gibson, V. C.; Gilles, M. R.; Walker, G. L. P. *Chem. Commun.* **1996**, 2229.

[290] Chen, W.; Liu, Y.; Chen, Z. *Eur. J. Org. Chem.* **2005**, *8*, 1665.

[291] Nicolaou, K. C.; King, N. P.; Finlay, M. R. V.; He, Y.; Roschangar, F.; Vourloumis, D.; Vallberg, H.; Sarabia, F.; Ninkovic, S.; Hepworth, D. *Bioorg. Med. Chem.* **1999**, *7*, 665.

[292] Ramachandran, P. V.; Burghardt, T. E.; Reddy, M. V. R. *J. Org. Chem.* **2005**, *70*, 2329.

[293] Nieczypor, P.; Mol, J. C.; Bespalova, N. B.; Bubnov, Y. N. *Eur. J. Org. Chem.* **2004**, 812.

[294] Bubnov, Y. N.; Pastukhov, F. V.; Yampolsky, I. V.; Ignatenko, A. V. *Eur. J. Org. Chem.* **2000**, 1503.

[295] Nicolaou, K. C.; Patron, A. P.; Ajito, K.; Richter, P. K.; Khatuya, H.; Bertinato, P.; Miller, R. A.; Tomaszewski, M. J. *Chem. Eur. J.* **1996**, *2*, 847.

[296] Aissa, C.; Riveiros, R.; Ragot, J.; Fürstner, A. *J. Am. Chem. Soc.* **2003**, *125*, 15512.

[297] Roush, W. R.; Palkowitz, A. D.; Ando, K. *J. Am. Chem. Soc.* **1990**, *112*, 6348.

[298] Judd, T. C.; Bischoff, A.; Kishi, Y.; Adusumilli, S.; Small, P. L. C. *Org. Lett.* **2004**, *6*, 4901.

[299] Hoffmann, R. W.; Froech, S. *Tetrahedron Lett.* **1985**, *26*, 1643.

[300] Matsushima, Y.; Nakayama, T.; Tohyama, S.; Eguchi, T.; Kakinuma, K. *J. Chem. Soc., Perkin Trans. 1* **2001**, 569.

[301] Chen, Z.; Ye, T. *Synlett* **2005**, 2781.

[302] Roush, W. R.; Straub, J. A.; Vannieuwenhze, M. S. *J. Org. Chem.* **1991**, *56*, 1636.

[303] Roush, W. R.; Straub, J. A. *Tetrahedron Lett.* **1986**, *27*, 3349.

[304] Roush, W. R.; Lin, X. F.; Straub, J. A. *J. Org. Chem.* **1991**, *56*, 1649.

[305] Steel, P. G.; Thomas, E. J. *J. Chem. Soc., Perkin Trans. 1* **1997**, 371.

[306] Merifield, E.; Steel, P. G.; Thomas, E. J. *J. Chem. Soc., Chem. Commun.* **1987**, 1826.

[307] Smith, A. B., III; Chen, S. S. Y.; Nelson, F. C.; Reichert, J. M.; Salvatore, B. A. *J. Am. Chem. Soc.* **1995**, *117*, 12013.

[308] Smith, A. B., III; Chen, S. S. Y.; Nelson, F. C.; Reichert, J. M.; Salvatore, B. A. *J. Am. Chem. Soc.* **1997**, *119*, 10935.

[309] Barth, M.; Bellamy, F. D.; Renaut, P.; Samreth, S.; Schuber, F. *Tetrahedron* **1990**, *46*, 6731.

[310] Paterson, I.; Anderson, E. A.; Dalby, S. M.; Loiseleur, O. *Org. Lett.* **2005**, *7*, 4125.

[311] White, J. D.; Kim, T. S.; Nambu, M. *J. Am. Chem. Soc.* **1995**, *117*, 5612.

[312] Nagle, D. G.; Geralds, R. S.; Yoo, H. D.; Gerwick, W. H.; Kim, T. S.; Nambu, M.; White, J. D. *Tetrahedron Lett.* **1995**, *36*, 1189.

[313] Xu, Z.; Chen, Z.; Ye, T. *Tetrahedron: Asymmetry* **2004**, *15*, 355.

[314] Ramachandran, P. V.; Krezminski, M. P.; Reddy, M. V. R.; Brown, H. C. *Tetrahedron: Asymmetry* **1999**, *10*, 11.

[315] Ghosh, A. K.; Wang, Y. *Tetrahedron Lett.* **2000**, *41*, 2319.

[316] Naito, H.; Kawahara, E.; Maruta, K.; Maeda, M.; Sasaki, S. *J. Org. Chem.* **1995**, *60*, 4419.

[317] Mulzer, J.; Hanbauer, M. *Tetrahedron Lett.* **2000**, *41*, 33.

[318] Ahmed, A.; Hoegenauer, E. K.; Enev, V. A. S.; Hanbauer, M.; Kaehlig, H.; Ohler, E.; Mulzer, J. *J. Org. Chem.* **2003**, *68*, 3026.

[319] Provencal, D. P.; Gardelli, C.; Lafontaine, J. A.; Leahy, J. W. *Tetrahedron Lett.* **1995**, *36*, 6033.

[320] Lafontaine, J. A.; Provencal, D. P.; Gardelli, C.; Leahy, J. W. *J. Org. Chem.* **2003**, *68*, 4215.

[321] Takahata, H.; Ouchi, H.; Ichinose, M.; Nemoto, H. *Org. Lett.* **2002**, *4*, 3459.

[322] Dinh, T. Q.; Du, X.; Smith, C. D.; Armstrong, R. W. *J. Org. Chem.* **1997**, *62*, 6773.

[323] Dinh, T. Q.; Smith, C. D.; Armstrong, R. W. *J. Org. Chem.* **1997**, *62*, 790.

[324] McAlpine, I. J.; Armstrong, R. W. *Tetrahedron Lett.* **2000**, *41*, 1849.

[325] Rychnovsky, S. D.; Griesgraber, G.; Schlegel, R. *J. Am. Chem. Soc.* **1995**, *117*, 197.

[326] Wender, P. A.; Clarke, M. O.; Horan, J. C. *Org. Lett.* **2005**, *7*, 1995.

[327] Wender, P. A.; Koehler, M. F. T.; Sendzik, M. *Org. Lett.* **2003**, *5*, 4549.

[328] Kang, S. H.; Kang, S. Y.; Kim, C. M.; Choi, H. W.; Jun, H. S.; Lee, B. M.; Park, C. M.; Jeong, J. W. *Angew. Chem., Int. Ed.* **2003**, *42*, 4779.

[329] Kang, S. H.; Kang, S. Y.; Choi, H. W.; Kim, C. M.; Jun, H. S.; Youn, J. H. *Synthesis* **2004**, 1102.

[330] Hayward, M. M.; Roth, R. M.; Duffy, K. J.; Dalko, P. I.; Stevens, K. L.; Guo, J.; Kishi, Y. *Angew. Chem., Int. Ed.* **1998**, *37*, 192.

[331] Kang, S. H.; Choi, H. W.; Kim, C. M.; Jun, H. S.; Kang, S. Y.; Jeong, J. W.; Youn, J. H. *Tetrahedron Lett.* **2003**, *44*, 6817.

[332] Murai, A. *J. Toxicol.: Toxin Rev.* **2003**, *22*, 617.

[333] Ishihara, J.; Yamamoto, Y.; Kanoh, N.; Murai, A. *Tetrahedron Lett.* **1999**, *40*, 4387.

[334] Grieco, P. A.; Speake, J. D.; Yeo, S. K.; Miyashita, M. *Tetrahedron Lett.* **1998**, *39*, 1125.

[335] Liu, B.; Zhou, W. S. *Org. Lett.* **2004**, *6*, 71.

[336] Ghosh, A. K.; Kim, J. H. *Tetrahedron Lett.* **2003**, *44*, 3967.

[337] Jirousek, M. R.; Cheung, A. W. H.; Babine, R. E.; Sass, P. M.; Schow, S. R.; Wick, M. M. *Tetrahedron Lett.* **1993**, *34*, 3671.

[338] Bhattacharjee, A.; Soltani, O.; De Brabander, J. K. *Org. Lett.* **2002**, *4*, 481.

[339] Barry, C. S.; Bushby, N.; Harding, J. R.; Hughes, R. A.; Parker, G. D.; Roe, R.; Willis, C. L. *Chem. Commun.* **2005**, 3727.

[340] Prahlad, V.; El-Ahl, A. A. S.; Donaldson, W. A. *Tetrahedron: Asymmetry* **2000**, *11*, 3091.

[341] White, J. D.; Hong, J.; Robarge, L. A. *Tetrahedron Lett.* **1999**, *40*, 1463.

[342] Roush, W. R.; Wada, C. K. *Tetrahedron Lett.* **1994**, *35*, 7347.

[343] Truesdale, L. K.; Swanson, D.; Sun, R. C. *Tetrahedron Lett.* **1985**, *26*, 5009.

[344] Barmann, H.; Prahlad, V.; Tao, C.; Yun, Y. K.; Wang, Z.; Donaldson, W. A. *Tetrahedron* **2000**, *56*, 2283.

[345] Prahlad, V.; Donaldson, W. A. *Tetrahedron Lett.* **1996**, *37*, 9169.

[346] Rychnovsky, S. D.; Rodriguez, C. *J. Org. Chem.* **1992**, *57*, 4793.

[347] Rychnovsky, S. D.; Hoye, R. C. *J. Am. Chem. Soc.* **1994**, *116*, 1753.

[348] Ahmed, M.; Barrett, A. G. M.; Beall, J. C.; Braddock, D. C.; Flack, K.; Gibson, V. C.; Procopiou, P. A.; Salter, M. M. *Tetrahedron* **1999**, *55*, 3219.

[349] Sebelius, S.; Szabó, K. J. *Eur. J. Org. Chem.* **2005**, 2539.

[350] Roush, W. R.; Banfi, L.; Park, J. C.; Hoong, L. K. *Tetrahedron Lett.* **1989**, *30*, 6457.

[351] Corey, E. J.; Kim, S. S. *Tetrahedron Lett.* **1990**, *31*, 3715.

[352] Ollivier, C.; Panchaud, P.; Renaud, P. *Synthesis* **2001**, 1573.

[353] Kabalka, G. W.; Venkataiah, B.; Das, B. C. *Green Chem.* **2002**, *4*, 472.

[354] Nicolaou, K. C.; Fylaktakidou, K. C.; Monenschein, H.; Li, Y. W.; Weyershausen, B.; Mitchell, H. J.; Wei, H. X.; Guntupalli, P.; Hepworth, D.; Sugita, K. *J. Am. Chem. Soc.* **2003**, *125*, 15433.

[355] Nicolaou, K. C.; Li, Y.; Fylaktakidou, K. C.; Mitchell, H. J.; Wei, H. X.; Weyershausen, B. *Angew. Chem., Int. Ed.* **2001**, *40*, 3849.

[356] Nicolaou, K. C.; Li, Y. W.; Weyershausen, B.; Wei, H. X. *Chem. Commun.* **2000**, 307.

[357] Smith, A. B., III; Barbosa, J.; Wong, W.; Wood, J. L. *J. Am. Chem. Soc.* **1996**, *118*, 8316.

[358] Hoemann, M. Z.; Agrios, K. A.; Aube, J. *Tetrahedron* **1997**, *53*, 11087.

[359] Vong, B. G.; Kim, S. H.; Abraham, S.; Theodorakis, E. A. *Angew. Chem., Int. Ed.* **2004**, *43*, 3947.

[360] Williams, D. R.; Ihle, D. C.; Plummer, S. V. *Org. Lett.* **2001**, *3*, 1383.

[361] White, J. D.; Smits, H. *Org. Lett.* **2005**, *7*, 235.

[362] Nadolski, G. T.; Davidson, B. S. *Tetrahedron Lett.* **2001**, *42*, 797.
[363] Ito, H.; Imai, N.; Tanikawa, S.; Kobayashi, S. *Tetrahedron Lett.* **1996**, *37*, 1795.
[364] Fujiwara, K.; Goto, A.; Sato, D.; Ohtaniuchi, Y.; Tanaka, H.; Murai, A.; Kawai, H.; Suzuki, T. *Tetrahedron Lett.* **2004**, *45*, 7011.
[365] Danieli, B.; Lesma, G.; Passarella, D.; Sacchetti, A.; Silvani, A.; Virdis, A. *Org. Lett.* **2004**, *6*, 493.
[366] Hilli, F.; White, J. M.; Rizzacasa, M. A. *Org. Lett.* **2004**, *6*, 1289.
[367] Ding, F.; Jennings, M. P. *Org. Lett.* **2005**, *7*, 2321.
[368] Fan, X.; Flentke, G. R.; Rich, D. H. *J. Am. Chem. Soc.* **1998**, *120*, 8893.
[369] Bonini, C.; Racioppi, R.; Viggiani, L. *Tetrahedron: Asymmetry* **1993**, *4*, 793.
[370] Bonini, C.; Giugliano, A.; Racioppi, R.; Righi, G. *Tetrahedron Lett.* **1996**, *37*, 2487.
[371] Schneider, C.; Rehfeuter, M. *Tetrahedron Lett.* **1998**, *39*, 9.
[372] Schneider, C.; Rehfeuter, M. *Chem. Eur. J.* **1999**, *5*, 2850.
[373] Schneider, C. *Synlett* **2001**, 1079.
[374] White, J. D.; Kranemann, C. L.; Kuntiyong, P. *Org. Lett.* **2001**, *3*, 4003.
[375] Le Flohic, A.; Meyer, C.; Cossy, J. *Org. Lett.* **2005**, *7*, 339.
[376] Yang, H. W.; Romo, D. *J. Org. Chem.* **1997**, *62*, 4.
[377] Yang, H. W.; Zhao, C. X.; Romo, D. *Tetrahedron* **1997**, *53*, 16471.
[378] Salmon, A.; Carboni, B. *J. Organomet. Chem.* **1998**, *567*, 31.
[379] Ramachandran, P. V.; Reddy, M. V. R.; Brown, H. C. *J. Indian Chem. Soc.* **1999**, *76*, 739.
[380] Pitts, M. R.; Mulzer, J. *Tetrahedron Lett.* **2002**, *43*, 8471.
[381] Kumar, V. S.; Wan, S.; Aubele, D. L.; Floreancig, P. E. *Tetrahedron: Asymmetry* **2005**, *16*, 3570.
[382] Ghosh, A. K.; Kim, J. H. *Tetrahedron Lett.* **2003**, *44*, 7659.
[383] Reddy, M. V. R.; Rearick, J. P.; Hoch, N.; Ramachandran, P. V. *Org. Lett.* **2001**, *3*, 19.
[384] Gingras, K.; Avedissian, H.; Thouin, E.; Boulanger, V.; Essagian, C.; McKerracher, L.; Lubell, W. D. *Bioorg. Med. Chem. Lett.* **2004**, *14*, 4931.
[385] Nagasawa, K. *J. Synth. Org. Chem. Jpn.* **2000**, *58*, 877.
[386] Liu, B.; Duan, S.; Sutterer, A. C.; Moeller, K. D. *J. Am. Chem. Soc.* **2002**, *124*, 10101.
[387] Liu, B.; Moeller, K. D. *Tetrahedron Lett.* **2001**, *42*, 7163.
[388] Ramachandran, P. V.; Liu, H.; Reddy, M. V. R.; Brown, H. C. *Org. Lett.* **2003**, *5*, 3755.
[389] Schneider, C.; Schuffenhauer, A. *Eur. J. Org. Chem.* **2000**, 73.
[390] Schneider, C. *Synlett* **1997**, 815.
[391] Garcia-Fortanet, J.; Murga, J.; Carda, M.; Marco, J. A. *Tetrahedron* **2004**, *60*, 12261.
[392] Murga, J.; Garcia-Fortanet, J.; Carda, M.; Marco, J. A. *Tetrahedron Lett.* **2003**, *44*, 1737.
[393] Wolbers, P.; Hoffmann, H. M. R. *Tetrahedron* **1999**, *55*, 1905.
[394] Yadav, J. S.; Prakash, S. J.; Gangadhar, Y. *Tetrahedron: Asymmetry* **2005**, *16*, 2722.
[395] Trost, B. M.; Yeh, V. S. C. *Org. Lett.* **2002**, *4*, 3513.
[396] Falomir, E.; Murga, J.; Ruiz, P.; Carda, M.; Marco, J. A. *J. Org. Chem.* **2003**, *68*, 5672.
[397] Falomir, E.; Murga, J.; Carda, M.; Marco, J. A. *Tetrahedron Lett.* **2003**, *44*, 539.
[398] Diaz-Oltra, S.; Murga, J.; Falomir, E.; Carda, M.; Marco, J. A. *Tetrahedron* **2004**, *60*, 2979.
[399] Wender, P. A.; De Brabander, J.; Harran, P. G.; Hinkle, K. W.; Lippa, B.; Pettit, G. R. *Tetrahedron Lett.* **1998**, *39*, 8625.
[400] Fürstner, A.; Mathes, C.; Lehmann, C. W. *Chem. Eur. J.* **2001**, *7*, 5299.
[401] Taylor, R. E.; Chen, Y. *Org. Lett.* **2001**, *3*, 2221.
[402] Meng, D. F.; Bertinato, P.; Balog, A.; Su, D. S.; Kamenecka, T.; Sorensen, E. J.; Danishefsky, S. J. *J. Am. Chem. Soc.* **1997**, *119*, 10073.
[403] Denmark, S. E.; Yang, S. M. *J. Am. Chem. Soc.* **2004**, *126*, 12432.
[404] Ritter, T.; Kvaerno, L.; Werder, M.; Hauser, H.; Carreira, E. M. *Org. Biomol. Chem.* **2005**, *3*, 3514.
[405] Jennings, M. P.; Clemens, R. T. *Tetrahedron Lett.* **2005**, *46*, 2021.
[406] Dreher, S. D.; Leighton, J. L. *J. Am. Chem. Soc.* **2001**, *123*, 341.
[407] Ramachandran, P. V.; Chandra, J. S.; Reddy, M. V. R. *J. Org. Chem.* **2002**, *67*, 7547.

[408] Banwell, M. G.; Coster, M. J.; Edwards, A. J.; Karunaratne, O. P.; Smith, J. A.; Welling, L. L.; Willis, A. C. *Aust. J. Chem.* **2003**, *56*, 585.

[409] Greer, P. B.; Donaldson, W. A. *Tetrahedron* **2002**, *58*, 6009.

[410] Greer, P. B.; Donaldson, W. A. *Tetrahedron Lett.* **2000**, *41*, 3801.

[411] Kadota, I.; Oguro, N.; Yamamoto, Y. *Tetrahedron Lett.* **2001**, *42*, 3645.

[412] Chattopadhyay, S. K.; Pattenden, G. *Tetrahedron Lett.* **1995**, *36*, 5271.

[413] Tan, C. H.; Kobayashi, Y.; Kishi, Y. *Angew. Chem., Int. Ed.* **2000**, *39*, 4282.

[414] Kadota, I.; Park, C. H.; Sato, K.; Yamamoto, Y. *Tetrahedron Lett.* **2001**, *42*, 6195.

[415] Kadota, I.; Kadowaki, C.; Park, C. H.; Takamura, H.; Sato, K.; Chan, P. W. H.; Thorand, S.; Yamamoto, Y. *Tetrahedron* **2002**, *58*, 1799.

[416] Nicolaou, K. C.; Kim, D. W.; Baati, R. *Angew. Chem., Int. Ed.* **2002**, *41*, 3701.

[417] Nicolaou, K. C.; Kim, D. W.; Baati, R.; O'Brate, A.; Giannakakou, P. *Chem. Eur. J.* **2003**, *9*, 6177.

[418] Duplantier, A. J.; Masamune, S. *J. Am. Chem. Soc.* **1990**, *112*, 7079.

[419] Sun, L. Q.; Takaki, K.; Chen, J.; Bertenshaw, S.; Iben, L.; Mahle, C. D.; Ryan, E.; Wu, D. D.; Gao, Q.; Xu, C. *Bioorg. Med. Chem. Lett.* **2005**, *15*, 1345.

[420] Dinh, T. Q.; Du, X.; Armstrong, R. W. *J. Org. Chem.* **1996**, *61*, 6606.

[421] Nishiyama, S. *J. Synth. Org. Chem. Jpn.* **2001**, *59*, 938.

[422] Dinh, T. Q.; Armstrong, R. W. *J. Org. Chem.* **1995**, *60*, 8118.

[423] Fuhrhop, J. H.; Bedurke, T.; Hahn, A.; Grund, S.; Gatzmann, J.; Riederer, M. *Angew. Chem., Int. Ed. Engl.* **1994**, *33*, 350.

[424] Hoeper, F.; Montforts, F. P. *Tetrahedron: Asymmetry* **1993**, *4*, 1439.

[425] Paterson, I.; Fessner, K.; Finlay, M. R. V. *Tetrahedron Lett.* **1997**, *38*, 4301.

[426] Crimmins, M. T.; Stanton, M. G.; Allwein, S. P. *J. Am. Chem. Soc.* **2002**, *124*, 5958.

[427] Pietruszka, J.; Wilhelm, T. *Synlett* **2003**, 1698.

[428] Sarraf, S. T.; Leighton, J. L. *Org. Lett.* **2000**, *2*, 3205.

[429] Vitale, J. P.; Wolckenhauer, S. A.; Do, N. M.; Rychnovsky, S. D. *Org. Lett.* **2005**, *7*, 3255.

[430] Goulet, M. T.; Boger, J. *Tetrahedron Lett.* **1990**, *31*, 4845.

[431] Koert, U.; Stein, M.; Harms, K. *Angew. Chem., Int. Ed. Engl.* **1994**, *33*, 1180.

[432] Koert, U.; Stein, M.; Wagner, H. *Chem. Eur. J.* **1997**, *3*, 1170.

[433] Schreiber, S. L.; Goulet, M. T. *J. Am. Chem. Soc.* **1987**, *109*, 8120.

[434] Crimmins, M. T.; Emmitte, K. A.; Choy, A. L. *Tetrahedron* **2002**, *58*, 1817.

[435] Mascavage, L. M.; Lu, Q.; Vey, J.; Dalton, D. R.; Carroll, P. J. *J. Org. Chem.* **2001**, *66*, 3621.

[436] Andrus, M. B.; Hicken, E. J.; Meredith, E. L.; Sommons, B. L.; Cannon, J. F. *Org. Lett.* **2003**, *5*, 3859.

[437] Muir, J. C.; Pattenden, G.; Ye, T. *J. Chem. Soc., Perkin Trans. 1* **2002**, 2243.

[438] Muir, J. C.; Pattenden, G.; Ye, T. *Tetrahedron Lett.* **1998**, *39*, 2861.

[439] Wipf, P.; Xu, W. J. *J. Org. Chem.* **1996**, *61*, 6556.

[440] Rychnovsky, S. D.; Khire, U. R.; Yang, G. *J. Am. Chem. Soc.* **1997**, *119*, 2058.

[441] White, J. D.; Ohba, Y.; Porter, W. J.; Wang, S. *Tetrahedron Lett.* **1997**, *38*, 3167.

[442] Williams, D. R.; Brugel, T. A. *Org. Lett.* **2000**, *2*, 1023.

[443] Mulzer, J.; Mantoulidis, A. *Tetrahedron Lett.* **1996**, *31*, 9179.

[444] Bolshakov, S.; Leighton, J. L. *Org. Lett.* **2005**, *7*, 3809.

[445] Kang, E. J.; Cho, E. J.; Ji, M. K.; Lee, Y. E.; Shin, D. M.; Choi, S. Y.; Chung, Y. K.; Kim, J. S.; Kim, H. J.; Lee, S. G.; Lah, M. S.; Lee, E. *J. Org. Chem.* **2005**, *70*, 6321.

[446] Ruiz, P.; Murga, J.; Carda, M.; Marco, J. A. *J. Org. Chem.* **2005**, *70*, 713.

[447] Hanessian, S.; Tehim, A.; Chen, P. *J. Org. Chem.* **1993**, *58*, 7768.

[448] Jadhav, P. K. *Tetrahedron Lett.* **1989**, *30*, 4763.

[449] Dirat, O.; Kouklovsky, C.; Langlois, Y.; Lesot, P.; Courtieu, J. *Tetrahderon: Asymmetry* **1999**, *10*, 3197.

[450] Crimmins, M. T.; Powell, M. T. *J. Am. Chem. Soc.* **2003**, *125*, 7592.

[451] Wipf, P.; Reeves, J. T.; Balachandran, R.; Day, B. W. *J. Med. Chem.* **2002**, *45*, 1901.

[452] Beumer, R.; Bayon, P.; Bugada, P.; Ducki, S.; Mongelli, N.; Sirtori, F. R.; Telser, J.; Gennari, C. *Tetrahedron* **2003**, *59*, 8803.

[453] Beumer, R.; Bayon, P.; Bugada, P.; Ducki, S.; Mongelli, N.; Sirtori, F. R.; Telser, J.; Gennari, C. *Tetrahedron Lett.* **2003**, *44*, 681.

[454] Telser, J.; Beumer, R.; Bell, A. A.; Ceccarelli, S. M.; Monti, D.; Gennari, C. *Tetrahedron Lett.* **2001**, *42*, 9187.

[455] Murga, J.; Garcia-Fortanet, J.; Carda, M.; Marco, J. A. *Tetrahedron Lett.* **2004**, *45*, 7499.

[456] Roush, W. R.; Palkowitz, A. D. *J. Am. Chem. Soc.* **1987**, *109*, 953.

[457] Wang, Z.; Deschenes, D. *J. Am. Chem. Soc.* **1992**, *114*, 1090.

[458] Liao, X. B.; Wu, Y. S.; De Brabander, J. K. *Angew. Chem., Int. Ed.* **2003**, *42*, 1648.

[459] Nakata, T.; Suenaga, T.; Nakashima, K.; Oishi, T. *Tetrahedron Lett.* **1989**, *30*, 6529.

[460] Lemarchand, A.; Bach, T. *Synthesis* **2005**, 1977.

[461] Murga, J.; García-Fortanet, J.; Carda, M.; Marco, J. A. *Tetrahedron Lett.* **2003**, *44*, 7909.

[462] Du, Y.; Wiemer, D. F. *J. Org. Chem.* **2002**, *67*, 5701.

[463] Vong, B. G.; Abraham, S.; Xiang, A. X.; Theodorakis, E. A. *Org. Lett.* **2003**, *5*, 1617.

[464] Hornberg, K. R.; Hamblett, C. L.; Leighton, J. L. *J. Am. Chem. Soc.* **2000**, *122*, 12894.

[465] Lewis, A.; Stefanuti, I.; Swain, S. A.; Smith, S. A.; Taylor, R. J. K. *Org. Biomol. Chem.* **2003**, *1*, 104.

[466] Bhattacharjee, A.; De Brabander, J. K. *Tetrahedron Lett.* **2000**, *41*, 8069.

[467] Vorontsova, N. V.; Rozenberg, V. I.; Vorontsov, E. V.; Antonov, D. Y.; Starikova, Z. A.; Bubnov, Y. N. *Russ. Chem. Bull., Int. Ed.* **2002**, *51*, 1483.

[468] Castoldi, D.; Caggiano, L.; Bayon, P.; Costa, A. M.; Cappella, P.; Sharon, O.; Gennari, C. *Tetrahedron* **2005**, *61*, 2123.

[469] Sutherlin, D. P.; Armstrong, R. W. *J. Org. Chem.* **1997**, *62*, 5267.

[470] Sutherlin, D. P.; Armstrong, R. W. *J. Am. Chem. Soc.* **1996**, *118*, 9802.

[471] Carter, R. G.; Bourland, T. C.; Zhou, X. T.; Gronemeyer, M. A. *Tetrahedron* **2003**, *59*, 8963.

[472] Hwang, C. H.; Keum, G.; Sohn, K. I.; Lee, D. H.; Lee, E. *Tetrahedron Lett.* **2005**, *46*, 6621.

[473] Yamamoto, Y.; Nishii, S.; Yamada, J. I. *J. Am. Chem. Soc.* **1986**, *108*, 7116.

[474] Mori, Y.; Kohchi, Y.; Noguchi, H.; Suzuki, M.; Carmeli, S.; Moore, R. E.; Patterson, G. M. L. *Tetrahedron* **1991**, *47*, 4889.

[475] Hanessian, S.; Giroux, S.; Buffat, M. *Org. Lett.* **2005**, *7*, 3989.

[476] Takahashi, S.; Ogawa, N.; Koshino, H.; Nakata, T. *Org. Lett.* **2005**, *7*, 2783.

[477] Hopner, F.; Montforts, F. P. *Liebigs Ann. Chem.* **1995**, 1033.

[478] Racherla, U. S.; Liao, Y.; Brown, H. C. *J. Org. Chem.* **1992**, *57*, 6614.

[479] Felpin, F. X.; Bertrand, M. J.; Lebreton, J. *Tetrahedron* **2002**, *58*, 7381.

[480] Felpin, F. X.; Girard, S.; Vo-Thanh, G.; Robins, R. J.; Villieras, J.; Lebreton, J. *J. Org. Chem.* **2001**, *66*, 6305.

[481] Felpin, F. X.; Vo-Thanh, G.; Robins, R. J.; Villieras, J.; Lebreton, J. *Synlett* **2000**, 1646.

[482] Chen, G. M.; Brown, H. C.; Ramachandran, P. V. *J. Org. Chem.* **1999**, *64*, 721.

[483] Kimura, M.; Kiyama, I.; Tomizawa, Y.; Horino, Y.; Tanaka, S.; Tamaru, Y. *Tetrahedron Lett.* **1999**, *40*, 6795.

[484] Felpin, F. X.; Lebreton, J. *J. Org. Chem.* **2002**, *67*, 9192.

[485] Mears, R. J.; DeSilva, H.; Whiting, A. *Tetrahedron* **1997**, *53*, 17395.

[486] Thormeier, S.; Carboni, B.; Kaufmann, D. E. *J. Organomet. Chem.* **2002**, *657*, 136.

[487] Vorontsova, N. V.; Rozenberg, V. I.; Vorontsov, E. V.; Tok, O. L.; Bubnov, Y. N. *Russ. Chem. Bull.* **2000**, *49*, 912.

[488] Crosby, S. R.; Harding, J. R.; King, C. D.; Parker, G. D.; Willis, C. L. *Org. Lett.* **2002**, *4*, 577.

[489] Koide, K.; Finkelstein, J. M.; Ball, Z.; Verdine, G. L. *J. Am. Chem. Soc.* **2001**, *123*, 398.

[490] Crosby, S. R.; Harding, J. R.; King, C. D.; Parker, G. D.; Willis, C. L. *Org. Lett.* **2002**, *4*, 3407.

[491] Ramachandran, P. V.; Padiya, K. J.; Rauniyar, V.; Reddy, M. V. R.; Brown, H. C. *J. Fluorine Chem.* **2004**, *125*, 615.

[492] Wang, X. Z.; Burke, J., T. R. *Synlett* **2004**, 469.

[493] Couladouros, E. A.; Plyta, Z. F.; Strongilos, A. T.; Papageorgiou, V. P. *Tetrahedron Lett.* **1997**, *38*, 7263.

[494] Couladouros, E. A.; Strongilos, A. T.; Papageorgiou, V. P.; Plyta, Z. F. *Chem. Eur. J.* **2002**, *8*, 1795.

[495] Ku, Y. Y.; Grieme, T.; Raje, P.; Sharma, P.; King, S. A.; Morton, H. E. *J. Am. Chem. Soc.* **2002**, *124*, 4282.

[496] Ku, Y. Y.; Grieme, T. *Curr. Opin. Drug Discov. Devel.* **2002**, *5*, 852.

[497] Brown, H. C.; Randad, R. S. *Tetrahedron* **1990**, *46*, 4457.

[498] Hoffmann, R. W.; Ladner, W. *Tetrahedron Lett.* **1979**, 4653.

[499] Hoffmann, R. W.; Ladner, W.; Steinbach, K.; Massa, W.; Schmidt, R.; Snatzke, G. *Chem. Ber.* **1981**, *114*, 2786.

[500] Richter, P. K.; Tomaszewski, M. J.; Miller, R. A.; Patron, A. P.; Nicolaou, K. C. *J. Chem. Soc., Chem. Commun.* **1994**, 1151.

[501] N'Zoutani, M. A.; Pancrazi, A.; Ardisson, J. *Synlett* **2001**, 769.

[502] Ireland, R. E.; Highsmith, T. K.; Gegnas, L. D.; Gleason, J. L. *J. Org. Chem.* **1992**, *57*, 5071.

[503] Ramachandran, P. V.; Prabhudas, B.; Pratihar, D.; Chandra, J. S.; Reddy, M. V. R. *Tetrahedron Lett.* **2003**, *44*, 3745.

[504] La Cruz, T. E.; Rychnovsky, S. D. *Org. Lett.* **2005**, *7*, 1873.

[505] Ramachandran, P. V.; Chandra, J. S.; Prabhudas, B.; Pratihar, D.; Reddy, M. V. R. *Org. Biomol. Chem.* **2005**, *3*, 3812.

[506] Bates, R. B.; Gangwar, S. *Tetrahedron: Asymmetry* **1993**, *4*, 69.

[507] Wu, B.; Liu, Q.; Sulikowski, G. A. *Angew. Chem., Int. Ed.* **2004**, *43*, 6673.

[508] Brimble, M. A.; Furkert, D. P. *Org. Biomol. Chem.* **2004**, *2*, 3573.

[509] Brimble, M. A.; Fares, F. A.; Turner, P. *Aust. J. Chem.* **2000**, *53*, 845.

[510] Boeckman, R. K.; Charette, A. B.; Asberom, T.; Johnston, B. H. *J. Am. Chem. Soc.* **1987**, *109*, 7553.

[511] Anderson, O. P.; Barrett, A. G. M.; Edmunds, J. J.; Hachiya, S. I.; Hendrix, J. A.; Horita, K.; Malecha, J. W.; Parkinson, C. J.; VanSickle, A. *Can. J. Chem.* **2001**, *79*, 1562.

[512] Berque, I.; Le Menez, P.; Razon, P.; Mahuteau, J.; Ferezou, J. P.; Pancrazi, A.; Ardisson, J.; Brion, J. D. *J. Org. Chem.* **1999**, *64*, 373.

[513] Barrett, A. G. M.; Edmunds, J. J.; Hendrix, J. A.; Horita, K.; Parkinson, C. J. *J. Chem. Soc., Chem. Commun.* **1992**, 1238.

[514] Meilert, K.; Brimble, M. A. *Org. Lett.* **2005**, *7*, 3497.

[515] Smith, A. B., III; Friestad, G. K.; Barbosa, J.; Bertounesque, E.; Hull, K. G.; Iwashima, M.; Qiu, Y. P.; Salvatore, B. A.; Spoors, P. G.; Duan, J. J. W. *J. Am. Chem. Soc.* **1999**, *121*, 10468.

[516] Vidya, R.; Eggen, M.; Georg, G. I.; Himes, R. H. *Bioorg. Med. Chem. Lett.* **2003**, *13*, 757.

[517] Statsuk, A. V.; Liu, D.; Kozmin, S. A. *J. Am. Chem. Soc.* **2004**, *126*, 9546.

[518] Schinzer, D.; Schulz, C.; Krug, O. *Synlett* **2004**, 2689.

[519] Roush, W. R.; Ando, K.; Powers, D. B.; Halterman, R. L.; Palkowitz, A. D. *Tetrahedron Lett.* **1988**, *29*, 5579.

[520] Giordano, A.; Spinella, A.; Sodano, G. *Tetrahedron: Asymmetry* **1999**, *10*, 1851.

[521] Brinkmann, H.; Hoffmann, R. W. *Chem. Ber.* **1990**, *123*, 2395.

[522] Nakata, M.; Ishiyama, T.; Akamatsu, S.; Hirose, Y.; Maruoka, H.; Suzuki, R.; Tatsuta, K. *Bull. Chem. Soc. Jpn.* **1995**, *68*, 967.

[523] Ireland, R. E.; Wipf, P.; Roper, T. D. *J. Org. Chem.* **1990**, *55*, 2284.

[524] Roush, W. R.; Koyama, K.; Curtin, M. L.; Moriarty, K. J. *J. Am. Chem. Soc.* **1996**, *118*, 7502.

[525] Niel, G.; Roux, F.; Maisonnasse, Y.; Maugras, I.; Poncet, J.; Jouin, P. *J. Chem. Soc., Perkin Trans. 1* **1994**, 1275.

[526] Guo-qiang, L.; Wei-chu, X. *Tetrahedron Lett.* **1993**, *34*, 5931.

[527] Boons, G. J.; Brown, D. S.; Clase, J. A.; Lennon, I. C.; Ley, S. V. *Tetrahedron Lett.* **1994**, *35*, 319.

[528] Boons, G. J.; Clase, J. A.; Lennon, I. C.; Ley, S. V.; Staunton, J. *Tetrahedron* **1995**, *51*, 5417.

[529] Zampella, A.; Sorgente, M.; D'Auria, M. V. *Tetrahedron: Asymmetry* **2002**, *13*, 681.

[530] Ley, S. V.; Brown, D. S.; Clase, J. A.; Fairbanks, A. J.; Lennon, I. C.; Osborn, H. M. I.; Stokes, E. S. E.; Wadsworth, D. J. *J. Chem. Soc., Perkin Trans. 1* **1998**, 2259.

[531] Francavilla, C.; Chen, W.; Kinder, J., F. R. *Org. Lett.* **2003**, *5*, 1233.

[532] Nakata, M.; Ishiyama, T.; Hirose, Y.; Maruoka, H.; Tatsuta, K. *Tetrahedron Lett.* **1993**, *34*, 8439.

533 Scarlato, G. R.; DeMattei, J. A.; Chong, L. S.; Ogawa, A. K.; Lin, M. R.; Armstrong, R. W. *J. Org. Chem.* **1996**, *61*, 6139.

534 Czuba, I. R.; Zammit, S.; Rizzacasa, M. A. *Org. Biomol. Chem.* **2003**, *1*, 2044.

535 Czuba, I. R.; Rizzacasa, M. A. *Chem. Commun.* **1999**, 1419.

536 Zampella, A.; Bassarello, C.; Bifulco, G.; Gomez-Paloma, L.; D'Auria, M. V. *Eur. J. Org. Chem.* **2002**, 785.

537 Roush, W. R.; Adam, M. A.; Harris, D. J. *J. Org. Chem.* **1985**, *50*, 2000.

538 Brown, H. C.; Bhat, K. S.; Randad, R. S. *J. Org. Chem.* **1987**, *52*, 3701.

539 Hoffmann, R. W.; Brinkmann, H.; Frenking, G. *Chem. Ber.* **1990**, *123*, 2387.

540 Hoffmann, R. W.; Zeiss, H. J. *Angew. Chem., Int. Ed. Engl.* **1980**, *19*, 218.

541 Roux, F.; Maugras, I.; Poncet, J.; Niel, G.; Jouin, P. *Tetrahedron* **1994**, *50*, 5345.

542 Terauchi, T.; Morita, M.; Kimijima, K.; Nakamura, Y.; Hayashi, G.; Tanaka, T.; Kanoh, N.; Nakata, M. *Tetrahedron Lett.* **2001**, *42*, 5505.

543 Kobayashi, Y.; Lee, J.; Tezuka, K.; Kishi, Y. *Org. Lett.* **1999**, *1*, 2177.

544 Monica, C. D.; Maulucci, N.; De Riccardis, F.; Izzo, I. *Tetrahedron: Asymmetry* **2003**, *14*, 3371.

545 Wuts, P. G.; Bigelow, S. S. *J. Org. Chem.* **1988**, *53*, 5023.

546 Moret, E.; Schlosser, M. *Tetrahedron Lett.* **1984**, *25*, 4491.

547 Lipomi, D. J.; Langille, N. F.; Panek, J. S. *Org. Lett.* **2004**, *6*, 3533.

548 Roush, W. R.; Brown, B. B.; Drozda, S. E. *Tetrahedron Lett.* **1988**, *29*, 3541.

549 Roush, W. R.; Brown, B. B. *J. Am. Chem. Soc.* **1993**, *115*, 2268.

550 Satoh, M.; Nomoto, Y.; Miyaura, N.; Suzuki, A. *Tetrahedron Lett.* **1989**, *30*, 3789.

551 Wang, X.; Porco, J., J. A. *Angew. Chem., Int. Ed.* **2005**, *44*, 3067.

552 Nakata, M.; Ishiyama, T.; Akamatsu, S.; Suzuki, R.; Tatsuta, K. *Synlett* **1994**, 601.

553 Chemler, S. R.; Roush, W. R. *J. Org. Chem.* **2003**, *68*, 1319.

554 Turk, J. A.; Visbal, G. S.; Lipton, M. A. *J. Org. Chem.* **2003**, *68*, 7841.

555 Burke, S. D.; Hong, J.; Lennox, J. R.; Mongin, A. P. *J. Org. Chem.* **1998**, *63*, 6952.

556 Cundy, D. J.; Donohue, A. C.; McCarthy, T. D. *J. Chem. Soc., Perkin Trans. 1* **1999**, 559.

557 Bauer, S. M.; Armstrong, R. W. *J. Am. Chem. Soc.* **1999**, *121*, 6355.

558 Cundy, D. J.; Donohue, A. C.; McCarthy, T. D. *Tetrahedron Lett.* **1998**, *39*, 5125.

559 Zheng, W. J.; DeMattei, J. A.; Wu, J. P.; Duan, J. J. W.; Cook, L. R.; Oinuma, H.; Kishi, Y. *J. Am. Chem. Soc.* **1996**, *118*, 7946.

560 Knapp, S.; Dong, Y. *Tetrahedron Lett.* **1997**, *38*, 3813.

561 Ireland, R. E.; Armstrong, J. D.; Lebreton, J.; Meissner, R. S.; Rizzacasa, M. A. *J. Am. Chem. Soc.* **1993**, *115*, 7152.

562 Andrus, M. B.; Schreiber, S. L. *J. Am. Chem. Soc.* **1993**, *115*, 10420.

563 Ireland, R. E.; Gleason, J. L.; Gegnas, L. D.; Highsmith, T. K. *J. Org. Chem.* **1996**, *61*, 6856.

564 White, J. D.; Toske, S. G.; Yakura, T. *Synlett* **1994**, 591.

565 Hermitage, S. A.; Roberts, S. M.; Watson, D. J. *Tetrahedron Lett.* **1998**, *39*, 3567.

566 Inoue, M.; Yamashita, S.; Tatami, A.; Miyazaki, K.; Hirama, M. *J. Org. Chem.* **2004**, *69*, 2797.

567 Uehara, H.; Oishi, T.; Inoue, M.; Shoji, M.; Nagumo, Y.; Kosaka, M.; Le Brazidec, J. Y.; Hirama, M. *Tetrahedron* **2002**, *58*, 6493.

568 Oishi, T.; Nagumo, Y.; Shoji, M.; Le Brazidec, J. Y.; Uehara, H.; Hirama, M. *Chem. Commun.* **1999**.

569 Tatami, A.; Inoue, M.; Uehara, H.; Hirama, M. *Tetrahedron Lett.* **2003**, *44*, 5229.

570 Inoue, M.; Hirama, M. *Synlett* **2004**, 577.

571 Jiang, X.; García-Fortanet, J. G.; De Brabander, J. K. *J. Am. Chem. Soc.* **2005**, *127*, 11254.

572 Smith, A. B., III; Adams, C. M.; Kozmin, S. A.; Paone, D. V. *J. Am. Chem. Soc.* **2001**, *123*, 5925.

573 Nicolaou, K. C.; Murphy, F.; Barluenga, S.; Ohshima, T.; Wei, H.; Xu, J.; Gray, D. L. F.; Baudoin, O. *J. Am. Chem. Soc.* **2000**, *122*, 3830.

574 Nicolaou, K. C.; Xu, J. Y.; Murphy, F.; Barluenga, S.; Baudoin, O.; Wei, H. X.; Gray, D. L. F.; Ohshima, T. *Angew. Chem., Int. Ed.* **1999**, *38*, 2447.

575 Paterson, I.; Cumming, J. G.; Smith, J. D.; Ward, R. A.; Yeung, K. S. *Tetrahedron Lett.* **1994**, *35*, 3405.

[576] Paterson, I.; Ward, R. A.; Smith, J. D.; Cumming, J. G.; Yeung, K. S. *Tetrahedron* **1995**, *51*, 9437.

[577] Yamamoto, Y.; Yatagai, H.; Maruyama, K. *J. Am. Chem. Soc.* **1981**, *103*, 1969.

[578] Oikawa, M.; Ueno, T.; Oikawa, H.; Ichihara, A. *J. Org. Chem.* **1995**, *60*, 5048.

[579] Oikawa, M.; Ueno, T.; Oikawa, H.; Ichihara, A. *J. Synth. Org. Chem. Jpn.* **1995**, *53*, 1123.

[580] Oikawa, M.; Oikawa, H.; Ichihara, A. *Tetrahedron Lett.* **1993**, *34*, 4797.

[581] Hoffmann, R. W.; Helbig, W. *Chem. Ber.* **1981**, *114*, 2802.

[582] Giordano, A.; Della Monica, C.; Landi, F.; Spinella, A.; Sodano, G. *Tetrahedron Lett.* **2000**, *41*, 3979.

[583] Zampella, A.; D'Auria, M. V. *Tetrahedron: Asymmetry* **2001**, *12*, 1543.

[584] Roush, W. R.; Hertel, L.; Schnaderbeck, M. J.; Yakelis, N. A. *Tetrahedron Lett.* **2002**, *43*, 4885.

[585] White, J. D.; Hong, J.; Robarge, L. A. *Tetrahedron Lett.* **1998**, *39*, 8779.

[586] Feutrill, J. T.; Holloway, G. A.; Hilli, F.; Hugel, H. M.; Rizzacasa, M. A. *Tetrahedron Lett.* **2000**, *41*, 8569.

[587] Dhokte, U. P.; Khau, V. V.; Hutchison, D. R.; Martinelli, M. J. *Tetrahedron Lett.* **1998**, *39*, 8771.

[588] Eggen, M.; Mossman, C. J.; Buck, S. B.; Nair, S. K.; Bhat, L.; Ali, S. M.; Reiff, E. A.; Boge, T. C.; Georg, G. I. *J. Org. Chem.* **2000**, *65*, 7792.

[589] Eggen, M. J.; Georg, G. I. *Bioorg. Med. Chem. Lett.* **1998**, *8*, 3177.

[590] Andrus, M. B.; Argade, A. B. *Tetrahedron Lett.* **1996**, *37*, 5049.

[591] McRae, K. J.; Rizzacasa, M. A. *J. Org. Chem.* **1997**, *62*, 1196.

[592] Paterson, I.; Ashton, K.; Britton, R.; Knust, H. *Org. Lett.* **2003**, *5*, 1963.

[593] Paterson, I.; Britton, R.; Delgado, O.; Meyer, A.; Poullennec, K. G. *Angew. Chem., Int. Ed.* **2004**, *43*, 4629.

[594] Cuzzupe, A. N.; Hutton, C. A.; Lilly, M. J.; Mann, R. K.; McRae, K. J.; Zammit, S. C.; Rizzacasa, M. A. *J. Org. Chem.* **2001**, *66*, 2382.

[595] Nicolaou, K. C.; Bertinato, P.; Piscopio, A. D.; Chakraborty, T. K.; Minowa, N. *J. Chem. Soc., Chem. Commun.* **1993**, 619.

[596] Cossy, J.; Bauer, D.; Bellosta, V. *Tetrahedron* **2002**, *58*, 5909.

[597] Yu, W. S.; Zhang, Y.; Jin, Z. D. *Org. Lett.* **2001**, *3*, 1447.

[598] Yamamoto, Y.; Maruyama, K.; Komatsu, T.; Ito, W. *J. Org. Chem.* **1986**, *51*, 886.

[599] White, J. D.; Wang, G. Q.; Quaranta, L. *Org. Lett.* **2003**, *5*, 4109.

[600] Hermitage, S. A.; Murphy, A.; Nielsen, P.; Roberts, S. M. *Tetrahedron* **1998**, *54*, 13185.

[601] Zweifel, G.; Shoup, T. M. *J. Am. Chem. Soc.* **1988**, *110*, 5578.

[602] Coe, J. W.; Roush, W. R. *J. Org. Chem.* **1989**, *54*, 915.

[603] Liu, B.; Zhou, W. *Tetrahedron Lett.* **2003**, *44*, 4933.

[604] Jyojima, T.; Katohno, M.; Miyamoto, N.; Nakata, M.; Matsumura, S.; Toshima, K. *Tetrahedron Lett.* **1998**, *39*, 6003.

[605] Merifield, E.; Thomas, E. J. *J. Chem. Soc., Perkin Trans. 1* **1999**, 3269.

[606] Merifield, E.; Thomas, E. J. *J. Chem. Soc., Chem. Commun.* **1990**, 464.

[607] Menager, E.; Merifield, E.; Smallridge, M.; Thomas, E. J. *Tetrahedron* **1997**, *53*, 9377.

[608] Khandekar, G.; Robinson, G. C.; Stacey, N. A.; Steel, P. G.; Thomas, E. J.; Vather, S. *J. Chem. Soc., Chem. Commun.* **1987**, 877.

[609] Kang, S. H.; Jeong, J. W. *Tetrahedron Lett.* **2002**, *43*, 3613.

[610] Paterson, I.; Yeung, K. S.; Watson, C.; Ward, R. A.; Wallace, P. A. *Tetrahedron* **1998**, *54*, 11935.

[611] Zampella, A.; D'Auria, M. V. *Tetrahedron: Asymmetry* **2002**, *13*, 1237.

[612] Cossy, J.; Bauer, D.; Bellosta, V. *Synlett* **2002**, 715.

[613] Toshima, H.; Ichihara, A. *Biosci. Biotechnol. Biochem.* **1998**, *62*, 599.

[614] Minguez, J. M.; Kim, S. Y.; Giuliano, K. A.; Balachandran, R.; Madiraju, C.; Day, B. W.; Curran, D. P. *Bioorg. Med. Chem.* **2003**, *11*, 3335.

[615] Archibald, S. C.; Barden, D. J.; Bazin, J. F. Y.; Fleming, I.; Foster, C. F.; Mandal, A. K.; Parker, D.; Takaki, K.; Ware, A. C.; Williams, A. R. B.; Zwicky, A. B. *Org. Biomol. Chem.* **2004**, *2*, 1051.

[616] Stenkamp, D.; Hoffmann, R. W.; Göttlich, R. *Eur. J. Org. Chem.* **1999**, 2929.

[617] Scheidt, K. A.; Bannister, T. D.; Tasaka, A.; Wendt, M. D.; Savall, B. M.; Fegley, G. J.; Roush, W. R. *J. Am. Chem. Soc.* **2002**, *124*, 6981.

[618] Spino, C.; Allan, M. *Can. J. Chem.* **2004**, *82*, 177.

[619] White, J. D.; Tiller, T.; Ohba, Y.; Porter, W. J.; Jackson, R. W.; Wang, S.; Hanselmann, R. *Chem. Commun.* **1998**, 79.

[620] Williams, D. R.; Rojas, C. M.; Bogen, S. L. *J. Org. Chem.* **1999**, *64*, 736.

[621] Smith, A. B., III; Zheng, J. *Tetrahedron* **2002**, *58*, 6455.

[622] Smith, A. B., III; Zheng, J. *Synlett* **2001**, 1019.

[623] Ooi, H.; Ishibashi, N.; Iwabuchi, Y.; Ishihara, J.; Hatakeyama, S. *J. Org. Chem.* **2004**, *69*, 7765.

[624] Kang, S. H.; Kim, C. M. *Synlett* **1996**, 515.

[625] Roush, W. R.; Bannister, T. D. *Tetrahedron Lett.* **1992**, *33*, 3587.

[626] Roush, W. R.; Bannister, T. D.; Wendt, M. D.; Jablonowski, J. A.; Scheidt, K. A. *J. Org. Chem.* **2002**, *67*, 4275.

[627] Kende, A. S.; Koch, K.; Dorey, G.; Kaldor, I.; Liu, K. *J. Am. Chem. Soc.* **1993**, *115*, 9842.

[628] Andersen, M. W.; Hildebrandt, B.; Hoffmann, R. W. *Angew. Chem., Int. Ed. Engl.* **1991**, *30*, 97.

[629] Oikawa, H.; Yoneta, Y.; Ueno, T.; Oikawa, M.; Wakayama, T.; Ichihara, A. *Tetrahedron Lett.* **1997**, *38*, 7897.

[630] Oikawa, H. *Curr. Med. Chem.* **2002**, *9*, 2033.

[631] Paterson, I.; Anderson, E. A.; Dalby, S. M.; Loiseleur, O. *Org. Lett.* **2005**, *7*, 4121.

[632] Fisher, M. J.; Myers, C. D.; Joglar, J.; Chen, S. H.; Danishefsky, S. J. *J. Org. Chem.* **1991**, *56*, 5826.

[633] Labrecque, D.; Charron, S.; Rej, R.; Blais, C.; Lamothe, S. *Tetrahedron Lett.* **2001**, *42*, 2645.

[634] Barrett, A. G. M.; Lebold, S. A. *J. Org. Chem.* **1991**, *56*, 4875.

[635] Barrett, A. G. M.; Lebold, S. A. *J. Org. Chem.* **1990**, *55*, 5818.

[636] Sebelius, S.; Wallner, O. A.; Szabó, K. J. *Org. Lett.* **2003**, *5*, 3065.

[637] Dineen, T. A.; Roush, W. R. *Org. Lett.* **2003**, *5*, 4725.

[638] Maurer, K. W.; Armstrong, R. W. *J. Org. Chem.* **1996**, *61*, 3106.

[639] Lin, G. Q.; Xu, W. C. *Tetrahedron* **1996**, *52*, 5907.

[640] Kang, S. H.; Jun, H. S.; Youn, J. H. *Synlett* **1998**, 1045.

[641] Zampella, A.; Sepe, V.; D'Orsi, R.; D'Auria, M. V. *Lett. Org. Chem.* **2004**, 308.

[642] Hori, K.; Kazuno, H.; Nomura, K.; Yoshii, E. *Tetrahedron Lett.* **1993**, *34*, 2183.

[643] Harada, T.; Kagamihara, Y.; Tanaka, S.; Sakamoto, K.; Oku, A. *J. Org. Chem.* **1992**, *57*, 1637.

[644] White, J. D.; Johnston, A. T. *J. Org. Chem.* **1994**, *59*, 3347.

[645] White, J. D.; Johnston, A. T. *J. Org. Chem.* **1990**, *55*, 5938.

[646] Hoffmann, R. W.; Göttlich, R.; Schopfer, U. *Eur. J. Org. Chem.* **2001**, 1865.

[647] Chen, S. H.; Horvath, R. F.; Joglar, J.; Fisher, M. J.; Danishefsky, S. J. *J. Org. Chem.* **1991**, *56*, 5834.

[648] May, S. A.; Grieco, P. A. *Chem. Commun.* **1998**, 1597.

[649] Tatsuta, K.; Ishiyama, T.; Tajima, S.; Koguchi, Y.; Gunji, H. *Tetrahedron Lett.* **1990**, *31*, 709.

[650] Andrus, M. B.; Turner, T. M.; Sauna, Z. E.; Ambudkar, S. V. *J. Org. Chem.* **2000**, *65*, 4973.

[651] Hu, H. S.; Jayaraman, S.; Oehlschage, A. C. *J. Org. Chem.* **1998**, *63*, 8843.

[652] Roush, W. R.; Lane, G. C. *Org. Lett.* **1999**, *1*, 95.

[653] Smith, A. B., III; Adams, C. M.; Barbosa, S. A. L.; Degnan, A. P. *Proc. Natl. Acad. Sci. U.S.A.* **2004**, *101*, 12042.

[654] Roush, W. R.; Coffey, D. S.; Madar, D.; Palkowitz, A. D. *J. Braz. Chem. Soc.* **1996**, *7*, 327.

[655] Sengoku, T.; Arimoto, H.; Uemura, D. *Chem. Commun.* **2004**, 1220.

[656] Andrus, M. B.; Lepore, S. D.; Turner, T. M. *J. Am. Chem. Soc.* **1997**, *119*, 12159.

[657] Akita, H.; Yamada, H.; Matsukura, H.; Nakata, T.; Oishi, T. *Tetrahedron Lett.* **1990**, *31*, 1735.

[658] Oishi, T.; Akita, H. *J. Synth. Org. Chem. Jpn.* **1991**, *49*, 657.

[659] Zeng, X.; Zeng, F.; Negishi, E. *Org. Lett.* **2004**, *6*, 3245.

[660] Andrus, M. B.; Lepore, S. D. *J. Am. Chem. Soc.* **1997**, *119*, 2327.

[661] Nagamitsu, T.; Sunazuka, T.; Tanaka, H.; Omura, S.; Sprengeler, P. A.; Smith, A. B., III *J. Am. Chem. Soc.* **1996**, *118*, 3584.

[662] Sunazuka, T.; Nagamitsu, T.; Matsuzaki, K.; Tanaka, H.; Omura, S.; Smith, A. B., III *J. Am. Chem. Soc.* **1993**, *115*, 5302.

[663] Kohyama, N.; Yamamoto, Y. *Synlett* **2001**, 694.

[664] Danishefsky, S. J.; Armistead, D. M.; Wincott, F. E.; Selnick, H. G.; Hungate, R. *J. Am. Chem. Soc.* **1989**, *111*, 2967.

[665] Danishefsky, S. J.; Armistead, D. M.; Wincott, F. E.; Selnick, H. G.; Hungate, R. *J. Am. Chem. Soc.* **1987**, *109*, 8117.

[666] Liu, B.; Zhou, W. S. *Tetrahedron* **2003**, *59*, 3379.

[667] Liu, B.; Zhou, W. *Tetrahedron Lett.* **2002**, *43*, 4187.

[668] Hoffmann, R. W.; Landmann, B. *Tetrahedron Lett.* **1983**, *24*, 3209.

[669] Hoffmann, R. W.; Landmann, B. *Angew. Chem., Int. Ed. Engl.* **1984**, *23*, 437.

[670] Mandal, A. K.; Schneekloth, J., J. S.; Crews, C. M. *Org. Lett.* **2005**, *7*, 3645.

[671] Yang, Z.; He, Y.; Vourloumis, D.; Vallberg, H.; Nicolaou, K. C. *Angew. Chem., Int. Ed. Engl.* **1997**, *36*, 166.

[672] Schlapbach, A.; Hoffmann, R. W. *Eur. J. Org. Chem.* **2001**, 323.

[673] Hoffmann, R. W.; Schlapbach, A. *Liebigs Ann. Chem.* **1991**, 1203.

[674] Midland, M. M.; Preston, S. B. *J. Am. Chem. Soc.* **1982**, *104*, 2330.

[675] Shimizu, M.; Kitagawa, H.; Kurahashi, T.; Hiyama, T. *Angew. Chem., Int. Ed.* **2001**, *40*, 4283.

[676] Andersen, M. W.; Hildebrandt, B.; Koster, G.; Hoffmann, R. W. *Chem. Ber.* **1989**, *122*, 1777.

[677] Zheng, B.; Srebnik, M. *J. Org. Chem.* **1995**, *60*, 486.

[678] Hoffmann, R. W.; Dahmann, G.; Andersen, M. W. *Synthesis* **1994**, 629.

[679] Hoffmann, R. W.; Dahmann, G. *Tetrahedron Lett.* **1993**, *34*, 1115.

[680] Lombardo, M.; Morganti, S.; Trombini, C. *J. Org. Chem.* **2000**, *65*, 8767.

[681] Hoffmann, R. W.; Stürmer, R. *Chem. Ber.* **1994**, *127*, 2511.

[682] Yamamoto, Y.; Yatagai, H.; Maruyama, K. *J. Am. Chem. Soc.* **1981**, *103*, 3229.

[683] Andemichael, Y. W.; Wang, K. K. *J. Org. Chem.* **1992**, *57*, 796.

[684] Wang, K. K.; Gu, G.; Liu, C. *J. Am. Chem. Soc.* **1990**, *112*, 4424.

[685] Yamaguchi, M.; Mukaiyama, T. *Chem. Lett.* **1980**, 993.

[686] Patron, A. P.; Richter, P. K.; Tomaszewski, M. J.; Miller, R. A.; Nicolaou, K. C. *J. Chem. Soc., Chem. Commun.* **1994**, 1147.

[687] Hoffmann, R. W.; Rolle, U.; Göttlich, R. *Liebigs Ann. Chem.* **1996**, 1717.

[688] Stürmer, R.; Ritter, K.; Hoffmann, R. W. *Angew. Chem., Int. Ed. Engl.* **1993**, *32*, 101.

[689] Tsai, D. J. S.; Matteson, D. S. *Organometallics* **1983**, *2*, 236.

[690] Carosi, L.; Lachance, H.; Hall, D. G. *Tetrahedron Lett.* **2005**, *46*, 8981.

[691] Beckmann, E.; Desai, V.; Hoppe, D. *Synlett* **2004**, 2275.

[692] Yamamoto, Y.; Saito, Y.; Maruyama, K. *J. Org. Chem.* **1983**, *48*, 5408.

[693] Hoffmann, R. W.; Ladner, W.; Ditrich, K. *Liebigs Ann. Chem.* **1989**, 883.

[694] Hoffmann, R. W.; Rolle, U. *Tetrahedron Lett.* **1994**, *35*, 4751.

[695] Hoffmann, R. W.; Haeberlin, E.; Rohde, T. *Synthesis* **2002**, 207.

[696] Stürmer, R.; Hoffmann, R. W. *Chem. Ber.* **1994**, *127*, 2519.

[697] Brown, H. C.; Jadhav, P. K.; Perumal, P. T. *Tetrahedron Lett.* **1984**, *25*, 5111.

[698] Bubnov, Y. N.; Etinger, M. Y. *Tetrahedron Lett.* **1985**, *26*, 2797.

[699] Brown, H. C.; Randad, R. S. *Tetrahedron* **1990**, *46*, 4463.

[700] Brown, H. C.; Randad, R. S. *Tetrahedron Lett.* **1990**, *31*, 455.

[701] Salunkhe, A. M.; Ramachandran, P. V.; Brown, H. C. *Tetrahedron Lett.* **1999**, *40*, 1433.

[702] Gurskii, M. E.; Golovin, S. B.; Ignatenko, A. V.; Bubnov, Y. N. *Russ. Chem. Bull.* **1993**, *42*, 139.

[703] Chataigner, I.; Lebreton, J.; Zammattio, F.; Villieras, J. *Tetrahedron Lett.* **1997**, *38*, 3719.

[704] van der Heide, T. A. J.; van der Baan, J. L.; Bijpost, E. A.; de Kanter, F. J. J.; Bickelhaupt, F.; Klumpp, G. W. *Tetrahedron Lett.* **1993**, *34*, 4655.

[705] Barrett, A. G. M.; Braddock, D. C.; de Koning, P. D. *Chem. Commun.* **1999**, 459.

[706] Gurskii, M. E.; Golovin, S. B.; Bubnov, Y. N. *Russ. Chem. Bull.* **1990**, *39*, 433.

[707] Barrett, A. G. M.; Wan, P. W. H. *J. Org. Chem.* **1996**, *61*, 8667.

[708] Williams, D. R.; Meyer, K. G.; Shamim, K.; Patnaik, S. *Can. J. Chem.* **2004**, *82*, 120.

[709] Hara, S.; Suzuki, A. *Tetrahedron Lett.* **1991**, *32*, 6749.
[710] Williams, D. R.; Brooks, D. A.; Meyer, K. G.; Clark, M. P. *Tetrahedron Lett.* **1998**, *39*, 7251.
[711] Yamamoto, Y.; Hara, S.; Suzuki, A. *Synth. Commun.* **1997**, *27*, 1029.
[712] Hara, S.; Yamamoto, Y.; Fujita, A.; Suzuki, A. *Synlett* **1994**, 639.
[713] Draillard, K.; Lebreton, J.; Villieras, J. *Tetrahedron: Asymmetry* **1999**, *10*, 4281.
[714] Brzezinski, L. J.; Leahy, J. W. *Tetrahedron Lett.* **1998**, *39*, 2039.
[715] Williams, D. R.; Clark, M. P.; Berliner, M. A. *Tetrahedron Lett.* **1999**, *40*, 2287.
[716] Williams, D. R.; Kiryanov, A. A.; Emde, U.; Clark, M. P.; Berliner, M. A.; Reeves, J. T. *Proc. Natl. Acad. Sci. U.S.A.* **2004**, *101*, 12058.
[717] Williams, D. R.; Kiryanov, A. A.; Emde, U.; Clark, M. P.; Berliner, M. A.; Reeves, J. T. *Angew. Chem., Int. Ed.* **2003**, *42*, 1258.
[718] Ishiyama, T.; Kitano, T.; Miyaura, N. *Tetrahedron Lett.* **1998**, *39*, 2357.
[719] Goujon, J. Y.; Zammattio, F.; Chretien, J. M.; Beaudet, I. *Tetrahedron* **2004**, *60*, 4037.
[720] Williams, D. R.; Clark, M. P.; Emde, U.; Berliner, M. A. *Org. Lett.* **2000**, *2*, 3023.
[721] Williams, D. R.; Meyer, K. G. *J. Am. Chem. Soc.* **2001**, *123*, 765.
[722] Zhai, H. B.; Luo, S. J.; Ye, C. F.; Ma, Y. X. *J. Org. Chem.* **2003**, *68*, 8268.
[723] Williams, D. R.; Patnaik, S.; Plummer, S. V. *Org. Lett.* **2003**, *5*, 5035.
[724] Williams, D. R.; Brooks, D. A.; Berliner, M. A. *J. Am. Chem. Soc.* **1999**, *121*, 4924.
[725] Corey, E. J.; Huang, H. C. *Tetrahedron Lett.* **1989**, *30*, 5235.
[726] Kim, S.; Sutton, S. C.; Guo, C.; LaCour, T. G.; Fuchs, P. L. *J. Am. Chem. Soc.* **1999**, *121*, 2056.
[727] Kim, S. K.; Sutton, S. C.; Fuchs, P. L. *Tetrahedron Lett.* **1995**, *36*, 2427.
[728] Thaper, R. K.; Zykov, A. Y.; Bubnov, Y. N.; Reshetova, I. G.; Kamernitsky, A. V. *Synth. Commun.* **1993**, *23*, 939.
[729] Jernelius, J. A.; Schrock, R. R.; Hoveyda, A. H. *Tetrahedron* **2004**, *60*, 7345.
[730] Barrett, A. G. M.; Seefeld, M. A. *Tetrahedron* **1993**, *49*, 7857.
[731] Barrett, A. G. M.; Seefeld, M. A. *J. Chem. Soc., Chem. Commun.* **1993**, 339.
[732] Gurskii, M. E.; Geiderickh, A. V.; Ignatenko, A. V.; Bubnov, Y. N. *Russ. Chem. Bull.* **1993**, *42*, 144.
[733] Narla, G.; Brown, H. C. *Tetrahedron Lett.* **1997**, *38*, 219.
[734] Roush, W. R.; Pinchuk, A. N.; Micalizio, G. C. *Tetrahedron Lett.* **2000**, *41*, 9413.
[735] Wang, K. K.; Liu, C. *J. Org. Chem.* **1986**, *51*, 4733.
[736] Wang, K. K.; Liu, C.; Gu, Y. G.; Burnett, F. N.; Sattsangi, P. D. *J. Org. Chem.* **1991**, *56*, 1914.
[737] Hoffmann, R. W.; Kemper, B. *Tetrahedron Lett.* **1980**, *21*, 4883.
[738] Brown, H. C.; Phadke, A. S. *Synlett* **1993**, 927.
[739] Hoffmann, R. W.; Metternich, R. *Liebigs Ann. Chem.* **1985**, 2390.
[740] Yamamoto, Y.; Hara, S.; Suzuki, A. *Synlett* **1996**, 883.
[741] Soundararajan, R.; Li, G. S.; Brown, H. C. *J. Org. Chem.* **1996**, *61*, 100.
[742] Barrett, A. G. M.; Beall, J. C.; Braddock, D. C.; Flack, K.; Gibson, V. C.; Salter, M. M. *J. Org. Chem.* **2000**, *65*, 6508.
[743] Micalizio, G. C.; Roush, W. R. *Org. Lett.* **2000**, *2*, 461.
[744] Soundararajan, R.; Li, G.; Brown, H. C. *Tetrahedron Lett.* **1995**, *36*, 2441.
[745] Coleman, R. S. *Synlett* **1998**, 1031.
[746] Barrett, A. G. M.; Seefeld, M. A.; White, A. J. P.; Williams, D. J. *J. Org. Chem.* **1996**, *61*, 2677.
[747] Ganesh, P.; Nicholas, K. M. *J. Org. Chem.* **1997**, *62*, 1737.
[748] Wang, X.; Porco, J. A. *J. Am. Chem. Soc.* **2003**, *125*, 6040.
[749] Schinzer, D.; Limberg, A.; Bohm, O. M. *Chem. Eur. J.* **1996**, *2*, 1477.
[750] Wuts, P. G. M.; Bigelow, S. S. *J. Org. Chem.* **1983**, *48*, 3489.
[751] Gaddoni, L.; Lombardo, M.; Trombini, C. *Tetrahedron Lett.* **1998**, *39*, 7571.
[752] Barrett, A. G. M.; Malecha, J. W. *J. Chem. Soc., Perkin Trans. 1* **1994**, 1901.
[753] Hertweck, C.; Boland, W. *J. Org. Chem.* **1999**, *64*, 4426.
[754] Hertweck, C.; Goerls, H.; Boland, W. *Chem. Commun.* **1998**, 1955.
[755] Roush, W. R.; Michaelides, M. R. *Tetrahedron Lett.* **1986**, *27*, 3353.
[756] Hu, S. J.; Jayaraman, S.; Oehlschlager, A. C. *J. Org. Chem.* **1998**, *63*, 8843.
[757] Hunt, J. A.; Roush, W. R. *Tetrahedron Lett.* **1995**, *36*, 501.

[758] Roush, W. R.; Grover, P. T.; Marron, T. G. In *Current Topics in the Chemistry of Boron*, Proceedings of the Eighth International Meeting, Knoxville, Tennessee, 1993; Kabalka, G. W., Ed.; Royal Society of Chemistry: Cambridge, U.K., 1994.

[759] Roush, W. R.; Follows, B. C. *Tetrahedron Lett.* **1994**, *35*, 4935.

[760] Heo, J. N.; Micalizio, G. C.; Roush, W. R. *Org. Lett.* **2003**, *5*, 1693.

[761] Chang, K. J.; Rayabarapu, D. K.; Yang, F. Y.; Cheng, C. H. *J. Am. Chem. Soc.* **2005**, *127*, 126.

[762] Jayaraman, S.; Hu, S.; Oehlschlager, A. C. *Tetrahedron Lett.* **1995**, *36*, 4765.

[763] Micalizio, G. C.; Pinchuk, A. N.; Roush, W. R. *J. Org. Chem.* **2000**, *65*, 8730.

[764] Barrett, A. G. M.; Bennett, A. J.; Menzer, S.; Smith, M. L.; White, A. J. P.; Williams, D. J. *J. Org. Chem.* **1999**, *64*, 162.

[765] Barluenga, S.; Lopez, P.; Moulin, E.; Winssinger, N. *Angew. Chem., Int. Ed.* **2004**, *43*, 3467.

[766] Roush, W. R.; Michaelides, M. R.; Tai, D. F.; Chong, W. K. M. *J. Am. Chem. Soc.* **1987**, *109*, 7575.

[767] Roush, W. R.; Michaelides, M. R.; Tai, D. F.; Lesur, B. M.; Chong, W. K. M.; Harris, D. J. *J. Am. Chem. Soc.* **1989**, *111*, 2984.

[768] Roush, W. R.; Harris, D. J.; Lesur, B. M. *Tetrahedron Lett.* **1983**, *24*, 2227.

[769] Smith, A. L.; Pitsinos, E. N.; Hwang, C. K.; Mizuno, Y.; Saimoto, H.; Scarlato, G. R.; Suzuki, T.; Nicolaou, K. C. *J. Am. Chem. Soc.* **1993**, *115*, 7612.

[770] Smith, A. L.; Hwang, C. K.; Pitsinos, E.; Scarlato, G. R.; Nicolaou, K. C. *J. Am. Chem. Soc.* **1992**, *114*, 3134.

[771] Burgess, K.; Henderson, I. *Tetrahedron Lett.* **1990**, *31*, 6949.

[772] Sato, M.; Yamamoto, Y.; Hara, S.; Suzuki, A. *Tetrahedron Lett.* **1993**, *34*, 7071.

[773] Duan, J. J. W.; Smith, A. B., III *J. Org. Chem.* **1993**, *58*, 3703.

[774] Heo, J. N.; Holson, E. B.; Roush, W. R. *Org. Lett.* **2003**, *5*, 1697.

[775] Burgess, K.; Chaplin, D. A.; Henderson, I.; Pan, Y. T.; Elbein, A. D. *J. Org. Chem.* **1992**, *57*, 1103.

[776] Jadhav, P. K.; Woemer, F. J. *Tetrahedron Lett.* **1994**, *35*, 8973.

[777] Gu, Y. G.; Wang, K. K. *Tetrahedron Lett.* **1991**, *32*, 3029.

[778] Sutherlin, D. P.; Armstrong, R. W. *Tetrahedron Lett.* **1993**, *34*, 4897.

[779] Kim, J. D.; Kim, I. S.; Hua, J. C.; Zee, O. P.; Jung, Y. H. *Tetrahedron Lett.* **2005**, *46*, 1079.

[780] Moriya, T.; Suzuki, A.; Miyaura, N. *Tetrahedron Lett.* **1995**, *36*, 1887.

[781] Suginome, M.; Nakamura, H.; Matsuda, T.; Ito, Y. *J. Am. Chem. Soc.* **1998**, *120*, 4248.

[782] Ramachandran, P. V.; Pratihar, D.; Biswas, D.; Srivastava, A.; Reddy, M. V. R. *Org. Lett.* **2004**, *6*, 481.

[783] Yamamoto, Y.; Yatagai, H.; Maruyama, K. *J. Chem. Soc., Chem. Commun.* **1980**, 1072.

[784] Kabalka, G. W.; Venkataiah, B.; Dong, G. *J. Org. Chem.* **2004**, *69*, 5807.

[785] Kabalka, G. W.; Venkataiah, B.; Dong, G. *Tetrahedron Lett.* **2005**, *46*, 4209.

[786] Coleman, R. S.; Gurrala, S. R.; Mitra, S.; Raao, A. *J. Org. Chem.* **2005**, *70*, 8932.

[787] Coleman, R. S.; Gurrala, S. R. *Org. Lett.* **2005**, *7*, 1849.

[788] Micalizio, G. C.; Roush, W. R. *Org. Lett.* **2001**, *3*, 1949.

[789] Hertweck, C.; Boland, W. *Tetrahedron* **1997**, *53*, 14651.

[790] Hertweck, C.; Boland, W. *J. Org. Chem.* **2000**, *65*, 2458.

[791] Suginome, M.; Yamamoto, Y.; Fujii, K.; Ito, Y. *J. Am. Chem. Soc.* **1995**, *117*, 9608.

[792] Jadhav, P. K.; Man, H. W. *Tetrahedron Lett.* **1996**, *37*, 1153.

[793] Taylor, R. E.; Hearn, B. R.; Ciavarri, J. P. *Org. Lett.* **2002**, *4*, 2953.

[794] Nakata, M.; Osumi, T.; Ueno, A.; Kimura, T.; Tamai, T.; Tatsuta, K. *Tetrahedron Lett.* **1991**, *32*, 6015.

[795] Nakata, M.; Osumi, T.; Ueno, A.; Kimura, T.; Tamai, T.; Tatsuta, K. *Bull. Chem. Soc. Jpn.* **1992**, *65*, 2974.

[796] Zhang, Q.; Lu, H.; Richard, C.; Curran, D. P. *J. Am. Chem. Soc.* **2004**, *126*, 36.

[797] Kabalka, G. W.; Venkataiah, B. *Tetrahedron Lett.* **2005**, *46*, 7325.

[798] Carter, K. D.; Panek, J. S. *Org. Lett.* **2004**, *6*, 55.

[799] Harried, S. S.; Lee, C. P.; Yang, G.; Lee, T. I. H.; Myles, D. C. *J. Org. Chem.* **2003**, *68*, 6646.

[800] Harried, S. S.; Yang, G.; Strawn, M. A.; Myles, D. C. *J. Org. Chem.* **1997**, *62*, 6098.

[801] Zakarian, A.; Batch, A.; Holton, R. A. *J. Am. Chem. Soc.* **2003**, *125*, 7822.
[802] Carrié, D.; Carboni, B.; Vaultier, M. *Tetrahedron Lett.* **1995**, *36*, 8209.
[803] Favre, E.; Gaudemar, M. *J. Organomet. Chem.* **1975**, *92*, 17.
[804] Kulkarni, S. V.; Brown, H. C. *Tetrahedron Lett.* **1996**, *37*, 4125.
[805] Brown, H. C.; Khire, U. R.; Narla, G. *J. Org. Chem.* **1995**, *60*, 8130.
[806] Hernandez, E.; Soderquist, J. A. *Org. Lett.* **2005**, *7*, 5397.
[807] Zweifel, G.; Backlund, S. J.; Leung, T. *J. Am. Chem. Soc.* **1978**, *100*, 5561.
[808] Brown, H. C.; Khire, U. R.; Racherla, U. S. *Tetrahedron Lett.* **1993**, *34*, 15.
[809] Brown, H. C.; Khire, U. R.; Narla, G.; Racherla, U. S. *J. Org. Chem.* **1995**, *60*, 544.
[810] Favre, E.; Gaudemar, M. *J. Organomet. Chem.* **1974**, *76*, 297.
[811] Wang, K. K.; Nikam, S. S.; Ho, C. D. *J. Org. Chem.* **1983**, *48*, 5376.
[812] Maddess, M. L.; Lautens, M. *Org. Lett.* **2005**, *7*, 3557.
[813] Ikeda, N.; Arai, I.; Yamamoto, H. *J. Am. Chem. Soc.* **1986**, *108*, 483.
[814] Favre, E.; Gaudemar, M. *J. Organomet. Chem.* **1974**, *76*, 305.
[815] Wang, K. K.; Liu, C. *J. Org. Chem.* **1985**, *50*, 2578.
[816] Corey, E. J.; Jones, G. B. *Tetrahedron Lett.* **1991**, *32*, 5713.
[817] Matsumoto, Y.; Naito, M.; Uozumi, Y.; Hayashi, T. *J. Chem. Soc., Chem. Commun.* **1993**, 1468.
[818] Hilt, G.; Luers, S.; Smolko, K. I. *Org. Lett.* **2005**, *7*, 251.
[819] Herberich, G. E.; Englert, U.; Wang, S. *Chem. Ber.* **1993**, *126*, 297.
[820] Brown, H. C.; Bhat, K. S.; Jadhav, P. K. *J. Chem. Soc., Perkin Trans. 1* **1991**, 2633.
[821] Lallemand, J. Y.; Six, Y.; Ricard, L. *Eur. J. Org. Chem.* **2002**, 503.
[822] Renard, J. Y.; Lallemand, J. Y. *Bull. Soc. Chim. Fr.* **1996**, *133*, 143.
[823] Renard, P. Y.; Lallemand, J. Y. *Tetrahedron: Asymmetry* **1996**, *7*, 2523.
[824] Ohanessian, G.; Six, Y.; Lallemand, J. Y. *Bull. Soc. Chim. Fr.* **1996**, *133*, 1143.
[825] Gao, X.; Hall, D. G. *Tetrahedron Lett.* **2003**, *44*, 2231.
[826] Schleth, F.; Vogler, T.; Harms, K.; Studer, A. *Chem. Eur. J.* **2004**, *10*, 4171.
[827] Brown, H. C.; Jayaraman, S. *Tetrahedron Lett.* **1993**, *34*, 3997.
[828] Deligny, M.; Carreaux, F.; Carboni, B.; Toupet, L.; Dujardin, G. *Chem. Commun.* **2003**, 276.
[829] Deligny, M.; Carreaux, F.; Carboni, B. *Synlett* **2005**, 1462.
[830] Grieco, P. A.; Dai, Y. J. *J. Am. Chem. Soc.* **1998**, *120*, 5128.
[831] Hoffmann, R. W.; Munster, I. *Tetrahedron Lett.* **1995**, *36*, 1431.
[832] Hoffmann, R. W.; Munster, I. *Liebigs Ann. Rec.* **1997**, 1143.
[833] Hoffmann, R. W.; Sander, T. *Liebigs Ann. Chem.* **1993**, 1185.
[834] Kruger, J.; Hoffmann, R. W. *J. Am. Chem. Soc.* **1997**, *119*, 7499.

Supplemental References for Table 1A.

[835] Allais, F.; Ducrot, P-H. *Synthesis* **2010**, *10*, 1649.
[836] Álvarez-Bercedo, P.; Falomir, E.; Murga, J.; Carda, M.; Marco, J. A. *Eur. J. Org. Chem.* **2008**, 4015.
[837] Barker, T. J.; Jarvo, E. R. *Org. Lett.* **2009**, *11*, 1047.
[838] Barnett, D. S.; Moquist, P. N.; Schaus, S. E. *Angew. Chem., Int. Ed.* **2009**, *48*, 8679.
[839] Boshra, R.; Doshi, A.; Jäkle, F. *Angew. Chem., Int. Ed.* **2008**, *47*, 1134.
[840] Brimble, M. A.; Bachu, P.; Sperry, J. *Synthesis* **2007**, *18*, 2887.
[841] Burgos, C. H.; Canales, E.; Matos, K.; Soderquist, J. A. *J. Am. Chem. Soc.* **2005**, *127*, 8044.
[842] Canales, E.; Prasad, K. G.; Soderquist, J. A. *J. Am. Chem. Soc.* **2005**, *127*, 11572.
[843] Cardona, W.; Quiñones, W.; Robledo, S.; Vélez, I. D.; Murga, J.; García-Fortanet, J.; Carda, M.; Cardona, D.; Echeverri, F. *Tetrahedron* **2006**, *62*, 4086.
[844] Castoldi, D.; Caggiano, L.; Bayón, P.; Costa, A. M.; Cappella, P.; Sharon, O.; Gennari, C. *Tetrahedron* **2005**, *61*, 2123.
[845] Chen, W; Liu, Y.; Chen, Z. *Eur. J. Org. Chem.* **2005**, 1665.
[846] Chen, Z.; Sinha, S. C. *Tetrahedron* **2008**, *64*, 1603.
[847] Chen, Z.; Ye, T. *Synlett* **2005**, *18*, 2781.
[848] Ding, F.; Jennings, M. P. *Org. Lett.* **2005**, *7*, 2321.

849 Fandrick, K. R.; Fandrick, D. R.; Gao, J. J.; Reeves, J. T.; Tan, Z.; Li, W.; Song, J. J.; Lu, B.; Yee, N. K.; Senanayake, C. H. *Org. Lett.* **2010**, *12*, 3748.

850 García-Fortanet, J.; Murga, J.; Carda, M.; Marco, J. A. *Arkivoc* **2005**, *ix*, 175.

851 García-Fortanet, J.; Murga, J.; Carda, M.; Marco, J. A. *Tetrahedron* **2004**, *60*, 12261.

852 Hiebel, M-A.; Pelotier, B.; Pia, O. *Tetrahedron Lett.* **2010**, *51*, 5091.

853 Jain, P.; Antilla, J. C. *J. Am. Chem. Soc.* **2010**, *132*, 11884.

854 Jennings, M. P.; Clemens, R. T. *Tetrahedron Lett.* **2005**, *46*, 2021.

855 Jirgensons, A.; Marinozzi, M.; Pellicciari, R. *Tetrahedron* **2005**, *61*, 373.

856 Kawai, N.; Hande, S. M.; Uenishi, J. *Tetrahedron* **2007**, *63*, 9049.

857 Liu, S.; Fan, Y.; Peng, X.; Wang, W.; Hua, W.; Akber, H.; Liao, L. *Tetrahedron Lett.* **2006**, *47*, 7681.

858 Lou, S.; Moquist, P. N.; Schaus, S. E. *J. Am. Chem. Soc.* **2006**, *128*, 12660.

859 Mamane, V.; García, A. B.; Umarye, J. D.; Lessmann, T.; Sommer, S.; Waldmann, H. *Tetrahedron* **2007**, *63*, 5754.

860 Martinez-Solorio, D.; Jennings, M. P. *J. Org. Chem.* **2007**, *72*, 6621.

861 Matsuoka, H.; Kondo, K. *Tetrahedron Lett.* **2009**, *50*, 2320.

862 Mitton-Fry, J. J.; Cullen, A. J.; Sammakia, T. *Angew. Chem., Int. Ed.* **2007**, *46*, 1066.

863 Nowrouzi, F.; Thadani, A. N.; Batey, R. A. *Org. Lett.* **2009**, *11*, 2631.

864 Nowrouzi, F.; Janetzko, J.; Batey, R. A. *Org. Lett.* **2010**, *12*, 5490.

865 Paterson, I.; Anderson, E. A.; Dalby, S. M.; Loiseleur, O. *Org. Lett.* **2005**, *7*, 4125.

866 Pattenden, G.; Ashweek, N. J.; Baker-Glenn, C. A. G.; Kempson, J.; Walker, G. M.; Yee, J. G. K. *Org. Biomol. Chem.* **2008**, *6*, 1478.

867 Peña-López, M.; Martínez, M. M.; Sarandeses, L. A.; Sestelo, J. P. *Org. Lett.* **2010**, *12*, 852.

868 Ramachandran, P. V.; Burghardt, T. E.; Reddy, M. V. R. *J. Org. Chem.* **2005**, *70*, 2329.

869 Rauniyar, V.; Hall, D. G. *Angew. Chem., Int. Ed.* **2006**, *45*, 2426.

870 Rauniyar, V.; Hall, D. G. *J. Org. Chem.* **2009**, *74*, 4236.

871 Rauniyar, V.; Hall, D. G. *Synthesis* **2007**, *21*, 3421.

872 Rauniyar, V.; Zhai, H.; Hall, D. G. *J. Am. Chem. Soc.* **2008**, *130*, 8481.

873 Reilly, M. K.; Rychnovsky, S. D. *Org. Lett.* **2010**, *12*, 4892.

874 Ruiz, P.; Murga, J.; Carda, M.; Marco, J. A. *J. Org. Chem.* **2005**, *70*, 713.

875 Schneider, U.; Dao, H. T.; Kobayashi, S. *Org. Lett.* **2010**, *12*, 2488.

876 Schneider, U.; Kobayashi, S. *Angew. Chem., Int. Ed.* **2007**, *46*, 5909.

877 Sebelius, S.; Szabó, K. J. *Eur. J. Org. Chem.* **2005**, 2539.

878 Shi, S-L.; Xu, L-W.; Oisaki, K.; Kanai, M.; Shibasaki, M. *J. Am. Chem. Soc.* **2010**, *132*, 6638.

879 Shimizu, M.; Kawanishi, M.; Mizota, I.; Hachiya, I. *Org. Lett.* **2010**, *12*, 3571.

880 Sieber, J. D.; Morken, J. P. *J. Am. Chem. Soc.* **2008**, *130*, 4978.

881 Singh, A. J.; Xu, C.-X.; Xu, X.; West, L. M.; Wilmes, A.; Chan, A.; Hamel, E.; Miller, J. H.; Northcote, P. T.; Ghosh, A. K. *J. Org. Chem.* **2010**, *75*, 2.

882 Soto-Cairoli, B.; Soderquist, J. A. *Org. Lett.* **2009**, *11*, 401.

883 Sugiura, M.; Mori, C.; Hirano, K.; Kobayashi, S. *Can. J. Chem.* **2005**, *83*, 937.

884 Takahata, H.; Saito, Y.; Ichinose, M. *Org. Biomol. Chem.* **2006**, *4*, 1587.

885 Umarye, J. D.; Leβmann, T.; García, A. B.; Mamane, V.; Sommer, S.; Waldmann, H. *Chem. Eur. J.* **2007**, *13*, 3305.

886 Yadav, J. S.; Prakash, S. J.; Gangadhar, Y. *Tetrahedron: Asymmetry* **2005**, *16*, 2722.

887 Yamaguchi, M.; Morita, N.; Schneider, U.; Kobayashi, S. *Adv. Synth. Catal.* **2010**, *352*, 1461.

888 Yin, X.; Zhao, G.; Schneller, S. W. *Tetrahedron Lett.* **2007**, 4809.

889 Zhang, P.; Morken, J. P. *J. Am. Chem. Soc.* **2009**, *131*, 12550.

890 Zhang, X.; Da, S.; Zhang, C.; Xie, Z.; Li. Y. *Tetrahedron Lett.* **2006**, *47*, 507.

Supplemental References for Table 1B.

837 Barker, T. J. Jarvo, E. R. *Org. Lett.* **2009**, *11*, 1047.

838 Barnett, D. S.; Moquist, P. N.; Schaus, S. E. *Angew. Chem., Int. Ed.* **2009**, *48*, 8679.

839 Boshra, R.; Doshi, A.; Jäkle, F. *Angew. Chem., Int. Ed.* **2008**, *47*, 1134.

[841] Burgos, C. H.; Canales, E.; Matos, K.; Soderquist, J. A. *J. Am. Chem. Soc.* **2005**, *127*, 8044.
[842] Canales, E.; Prasad, G.; Soderquist, J. A. *J. Am. Chem. Soc.* **2005**, *127*, 11572.
[845] Chen, W.; Liu, Y.; Chen, Z. *Eur. J. Org. Chem.* **2005**, 1665.
[849] Fandrick, K. R.; Fandrick, D. R.; Gao, J. J.; Reeves, J. T.; Tan, Z.; Li, W.; Song, J. J.; Lu, B.; Yee, N. K.; Senanayake, C. H. *Org. Lett.* **2010**, *12*, 3748.
[891] Friberg, A.; Sarvary, I.; Wendt, O. F.; Frejd, T. *Tetrahedron: Asymmetry* **2008**, *19*, 1765.
[853] Jain, P.; Antilla, J. C. *J. Am. Chem. Soc.* **2010**, *132*, 11884.
[858] Lou, S.; Moquist, P. N.; Schaus, S. E. *J. Am. Chem. Soc.* **2006**, *128*, 12660.
[892] Lou, S.; Moquist, P. N.; Schaus, S. E. *J. Am. Chem. Soc.* **2007**, *129*, 15398.
[859] Mamane, V.; García, A. B.; Umarye, J. D.; Lessmann, T.; Sommer, S.; Waldmann, H. *Tetrahedron* **2007**, *63*, 5754.
[861] Matsuoka, H.; Kondo, K. *Tetrahedron Lett.* **2009**, *50*, 2320.
[893] Melancon, B. J.; Perl, N. R.; Taylor, R. E. *Org. Lett.* **2007**, *9*, 1425.
[863] Nowrouzi, F.; Thadani, A. N.; Batey, R. A. *Org. Lett.* **2009**, *11*, 2631.
[869] Rauniyar, V.; Hall, D. G. *Angew. Chem., Int. Ed.* **2006**, *45*, 2426.
[843] Rauniyar, V.; Hall, D. G. *Synthesis* **2007**, *21*, 3421.
[844] Rauniyar, V.; Zhai, H.; Hall, D. G. *J. Am. Chem. Soc.* **2008**, *130*, 8481.
[845] Reilly, M. K.; Rychnovsky, S. D. *Org. Lett.* **2010**, *12*, 4892.
[894] Schneider, U.; Chen, I-H.; Kobayashi, S. *Org. Lett.* **2008**, *10*, 737.
[876] Schneider, U.; Kobayashi, S. *Angew Chem., Int. Ed.* **2007**, *46*, 5909.
[895] Schneider, U.; Ueno, M.; Kobayashi, S. *J. Am. Chem. Soc.* **2008**, *130*, 13824.
[875] Schneider, U.; Dao, H. T.; Kobayashi, S. *Org. Lett.* **2010**, *12*, 2488.
[877] Sebellus, S.; Szabó, K. J. *Eur. J. Org. Chem.* **2005**, 2539.
[896] Selander, N.; Sebelius, S.; Estay, C.; Szabó, K. J. *Eur. J. Org. Chem.* **2006**, 4085.
[878] Shi, S-L.; Xu, L-W.; Oisaki, K.; Kanai, M.; Shibasaki, M. *J. Am. Chem. Soc.* **2010**, *132*, 6638.
[897] Vorontsova, N. V.; Zhuravsky, R. P.; Sergeeva, E. V.; Vorontsov, E. V.; Starikova, Z. A.; Rozenberg, V. I. *Russ. Chem. Bull., Int. Ed.* **2007**, *56*, 2225.
[887] Yamaguchi, M.; Morita, N.; Schneider, U.; Kobayashi, S. *Adv. Synth. Catal.* **2010**, *352*, 1461.

Supplemental References for Table 2.

[838] Barnett, D. S.; Moquist, P. N.; Schaus, S. E. *Angew. Chem., Int. Ed.* **2009**, *48*, 8679.
[841] Burgos, C. H.; Canales, E.; Matos, K.; Soderquist, J. A. *J. Am. Chem. Soc.* **2005**, *127*, 8044.
[898] Domon, D.; Fujiwara, K.; Ohtaniuchi, Y.; Takezawa, A.; Takeda, S.; Kawasaki, H.; Murai, A.; Kawai, H.; Suzuki, T. *Tetrahedron Lett.* **2005**, *46*, 8279.
[899] Horneff, T.; Herdtweck, E.; Randoll, S.; Bach, T. *Bioorg. Med. Chem.* **2006**, *14*, 6223.
[853] Jain, P.; Antilla, J. C. *J. Am. Chem. Soc.* **2010**, *132*, 11884.
[858] Lou, S.; Moquist, P. N.; Schaus, S. E. *J. Am. Chem. Soc.* **2006**, *128*, 12660.
[859] Mamane, V.; García, A. B.; Umarye, J. D.; Lessmann, T.; Sommer, S.; Waldmann, H. *Tetrahedron* **2007**, *63*, 5754.
[863] Nowrouzi, F.; Thadani, A. N.; Batey, R. A. *Org. Lett.* **2009**, *11*, 2631.
[900] Ramachandran, P. V.; Chandra, J. S.; Prabhudas, B.; Pratihar, D.; Reddy, M. V. R. *Org. Biomol. Chem.* **2005**, *3*, 3812.
[901] Ramachandran, P. V.; Pratihar, D.; Biswas, D. *Chem. Commun.* **2005**, 1988.
[902] Ramachandran, P. V.; Srivastava, A.; Hazra, D. *Org. Lett.* **2007**, *9*, 157.
[869] Rauniyar, V.; Hall, D. G. *Angew. Chem., Int. Ed.* **2006**, *45*, 2426.
[871] Rauniyar, V.; Hall, D. G. *Synthesis* **2007**, *21*, 3421.
[872] Rauniyar, V.; Zhai, H.; Hall, D. G. *J. Am. Chem. Soc.* **2008**, *130*, 8481.
[873] Reilly, M. K.; Rychnovsky, S. D. *Org. Lett.* **2010**, *12*, 4892.
[875] Schneider, U.; Dao, H. T.; Kobayashi, S. *Org. Lett.* **2010**, 12, 2488.
[878] Shi, S-L.; Xu, L-W.; Oisaki, K.; Kanai, M.; Shibasaki, M. *J. Am. Chem. Soc.* **2010**, *132*, 6638.
[879] Shimizu, M.; Kawanishi, M.; Mizota, I.; Hachiya, I. *Org. Lett.* **2010**, *12*, 3571.
[903] Va, P.; Roush, W. R. *Tetrahedron* **2007**, *63*, 5768.

Supplemental References for Table 3.

[838] Barnett, D. S.; Moquist, P. N.; Schaus, S. E. *Angew. Chem., Int. Ed.* **2009**, *48*, 8679.

[841] Burgos, C. H.; Canales, E.; Matos, K.; Soderquist, J. A. *J. Am. Chem. Soc.* **2005**, *127*, 8044.

[904] Chakraborty, T. K.; Chattopadhyay, A. K. *J. Org. Chem.* **2008**, *73*, 3578.

[849] Fandrick, K. R.; Fandrick, D. R.; Gao, J. J.; Reeves, J. T.; Tan, Z.; Li, W.; Song, J. J.; Lu, B.; Yee, N. K.; Senanayake, C. H. *Org. Lett.* **2010**, *12*, 3748.

[905] Gao, X.; Hall, D. G. *J. Am. Chem. Soc.* **2005**, *127*, 1628.

[906] Ghosh, A. K.; Yuan, H. *Org. Lett.* **2010**, *12*, 3120.

[907] Hirose, T.; Sunazuka, T.; Yamamoto, D.; Mouri, M.; Hagiwara, Y.; Matsumaru, T.; Kaji, E.; Omura, S. *Tetrahedron Lett.* **2007**, *48*, 413.

[853] Jain, P.; Antilla, J. C. *J. Am. Chem. Soc.* **2010**, *132*, 11884.

[908] Hua, Z.; Jin, Z. *Tetrahedron Lett.* **2007**, *48*, 7695.

[909] Keck, G. E.; Giles, R. L.; Cee, V. J.; Wager, C. A.; Yu, T.; Kraft, M. B. *J. Org. Chem.* **2008**, *73*, 9675.

[858] Lou, S.; Moquist, P. N.; Schaus, S. E. *J. Am. Chem. Soc.* **2006**, *128*, 12660.

[910] Magnin-Lachaux, M.; Tan, Z.; Liang, B.; Negishi, E.-i. *Org. Lett.* **2004**, *6*, 1425.

[859] Mamane, V.; García, A. B.; Umarye, J. D.; Lessmann, T.; Sommer, S.; Waldmann, H. *Tetrahedron* **2007**, *63*, 5754.

[911] Morita, A.; Kuwahara, S. *Tetrahedron Lett.* **2007**, *48*, 3163.

[863] Nowrouzi, F.; Thadani, A. N.; Batey, R. A. *Org. Lett.* **2009**, *11*, 2631.

[912] Penner, M.; Rauniyar, V.; Kaspar, L. T.; Hall, D. G. *J. Am. Chem. Soc.* **2009**, *131*, 14216.

[868] Ramachandran, P. V.; Burghardt, T. E.; Reddy, M. V. R. *J. Org. Chem.* **2005**, *70*, 2329.

[902] Ramachandran, P. V.; Srivastava, A.; Hazra, D. *Org. Lett.* **2007**, *9*, 157.

[901] Ramachandran, P. V.; Pratihar, D.; Biswas, D. *Chem. Commun.* **2005**, 1988.

[869] Rauniyar, V.; Hall, D. G. *Angew. Chem., Int. Ed.* **2006**, *45*, 2426.

[870] Rauniyar, V.; Hall, D. G. *J. Org. Chem.* **2009**, *74*, 4236.

[871] Rauniyar, V.; Hall, D. G. *Synthesis* **2007**, *21*, 3421.

[872] Rauniyar, V.; Zhai, H.; Hall, D. G. *J. Am. Chem. Soc.* **2008**, *130*, 8481.

[873] Reilly, M. K.; Rychnovsky, S. D. *Org. Lett.* **2010**, *12*, 4892.

[875] Schneider, U.; Dao, H. T.; Kobayashi, S. *Org. Lett.* **2010**, *12*, 2488.

[877] Sebelius, S.; Szabó, K. J. *Eur. J. Org. Chem.* **2005**, 2539.

[878] Shi, S-L.; Xu, L-W.; Oisaki, K.; Kanai, M.; Shibasaki, M. *J. Am. Chem. Soc.* **2010**, *132*, 6638.

[879] Shimizu, M.; Kawanishi, M.; Mizota, I.; Hachiya, I. *Org. Lett.* **2010**, *12*, 3571.

[913] Suenaga, K.; Kimura, T.; Kuroda, T.; Matsui, K.; Miya, S.; Kuribayashi, S.; Sakakura, A.; Kigoshi, H. *Tetrahedron* **2006**, *62*, 8278.

[885] Umarye, J. D.; Leßmann, T.; García, A. B.; Mamane, V.; Sommer, S.; Waldmann, H. *Chem. Eur. J.* **2007**, *13*, 3305.

[914] Vogt, M.; Ceylan, S.; Kirschning, A. *Tetrahedron* **2010**, *66*, 6450.

[915] Zampella, A.; Sepe, V.; D'Orsi, R.; D'Auria, M. V. *Lett. Org. Chem.* **2004**, *1*, 308.

Supplemental References for Table 4.

[916] Althaus, J.; Mahmood, A.; Suárez, J. R.; Thomas, S. P.; Aggarwal, V. K. *J. Am. Chem Soc.* **2010**, *132*, 4025.

[917] Beckmann, E.; Hoppe, D. *Synthesis* **2005**, *2*, 217.

[918] Berrée, F.; Gernigon, N.; Hercouet, A.; Lin,C. H.; Carboni, B. *Eur. J. Org. Chem.* **2009**, 329.

[919] Bischop, M.; Doum, V.; Nordschild (née Rieche), A. C. M.; Pietruszka, J.; Sandkuhl, D. *Synthesis* **2010**, *3*, 527.

[920] Burks, H. E.; Kilman, L. T.; Morken, J. P. *J. Am. Chem. Soc.* **2009**, *131*, 9134.

[921] Carosi, L.; Hall, D. G. *Angew. Chem., Int. Ed.* **2007**, *46*, 5913.

[922] Carosi, L.; Hall, D. G. *Can. J. Chem.* **2009**, *87*, 650.

[923] Carosi, L.; Lachance, H.; Hall, D. G. *Tetrahedron Lett.* **2005**, *46*, 8981.

[924] Chen, M.; Ess, D. H.; Roush, W. R. *J. Am. Chem. Soc.* **2010**, *132*, 7881.

[925] Chen, M.; Roush, W. R. *Org. Lett.* **2010**, *12*, 2706.

926 (a) Cho, H. Y.; Morken, J. P. *J. Am. Chem. Soc.* **2008**, *130*, 16140. (b) Cho, H. Y.; Morken, J. P. *J. Am. Chem. Soc.* **2010**, *132*, 7576.

927 Cmrecki, V.; Eichenauer, N. C.; Frey, W.; Pietruszka, J. *Tetrahedron* **2010**, *66*, 6550.

928 Fang, G. Y.; Aggarwal, V. K. *Angew. Chem., Int. Ed.* **2007**, *46*, 359.

849 Fandrick, K. R.; Fandrick, D. R.; Gao, J. J.; Reeves, J. T.; Tan, Z.; Li, W.; Song, J. J.; Lu, B.; Yee, N. K.; Senanayake, C. H. *Org. Lett.* **2010**, *12*, 3748.

929 Fernández, E.; Pietruszka, J. *Synlett* **2009**, *9*, 1474.

930 Fernández, E.; Pietruszka, J.; Frey, W. *J. Org. Chem.* **2010**, *75*, 5580.

931 Flamme, E. M.; Roush, W. R. *Org. Lett.* **2005**, *7*, 1411.

932 González, J. R.; González, A. Z.; Soderquist, J. A. *J. Am. Chem. Soc.* **2009**, *131*, 9924.

933 Hicks, J. D.; Flamme, E. M.; Roush, W. R. *Org. Lett.* **2005**, *7*, 5509.

934 Ito, H.; Ito, S.; Sasaki, Y.; Matsuura, K.; Sawamura, M. *J. Am. Chem. Soc.* **2007**, *129*, 14856.

935 Ito, H.; Okura, T.; Matsuura, K.; Sawamura, M. *Angew. Chem., Int. Ed.* **2010**, *49*, 560.

936 Kister, J.; DeBaillie, A. C.; Lira, R.; Roush, W. R. *J. Am. Chem. Soc.* **2009**, *131*, 14174.

937 Kobayashi, S.; Endo, T.; Schneider, U.; Ueno M. *Chem. Commun.* **2010**, *46*, 1260.

938 Kobayashi, S.; Konishi, H.; Schneider, U. *Chem. Comm.* **2008**, 2313.

939 Li, F.; Roush, W. R. *Org. Lett.* **2009**, *11*, 2932.

940 Lira, R.; Roush, W. R. *Org. Lett.* **2007**, *9*, 533.

941 Owen, R. M.; Roush, W. R. *Org. Lett.* **2005**, *7*, 3941.

942 Pardo-Rodríguez, V.; Marco-Martínez, J.; Buñuel, E.; Cárdenas, D. J. *Org. Lett.* **2009**, *11*, 4548.

943 Peng, F.; Hall, D. G. *J. Am. Chem. Soc.* **2007**, *129*, 3070.

944 Peng, F.; Hall, D. G. *Tetrahedron Lett.* **2007**, *48*, 3305.

945 Pietruszka, J.; Rieche, A. C. M.; Schöne, N. *Synlett* **2007**, 2525.

946 Pietruszka, J.; Schöne, N. *Eur. J. Org. Chem.* **2004**, 5011.

947 Pietruszka, J.; Schöne, N. *Synthesis* **2006**, *1*, 24.

948 Pietruszka, J.; Schöne, N.; Frey, W.; Grundl, L. *Chem. Eur. J.* **2008**, *14*, 5178.

949 Possémé, F.; Deligny, M.; Carreaux, F.; Carboni, B. *J. Org. Chem.* **2007**, *72*, 984.

950 Ramachandran, P. V.; Chatterjee, A. *Org. Lett.* **2008**, *10*, 1195.

901 Ramachandran, P. V.; Pratihar, D.; Biswas, D. *Chem. Commun.* **2005**, 1988.

951 Ramachandran, P. V.; Pratihar, D.; Nair, H. N. G.; Walters, M.; Smith, S.; Yip-Schneider, M. T.; Wu, H.; Schmidt, C. M. *Bioorg. Med. Chem. Lett.* **2010**, *20*, 6620.

952 Robinson, A.; Aggarwal, V. K. *Angew. Chem., Int. Ed.* **2010**, *49*, 6673.

953 Ros, A.; Bermejo, A.; Aggarwal, V. K. *Chem. Eur. J.* **2010**, *16*, 9741.

875 Schneider, U.; Dao, H. T.; Kobayashi, S. *Org. Lett.* **2010**, *12*, 2488.

954 Sivasubramaniam, U.; Hall, D. G. *Heterocycles* **2010**, *80*, 1449.

955 Tang, S.; Xie, X.; Wang, X.; He, L.; Xu, K.; She, X. *J. Org. Chem.* **2010**, *75*, 8234.

956 Winbush, S. M.; Roush, W. R. *Org. Lett.* **2010**, *12*, 4344.

957 Woodward, A. R.; Burks, H. E.; Chan, L. M.; Morken, J. P. *Org. Lett.* **2005**, *7*, 5505.

887 Yamaguchi, M.; Morita, N.; Schneider, U.; Kobayashi, S. *Adv. Synth. Catal.* **2010**, *352*, 1461.

958 Yamamoto, A.; Ikeda, Y.; Suginome, M. *Tetrahedron Lett.* **2009**, *50*, 3168.

Supplemental References for Table 5.

959 Arndt, M.; Reinhold, A.; Hilt, G. *J. Org. Chem.* **2010**, *75*, 5203.

960 Chataigner, I.; Zammattio, F.; Lebreton, J.; Villiéras, J. *Tetrahedron* **2008**, *64*, 2441.

845 Chen, W.; Liu, Y.; Chen, Z. *Eur. J. Org. Chem.* **2005**, 1665.

961 Elford, T. G.; Ulaczyk-Lesanko, A.; De Pascale, G.; Wright, G. D.; Hall, D. G. *J. Comb. Chem.* **2009**, *11*, 155.

849 Fandrick, K. R.; Fandrick, D. R.; Gao, J. J.; Reeves, J. T.; Tan, Z.; Li, W.; Song, J. J.; Lu, B.; Yee, N. K.; Senanayake, C. H. *Org. Lett.* **2010**, *12*, 3748.

962 Faveau, C.; Mondon, M.; Gesson, J-P.; Mahnke, T.; Gebhardt, S.; Koert, U. *Tetrahedron Lett.* **2006**, *47*, 8305.

963 Lachance, H.; Marion, O.; Hall, D. G. *Tetrahedron Lett.* **2008**, *49*, 6061.

964 Ohmura, T.; Taniguchi, H.; Kondo, Y.; Suginome, M. *J. Am. Chem. Soc.* **2007**, *129*, 3518.

[951] Ramachandran, P. V.; Pratihar, D.; Nair, H. N. G.; Walters, M.; Smith, S.; Yip-Schneider, M. T.; Wu, H.; Schmidt, C. M. *Bioorg. Med. Chem. Lett.* **2010**, *20*, 6620.

[870] Rauniyar, V.; Hall, D. G. *J. Org. Chem.* **2009**, *74*, 4236.

[871] Rauniyar, V.; Hall, D. G. *Synthesis* **2007**, *21*, 3421.

[872] Rauniyar, V.; Zhai, H.; Hall, D. G. *J. Am. Chem. Soc.* **2008**, *130*, 8481.

[965] Román, J. G.; Soderquist, J. A. *J. Org. Chem.* **2007**, *72*, 9772.

[966] Sánchez, C.; Ariza, X.; Cornellà, J.; Farràs, J.; Garcia, J.; Ortiz, J. *Chem. Eur. J.* **2010**, *16*, 11535.

[967] Selander, N.; Kipke, A.; Sebelius, S.; Szabó, K. J. *J. Am. Chem. Soc.* **2007**, *129*, 13723.

[955] Tang, S.; Xie, X.; Wang, X.; He, L.; Xu, K.; She, X. *J. Org. Chem.* **2010**, *75*, 8234.

Supplemental References for Table 6.

[844] Castoldi, D.; Caggiano, L.; Bayón, P.; Costa, A. M.; Cappella, P.; Sharon, O.; Gennari, C. *Tetrahedron* **2005**, *61*, 2123.

[968] Chang, K-J.; Rayabarapu, D. K.; Yang, F-Y.; Cheng, C-H. *J. Am. Chem. Soc.* **2005**, *127*, 126.

[960] Chataigner, I.; Zammattio, F.; Lebreton, J.; Villiéras, J. *Tetrahedron* **2008**, *64*, 2441.

[969] Chen, M.; Handa, M.; Roush, W. R. *J. Am. Chem. Soc.* **2009**, *131*, 14602.

[970] Crotti, S.; Bertolini, F.; Macchia, F.; Pineschi, M. *Org. Lett.* **2009**, *11*, 3762.

[971] Elford, T. G.; Arimura, Y.; Yu, S. H.; Hall, D. G. *J. Org. Chem.* **2007**, *72*, 1276.

[961] Elford, T. G.; Ulaczyk-Lesanko, A.; De Pascale, G.; Wright, G. D.; Hall, D. G. *J. Comb. Chem.* **2009**, *11*, 155.

[972] Ely, R. J.; Morken, J. P. *J. Am. Chem. Soc.* **2010**, *132*, 2534.

[973] Ess, D. H.; Kister, J.; Chen, M.; Roush, W. R. *Org. Lett.* **2009**, *11, 5538*.

[931] Flamme, E. M.; Roush, W. R. *Org. Lett.* **2005**, *7*, 1411.

[974] Fürstner, A.; Bonnekessel, M.; Blank, J. T.; Radkowski, K.; Seidel, G.; Lacombe, F.; Gabor, B.; Mynott, R. *Chem. Eur. J.* **2007**, *13*, 8762.

[975] González, A. Z.; Román, J. G.; Alicea, E.; Canales, E.; Soderquist, J. A. *J. Am. Chem. Soc.* **2009**, *131*, 1269.

[933] Hicks, J. D.; Flamme, E. M.; Roush, W. R. *Org. Lett.* **2005**, *7*, 5509.

[976] Hoffmann, R. W.; Menzel, K.; Harms, K. *Eur. J. Org. Chem.* **2002**, 2603.

[977] Kabalka, G. W.; Venkataiah, B. *Tetrahedron Lett.* **2005**, *46*, 7325.

[978] Kabalka, G. W.; Venkataiah, B.; Dong, G. *Tetrahedron Lett.* **2005**, 46, 4209.

[936] Kister, J.; DeBaillie, A. C.; Lira, R.; Roush, W. R. *J. Am. Chem. Soc.* **2009**, *131*, 14174.

[939] Li, F.; Roush, W. R. *Org. Lett.* **2009**, *11*, 2932.

[979] Lira, R.; Roush, W. R. *Org. Lett.* **2007**, *9*, 4315.

[940] Lira, R.; Roush, W. R. *Org. Lett.* **2007**, *9*, 533.

[980] Mitra, S.; Gurrala, S. R.; Coleman, R. S. *J. Org. Chem.* **2007**, *72*, 8724.

[941] Owen, R. M.; Roush, W. R. *Org. Lett.* **2005**, *7*, 3941.

[981] Purser, S.; Wilson, C.; Moore, P. R.; Gouverneur, V. *Synlett* **2007**, *7*, 1166.

[868] Ramachandran, P. V.; Burghardt, T. E.; Reddy, M. V. R. *J. Org. Chem.* **2005**, *70*, 2329.

[900] Ramachandran, P. V.; Chandra, J. S.; Prabhudas, B.; Pratihar, D.; Reddy, M. V. R. *Org. Biomol. Chem.* **2005**, *3*, 3812.

[982] Ramachandran, P. V.; Chatterjee, A. *J. Fluorine Chem.* **2009**, *130*, 144.

[983] Ramachandran, P. V.; Pratihar, D. *Org. Lett.* **2007**, *9*, 2087.

[984] Ramachandran, P. V.; Pratihar, D.; Biswas, D. *Org. Lett.* **2006**, *8*, 3877.

[951] Ramachandran, P. V.; Pratihar, D.; Nair, H. N. G.; Walters, M.; Smith, S.; Yip-Schneider, M. T.; Wu, H.; Schmidt, C. M. *Bioorg. Med. Chem. Lett.* **2010**, *20*, 6620.

[877] Sebelius, S.; Szabó, K. J. *Eur. J. Org. Chem.* **2005**, 2539.

[967] Selander, N.; Kipke, A.; Sebelius, S.; Szabó, K. J. *J. Am. Chem. Soc.* **2007**, *129*, 13723.

[896] Selander, N.; Sebelius, S.; Estay, C.; Szabó, K. J. *Eur. J. Org. Chem.* **2006**, 4085.

[985] Selander, N.; Szabó, K. J. *Adv. Synth. Catal.* **2008**, *350*, 2045.

[986] Selander, N.; Szabó, K. J. *Chem. Commun.* **2008**, 3420.

[987] Selander, N.; Szabó, K. J. *J. Org. Chem.* **2009**, *74*, 5695.

[879] Shimizu, M.; Kawanishi, M.; Mizota, I.; Hachiya, I. *Org. Lett.* **2010**, *12*, 3571.

[988] Sumida, Y.; Yorimitsu, H.; Oshima,K. *J. Am. Chem. Soc.* **2006**, *128*, 15960.
[989] Sumida, Y.; Yorimitsu, H.; Oshima,K. *Org. Lett.* **2008**, *10*, 4677.
[903] Va, P.; Roush, W. R. *Tetrahedron* **2007**, *63*, 5768.
[914] Vogt, M.; Ceylan, S.; Kirschning, A. *Tetrahedron* **2010**, *66*, 6450.
[990] Wu. J. Y.; Moreau, B.; Ritter, T. *J. Am. Chem. Soc.* **2009**, *131*, 12915.
[991] Yu, S. H.; Ferguson, M. J.; McDonald, R.; Hall, D. G. *J. Am. Chem. Soc.* **2005**, *127*, 12808.

Supplemental References for Table 7.

[992] Canales, E.; Gonzalez, A. Z.; Soderquist, J. A. *Angew. Chem., Int. Ed.* **2007**, *46*, 397.
[993] Fandrick, D. R.; Fandrick, K. R.; Reeves, J. T.; Tan, Z.; Johnson, C. S.; Lee, H.; Song, J. J.; Yee, N. K.; Senanayake, C. H. *Org. Lett.* **2010**, *12*, 88.
[994] Fandrick, D. R.; Reeves, J. T.; Tan, Z.; Lee, H.; Song, J. J.; Yee, N. K.; Senanayake, C. H. *Org. Lett.* **2009**, *11*, 5458.
[995] Hernandez, E.; Burgos, C. H.; Alicea, E.; Soderquist, J. A. *Org. Lett.* **2006**, *8*, 4089.
[996] Ito, H.; Sasaki, Y.; Sawamura, M. *J. Am. Chem. Soc.* **2008**, *130*, 15774.
[997] Lai, C.; Soderquist, J. A. *Org. Lett.* **2005**, *7*, 799.
[998] Maddess, M. L.; Lautens, M. *Org. Lett.* **2005**, *7*, 3557.
[875] Schneider, U.; Dao, H. T.; Kobayashi, S. *Org. Lett.* **2010**, *12*, 2488.
[878] Shi, S-L.; Xu, L-W.; Oisaki, K.; Kanai, M.; Shibasaki, M. *J. Am. Chem. Soc.* **2010**, *132*, 6638.

Supplemental References for Table 8.

[999] Bouziane, A.; Régnier, T.; Carreaux, F.; Carboni, B.; Bruneau, C.; Renaud, J-L. *Synlett* **2010**, *2*, 207.
[1000] Carreaux, F.; Favre, A.; Carboni, B.; Rouaud, I.; Boustie, J. *Tetrahedron Lett.* **2006**, *47*, 4545.
[970] Crotti, S.; Bertolini, F.; Macchia, F.; Pineschi, M. *Org. Lett.* **2009**, *11*, 3762.
[1001] Deligny, M.; Carreaux, F.; Carboni, B. *Synlett* **2005**, *9*, 1462.
[1002] Favre, A.; Carreaux, F.; Deligny, M.; Carboni, B. *Eur. J. Org. Chem.* **2008**, 4900.
[905] Gao, X.; Hall, D. G. *J. Am. Chem. Soc.* **2005**, *127*, 1628.
[1003] Gao, X.; Hall, D. G.; Deligny, M.; Favre, A.; Carreaux, F.; Carboni, B. *Chem. Eur. J.* **2006**, *12*, 3132.
[1004] Gerdin, M.; Penhoat, M.; Zalubovskis, R.; Pétermann, C.; Moberg, C. *J. Organomet. Chem.* **2008**, *693*, 3519.
[1005] Hercouet, A.; Berrée, F.; Lin, C. H.; Toupet, L.; Carboni, B. *Org. Lett.* **2007**, *9*, 1717.
[1006] Hilt, G.; Hess, W.; Harms, K. *Org. Lett.* **2006**, *8*, 3287.
[1007] Hilt, G.; Lüers, S.; Smolko, K. I. *Org. Lett.* **2005**, *7*, 251.
[1008] Ito, H.; Kunii, S.; Sawamura, M. *Nature Chem.* **2010**, *2*, 972.
[935] Ito, H.; Okura, T.; Matsuura, K.; Sawamura, M. *Angew. Chem., Int. Ed.* **2010**, *49*, 560.
[963] Lachance, H.; Marion, O.; Hall, D. G. *Tetrahedron Lett.* **2008**, *49*, 6061.
[1009] Lessard, S.; Peng, F. *J. Am. Chem. Soc.* **2009**, *131*, 9612.
[1010] Marion, O.; Gao, X.; Marcus, S.; Hall, D. G. *Bioorg. Med. Chem.* **2009**, *17*, 1006.
[1011] Nakagawa, D.; Middyashita, M.; Tanino, K. *Tetrahedron Lett.* **2010**, *51*, 2771.
[1012] Olsson, V. J.; Szabó, K. J. *Angew. Chem., Int. Ed.* **2007**, *46*, 6891.
[1013] Olsson, V. J.; Szabó, K. J. *J. Org. Chem.* **2009**, *74*, 7715.
[967] Selander, N.; Kipke, A.; Sebelius, S.; Szabó, K. J. *J. Am. Chem. Soc.* **2007**, *129*, 13723.
[896] Selander, N.; Sebelius, S.; Estay, C.; Szabó, K. J. *Eur. J. Org. Chem.* **2006**, 4085.
[986] Selander, N.; Szabó, K. J. *Chem. Commun.* **2008**, 3420.
[1014] Ulaczyk-Lesanko, A.; Pelletier, E.; Lee, M.; Prinz, H.; Waldmann, H.; Hall, D. G. *J. Comb. Chem.* **2007**, *9*, 695.

Supplemental References for Table 9.

[1015] Bandur, N. G.; Brükner, D.; Hoffmann, R. W.; Koert, U. *Org. Lett.* **2006**, *8*, 3829.

Selected Supplemental References on New Preparative Methods for Allylic Boronates

[920] Burks, H. E.; Kilman, L. T.; Morken, J. P. *J. Am. Chem. Soc.* **2009**, *131*, 9134.

[921] Carosi, L.; Hall, D. G. *Angew. Chem., Int. Ed.* **2007**, *46*, 5913.

[1016] Dutheuil, G.; Selander, N.; Szabó, K. J.; Aggarwal, V. K. *Synthesis* **2008**, 2293.

[972] Ely, R. J.; Morken, J. P. *J. Am. Chem. Soc.* **2010**, *132*, 2534.

[929] Fernández, E.; Pietruszka, J. *Synlett* **2009**, *9*, 1474.

[1017] Guzman-Martinez, A.; Hoveyda, A. H. *J. Am. Chem. Soc.* **2010**, *132*, 10634.

[934] Ito, H.; Ito, S.; Sasaki, Y.; Matsuura, K.; Sawamura, M. *J. Am. Chem. Soc.* **2007**, *129*, 14856.

[1018] Ito, H.; Kawakami, C.; Sawamura, M. *J. Am. Chem. Soc.* **2005**, *127*, 16034.

[1009] Lessard, S.; Peng, F. *J. Am. Chem. Soc.* **2009**, *131*, 9612.

[1019] Sasaki, Y.; Zhong, C.; Sawamura, M.; Ito, H. *J. Am. Chem. Soc.* **2010**, *132*, 1226.

[1020] Sumida, Y.; Yorimitsu, H.; Oshima, K. *Org. Lett.* **2008**, *10*, 4677.

[990] Wu. J. Y.; Moreau, B.; Ritter, T. *J. Am. Chem. Soc.* **2009**, *131*, 12915.

Selected New Reviews on Carbonyl Allylboration

[1021] Hall, D. G. *Synlett* **2007**, 1644.

[1022] Hall, D. G. *Pure Appl. Chem.* **2008**, *80*, 913.

Selected References on the Mechanism of Catalytic Allylborations

[1023] Sakata, K.; Fujimoto, H. *J. Am. Chem. Soc.* **2008**, *130*, 12519.

INDEX

Allylboration of Carbonyl Compounds, by Dennis G. Hall and Hugo Lachance.
© 2012 Organic Reactions, Inc. Published 2012 by John Wiley & Sons, Inc.